Handbook of Formal Languages

Editors: G. Rozenberg   A. Salomaa

T0188879

Springer
Berlin
Heidelberg
New York
Barcelona
Budapest
Hong Kong
London
Milan
Paris
Santa Clara
Singapore
Tokyo

G. Rozenberg A. Salomaa (Eds.)

Handbook of Formal Languages

Volume 2
Linear Modeling:
Background and Application

With 87 Figures

Springer

Prof. Dr. Grzegorz Rozenberg
Leiden University
Department of Computer Science
P.O. Box 9512
NL-2300 RA Leiden
The Netherlands

Prof. Dr. Arto Salomaa
Turku Centre
for Computer Science
Data City
FIN-20520 Turku
Finland

Library of Congress Cataloging-in-Publication Data
Handbook of formal languages / G. Rozenberg, A. Salomaa, (eds.).
p. cm.
Includes bibliographical references and index.
Contents: v. 1. Word, language, grammar – v.2. Linear modeling:
background and application – v. 3. Beyond words.
ISBN 3-540-61486-9 (set). – ISBN 3-540-60420-0 (alk. paper: v. 1). –
ISBN 3-540-60648-3 (alk. paper: v. 2). – ISBN 3-540-60649-1
(alk. paper: v. 3)
1. Formal languages.   I. Rozenberg, Grzegorz.   II. Salomaa, Arto.
QA267.3.H36   1997
511.3 – DC21   96-47134
                                              CIP

CR Subject Classification (1991): F.4 (esp. F.4.2-3), F.2.2, A.2, G.2, I.3, D.3.1, E.4

ISBN 3-540-60648-3 Springer-Verlag Berlin Heidelberg New York
ISBN 3-540-61486-9 (set) Springer-Verlag Berlin Heidelberg New York

©Springer-Verlag Berlin Heidelberg 1997
Springer is a part of Springer Science+Business Media

Cover design: MetaDesign, Berlin
Typesetting: Data conversion by Lewis & Leins, Berlin
SPIN: 11326724   45/3111 - 5 4 3 2 1 - Printed on acid-free paper

Preface

The need for a comprehensive survey-type exposition on formal languages and related mainstream areas of computer science has been evident for some years. In the early 1970s, when the book *Formal Languages* by the second-mentioned editor appeared, it was still quite feasible to write a comprehensive book with that title and include also topics of current research interest. This would not be possible anymore. A standard-sized book on formal languages would either have to stay on a fairly low level or else be specialized and restricted to some narrow sector of the field.

The setup becomes drastically different in a collection of contributions, where the best authorities in the world join forces, each of them concentrating on their own areas of specialization. The present three-volume Handbook constitutes such a unique collection. In these three volumes we present the current state of the art in formal language theory. We were most satisfied with the enthusiastic response given to our request for contributions by specialists representing various subfields. The need for a Handbook of Formal Languages was in many answers expressed in different ways: as an easily accessible historical reference, a general source of information, an overall course-aid, and a compact collection of material for self-study. We are convinced that the final result will satisfy such various needs.

The theory of formal languages constitutes the stem or backbone of the field of science now generally known as theoretical computer science. In a very true sense its role has been the same as that of philosophy with respect to science in general: it has nourished and often initiated a number of more specialized fields. In this sense formal language theory has been the origin of many other fields. However, the historical development can be viewed also from a different angle. The origins of formal language theory, as we know it today, come from different parts of human knowledge. This also explains the wide and diverse applicability of the theory. Let us have a brief look at some of these origins. The topic is discussed in more detail in the introductory Chapter 1 of Volume 1.

The main source of the theory of formal languages, most clearly visible in Volume 1 of this Handbook, is *mathematics*. Particular areas of mathematics important in this respect are combinatorics and the algebra of semigroups and monoids. An outstanding pioneer in this line of research was

Axel Thue. Already in 1906 he published a paper about avoidable and unavoidable patterns in long and infinite words. Thue and Emil Post were the two originators of the formal notion of a rewriting system or a grammar. That their work remained largely unknown for decades was due to the difficult accessibility of their writings and, perhaps much more importantly, to the fact that the time was not yet ripe for mathematical ideas, where noncommutativity played an essential role in an otherwise very simple setup.

Mathematical origins of formal language theory come also from mathematical logic and, according to the present terminology, computability theory. Here the work of Alan Turing in the mid-1930s is of crucial importance. The general idea is to find models of computing. The power of a specific model can be described by the complexity of the language it generates or accepts. Trends and aspects of mathematical language theory are the subject matter of each chapter in Volume 1 of the Handbook. Such trends and aspects are present also in many chapters in Volumes 2 and 3.

Returning to the origins of formal language theory, we observe next that much of formal language theory has originated from *linguistics*. In particular, this concerns the study of grammars and the grammatical structure of a language, initiated by Noam Chomsky in the 1950s. While the basic hierarchy of grammars is thoroughly covered in Volume 1, many aspects pertinent to linguistics are discussed later, notably in Volume 2.

The *modeling* of certain objects or phenomena has initiated large and significant parts of formal language theory. A model can be expressed by or identified with a language. Specific tasks of modeling have given rise to specific kinds of languages. A very typical example of this are the L systems introduced by Aristid Lindenmayer in the late 1960s, intended as models in developmental biology. This and other types of modeling situations, ranging from molecular genetics and semiotics to artificial intelligence and artificial life, are presented in this Handbook. Words are one-dimensional, therefore linearity is a feature present in most of formal language theory. However, sometimes a linear model is not sufficient. This means that the language used does not consist of words (strings) but rather of trees, graphs, or some other nonlinear objects. In this way the possibilities for modeling will be greatly increased. Such extensions of formal language theory are considered in Volume 3: languages are built from nonlinear objects rather than strings.

We have now already described the contents of the different volumes of this Handbook in brief terms. Volume 1 is devoted to the mathematical aspects of the theory, whereas applications are more directly present in the other two volumes, of which Volume 3 also goes into nonlinearity. The division of topics is also reflected in the titles of the volumes. However, the borderlines between the volumes are by no means strict. From many points of view, for instance, the first chapters of Volumes 2 and 3 could have been included in Volume 1.

We now come to a very important editorial decision we have made. Each of the 33 individual chapters constitutes its own entity, where the subject matter is developed from the beginning. References to other chapters are only occasional and comparable with references to other existing literature. This style of writing was suggested to the authors of the individual chapters by us from the very beginning. Such an editorial policy has both advantages and disadvantages as regards the final result. A person who reads through the whole Handbook has to get used to the fact that notation and terminology are by no means uniform in different chapters; the same term may have different meanings, and several terms may mean the same thing. Moreover, the prerequisites, especially in regard to mathematical maturity, vary from chapter to chapter. On the positive side, for a person interested in studying only a specific area, the material is all presented in a compact form in one place. Moreover, it might be counterproductive to try to change, even for the purposes of a handbook, the terminology and notation already well-established within the research community of a specific subarea. In this connection we also want to emphasize the diversity of many of the subareas of the field. An interested reader will find several chapters in this Handbook having almost totally disjoint reference lists, although each of them contains more than 100 references.

We noticed that guaranteed timeliness of the production of the Handbook gave additional impetus and motivation to the authors. As an illustration of the timeliness, we only mention that detailed accounts about DNA computing appear here in a handbook form, less than two years after the first ideas about DNA computing were published.

Having discussed the reasons behind our most important editorial decision, let us still go back to formal languages in general. Obviously there cannot be any doubt about the mathematical strength of the theory – many chapters in Volume 1 alone suffice to show the strength. The theory still abounds with challenging problems for an interested student or researcher. Mathematical strength is also a necessary condition for applicability, which in the case of formal language theory has proved to be both broad and diverse. Some details of this were already mentioned above. As the whole Handbook abounds with illustrations of various applications, it would serve no purpose to try to classify them here according to their importance or frequency. The reader is invited to study from the Handbook older applications of context-free and contextual grammars to linguistics, of parsing techniques to compiler construction, of combinatorics of words to information theory, or of morphisms to developmental biology. Among the newer application areas the reader may be interested in computer graphics (application of L systems, picture languages, weighted automata), construction and verification of concurrent and distributed systems (traces, omega-languages, grammar systems), molecular biology (splicing systems, theory of deletion), pattern matching, or cryptology, just to mention a few of the topics discussed in the Handbook.

About Volume 2

Some brief guidelines about the contents of the present Volume 2 follow. Problems about complexity occur everywhere in language theory; Chapter 1 gives an overall account. Parsing techniques are essential in applications, both for natural and programming languages. They are dealt with in Chapter 2, while Chapters 3–6 study extensions and variations of classical language theory. While Chapter 3 continues the general theory of context-free languages, Chapters 5 and 6 are motivated by linguistics, and Chapter 4, motivated by artificial intelligence, is also applicable to distributed systems. DNA computing has been an important recent breakthrough – some language-theoretic aspects are presented in Chapter 7. Chapter 8 considers the string editing problem which in various settings models a variety of problems arising from DNA and protein sequences. Chapter 9 considers several methods of word matching that are based on the use of automata. Chapter 10 discusses the relationship between automata theory and symbolic dynamics (the latter area has originated in topology). By its very nature the whole of cryptology can be viewed as a part of language theory. Chapter 11 gives an account of language-theoretic techniques that have turned out to be especially useful in cryptology.

Acknowledgements

We would like to express our deep gratitude to all the authors of the Handbook. Contrary to what is usual in case of collective works of this kind, we hoped to be able to keep the time schedule – and succeeded because of the marvellous cooperation of the authors. Still more importantly, thanks are due because the authors were really devoted to their task and were willing to sacrifice much time and energy in order to achieve the remarkable end result, often exceeding our already high expectations.

The Advisory Board consisting of J. Berstel, C. Calude, K. Culik II, J. Engelfriet, H. Jürgensen, J. Karhumäki, W. Kuich, M. Nivat, G. Păun, A. Restivo, W. Thomas, and D. Wood was of great help to us at the initial stages. Not only was their advice invaluable for the planning of the Handbook but we also appreciate their encouragement and support.

We would also like to extend our thanks from the actual authors of the Handbook to all members of the scientific community who have supported, advised, and helped us in many ways during the various stages of work. It would be a hopeless and maybe also unrewarding task to try to single out any list of names, since so many people have been involved in one way or another.

We are grateful also to Springer-Verlag, in particular Dr. Hans Wössner, Ingeborg Mayer, J. Andrew Ross, and Gabriele Fischer, for their cooperation, excellent in every respect. Last but not least, thanks are due to Marloes Boon-van der Nat for her assistance in all editorial stages, and in particular, for keeping up contacts to the authors.

September 1996   Grzegorz Rozenberg, Arto Salomaa

Contents of Volume 2

Chapter 3. **Grammars with Controlled Derivations**

Chapter 4. **Grammar Systems**

Contents of Volume 1

Contents of Volume 3

Authors' Addresses

Alberto Apostolico
Dipartimento di Elettronica e Informatica, Università di Padova
Via Gradenigo 6/a, I-35131 Padova, Italy
and Department of Computer Science, Purdue University
1398 Computer Science Building, West Lafayette, IN 47907-1398, U.S.A.
axa@cs.purdue.edu

Marie-Pierre Béal
Institut Gaspard Monge, Université de Marne-la-Vallée
2, rue de la Butte verte, F-93166 Noisy-le-Grand, France
beal@monge.univ-mlv.fr

Cristian Calude
Centre for Discrete Mathematics and Theoretical Computer Science
The University of Auckland, Private Bag 92019, Auckland, New Zealand
c_calude@cs.aukuni.ac.nz

Maxime Crochemore
Institut Gaspard Monge, Université de Marne-la-Vallée
2, rue de la Butte verte, F-93166 Noisy-le-Grand, France
mac@univ-mlv.fr

Jürgen Dassow
Faculty of Computer Science, Otto-von-Guericke-University of Magdeburg
P.O. Box 4120, D-39016 Magdeburg, Germany
dassow@cs.uni-magdeburg.de

Andrzej Ehrenfeucht
Department of Computer Science, University of Colorado at Boulder
Campus 430, Boulder, CO 80309, U.S.A.
andrzej@piper.cs.colorado.edu

Christophe Hancart
Laboratoire d'Informatique de Rouen, Faculté des Sciences et Techniques
Université de Rouen, F-76821 Mont-Saint-Aignan Cedex, France
hancart@dir.univ-rouen.fr

Thomas Head
Department of Mathematics, University of Binghamton
P.O. Box 6000, Binghamton, NY 13902, U.S.A.
tom@math.binghamton.edu

Juraj Hromkovič
Institut für Informatik und Praktische Mathematik, Universität Kiel
Olshausenstrasse 40, D-24098 Kiel, Germany
jhr@informatik.uni-kiel.d400.de

Solomon Marcus
Faculty of Mathematics, University of Bucharest
Str. Academiei, RO-70109 Bucharest, Romania
solomon@imar.ro

Valtteri Niemi
Department of Mathematics and Statistics, University of Vaasa
FIN-65101 Vaasa, Finland
vni@uwasa.fi

Anton Nijholt
Computer Science Department, University of Twente
P.O. Box 217, NL-7500 AE Enschede, The Netherlands
anijholt@cs.utwente.nl

Gheorghe Păun
Institute of Mathematics of the Romanian Academy
P.O. Box 1-764, RO-70700 Bucharest, Romania
gpaun@imar.ro

Dominique Perrin
Institut Gaspard Monge, Université de Marne-la-Vallée
2, rue de la Butte verte, F-93166 Noisy-le-Grand, France
perrin@univ-mlv.fr

Dennis Pixton
Department of Mathematics, University of Binghamton
P.O. Box 6000, Binghamton, New York 13902, U.S.A.
dennis@math.binghamton.edu

Grzegorz Rozenberg
Department of Computer Science, Leiden University
P.O. Box 9512, NL-2300 RA Leiden, The Netherlands
and Department of Computer Science, University of Colorado at Boulder
Campus 430, Boulder, CO 80309, U.S.A.
rozenber@wi.leidenuniv.nl

Arto Salomaa
Academy of Finland and Turku Centre for Computer Science (TUCS)
Lemninkäisenkatu 14 A, FIN-20520 Turku, Finland
asalomaa@sara.cc.utu.fi

Klaas Sikkel
FIT, CSCW, German National Research Centre for Information Technology (GMD)
Schloß Birlinghoven, D-53757 Sankt Augustin, Germany
sikkel@gmd.de

Complexity:
A Language-Theoretic Point of View

Cristian Calude and Juraj Hromkovič

1. Introduction

The theory of computation and complexity theory are fundamental parts of current theoretical computer science. They study the borders between possible and impossible in information processing, quantitative rules governing discrete computations (how much work (computational resources) has to be done (have to be used) and suffices (suffice) to algorithmically solve various computing problems), algorithmical aspects of complexity, optimization, approximation, reducibility, simulation, communication, knowledge representation, information, etc. Historically, theoretical computer science started in the 1930s with the theory of computation (computability theory) giving the exact formal border between algorithmically solvable computing problems and problems which cannot be solved by any program (algorithm). The birth of complexity theory can be set in the 1960s when computers started to be widely used and the inner difficulty of computing problems started to be investigated. At that time people defined quantitative complexity measures enabling one to compare the efficiency of computer programs and to study the computational hardness of computing problems as an inherent property of problems. The abstract part of complexity theory has tried to classify computing problems according to their hardness (computational complexity) while the algorithmic part of complexity theory has dealt with the development of methods for the design of effective algorithms for concrete problems.

The theory of computation and complexity theory provide a variety of concepts, methods, and tools building the fundamentals of theoretical computer science. The goal of this chapter is to concentrate on the intersection of formal language theory and computation (complexity) theory, on the methods developed in formal language theory and used in complexity theory as well as on the complexity of language recognition and generation. An effort in this direction is reasonable because the core formalism used in complexity and computation theory is based on words and languages, the class of algorithmically solvable problems is usually defined as a class of languages, the fundamental complexity classes are defined as classes of languages, etc.

In what follows we assume that the reader is familiar with the elementary notions and concepts of formal language theory (words, languages, automata,

Turing machines, grammars and rewriting systems, etc.) and we review the basic concepts, results, and proof methods of the computation and complexity theory using the formalism of formal language theory.

This chapter is organized as follows. In Section 2 the fundamentals of computability theory (theory of computation) are given.

Section 3 is devoted to abstract (structural) complexity theory. First the definitions of time and space complexity as basic complexity measures are given and the corresponding complexity classes are defined. Using proof methods from computability and formal language theory, strong hierarchies of complexity measures (more time/space helps to recognize more languages) are proven. The problem of proving nontrivial lower bounds on the complexity of concrete problems is discussed and nondeterminism is used in order to obtain a new insight on the classification of the hardness of computing problems. Finally, probabilistic Turing machines and the corresponding probabilistic complexity classes are introduced.

Section 4 is devoted to program-size (or descriptional) complexity. We begin by contrasting dynamic and descriptional complexities, then revisit the halting problem; random strings and random languages will be introduced and studied. Recursive and regular languages are characterized in terms of descriptive complexity and, at the end, we review a few results – obtained by program-size methods – concerning the problem **P** versus **NP**.

The last section of this chapter is devoted to parallel data processing. Alternating Turing machines are used to represent a parallel computing model that enables one to relate sequential complexity measures to parallel ones. A further extension to synchronized alternating Turing machines shows the importance of communication facilities in parallel computing. A formal language approach enabling one to study and to compare the power of different communication structures as candidates for parallel architectures (interconnection networks) closes this section.

2. Theory of computation

2.1 Computing fallibilities

This section will describe a few tasks which appear to be beyond the capabilities of computers.

From minimal art to minimal programs

According to Gardner [66], minimal art[1] – painting[2], sculpture[3], music[4] – appears to be *minimal* in at least two senses:

- it requires minimal resources, i.e., time, space, cost, thought, talent, to produce, and
- it has *some*, but rather minimal, aesthetic value.

Let's imagine with [110] that we find ourselves in a large and crowded hall where ten thousand people are talking on a large variety of subjects. The loud hubbub generated by this environment is certainly very rich in information. However, it is totally beyond human feasibility to extract one single item from it. Pushing this experiment to the extreme we reach "white noise" where all sounds that have been made, that are being made, or that will be ever made are put together. Similar experiments would consist in

- considering a canvas on which all colours are mixed to the extent that the whole painting becomes a uniform shade of grey, or
- mixing matter and anti-matter until one reaches the quantum vacuum, or
- considering a lexicon containing all writings that have been written, that are being written or that will be ever written.

In all these experiments information tends to be so "dense" and "large" that it is impossible to conceive it as a human creation: it reaches the level

---

[1] In painting, minimalism was characterized chiefly by the minimal presence of such standard "artistic" means as form and color and by the use of components that in themselves have no emotive or aesthetic significance. Minimal sculpture is often constructed by others, from the artist's plans, in commonplace industrial materials such as plastic or concrete. Music minimalism is based on the repetition of a musical phrase with subtle, slowly shifting tonalities and a rhythmic structure – if there is one. Minimal art works are not intended to embody any representational or emotional qualities but must be seen simply as what they are. See [5, 137].

[2] P. Mondrian (Composition 2, 1922), R. Tuttle (Silver Picture, 1964).

[3] C. Brâncuşi (Endless Column, 1937-8), C. Andre (Cedar Piece, 1959), Picasso (Chicago statue, 1967).

[4] La Monte Young (Trio for Strings, 1958), T. Riley (In C, 1964), S. Reich (Drumming, 1971), P. Glass (Akhnateon, 1984), J. Adams (Nixon in China, 1987).

of *randomness*.[5] What about computers and their "minimal programs"? Any computation can be done in infinitely many different ways. However, the programs of greatest interest are the smallest ones, i.e., the minimal ones.

What is the typical "behaviour" of such a program? Can we "compute" the smallest minimal programs?

To answer the first question we claim that *minimal programs should be random*. But what is a random program? According to the point of view of Algorithmic Information Theory (see [37, 24]), a random program is a program whose minimal program has roughly the same length as the generating program.[6]

Now, assume that x is a minimal program generating y. If x is not random, then there exists a program z generating x which is substantially smaller than x. To conclude, let us consider the program

from z calculate x, then from x calculate y.

This program is only a few letters longer than z, and thus it should be much shorter than x, so x is not minimal.

The answer to the second question is *negative* and the argument is related to the answer of the first question: minimal programs cannot be computed because they are random.

Like minimal art works, minimal programs tend to display an inner "randomness"; in contrast, minimal programs cannot be computed at all while minimal work arts appear to require very little resources.

Word problems

Suppose that we have fixed a finite set of strings (words) over a fixed alphabet. These strings do not need to have in themselves any meaning, but a meaning will be assigned by considering certain "equalities" between strings. These equalities will be used to derive further equalities by making substitutions of strings from the initial list into other, normally much longer, strings which contain them as "portions". Each portion may be in turn replaced by another portion which is deemed to be equal to it according to the list.

For example, from the list

$$ATE = A$$
$$CARP = ME$$
$$EAT = AT$$
$$PAN = PILLOW$$

---

[5] Xenakis [147] says that the amount of information conveyed in sounds gives the real value of music, and Eco [57] observes that a lexicon, no matter how complete or well constructed, has no poetic value.

[6] See Section 1.3.3.

we can derive, by successively substitutions,

$$LAP=LEAP$$

as

$$LAP=LATEP=LEATEP=LEAP.[7]$$

Is it possible to derive from the string CARPET the string MEAT? A possible way to get a negative answer is to notice that in every equality in our initial list the number of As plus the number of Ws plus the number of Ms is constant in each side. Computing this "invariant" for the strings above we get 1 for CARPET and 2 for MEAT, so we cannot get MEAT from CARPET.

What about the general decision problem, i.e. one in which we have an arbitrary initial (finite) list of strings, two fixed arbitrary strings x, y and we ask whether we can get from x to y by allowed substitutions? Clearly, by generating quasi-lexicographically all possible finite sequences of strings starting with x and ending with y, and then checking if such a list satisfy the required rules, one can establish equality between strings which are indeed equal. For some lists (e.g., the list displayed above) it is possible to design an algorithm to test whether two arbitrary strings are or are not equal.[8] Is it possible to do this in general? The answer is *negative* and here is an example (discovered by G. S. Tseitin and D. Scott in 1955 and modified by Gardner [65]) of an instance for which there is no single algorithm to test whether, for arbitrary strings x, y we can get from x to y:

$$
\begin{aligned}
ac &= ca \\
ad &= da \\
bc &= cd \\
bd &= db \\
abac &= abacc \\
eca &= ae \\
edb &= be.
\end{aligned}
$$

For more information on word problems see [2, 50].

Tilings

Consider a positive integer n and a $2^n \times 2^n$ square grid with only one square removed. Let us define an L-shaped tile to be a figure consisting of three squares arranged in the form of the letter L. Is it possible to cover the square grid with L-shaped tiles? The answer is *affirmative* and here is an

---

[7] We have used, in order, the first, third, and again first equality.

[8] The reader is encouraged to find such an algorithm.

inductive argument (see [152]). If $n = 1$, then the grid has the dimension 2×2 and one square has been removed: the figure is exactly an L-shaped tile. Suppose now that for a positive integer n, every $2^n \times 2^n$ square grid in which one square has been removed can be covered with L-shaped tiles. Let us consider a $2^{n+1} \times 2^{n+1}$ grid with one square removed. Cutting this grid in half both vertically and horizontally we get four $2^n \times 2^n$ square subgrids. The missing square comes from one of these four subgrids, so by applying the inductive hypothesis for that subgrid we deduce that it can be covered with L-shaped tiles. To cover the remainder subgrids, first place one L-shaped tile in the center so that it covers one square from each of the remaining subgrids. We have to cover an area which contains every square except one in each of the subgrids, so applying again the inductive hypothesis we get the desired result. Notice that the above proof can be easily converted in an algorithm constructing the required cover.

We can go one further step and ask if it possible to cover the Euclidean plane with polygonal shapes, i.e., we are given a finite number of shapes and we ask whether it is possible to cover the plane *completely, without gaps or overlaps*, with just the selected shapes. Choosing only squares, or equilateral triangles, or regular hexagons, the answer is *affirmative*. In all these cases the tilings are *periodic*, in the sense that they are exactly repetitive in two independent directions. However, H. Wang has shown the existence of non-periodic tilings. In 1961 he addressed the following question: Is there an algorithm for deciding whether or not a given finite set of different polygonal shapes will tile the Euclidean plane? Five years later, R. Berger proven that the answer is *negative*.[9]

The world of polynomials

We shall use Diophantine equations, that is equations of the form

$$P(x_1, \ldots, x_n) = 0,$$

where P is a polynomial with integer coefficients in the variables $x_1, \ldots, x_n$, to define sets of positive integers. To every polynomial $P(x, y_1, y_2, \ldots, y_m)$ with integer coefficients one associates the set

$$D = \{x \in \mathbb{N} \mid P(x, y_1, y_2, \ldots, y_m) = 0, \text{for some } y_1, y_2, \ldots, y_m \in \mathbb{Z}\}.$$

Call a set *Diophantine* if it is of the above form.

For example, the set of composite numbers is Diophantine as it can be written in the form

$$\{x \in \mathbb{N} \mid x = (y + 2)(z + 2), \text{ for some } y, z \in \mathbb{Z}\}.$$

---

[9] Berger used a set of 20 426 tiles; R. Robinson was able to reduce this number to six, and R. Penrose to two. See more in [74].

The language of Diophantine sets permits the use of existential quantifiers (by definition), as well as the logical connectives *and* and *or* (as the system $P_1 = 0$ and $P_2 = 0$ is equivalent to the equation $P_1^2 + P_2^2 = 0$, and the condition $P_1 = 0$ or $P_2 = 0$ can equivalently be written as $P_1 P_2 = 0$). Many complicated sets, including the set of all primes or the exponential set

$$\{2^{3^{\cdot^{\cdot^{\cdot^n}}}} \mid n > 1\},$$

are Diophantine.

Actually, the work of J. Robinson, M. Davis and Y. Matijasevič (see the recent book [109]) has shown that combining the fact that every possible computation can be represented by a suitable polynomial with the fact that the language of Diophantine sets permits neither the use of universal quantification nor the use of the logical negation, proves that the famous *Hilbert's tenth problem* "Does there exist an algorithm to tell whether or not an arbitrary given Diophantine equation has a solution" has a *negative* answer.

2.2 Turing machines, Chaitin computers, and Chomsky grammars

Before the work of A. Church, S. Kleene, K. Gödel, and A. Turing there was a great deal of uncertainty about whether the informal notion of an algorithm – used since Euclid and Archimedes – could ever be made mathematically rigorous.

Turing's approach was to think of algorithms as procedures for manipulating symbols in a purely deterministic way. He imagined a device, a *Turing machine*, as it has come to be called later, having a finite number of states. At any given moment the machine is scanning a single square on a long, thin tape and, depending upon the state and the scanned symbol, it writes a symbol (chosen from a finite alphabet), moves left or right, and enters a new, possibly the same, state. Despite their extreme conceptual simplicity Turing machines are very powerful.[10]

There are many different models of Turing machines. At this point we use the following variant: a Turing machine TM will have three tapes, an input tape, an output tape, and a scratch tape. Such a machine determines a partial function, φ_{TM} from the set of strings over an alphabet Σ into itself: $\varphi_{TM}(x) = y$ if TM started in its initial state, with scratch and output tapes blank, and x on its input tape, writes y on its output tape and then halts. The class of partial functions computed by Turing machines coincides with the class of partial recursive functions; the languages computed by Turing machines are the recursively enumerable languages, [21, 23, 84, 114, 128].

For information-theoretical reasons Chaitin [35] (and, independently, Levin [103]) has modified the standard notion of Turing machine by requiring

---

[10] In fact, Turing conjectured that a symbolic procedure is algorithmically computable just in case we can design a Turing machine to carry on the procedure. This claim, known as the *Church–Turing Thesis*, will be discussed later.

that as we are reading a string, we are able to tell when we have read the entire string. More precisely, we require that the input tape reading head cannot move to the left: at the start of the computation, the input tape is positioned at the leftmost binary digits of x, and at the end of the computation, for $\varphi_{TM}(x)$ to be defined, we now require that the input head be positioned on the last digit of x. Thus, while reading x, TM was able to detect at which point the last digit of x has occurred. Such a special Turing machine is called *self-delimiting Turing machine* or *Chaitin computer* (cf. [135, 24]). The class of partial recursive functions having a prefix-free domain[11] is exactly the class of partial functions computed Chaitin computers. In terms of languages, Chaitin computers have the same capability as Turing machines.

Chomsky type-0 grammars offer another way to generate languages. Before going to some details let us a fix a piece of notation. For an alphabet V we denote by V^* the free monoid generated by V (λ is the empty string); the elements of the Cartesian product $V^* \times V^*$ will be written in the form $\alpha \to \beta$, $\alpha, \beta \in V^*$.

A *Chomsky (type-0) grammar* is a system $G = (V_N, V_T, w, P)$, where V_N and V_T are disjoint alphabets, $w \in V^*$ and P is a finite subset of $V^* V_N V^* \times V^*$, where $V = V_N \cup V_T$. The elements of V_N and V_T are referred to as nonterminal and terminal letters, respectively, and w is called the start, or axiom, string.

A *derivation* of length n in G with domain a_1 and codomain a_{n+1} is a triple of finite sequences $x = (pr^x, r^x, k^x)$ such that

1. $pr^x = (a_1, a_2, \dots, a_{n+1})$ is a sequence of $n+1$ strings over V,
2. $r^x = (r_1, r_2, \dots, r_n)$ is a sequence of n elements in P,
3. $k^x = (\langle u_1, v_1 \rangle, \dots, \langle u_n, v_n \rangle)$ is a sequence of n pairs of strings over V, and
4. for each $1 \le i \le n$, $a_i = u_i \alpha v_i$, $a_{i+1} = u_i \beta v_i$, and $r_i = \alpha \to \beta$.

A derivation with domain w and codomain in V_T^* is called *terminal*. The language generated by G is defined to be the set of all codomains of terminal derivations. The languages generated by Chomsky grammars are exactly the recursively enumerable languages.[12]

2.3 Universality

Is it possible to design a "machine" capable to simulate any other "machine"? If we replace the word "machine" by Turing machine or Chaitin computer or Chomsky grammar, then the answer is *affirmative*. Proofs of this extremely

---

[11] No string in the domain is a proper prefix of another string in the domain.

[12] Traditionally, see [128], the axiom is a nonterminal letter; using a string in V^* instead of a single letter in V_T does not modify the generative capacity of grammars.

important result[13] can be found in Turing's seminal paper [144], and in many monographs and textbooks (e.g., [84, 128]). The result is true also for Chomsky grammars as well, and in the next paragraph we shall illustrate the universality with a construction of the universal Chomsky grammar.

The universal Chomsky grammar

First we have to make precise the notion of "simulation".

A *universal Chomsky grammar* is a Chomsky grammar $U = (V_N, V_T, w, P)$ with the following property: for every recursively enumerable language L over V_T there exists a string $w(L)$ (depending upon L) such that the language generated by the grammar $(V_N, V_T, w(L), P)$ coincides with L.

Theorem 2.1. [27] *There exists a universal Chomsky grammar.*

Proof. Let

$$V_N = \{A, B, C, D, E, F, H, R, S, T, X, Y\} \cup V_T \times \{1, 2, 3, 4, 5, 6, 7, 8, 9\},$$

P consists of the following rules (we have used also the set $\Sigma = \{S, X, Y\} \cup V_T$):

1. $C \to BT$,
2. $Tx \to xT$, $x \in \Sigma \cup \{D, E\}$,

3. $TDx \to (x, 2)D(x, 1)$, $x \in \Sigma$,
4. $y(x, 2) \to (x, 2)y$, $x \in \Sigma\ y \in \Sigma \cup \{D, E\}$,
5. $B(x, 2) \to (x, 3)B$, $x \in \Sigma$,
6. $y(x, 3) \to (x, 3)y$, $x, y \in \Sigma$,
7. $x(x, 3) \to (x, 4)$, $x \in \Sigma$,

8. $(x, 1)y \to (y, 5)(x, 1)(y, 1)$, $x, y \in \Sigma$,
9. $x(y, 5) \to (y, 5)x$, $y \in \Sigma, x \in \Sigma \cup \{D, E\}$,
10. $B(y, 5) \to (y, 6)B$, $y \in \Sigma$,
11. $x(y, 6) \to (y, 6)x$, $x, y \in \Sigma$,
12. $(x, 4)y(y, 6) \to (x, 4)$, $x, y \in \Sigma$,

13. $(x, 1)Ey \to (x, 7)E(y, 9)$, $x, y \in \Sigma$,
14. $(x, 1)(y, 7) \to (x, 7)y$, $x, y \in \Sigma$,
15. $D(x, 7) \to DX$, $x \in \Sigma$,
16. $(x, 9)y \to (y, 8)(x, 9)(y, 9)$, $x, y \in \Sigma$,
17. $x(y, 8) \to (y, 8)x$, $y \in \Sigma, x \in \Sigma \cup \{D, E, B\}$,
18. $(x, 9)(y, 8) \to (y, 8)(x, 9)$, $x, y \in \Sigma$,
19. $(x, 4)(y, 8) \to y(x, 4)$, $x, y \in \Sigma$,

---

[13] Which represents the mathematical fact justifying the construction of present day computers.

20. $(x,1)ED \rightarrow (x,7)RED$, $x,y \in \Sigma$,

21. $(x,9)D \rightarrow RxD$, $x \in \Sigma$,
22. $(x,9)R \rightarrow Rx$, $x \in \Sigma$,
23. $xR \rightarrow Rx$, $x \in \Sigma \cup \{D,E\}$,
24. $BR \rightarrow RC$,
25. $(x,4)R \rightarrow \lambda$, $x \in \Sigma$,

26. $Ax \rightarrow xA$, $x \in V_T$,
27. $AC \rightarrow H$,
28. $Hx \rightarrow H$, $x \in \Sigma \cup \{D,E\}$,
29. $HF \rightarrow \lambda$.

First we notice that every recursively enumerable language can be generated by a Chomsky grammar having at most three nonterminals. Indeed, if L is generated by the grammar $G = (V_N, V_T, S, P)$ and V_N contains more than three elements, say $V_N = \{S, X_1, \ldots, X_m\}$ with $m > 2$, then we define the morphism $h : V^* \rightarrow (V_T \cup \{S, A, B\})^*$ (here A, B are symbols not contained in V) by $h(S) = S$, $h(a) = a$, for all $a \in V_T$, and $h(X_i) = AB^i$, $1 \le i \le m$. Let $h(P') = \{h(u) \rightarrow h(v) \mid u \rightarrow v \in P\}$. It is easy to check that L is generated by the grammar $G' = (\{S, A, B\}, V_T, S, P')$.

To complete the proof we consider the language L generated by the grammar $G = (\{S, X, X\}, V_T, S, Q)$ and we put

$$w(L) = ASCD\alpha_1 D\alpha_2 E\beta_2 D \ldots D\alpha_k R\beta_k DF,$$

where the set of productions is $Q = \{\alpha_i \rightarrow \beta_i \mid 1 \le i \le k\}$.

We analyse now a derivation from a string

$$A\gamma CD\alpha_1 D\alpha_2 E\beta_2 D \ldots D\alpha_k R\beta_k DF. \tag{2.1}$$

The first group of rules[14] constructs the nonterminal T which selects the rule $\alpha_i \rightarrow \beta_i$ occurring in the right hand side of a nonterminal D (see rule 3). By the second group of rules, the first symbol x in α_i goes into $(x,1)$ and the nonterminal $(x,2)$ is translated into the left hand side of B, where it becomes $(x,3)$. If in γ there exists a symbol x, then by the rule 7 we construct the nonterminal $(x,4)$.

The third group of rules transforms all symbols x from α_i in $(x,1)$; then, every such x is removed from the right hand side of $(x,4)$ in case such elements do appear in the same order.

The fourth group transforms every symbol y from $\beta_i \ne \lambda$ into $(y,9)$ and translates the symbol $(y,8)$ on the left hand side. When $(y,8)$ reaches $(x,4)$ the symbol y is introduced. In this way the string α_i, erased by the third group of rules, is replaced by β_i. In case $\beta_i = \lambda$ the rule 20 is used instead of the fourth group. Accordingly, we have reconstructed the

---

[14] There are seven groups of rules separated by empty lines.

derivation from γ to γ' using the rule $\alpha_i \to \beta_i$. This procedure can be iterated by means of rules in the sixth group. If γ' does not contain any nonterminal, then by rules in the last group the string can be reduced to γ'.

So, every string in L can be generated by the universal grammar with the axiom $w(L)$.

For the converse relation we notice that the nonterminal A can be eliminated only in the case when between A and C there exists a terminal string. Every derivation has to begin with the introduction of the nonterminal T. Erasing T determines the introduction of a nonterminal $(x, 1)$, which, in turn, can be eliminated by the nonterminal $(y, 9)$. These operations are possible if and only if the string α_i has been removed from γ. One can erase $(y, 9)$ after the translation of β_i in the place occupied by α_i. The symbol R constructed in this way can be eliminated when the given string is reduced to the form (2.1). In this way we have constructed a derivation using the rule $\alpha_i \to \beta_i$. All derivations which are not of this from will be eventually be blocked.   □

2.4 Silencing a universal computer

A universal machine, be it Turing, Chomsky, or Chaitin, despite all the clever things it can do, is not *omniscient*.[15]

Let P be program whose intended behaviour is to input a string, and then output another string. The "meaning" of P may be regarded as a function from strings over an alphabet V into strings over the same alphabet, i.e. a function $f : V^* \to V^*$. Such a program P on input x

- may eventually stop, in which case it prints a string, or
- it may run forever, in which case we say, following [134], that the program has been "silenced" by x.

We shall prove that *every universal program can be silenced by some input.* Here is the argument.

List all valid programs, $P_1, P_2, \ldots, P_n, \ldots$.[16]

Now suppose, by absurdity, that we have a universal program that cannot be silenced, so, by universality, we have a universal programming language such that none of its programs can be silenced. Let g_n by the function computed by P_n and construct, by *diagonalization*, the function $f(x) = x g_{string(x)}(x)$, where $string(x)$ is the position of the string x in the quasi-lexicographical enumeration of all strings in V^*.

- The function f is computable by the following procedure: given the string x, first compute $string(x)$, then generate all programs until we reach the

---

[15] As Leibniz might have hoped.

[16] There are many ways to do it, for instance, by listing all valid programs in quasi-lexicographical order. Such an enumeration is referred to as a *gödelization* of programs, as Gödel used first this method in his famous proof of the Incompleteness Theorem [59, 60].

program $P_{string(x)}$, run $P_{string(x)}$ on the input x and then concatenate x with the result of the above computation.

- In view of the universality it follows that f has to be computed by some program P_n. Now, take the string x such that $string(x) = n$ and the input x: $f(x) = g_n(x)$, which false.

This concludes our proof.

At a first glance the contradiction in the above argument can be avoided, and here are two possibilities. First add to the initial programming language the program computing f. This construction does not work as we can use again the diagonalization method to derive a contradiction for the new language. Another possibility would be to say that f is "pathological" and we don't really want it computable. The last option is perfectly legitimate;[17] however, if we want to preserve universality we have to admit non-terminating programs.

2.5 *Digression: A simple grammatical model of brain behaviour

A universal grammar can be used in describing a model of human brain behaviour.[18] The model has four grammatical components:

- the "grammar generator",
- the "parser",
- the "semantic analyzer", and
- the universal grammar.

The "grammar generator" can be realized as a regular grammar, but the "parser" and the universal grammar are type-0 Chomsky grammars.[19] The model works as follows (see [26]): an external stimulus, a question, for example, comes to the brain under the form of a string x in some language. Then an "excitatory relay" activates the "grammar generator" which produces strings of the form

$$x_1 \to y_1/x_2 \to y_2/\cdots/x_k \to y_k$$

corresponding to the rules of that specific grammar. In fact, we assume that the grammar generator is producing exactly a string w of the form (2.1) required by the universal grammar. The string w and the input string come to a "syntactic analyzer" which decides whether x belongs or not

---

[17] The language of primitive recursive functions is such a language.

[18] We adopt here the *Computabilism Thesis for Brains* according to which the "brain functions basically like a digital computer". This thesis has been formulated, without elaboration, by Gödel in 1972, cf. [153]. We also distinguish, again with Gödel, the mind from the brain: *The mind is the user of the brain functioning as a computer.* For more details we refer to Section 1.2.8.

[19] An attempt to implement this model at the level of context-sensitive grammars is discussed in [28].

to the language identified by w. If the answer is affirmative, that is the right "competence" corresponding to x has been found, then x and w come in the universal grammar which is the core of the answering device. The universal grammar behaves now as the specific grammar and is used to "answer" the stimulus. The answering mechanism, using some information provided by a "semantic analyzer" and some memory files, generates the answer and sends it to the environment.

The full information capability of the brain can be accomplished by using only these "few" components of the model, without actually retaining all grammars corresponding to the competences of the human brain.

2.6 The halting problem

We have seen that each universal machine can be silenced by some input. Is it natural to ask the following question: Can a machine test whether an arbitrary input will silence the universal machine? The answer is again *negative*.

For the proof we will assume, without restricting generality, that all valid programs incorporate inputs – which are coded as natural numbers. So, a program may be silenced or may just eventually stop, in which case it prints a natural number. Assume further that there exists a **halting program** deciding whether the universal program is silenced by an arbitrary input, that is, by universality, whether an arbitrary program is going to be silenced or not. Construct the following program:

1. read a natural N;
2. generate all programs up to N bits in size;
3. use the **halting program** to check for each generated program whether it halts;
4. simulate the running of the above generated programs, and
5. output double the biggest value output by these programs.

The above program halts for every natural N. How long is it? It is about $\log N$ bits. Reason: to know N we need $\log N$ bits (in binary); the rest of the program is a constant, so our program is $\log N + O(1)$[20] bits.

Now observe that *there is a big difference between the size – in bits – of our program and the size of the output produced by this program*. Indeed, for large enough N, our program will belong to the set of programs having less than N bits (because $\log N + O(1) < N$). Accordingly, the program will be generated by itself – at some stage of the computation. In this case we have got a contradiction since our program will output a natural number two times bigger than the output produced by itself!

---

[20] $f(n) = O(g(n))$ means that there exists a constant c such that $|f(n)| \leq c|g(n)|$, for all n.

The following two questions:

Does the Diophantine equation $P = 0$ have a solution?
Does the Diophantine equation $P = 0$ have an infinity of solutions?

are algorithmically unsolvable, but, they have a *different degree of unsolvability*!

If one considers a Diophantine equation with a parameter n, and asks whether or not there is a solution for $n = 0, 1, 2, \ldots, N - 1$, then the N answers to these N questions really constitute only $\log_2 N$ bits of information, as we can determine which equation has a solution if we know *how many* of them are solvable. These answers are not independent. On the other hand, if we ask the second question, then the answers can be independent, if the equation is constructed properly.[21] The first question never leads to purely chaotic, random behaviour, while the second question may sometimes lead to randomness.

2.7 The Church–Turing Thesis

The Church–Turing Thesis, a prevailing paradigm in computation theory, states that no realizable computing device can be "globally" more powerful, that is, aside from relative speedups, than a universal Turing machine. It is a *thesis*, and not a theorem, as it relates an informal notion – a realizable computing device – to a mathematical notion. Re-phrasing, the Church–Turing Thesis states that the universal Turing machine is an *adequate* model for discrete computation. Here are some reasons why the Church–Turing Thesis is universally accepted:

- *Philosophical argument*: Due to Turing's analysis it seems very difficult to imagine some other method which falls outside the scope of his description.
- *Mathematical evidence*: Every mathematical notion of computability which has been proposed was proven equivalent to Turing computability.
- *Sociological evidence*: No example of computing device which cannot be simulated by a Turing machine has been given, i.e., the thesis has not been disproved despite having proposed over 60 years ago.

The Church–Turing Thesis includes a syntactic as well a physical claim. In particular, it specifies which types of computations are physically realisable. According to Deutsch [52], p. 101:

> The reason why we find it possible to construct, say, electronic calculators, and indeed why we can perform mental arithmetic, cannot be found in mathematics or logic. *The reason is that the laws of physics*

---

[21] Chaitin [37] has effectively constructed such an equation; the result is a 900 000–character 17 000–variable universal exponential Diophantine equation.

"happen" to permit the existence of physical models for the operations of arithmetic such as addition, subtraction and multiplication. If they did not, these familiar operations would be non-computable functions. We might still know *of* them and invoke them in mathematical proofs (which would presumably be called "non-constructive") but we could not perform them.

As physical statements may change in time, so may our concept of computation. Indeed, the Church–Turing Thesis has been recently re-questioned; for instance, [132] has proposed an alternative model of computation, which builds on a particular chaotic dynamical system [55] and surpasses the computational power of the universal Turing machine. See [146, 72, 73, 145, 14, 123, 81, 56, 140, 141, 113] for related ideas.

2.8 *Digression: mind, brain, and computers

Thinking is an essential, if not the most essential, component of human life – it is a mark of "intelligence".[22] In the intervening years the Church–Turing Thesis has been used to approach formally the notion of "intelligent being". In simple terms, the Church–Turing Thesis was stated as follows: *What is humanly computable is computable by a universal Turing machine.* Thus, it equates information-processing capabilities of a human being with the "intellectual capacities" of a universal Turing machine.[23] This discussion leads directly to the traditional problem of mind and matter, which exceeds the aim of this paper (see the discussion in [51, 58, 70, 126, 127, 120, 121, 130]); in what follows we shall superficially review this topic in connection with the related question: can computers think?

The responses to the mind-body problem are very diverse; however, there are two main trends, *monism*, which claims that the distinction between mind and matter is only apparent, simply, the mind is identical with the brain and its function, and *dualism*, which maintains the we have a real distinction.

The dualism can be traced to Descartes. There are many types of dualism, among them being:

- "categorical dualism" (the mind and the body are different logical entities);
- "substance dualism" illustrated by Popper or Gödel, and claiming that mind exists in a mental space outside space or time, and the brain is just a complex organ which "translates" thoughts into the corporeal movements of the body. Gödel rejected monism by saying (in Wang's words, [153], p. 164) that monism *is a prejudice of our time which will be disproved scientifically – perhaps by the fact that there aren't enough nerve cells to perform the observable operations of the mind.* This is a challenging *scientific conjecture* – indeed, the capacity of nerve cells is a scientific research topic for

---

[22] Descartes placed the essence of being in thinking.
[23] This may create the false impression of "EOE policy": *all brains are equal.*

neuroscience and the observable operations of mind are also things subject to scientific analysis. According to same author ([153], p. 169) Gödel asserted that *the brain functions basically like a digital computer*. The user of the brain functioning as a computer is just the *mind*.

- "property dualism" maintaining that the mind and our experiences are "emergent" properties of the material brain;
- "epistemic dualism", illustrated by Kant, saying that for "theoretical reason" the states of the mind are reducible to the states of the brain, but for "practical reason" such a reduction is not possible.

Von Neumann remarked that in 1940s two outstanding problems were confronting science: weather prediction and the brain's operation. Today we have a somewhat better understanding of the complexity of weather,[24] but the brain still remains a mystery. Perhaps the brain, and accordingly, the mind, are simply unsimulatable and the reason is as von Neumann remarked: *the simplest model of a neuron may be a neuron itself*. This property suggests the notion of *randomness*, which will be discussed in a separate section.

3. Computational complexity measures and complexity classes

3.1 Time and space complexities and their properties

In the early 1960s the border between algorithmically solvable problems and algorithmically unsolvable ones was already well-defined and understood and scientists knew methods powerful enough to decide whether a given computing problem is decidable (algorithmically solvable) or undecidable. Because of the growth of computer use in many areas of everyday life the interest of researchers has moved to questions like how to measure the effectiveness of computer programs (algorithms), how to compare the effectiveness of two algorithms, and how to measure the computational difficulty of computing problems. Dealing with these questions two fundamental complexity measures, time and space, have been introduced by Hartmanis et. al. [78, 77]. Both these complexity measures are considered as functions of the input. Informally, the time complexity of an algorithm working on an input is the number of "elementary" operations executed by the algorithm processing the given input. In another words, it is the amount of work done to come from the input to the corresponding output. The space complexity of an algorithm is the number of "elementary" cells of the memory used in the computing process. Obviously, what does "elementary" mean depends on the formal model of algorithms one chooses (machine models, axiomatic models, programming languages, etc.). Since the theory always tries to establish results

---

[24] Even if the weather equations behave chaotically, that is a small difference in the data can cause wildly different weather patters.

(assertions) which are independent of the formalism used and have a general validity, this dependence of the measurement on the model does not seem welcome. Fortunately, all reasonable computing models used are equivalent, in the sense that the differences in the complexity measurement are negligible for the main concepts and statements of the complexity theory.

Here, we use Turing machine (TM) as the standard computing model of computation theory to define the complexity measures. From the several versions of Turing machine we consider the off-line multitape Turing machine (MTM) consisting of a finite state control, one two-way read-only input tape with one read-only head and a finite number of infinite working tapes, each with one read/write head. This is the standard model used for the definitions of time and space complexities because it clearly separates the input data (input tape) from the computer memory (working tapes) [78]. A **computing step** (shortly, step) of a MTM is considered to be the elementary operation of this computing model. In one step a MTM M reads one symbol from each of its tapes (exactly those symbols are read which are on the positions where the heads are adjusted) and depending on them and the current state of the machine M, M possibly changes its state, rewrites the symbols read from the working tapes and moves the heads at most one position to the left or to the right. A configuration of a MTM M is the complete description of the global state of the machine M including the state of the finite control, the current contents of all tapes, and the positions of all heads on the tapes. A computation of M is a sequence of configurations $C_1, C_2, \ldots, C_m$ such that $C_i \vdash C_{i+1}$ (C_{i+1} is reached in one step from C_i). A formal description of $MTMs$ can be found in all textbooks on this topic and we omit it here. Now, we are ready to define the complexity measures.

Definition 3.1. *Let M be a MTM recognizing a language $L(M)$ and let $x \in Z^*$, where Z is the input alphabet of M. If $C = C_1, C_2, \ldots, C_k$ is the finite computation of M on x, then the* **time complexity of the computation of M on x** *is*

$$T_M(x) = k - 1 .$$

If the computation of M on x is infinite, then

$$T_M(x) = \infty .$$

The **time complexity of** M *is the partial function from* $\mathbb{N}$ *to* $\mathbb{N}$,

$$T_M(n) = \max\{T_M(x) \mid x \in \Sigma^n \cap L(M)\}.$$

The space complexity of one configuration C of M, $S_M(C)$, is the length of the longest word over the working alphabet stored on the working tapes in C. The **space complexity of a computation** $D = C_1, \ldots, C_m$ *is*

$$S_M(D) = \max\{S_M(C_i) \mid i = 1, \ldots, m\}.$$

The **space complexity of** M **on a word** $x \in L(M)$ *is* $S_M(x) = S_M(D_x)$, *where* D_x *is the computation of* M *on* x.
The **space complexity of** M *is the partial function from* $\mathbb{N}$ *to* $\mathbb{N}$,

$$S_M(n) = \max\{S_M(x) \mid x \in \Sigma^n \cap L(M)\}.$$

One can observe that the function $T_M(x)(S_M(x))$, as a function from $L(M)$ to $\mathbb{N}$ is the most precise (complete) description of the complexity behaviour of the machine M. But this description is not useful for the comparison of complexities of different algorithms ($MTMs$). It is so complex that one can have trouble to find a reasonable description of it. Thus, we prefer to consider the complexity measures as functions of the input sizes rather than of specific inputs. The time and space complexities are so called **worst case** complexities because $T_M(n)(S_M(n))$ is maximum of the complexities over all words of length n from $L(M)$ (i.e., M can recognize every input from $L(M) \cap \Sigma^n$ in time $T_M(n)$ and there exists $x \in L(M) \cap \Sigma^n$ such that $T_M(n) = T_M(x)$). In abstract complexity theory we usually prefer this worst–case approach to define the complexity measures, but in the analysis of concrete algorithms we are often interested to learn the average behaviour of the complexity on inputs of a fixed length. In such case one has to make a probabilistic analysis of the behaviour of $T_M(x)$ according to the probability distribution of inputs of a fixed length.

Another interesting point in the definition above is that to define $T_M(n)$ and $S_M(n)$ we have considered $T_M(x)$ and $S_M(x)$ only for words from $L(M)$. This is because we allow infinite (or very complex) computations of M on words in $\Sigma^* - L(M)$. In what follows we show that if $T_M(S_M)$ is a "nice" function, then it does not matter for the complexity of the recognition of the language $L = L(M)$ whether one defines T_M as above or as $T_M(n) = \max\{T_M(x) \mid x \in \Sigma^n\}$. The idea is to construct a machine M' simulating M in such a way that if M' simulates more than $T_M(n)$ steps of M on an input y, then M' halts and rejects the input. Obviously, M' is able to do it in this way if M' is able to compute the number $T_M(|y|)$ in time $O(T_M(|y|))$. Before starting to describe this we show that it is sufficient to study the asymptotical behaviour of $T_M(n)$ and $S_M(n)$.

Definition 3.2. [78, 77] *Let* t *and* s *be two functions from* $\mathbb{N}$ *to* $\mathbb{N}$. *Then*

$$TIME(t(n)) = \{L \mid L = L(M) \text{ for a MTM } M \text{ with } T_M(n) \leq t(n)\},$$

and

$$SPACE(s(n)) = \{L \mid L = L(M) \text{ for a MTM } M \text{ with } S_M(n) \leq s(n)\}.$$

Theorem 3.1. [77] *Let* c *be a positive real number and let* $s : \mathbb{N} \to \mathbb{N}$ *be a function. Then*

$$SPACE(s(n)) = SPACE(c \cdot s(n)).$$

Idea of proof. Without loss of generality we assume $c < 1$. Then, $SPACE(c \cdot s(n)) \subseteq SPACE(s(n))$ is obvious. To show that $SPACE(s(n)) \subseteq SPACE(c \cdot s(n))$ we use a compression of the contents of the tapes. Let M be a MTM recognizing a language $L(M) \in SPACE(s(n))$ with $S_M(n) \leq s(n)$. Let $k = \lceil 1/c \rceil$ and let Γ be the working alphabet of M. Then one constructs a MTM M' with the working alphabet Γ^k. This enables to store in one cell of the tapes of M' the contents of k adjacent cells of tapes of M. The detailed construction of the rules of M' is left to the reader. $\square$

The same idea of the compression can be used to reach the following speed-up theorem.

Theorem 3.2. [78] *For every constant* $c > 0$ *and every function* $t : \mathbb{N} \to \mathbb{N}$ *such that* $\liminf\limits_{n \to \infty} t(n)/n = \infty$

$$TIME(t(n)) = TIME(c \cdot t(n)).$$

Obviously, the compression of tapes in the theorems above is not completely fair because we save space and time according to the complexity definition by storing larger data in one memory cell and executing more complicated operations in a step of the simulating machine. If one fixes the working alphabet to $\{0, 1\}$ this effect will be impossible. So, the lack of difference between $c \cdot f(n)$ and $f(n)$ for complexity considerations on Turing machines is rather a property of the computing model than a complexity property of algorithms. On the other hand, in studying the complexity of concrete problems one primarily estimates the asymptotical behaviour (polynomial growth, for instance) of complexity functions. Thus, we are satisfied with the study of asymptotical behaviours of complexity functions and, accordingly, we can accept Turing machines as a computing model for the complexity measurement.

Definition 3.3. *A function* $s : \mathbb{N} \to \mathbb{N}$ *is called* **space-constructible** *if there exists a* MTM M *having the following two properties:*

(i) $S_M(n) \leq s(n)$ *,for all* $n \in \mathbb{N}$, *and*

(ii) for every $n \in \mathbb{N}$, M *starting on the input* 0^n *computes* $0^{s(n)}$ *on the first working tape and halts in a final state.*

A function $t : \mathbb{N} \to \mathbb{N}$ *is called* **time-constructible** *if there exists a* MTM A *having the following two properties:*

(iii) $T_M(n) = O(t(n))$, *and*

(iv) for every $n \in \mathbb{N}$, M *starting on the input* 0^n *computes* $0^{t(n)}$ *on the first working tape and halts in a final state.*

Now, we can show that for constructible functions it does not matter whether $T_M(n)$ is defined as $\max\{T_M(x) \mid x \in L(M) \cap \Sigma^n\}$ or as $\max\{T_M(x) \mid x \in \Sigma^n\}$.

Lemma 3.1. *Let* $t : \mathbb{N} \to \mathbb{N}$ $(s : \mathbb{N} \to \mathbb{N})$ *be a time-constructible (space-constructible) function. Then for every* $L \in TIME(t(n))$ $(L \in SPACE(s(n)))$, $L \subseteq \Sigma^*$, *there exists a MTM* M *such that*

(i) $L(M) = L$, *and*
(ii) for every $x \in \Sigma^*, T_M(x) = O(t(|x|))$ $(S_M(x) \leq S(|x|))$.

Proof. We give the proof for the time complexity. Since $L \in TIME(t(n))$ we may assume there is a MTM A with $L(A) = L$ and $T_A(n) \leq t(n)$. If A has k working tapes, then we construct M with $k+1$ working tapes acting on the input x as follows:

1. M computes $0^{t(|x|)}$ on the first working tape in time $O(t(n))$.
2. M simulates $t(|x|)$ steps of the computation of A on x.
3. If A accepts x, then M accepts x too. If A halts and rejects x, then M rejects, too. If A does not halt after $t(|x|)$ steps, then M halts and rejects x.

Note that step 1 can be done because t is time-constructible. Since M uses k free working tapes to simulate A with k working tapes, M can simulate one step of A in one step. Rewriting 0 to 1 in each simulation step M one can check that $t(|x|)$ simulation steps were executed. Since each accepting computation of A on a word $x \in L(A)$ fulfills $T_A(x) \leq T_A(|x|)$ one sees that every computation longer than $T_A(|x|)$ cannot lead to acceptance of x (i.e. $x \notin L(A)$). Thus $L = L(A) = L(M)$ and $T_M(|x|) = O(t(|x|))$ for every $x \in \Sigma^*$. □

As we have seen the complexity classes bounded by two functions f and g are the same if $f(n) = \Theta(g(n))$.[25] A natural and important question has arisen. Which increase of the growth of g in the comparison to the growth of f is sufficient and necessary to reach $TIME(f(n)) \subset TIME(g(n))$ or $SPACE(f(n)) \subset SPACE(g(n))$? The answer to this question is provided by the following theorems which have been established by generalizing the well-known diagonalization method from computability theory.

Before starting to formulate them, we observe that every MTM A can be simulated by a MTM B with one working tape only in the same space [77]. This is because the i-th cell of the working tape of B can save the tuple containing all symbols on the i-th positions of all working tapes of A.

---

[25] If there are two constants c, N such that for all $n \geq N$, $f(n) \geq cg(n)$, then we say that $f(n) = \Omega(g(n))$. We say that $f(n) = \Theta(g(n))$ in case $f(n) = O(g(n))$ and $f(n) = \Omega(g(n))$.

Theorem 3.3. [77] *Let* s_1 *and* s_2 *be two functions having the following properties:*

(i) $s_2(n) \geq s_1(n) \geq \log_2 n$ *for every* $n \in \mathbb{N}$,
(ii) s_2 *is space-constructible,*
(iii) $\liminf\limits_{n \to \infty} \frac{s_1(n)}{s_2(n)} = 0$.

Then, $SPACE(s_1(n)) \subset SPACE(s_2(n))$.

Proof. The proof idea is based on the diagonalization technique. The simple version of this technique orders all Turing machines in an enumerable sequence $T_1, T_2, \ldots, T_i, \ldots$ and, for any $i \in \mathbb{N}$, it chooses a word $x_i \in \{0,1\}^*$ ($x_i \neq x_j$ for $i \neq j$). Then, the diagonal language L_d is defined as $\{x_i \mid x_i \notin L(T_i),\ i \in \mathbb{N}\}$. Since, for every $i \in \mathbb{N}$, $L(T_i)$ and L_d differs in the behaviour on x_i we call x_i the **candidate** for the difference between $L(T_i)$ and L_d in the process of the choice of x_i.

There are two reasons why this simple approach fails to work directly for our theorem. We cannot enumerate the sequence of $MTMs$ working in space s_1 because it is not decidable whether a given MTM is $s_1(n)$-space bounded. This means we have to take the sequence of all $MTMs$, and for every MTM M of this sequence (even in case M is not $s_1(n)$-space bounded), we have to try to find a candidate for the difference between $L(M)$ and the diagonal language. Obviously, if a machine M is not $s_1(n)$-space bounded the candidate may fail because we do not need to have the difference between $L(M)$ and L_d. We must be only sure that our candidates will work for $s_1(n)$-space bounded machines.

A second, more serious problem is the following one. L_d has to be in $SPACE(s_2(n))$, i.e., there must exist a MTM M_2 such that $L(M_2) = L_d$ and $S_{M_2}(n) \leq s_2(n)$. Since M_2 has a working alphabet of a fixed size and we want to simulate the work of other $s_1(n)$-space bounded machines on the chosen candidates we need more than $s_1(n)$ space to do it if the working alphabet Γ of the simulated machine is larger than the working alphabet of M_2. More precisely, M_2 needs $\lceil \log_2 |\Gamma| \rceil \cdot s_1(n)$ space for this simulation which may be greater than $s_2(n)$ for finitely many small n's. Thus, for words of small lengths M_2 is not able to simulate the machine with large Γ. It means we cannot fix the candidates x_i's for T_i's without knowing that they are large enough to have $s_2(|x_i|) \geq (\log_2 |\Gamma_i|) \cdot s_1(|x_i|)$, where Γ_i is the working alphabet of T_i.

To overcome this difficulty we choose for every MTM T_i an infinite set of candidates for the difference between $L(T_i)$ and $L(M_2)$. Then, we can be sure that one candidate y will be large enough for $s_2(|y|) \geq (\log_2 |\Gamma_i|) \cdot s_1(|y|)$ and so M_2 can simulate T_i on y and accepts (rejects) if T_i has rejected (accepted).

Now, we are prepared to give the formal proof. As we have already observed it is sufficient to consider $MTMs$ with one working tape only. Let $T_1, T_2, \ldots$ be sequence of all $MTMs$ with one working tape and the input alphabet $\{0,1\}$, and let $\tilde{T}_1, \tilde{T}_2, \ldots$ be their binary codes. The infinite set $X_i = \{x \in \{0,1\}^* \mid x = 0^n 1 \tilde{T}_i\}$ will be considered as the set of candidates for the difference between $L(T_i)$ and $L(M_2)$. Obviously $X_i \cap X_j = \emptyset$ for $i \neq j$.

We define the diagonal language as the language $L(M_2)$ accepted by M_2 acting on every input w as follows:

1. M_2 computes $0^{S_2(|w|)}$ on the first working tape.
2. M_2 considers w as $0^r 1 \tilde{T}$ for some $r \in \mathbb{N}$ and $\tilde{T} \in \{0,1\}^*$. If $|\tilde{T}| > S_2(|w|)$, then M_2 halts and rejects w, else M_2 continues with the step 3.
3. M_2 decides whether $\tilde{T}$ is a code of a MTM T with one working tape and the input alphabet $\{0,1\}$. If not, M_2 halts and rejects w. If the answer is yes, then M_2 continues with the step 4.
4. M_2 writes the number $2^{S_2(|w|)}$ in binary on the second working tape. On the third working tape M_2 simulates step by step the computation of T on w. The working alphabet Γ of T is coded by the binary working alphabet of M_2.
    (a) If M_2 needs more than $s_2(|w|)$ space on the third tape (i.e., if $s_2(|w|) < \lceil \log_2(|\Gamma|) \rceil \cdot s_1(|w|)$ or T is not s_1-space bounded), then M_2 halts and rejects w.
    (b) If M_2 has successfully simulated $2^{s_2(|w|)}$ steps of T and T does not halt, then M halts and accepts w.
    (c) If T halts on w in fewer than $2^{s_2(|w|)}$ steps and M_2 succeeds to simulate all these steps, then M_2 accepts w iff T rejects w.

We observe that M_2 always halts and that $S_{M_2}(n) \leq s_2(n)$ for every $n \in \mathbb{N}$.

Now, we have to show that $L(M_2) \notin SPACE(s_1(n))$. We assume that $L(M_2) \in SPACE(s_1(n))$. Then there exists a MTM M with one working tape such that $L(M) = L(M_2)$ and $S_M(n) \leq s_1(n)$. Let Q be the set of states of M, and let Γ be the working alphabet of M. Obviously, for a given input w, there are at most

$$r(|w|) = |Q| \cdot s_1(|w|) \cdot (|\Gamma| + 1)^{s_1(|w|)} \cdot |w|$$

different configurations of M. This means that any computation of M on w consisting of more than $r(|w|)$ steps is infinite because M is deterministic and some configuration has occurred twice in the computation. Since $\liminf_{n \to \infty} s_1(n)/s_2(n) = 0$, there exists a positive integer n_0 such that:

(i) $|\tilde{M}| \leq s_2(n_0)$,
(ii) $\lceil \log_2(|\Gamma|) \rceil \cdot s_1(n_0) \leq s_2(n_0)$, and
(iii) $|Q| \cdot s_1(n_0) \cdot (|\Gamma| + 1)^{s_1(n_0)} \cdot n_0 < 2^{s_2(n_0)}$.

Set $x = 0^j 1 \tilde{M}$ for a $j \in \mathbb{N}$ such that $|x| = n_0$. If $x \in L(M)$, then there exists an accepting computation of M on x of length at most $r(|x|)$. Because (i), (ii), (iii) are fulfilled, M_2 succeeds in simulating the whole computation of M on x. Thus, M_2 rejects x according to 4(c). If $x \notin L(M)$, then two different reasons may determine it:

1. M halts in at most $r(|x|)$ steps and rejects x. Then, following the step 4(c), M accepts x.
2. The computation of M on x is infinite. Then according to (i), (ii), (iii), M_2 succeeds in simulating the first $2^{s_2(|x|)} = 2^{s_2(n_0)}$ steps of M on x. Because of step 4(b) M_2 halts and accepts x.

Thus, we have showed "$x \in L(M)$ iff $x \notin L(M_2)$" which is a contradiction.
$\square$

The assertion above shows that the power of Turing machines strongly grows with the asymptotical growth of the space complexity used. For space bounds growing slower than $\log_2 n$ the situation is a little bit different. A survey of such small space-bounded classes can be found in [67, 139]. We only mention that $SPACE(0(1))$ [constant space] is exactly the class of regular languages and that if $SPACE(f(n))$ contains at least one non-regular language, then $\limsup\limits_{n\to\infty} f(n)/\log\log_2 n > 0$. Thus the space classes $SPACE(s(n))$ with $s(n) = o(\log\log_2 n)$ [26] contain only regular languages.

A similar hierarchy can be achieved for time complexity, but it is not so strong as the space hierarchy. The reason is that we do not have only the problem to simulate different working alphabets of arbitrarily large cardinalities by one working alphabet of the "diagonalizing" MTM, but we have to fix the number of working tapes for the diagonalizing MTM while the simulated MTMs may have any number of working tapes. Since, for any $k \in \mathbb{N}$, we do not know a linear time simulation of an arbitrary number of tapes by k working tapes we formulate the hierarchy result as follows.

Theorem 3.4. [78] *Let there exist a positive integer k, a function $f : \mathbb{N} \to \mathbb{N}$, and a simulation of an arbitrary MTM A by a MTM with k working tapes in $O(T_A(n) \cdot f(n))$ time. Let t_1, t_2 be two functions from $\mathbb{N}$ to $\mathbb{N}$ such that*

(i) $t_2(n) \cdot f(n) \geq t_1(n)$ for every $n \in \mathbb{N}$,
(ii) t_2 is time-constructible, and
(iii) $\liminf\limits_{n\to\infty} \frac{t_1(n)\cdot f(n)}{t_2(n)} = 0$.

Then $TIME(t_1(n)) \subset TIME(t_2(n))$.

Due to the fact that every $t(n)$ time-bounded MTM can be simulated by a MTM with two working tapes in $O(t(n) \cdot \log_2(t(n)))$ [80], we deduce the relation

$$TIME(t_1(n)) \subset TIME(t_1(n) \cdot \log_2(t_1(n)) \cdot q(n)),$$

for any monotone, unbounded $q : \mathbb{N} \to \mathbb{N}$.

---

[26] $f(n) = o(g(n))$ in case $\lim_{n\to\infty} \frac{f(n)}{g(n)} = 0$.

The hierarchy result above shows that there are languages of arbitrarily large complexity, i.e., we have an infinite number of levels for the classification of the computational difficulty of language recognition (computing problems). If $L \in TIME(t_1(n)) - TIME(t_2(n))$, then we say that $t_1(n)$ is the **upper bound** on the time complexity of L and $t_2(n)$ is the **lower bound** on the time complexity of L. So, the upper bound t_1 for L means that there is an algorithm (MTM) A recognizing L with $T_A(n) \leq t_1(n)$ and the lower bound $t_2(n)$ means that there is no algorithm (MTM) B recognizing L with $T_B(n) \leq t_2(n)$. At this point one may wish to define the time (space) complexity of a language (computing problem) as the complexity of the best (asymptotically optimal) algorithm (MTM) recognizing L. But this is impossible because there are languages with no best MTM. This fact is more precisely formulated for time complexity in the following version of Blum's Speed-up Theorem (this result works for an arbitrary Blum space, cf. [23]).

Theorem 3.5. [17] *There exists a recursive language L such that for any MTM M_1 accepting L, there exists a MTM M_2 such that*

(i) $L(M_2) = L(M_1)$, and
(ii) $T_{M_2}(n) \leq \log_2(T_{M_1}(n))$ for almost all $n \in \mathbb{N}$.

Idea of proof. The assertion of Blum's Speed-up Theorem may seem curious and surprising at first sight. But the following idea shows an explanation of this phenomenon. First, one can observe that for every language L' there are infinitely many $MTMs$ accepting it. Then, one has to construct a language L which cannot be recognized efficiently by any MTM (program) of small size (small index), but as MTM (program) sizes increase faster and faster $MTMs$ accepting L more and more efficiently exist. The formal construction of L can be done by diagonalization.   □

We see that the theorem above can be applied infinitely many times to L and so there is no optimal MTM accepting L. For every concrete MTM M there exists a larger MTM M' working more efficiently than M.[27]

This is the reason why we cannot generally define the complexity of a computing problem (language) as the complexity of the optimal algorithm for it. But, we can always speak about lower and upper bounds on the problem complexity which is sufficient for the classification of computing problems according to their computational difficulty.

We close this section studying the relation between time complexity and space complexity. We observe that $TIME(t(n)) \subseteq SPACE(t(n))$ for any function $t(n) \geq n$ because no MTM can use more cells of the working

---

[27] The speed-up phenomenon is non-constructive; for instance, there is no recursive function of the initial index that gives a bound for the exceptional values in Blum speed-up, but that there is a recursive bounding function of the speed-up index, [22]. The class of speedable functions is large, [31].

tapes than the number of executed steps of M is. On the other hand we have $SPACE(s(n)) \subseteq \bigcup_{c \in \mathbf{N}} TIME(c^{s(n)})$ for any $s(n) \geq \log_2 n$. This is because for every $s(n)$-space bounded MTM M there is a constant c such that the number of different configurations of M with a fixed content of the input tape is bounded by $c^{s(n)}$. These above two relations suggest that space may be much more powerful than time. The best result currently known in this direction is contained in the following assertion.

Theorem 3.6. [82] *For any function* $t : \mathbf{N} \to \mathbf{N}$ *such that* $t(n)/\log(t(n)) = \Omega(n)$:

$$TIME(t(n)) \subseteq SPACE(t(n)/\log(t(n))).$$

3.2 Classification of problems according to computational difficulty and nondeterminism

The hierarchy theorems show that there are problems of arbitrarily large computational difficulty. The main practical interest is in the classification of concrete computing problems according to their computational difficulty. To see the importance of this let us consider the following classical example [79]. Let us have four algorithms $(MTMs)$ A_1, A_2, A_3, and A_4. Let $T_{A_1}(n) = 5n, T_{A_2}(n) = n \cdot \log n, T_{A_3}(n) = n^2$, and $T_{A_4}(n) = 2^n$. Then for the realistic input size $n = 1000$ we have $T_{A_1}(1000) = 5000, T_{A_2}(1000) = 9966, T_{A_3}(1000) = 1000000$, and $T_{A_4}(1000)$ is a 302-digit number. For comparison, the number of protons in the known universe has 126 digits, and the number of microseconds since the "Big Bang" has 24 digits. Thus, A_4 is not useful for any practical purposes because already $T_{A_4}(100)$ is a 31-digit number. If A_4 is an optimal algorithm for some problem L, then we cannot algorithmically solve L in general. We can see this also from the opposite side. Every computer has a finite memory and we always have a bound on the time we can wait for a result. These two constants together with the complexity functions of the given algorithm A bound the size of inputs which can be processed by the algorithm A on a given computer. We observe that, for algorithms with exponential complexity, the bounds are very small because, for $f(n) = 2^n, f(n + 10) \geq 10^3 \cdot f(n)$. Considerations similar to those made above have lead to the classification of computing problems into **tractable** (feasible) problems admitting a polynomial-time solution [46] and **intractable** problems for which no polynomial algorithm exists. The basic class for the classification of problems (languages) is the class

$$P = \bigcup_{k \in \mathbf{N}} TIME(n^k).$$

An important observation is that the definition of P is invariant with respect to all "reasonable" computing models used, i.e., if one finds a polynomial algorithm for a problem in one formalism than we can be sure that there exist polynomial algorithms for this problem in all other formalisms, and if one proves $L \notin P$ in one of the computing model formalisms then this is true for all.

Some further fundamental complexity classes are:

$$DLOG = SPACE(\log_2 n),$$
$$PSPACE = \bigcup_{k \in \mathbb{N}} SPACE(n^k),$$
$$EXPTIME = \bigcup_{k \in \mathbb{N}} TIME(2^{n^k}).$$

They satisfy the following relations:

$$DLOG \subseteq P \subseteq PSPACE \subseteq EXPTIME.$$

As we have already mentioned above, the theory of computation provides successful methods for the classification of languages (problems) into recursive (decidable) and nonrecursive (undecidable) ones. Unfortunately, this is not true for the decision whether a language L is in P or not. Usually it is not very hard to prove that $L \in P$, because it is sufficient to find a polynomial MTM accepting L. So, if $L \in P$, then it is rarely a problem to find a polynomial algorithm for L. An exception is the problem of linear programming which was not known to be in P, nor to be not in P for a longer time. In 1979 an ingenious polynomial-time algorithm was found for it [98].

The main difficulty is in proving lower bounds on concrete computing problems. We do not have any mathematical method enabling to prove higher than quadratic lower bounds on the time of restricted computing models (for instance, one-tape TM) or superlinear (higher than linear) lower bounds on the time of general computing models (register machines). Thus, we cannot really classify computing problems because we are unable to obtain tight lower bounds on their complexity.

To overcome the absence of absolute lower bounds (i.e., the evidence that some computing problems are difficult) Cook [47] has introduced a method enabling one to prove so-called **relative** lower bounds which are viewed as a strong implication of hardness. The idea of this method is based on the reduction between computing problems – a classical mathematical approach to problem solving. The novelty consists in connecting the reduction, of one problem to another one, with the complexity.

Definition 3.4. [47] *Let $L_1 \subseteq \Sigma_1^*$ and $L_2 \subseteq \Sigma_2^*$ be two languages. We say, that L_1 is **polynomial-time reducible** to L_2 (denoted $L_1 \leq L_2$) if there is a polynomial-time bounded MTM that for every word $x \in \Sigma_1^*$ writes x_2 on the first working tape (considered as an output tape here) such that $x \in L_1 \iff y \in L_2$. We say that L_1 and L_2 are **polynomial-time equivalent** if $L_1 \leq L_2$ and $L_2 \leq L_1$.*

We observe that "L_1 is polynomial-time reducible to L_2" means that L_1 cannot be "much more difficult" than L_2, i.e. if $L_2 \in P$ then L_1 must be in P, too. Thus, if L_1 is **polynomial-time equivalent** to L_2, then $L_1 \in P$ iff $L_2 \in P$.

The first idea behind the relative lower bounds (a strong implication to be difficult) is that if one finds many computing problems polynomial-time equivalent (reducible) each to the other, and no polynomial-time algorithm is known for any of these problems, then we have a large experience (strong implication) that each of these problems is computationally difficult. This idea is still strengthen by considering nondeterminism as follows (we assume that the reader is familiar with the concept of nondeterminism and with nondeterministic $MTMs$ [116]).

Definition 3.5. *Let M be a nondeterministic MTM with an input alphabet Σ. Let $x \in \Sigma^*$ and C be a computation of M on x. We denote by $T_M(C)$ the length of C minus 1. For every $x \in L(M)$, the* **time complexity of M on x** *is*

$$T_M(x) = \min\{T_M(C) \mid C \text{ is an accepting computation of } M \text{ on } x\}.$$

The **time complexity of M** *is a partial function $T_M : \mathbb{N} \to \mathbb{N}$ such that*

$$T_M(n) = \max\{T_M(x) \mid x \in L(M) \cap \Sigma^n\}.$$

Further, for any $f : \mathbb{N} \to \mathbb{N}, f(n) \geq n$,

$$NTIME(f(n)) = \{L \mid L = L(M) \text{ for a nondeterministic } MTM \text{ } M$$
$$\text{with } T_M(n) \leq f(n)\}.$$

We note that one can represent all possible computations of a nondeterministic MTM M on a given word as a directed possibly infinite tree $Tr_M(w)$ whose nodes are labeled by configurations of M. The root of $Tr_M(w)$ is labeled by the initial configuration $C_0(w)$ of M on w. Each node labeled by C has indegree 1 and outdegree equal to the number of all possible nondeterministic actions from the configuration C. M accepts w if and only if $Tr_M(w)$ contains at least one accepting configuration. We observe that $T_M(x)$ is bounded by the depth of $Tr_M(x)$ because $T_M(x)$ is the minimum of the distances between the root of $Tr_M(x)$ and accepting configurations of $Tr_M(x)$. A deterministic simulation of the work of M on an input w is usually a search for an accepting configuration in $Tr_M(w)$. Since the size of $Tr_M(w)$ may be exponential in the depth of $Tr_M(w)$ (and so exponential in $T_M(w)$) this deterministic search for an accepting computation consists of exponentially many steps according to $T_M(w)$. All simulations of nondeterministic machines by deterministic ones cause an exponential increase of time complexity. This is one reason to believe that

$$P \subset NP = \bigcup_{k \in \mathbb{N}} NTIME(n^k).$$

Another reason for that is that for some mathematical problems the deterministic time corresponds to the complexity of the search for a solution while the nondeterministic time corresponds to the complexity of verifying whether a given candidate for the solution is a consistent solution. This is because a nondeterministic algorithm can guess the solution (for instance the values of variables over $\{0, 1\}$ in a system of equations) in real time and then check whether this guess was correct. For many such problems we cannot find better deterministic algorithms than those searching for a solution by generating exponentially many candidates for it. Thus, we have enough experience to believe $P \subset NP$.

Now, we add these two ideas together by saying that a problem is relatively hard if it is one of the hardest in NP in the following sense.

Definition 3.6. [47] *A language L is called NP-complete if*

(i) $L \in NP$, and
(ii) $\forall L' \in NP$ L' is polynomially reducible to L.

Clearly,

(i) If $P \subset NP$ then every NP-complete language is in $NP - P$ (i.e., intractable).
(ii) If an NP-complete language $L \in P$, then $P = NP$.

A few thousands NP-complete problems are known, and for none of them we have a polynomial-time algorithm. Moreover, we do not know any deterministic time-effective simulation of nondeterministic computations and we do not believe that to find a solution is not much harder than to verify whether a given candidate for the solution is correct. All these facts together provide a large experience supporting the opinion that NP-complete problems do not have polynomial algorithms.

We omit examples and proofs of NP-completeness in this short survey. A nice overview on this topic can be found in [64]. We note only that we can also define complete problems in other classes like $P, PSPACE$ too. If one defines P-complete problems according to the $\log_2 n$-space reduction, then such P-complete problems are candidates for membership in $P - DLOG$. This question has been found interesting from the very beginning of the complexity theory because each polynomial time computation contains at most a polynomial number of different configurations. To generate any polynomial number of configuration the space $O(\log_2 n)$ is sufficient. Thus $P = DLOG$ would mean that the memory of every polynomial-time algorithm can be optimized within the $\log_2 n$-space bound. But we conjecture that there are problems in P (exactly the P-complete problems according to $\log_2 n$-space reduction) which require polynomial time as well as polynomial space to be solved.

We conclude this section by giving the fundamental relations among the basic complexity classes.

Definition 3.7. *Let M be a nondeterministic MTM with an input alphabet Σ. Let $x \in \Sigma^*$ and $C = C_1, \ldots, C_k$ be a computation of M on x. We denote by $S_M(C_i)$ the length of the longest content of the working tapes of M in the configuration C_i. $S_M(C) = \max\{S_M(C_i) \mid i = 1, \ldots, k\}$. For every $x \in L(M)$, the* **space complexity of M on x** *is*

$$S_M(x) = \min\{S_M(C) \mid C \text{ is an accepting computation of } M \text{ on } x\}.$$

The **space complexity of M** *is a function $S_M : \mathbb{N} \to \mathbb{N}$ such that*

$$S_M(n) = \max\{S_M(x) \mid x \in L(M) \cap \Sigma^n\}.$$

Further, for any $f : \mathbb{N} \to \mathbb{N}, f(n) \geq n$,

$$NSPACE(f(n)) = \{L \mid L(M) \text{ for a nondeterministic MTM } M \text{ with}$$

$$S_M(n) \leq f(n)\}.$$

$$NLOG = NSPACE(\log_2 n),$$

$$NPSPACE = \bigcup_{k \in \mathbb{N}} NSPACE(n^k).$$

First, we observe that there exists a space-efficient simulation of nondeterministic computations.

Theorem 3.7. [129] *Let $f(n) \geq \log_2 n$ is space-constructible. Then*

$$NSPACE(s(n)) \subseteq SPACE((s(n))^2).$$

Idea of proof. Let $L = L(M)$, $S_M(n) \leq s(n)$. Then there exists a constant c such that the number of different configurations of the nondeterministic MTM M on an input of length n is bounded by $c^{s(n)}$, for every $n \in \mathbb{N}$. So, if M accepts a word w, then $Tr_M(w)$ contains an accepting configuration in the distance at most $c^{s(|w|)}$ from the root.

Let $C_0(w)$ be the initial configuration of M on w. Without loss of generality we can assume that there is a unique accepting configuration C'. We have to test whether C' is reachable from $C_0(w)$ in exactly $c^{s(|w|)}$ steps. We can do it by testing for every configuration C whether C is reachable from $C_0(w)$ in $c^{s(|w|)/2}$ steps and C' is reachable from C in $c^{s(|w|)/2}$ steps. By this approach the space needed to determine whether C_1 is reachable from C_2 in 2^i steps is equal to the space needed to record the configuration C plus the space needed to test if one can reach one configuration from another one in 2^{i-1} steps. Using this approach recursively until one has to test whether one configuration can be reached from another one in one step we have to store at most $\log_2(c^{s(n)}) = O(s(n))$ configurations. Since each configuration is of size $s(n)$, the simulation uses $O((s(n))^2)$ space. $\qquad\square$

Corollary 3.1. $PSPACE = NPSPACE$

As we have already mentioned, the best known time-bounded simulation causes an exponential increase of time.

Theorem 3.8. *Let* t *be a time-constructible function. Then*

$$NTIME(t(n)) \subseteq \bigcup_{c \in \mathbb{N}} TIME(c^{t(n)}).$$

Idea of proof. Let $L \in NTIME(t(n))$, i.e., $L = L(M)$ for some $t(n)$-time bounded nondeterministic NTM. Obviously, M is $t(n)$-space bounded, too. As we already know, there exists a constant d such that the number of configurations of M on an input of the length n is bounded by $d^{t(n)}$. A deterministic MTM can generate all configurations of M reachable from the given initial configuration in time $c^{t(n)}$, for some suitable constant c.   □

The fundamental complexity hierarchy is as follows.

Theorem 3.9.

$$DLOG \subseteq NLOG \subseteq P \subseteq NP \subseteq PSPACE \subseteq EXPTIME$$

Proof. The following inclusions $DLOG \subseteq NLOG$, $P \subseteq NP$, $PSPACE \subseteq EXPTIME$ are obvious. Since $NP \subseteq NPSPACE = PSPACE$ we obtain $NP \subseteq PSPACE$. $NLOG \subseteq P$, because every nondeterministic $\log_2 n$-space bounded MTM has at most a polynomial number of configurations in any computation tree for an input of length n.   □

All inclusions in Theorem 3.9 are believed to be proper, but nobody has proven it for any of them. To prove at least one proper inclusion or equality in this hierarchy is one of the central open problems of theoretical computer science. One supposes that this problem is very hard because it can be proven that the usual machine simulations, as a method for proving equality between two language (complexity) classes, and the diagonalization, as a method for proving non-equality among complexity classes, do not work for any of the inclusions of the fundamental complexity hierarchy. How can one prove that these methods do not work? We illustrate this approach on the most important problem $P =?NP$ (see, for instance, [64, 7, 8, 116]).

An **oracle machine** is a pair (M, L), denoted M^L too, where M is a MTM with one additional oracle tape and L is an oracle. M has special states $q_?, q_Y, q_N$, and if M is in the state $q_?$ then M without looking on the content of its tapes, enters the state $q_Y(q_N)$ if the content of the oracle tape is (not) in L. A move from $q_?$ to q_Y or q_N according to L is considered as one step of M^L. After this, the content of the oracle tape is erased. We denote $L(M^A)$ the language accepted by the oracle machine M^A and

$P^A = \{L(M^A) \mid M^A$ is a polynomial-time bounded oracle machine$\}$,

$NP^A = \{L(M^A) \mid M^A$ is a nondeterministic polynomial-time bounded oracle machine$\}$.

Theorem 3.10. [6] *There exist two languages* A, B *such that*

(i) $P^A = NP^A$
(ii) $P^B \subset NP^B$

Idea of proof. To prove (i) it is sufficient to take a $PSPACE$-complete language A. For this choice of A one can easily observe that $P^A = PSPACE = NPSPACE = NP^A$. The proof of (ii) is much more involved. The idea is to find an oracle B and a language $L \in NP^B - P^B$ in such a way $L = \{0^i \mid B$ contains a word of the length $i\}$ and $|B \cap \{0,1\}^n| \le 1$, for any $n \in \mathbb{N}$. A nondeterministic MTM M^B can accept L in linear time by guessing the word x from $B \cap \{0,1\}^n$, for the input 0^n, and by verifying its nondeterministic decisions by asking the oracle B whether $x \in B$. If the oracle B is cleverly constructed then no polynomial-time bounded MTM can find in polynomial time the candidate $x \in B \cap \{0,1\}^n$ among the 2^n words of the length n. □

The above theorem shows that we cannot prove $P = NP$ by any usual simulation and $P \ne NP$ by any typical diagonalization. Suppose one could "simulate" nondeterministic polynomial-time bounded $MTMs$ by polynomial-time bounded $MTMs$. All known simulations of a machine M_1 by a machine M_2 remain valid if one attaches the same oracle to both machines M_1 and M_2. This means that $P = NP$, proven by such simulation implies $P^C = NP^C$, for every oracle C. But (ii) of Theorem 3.10 contradicts this. The same is true for the diagonalization method. All usual diagonalization proofs for the non-equality $G \ne H$ between two language classes G and H would also work when oracles are attached ($G^L \ne H^L$, for every language L). But we have an oracle A such that $P^A = NP^A$ which means that usual diagonalizations cannot help to separate P from NP.

Finishing this section we note that we omit to present two large intersections of formal language theory and complexity theory here. One is devoted to the characterizations of basic complexity classes by different kind of automata (multihead automata, pushdown machines, etc.). A nice overview about this topic can be found by Lange [100]. Another topic omitted is devoted to the complexity of classical problems of formal language theory like string matching, parsing, etc. Each one of this research areas involves enough result for a monograph and according to the size restrictions of this chapter we do not try to give a survey of them.

3.3 Hard problems and probabilistic computations

What to do if one has proven the evidence or a strong implication that a given computing problem is computationally difficult? One can stop any attempt to solve the problem by giving the mathematical arguments justifying why the problem cannot be solved on a computer. This may be good for pure theory but quite unsatisfactory if there is a large practical interest to have a program solving the problem. In this case one uses one of the following approaches.

1) **Deterministic Algorithms**

Let us assume that our problem has exponential time complexity. Then one can still try to find a MTM M (algorithm, program) which solves the problem in time complexity close to $T_M(n) = \frac{1}{100} \cdot 2^{n/50}$. It means, that in spite of the fact that M has an exponential complexity, one can still effectively compute results for inputs of several hundreds bits. Thus, if the lengths of the input from practical applications are in the range where $T_M(n)$ is not too large, then one has an useful algorithmical solution for a hard problem.

2) **Problem Restrictions**

Usually, you do not need to solve the given computing problem in the generality of its original formulation. In several cases some additional restrictions essentially decrease the computational difficulty of the problem. The most typical case is to restrict the set of potential inputs only (for instance, the Post Correspondence Problem starts to be decidable if one considers inputs over one-letter alphabet only, the Satisfiability Problem is in P if one restricts the inputs to conjunctive normal formulas in which the length of the elementary disjunctions are bounded by two).

3) **Approximate Algorithms**

If our difficult problem is an optimization problem, one can make it easier searching for a solution which is not too far from the optimum, but it is not necessarily the optimum. There are NP-complete optimization problems for which good approximations of optimal solutions can be found in polynomial time. Since the practice is often satisfied with approximate solutions this is one of the most successful methods for the design of algorithms for hard problems.

4) **Probabilistic Algorithms**

Probabilistic algorithms are based on nondeterministic ones. Each nondeterministic step (branching) is interpreted as a random decision (tossing coin). While for nondeterministic computations an accepting configuration in the computation tree has been sufficient to accept the input, for probabilistic computations we require a "sufficiently large" probability to reach an accepting configuration (correct output). Thus, instead of surely correct outputs produced by deterministic algorithms one computes an output whose probability to be correct is large enough. Usually the probabilistic algorithms allow repeated runs on the same input due to which the probability to get the right answer tends to 1 with the size of the input.

5) **Probabilistic Approximate Algorithms**

A combination of the two previous methods leads to algorithms providing with high probability solutions of optimization problems which are very close to optimum.

6) **Heuristics**

Heuristic algorithms are similar to probabilistic ones in that they make random decision during their computations, too. The difference is in that we are unable to prove that heuristic algorithms provide good solutions (outputs) despite of the fact that they successfully work in some practical applications (at least for most of the inputs generated in practice). The ideas of the algorithm design are based on some analogies to efficient optimization processes running in the nature (biology, physics). Two of the most popular representants are genetic algorithms and simulated annealing. But theoretical as well as experimental results show that the methods previously described are much more reliable than heuristic ones and the use of heuristic methods is recommended only in cases in which one is not able to find an efficient solution by the previous five approaches.

Note that the above approaches may be also mixed. For instance, we have polynomial-time approximate algorithms for Traveling Salesman Problem in Euclidean space.

Out of the six approaches mentioned above we shall give more details about the probabilistic one only. There are two reasons for this: i) one can use the formalism of formal language theory to describe probabilistic algorithms, and ii) the power of randomized computations is one of the topics of main interest in the recent complexity theory.

We distinguish two types of probabilistic MTMs (algorithms) called Las Vegas MTMs and Monte Carlo MTMs. Las Vegas algorithms always give the right result and their complexity on an input x is measured as the "weighted average" over all computations on x. Monte Carlo algorithms give the right result with some probability greater than $1/2$ (i.e., some computations may lead to wrong results) and one usually considers that the time complexity is measured by the depth of the computation tree. In what follows we give more details about these two models of probabilistic computations in the formalism of formal language theory.

Definition 3.8. *A nondeterministic MTM M is called a* **Las Vegas MTM M** *if*

(i) the degree of nondeterminism is bounded by two (i.e. all computation trees are binary trees), and

(ii) for every $x \in L(M)$ [$x \notin L(M)$] all leaves of $Tr_M(x)$ are accepting [rejecting] configurations.

Let $D \in Tr_M(x)$ denote the fact that D is a computation of M on x corresponding to a path in $Tr_M(x)$ leading from the root to a leaf of $Tr_M(x)$. The probability of executing $D \in Tr_M(x)$ is $Prob(D) = 2^k$, where k is the number of nondeterministic choices of D. Let $|D|$ denote the length of D. **Las Vegas time complexity of M on an input x** *is*

$$LVT_M(x) = \sum_{D \in Tr_M(x)} Prob(D) \cdot |D|.$$

Las Vegas time complexity of M *is*

$$LVT_M(n) = \max\{LTV_M(x) \mid |x| = n\}.$$

We observe that polynomial Las Vegas algorithms are very useful in practical applications because they are reliable (they compute the right outputs), and the outputs are computed in polynomial time with a high probability.

Definition 3.9. *A nondeterministic MTM M is called* **one-sided** *error Monte Carlo MTM if*

(i) *the degree of nondeterminism is bounded by two,*

(ii) *there is a function* $f : \mathbb{N} \to \mathbb{N}$ *such that all computations on the inputs of the length* n *have length* $f(n)$ *and each one uses exactly* $f(n) - 1$ *nondeterministic guesses (i.e. all* $2^{f(n)-1}$ *computations have the same probability)*

(iii) *if* $x \in L(M)$, *then at least* $2^{f(n)-2}$ *computations on* x *finish in accepting states, and*

(iv) *if* $x \notin L(M)$, *then all computations on* x *finish in rejecting states.*

A nondeterministic MTM M is called **two-sided error Monte Carlo MTM** *if it satisfies* (i), (ii), *and*

(v) *there exists a constant* ϵ *such that if* $x \in L(M)(x \notin L(M))$, *then more than* $(2^{f(n)-1}/2)(1 + \epsilon)$ *computations finish in accepting (rejecting) states.*

We observe that Monte Carlo algorithms are not reliable as Las Vegas ones, but one-sided error Monte Carlo algorithms are still extremely useful in solving decision problems. The reason for this claim is that we can iterate the algorithm on the same input a certain number of times with independent random choices in order to make the error probability arbitrarily small. For instance, k runs of a one-sided error Monte Carlo algorithm on the same input produce k answers. If at least one is "yes", then the answer "yes" is certainly correct. If all answers are "no", then we can conclude that the answer is "no" with probability $1 - (\frac{1}{2})^k$. For sufficiently large k the error 2^{-k} is smaller than the probability that our hardware/software fails during the execution of the k runs of the algorithm.

A nice example of the Monte Carlo approach is the recognition of the language

$$Com = \{w \in 1\{0,1\}^* \mid w \text{ is the binary code of a composite number}\}.$$

Obviously, the simple approach based on trying to divide the given input number n by all numbers $n \in \{1, 2, \ldots, \lceil \sqrt{n} \rceil\}$ has exponential complexity in the input length $\lceil \log_2 n \rceil$. We do not know any polynomial-time deterministic algorithm for Com, but one can construct an $O((\log_2 n)^5)$ algorithm assuming the extended Riemann hypothesis [111]. In what follows we present a polynomial time one-sided error Monte Carlo algorithm for Com [136, 124]. It is based on the following two results of the number theory. Let $GCD(a, b)$ denote the greatest common divisor of the numbers a and b.

Lemma 3.2. *If a positive integer* $n > 2$ *is composite, then there exists an integer* $a, 1 \leq a \leq n$, *such that*

(i) $a^{n-1} \not\equiv 1$ *(mod n) or*
(ii) there exists an integer i *such that* 2^i *divides* $n - 1$ *and* $1 < GCD(a^{(n-1)/2^i} - 1, n) < n$.

Definition 3.10. *Let* n *be a positive integer. An integer* $a, 1 \leq a \leq n$, *is called a* **compositeness witness for n** *if (i) or (ii) of Lemma 1.22 hold.*

Lemma 3.3. *If* $n \geq 3$ *is an odd integer, then at least* $(n - 1)/2$ *distinct integers from* $\{1, 2, \ldots, n - 1\}$ *are compositeness witnesses for* n.

So, if for a fixed composite number n one randomly chooses a number $a \in \{1, 2, \ldots, n - 1\}$, then the probability that a is a compositeness witness for n is at least $1/2$. On the other hand, if n is a prime, then no $a \in \{1, 2, \ldots, n - 1\}$ is a compositeness witness for n. This two facts yield a Monte Carlo MTM (algorithm) M working as follows:

1) If n is even, then M accepts.
2) If n is odd, then M randomly chooses a number $a \in \{1, 2, \ldots, n - 1\}$. If a is a compositeness witness for n then M accepts. Otherwise, M rejects.

The quwstion whether a is a compositeness witness for n can be checked in polynomial time. Obviously, if M accepts, then the input is certainly composite. If M rejects, then one can state that n is prime with probability at least $1/2$. As we have already seen, if one wishes a smaller error probability, the algorithm has to be executed repeatedly with the same input n.

We note that there even exists a Las Vegas algorithm deciding whether a given number is prime or not [3], but it is too technical to be presented here.

Probabilistic algorithms are very efficient, but only "probably correct". However, many probabilistic algorithms – in particular, the above primality probabilistic algorithm – can be "theoretically" converted into rigorous, deterministic algorithms provided a sufficiently long random string[28] input is supplied [44, 30]; this possibility is only "theoretical" as the set of random strings is not recursively enumerable [24].

A recent quantum algorithm[29] (based on Fourier transformation) proposed by Shor [131] seems to indicate that primality can be checked in polynomial time on a quantum computer.

---

[28] See Section 1.4.3.
[29] One decisive feature of quantum computation is *parallelism*: during a computation cycle a quantum computer is processing all coherent paths *at once*. See more in [4, 9, 10, 32, 61, 62, 52, 53, 54, 15, 16, 12, 13, 142, 143].

From a practical point of view one can consider the class of languages (problems) recognized (solved) in polynomial time by some probabilistic algorithms to be the class of practically (solvable) languages (problems). Because of this it is reasonable to consider the following language classes, introduced by Gill [68]:

$$ZPP = \{L \mid L = L(M), \quad \text{for a Las Vegas } MTM \ M \text{ working in}$$
$$\text{polynomial time}\},$$
$$R = \{L \mid L = L(M), \quad \text{for a one-sided error Monte Carlo } MTM$$
$$\text{working in polynomial time}\},$$
$$BPP = \{L \mid L = L(M), \quad \text{for a two-sided error Monte Carlo } MTM$$
$$\text{working in polynomial time}\}.$$

The following relations can be easily proven.

Theorem 3.11.

$$P \subseteq ZPP = R \cap coR \subseteq R \subseteq NP,$$

$$R \cup coR \subseteq BPP \subseteq PSPACE.$$

Whether some of the inclusions above are proper or not is unknown. So, we do not know any NP-complete language recognizable by one-sided error Monte Carlo in polynomial time. But we have either examples of problems in P for which probabilistic algorithms are more effective than the best deterministic algorithms or we have polynomial-time probabilistic algorithms for problems whose membership in P is unknown. Thus, we do not conjecture that a NP-complete problem may have a polynomial-time probabilistic solution.

4. Program-size complexity

4.1 Dynamic versus program-size complexities

Let us recall that a Chaitin computer C is a partial recursive function carrying strings (on A) into strings such that the domain of C is prefix-free. If C is a computer, then T_C denotes its time complexity, i.e., $T_C(x)$ is the running time of C on the entry x, if x is in the domain of C; $T_C(x)$ is undefined in the opposite case.

The property of universality discussed in Section 1.2.3 can be presented, in a stronger form, as follows:

Theorem 4.1. [Invariance Theorem] *There exists a universal Chaitin computer* U *with the property that for every computer* C *there exists a con-*

stant $sim(U, C)$ - *which depends upon* U, C - *such that in case* $C(x) = y$, *there exists*[30] *a string* x' *satisfying the following conditions:*

$$U(x') = y, \tag{4.1}$$

$$|x'| \leq |x| + sim(U, C). \tag{4.2}$$

Indeed, let (C_i) be a gödelization of all Chaitin computers and define $U(a_1^i a_2 u) = C_i(u)$, for all strings u (here a_1, a_2 are two distinct letters from the alphabet A).

The *program-size* or *Chaitin complexity* associated with the Chaitin computer C is the partial function H_C defined by $H_C(x) = \min\{|u| \in A^* \mid C(u) = x\}$ iff such a u does exist. The above result can be rephrased as follows:

Theorem 4.2. *There exists a Chaitin computer* U *such that for every Chaitin computer* C *there effectively exists a constant* c - *depending upon* U *and* C - *such that for all strings* x,

$$H_U(x) \leq H_C(x) + c.$$

So, U not only simulates every Chaitin computer C, but the simulation is asymptotically optimal. Is it possible to prove a similar result for a dynamical complexity, e.g., time complexity? The answer is *negative* and here is a proof (cf. [25]).

Assume, for the sake of a contradiction, that U can simulate every other computer (4.1), in a shorter time. Formally, to equation (4.1) we add the constraint:

$$T_U(x') < T_C(x). \tag{4.3}$$

For every string x in the domain of U let

$$t(x) = \min\{T_U(z) \mid z \in A^*, U(z) = U(x)\}, \tag{4.4}$$

i.e., $t(x)$ is the minimal running time necessary for U to produce $U(x)$.[31]

Next define the *temporal canonical program (input)* associated with x to be the first (in quasi-lexicographical order) string $x^\#$ satisfying the equation (4.4):

$$x^\# = \min\{z \in dom(U) \mid U(z) = U(x), T_U(z) = t(x)\}.$$

So,

$$U(x^\#) = U(x), \text{ and } T_U(x^\#) = t(x).$$

As the universal computer U can simulate itself, it follows from (4.3) that there exists a string x' such that $U(x') = U(x^\#) = U(x)$, and $T_U(x') < T_U(x^\#) = t(x)$, which is false.

---

[30] And can be effectively constructed.
[31] Actually, $t(x)$ is not computable.

The reason for the above phenomenon can be illustrated by showing the existence of "small-sized" computers requiring "very large" running times. To this aim we use the information-theoretic version of the Busy Beaver function Σ. For every natural m let us denote by $string(m)$ the mth string in quasi-lexicographical order, and let $\Sigma(n)$ be the largest natural number whose algorithmic information content is less than or equal to n, i.e.

$$\Sigma(n) = \max\{m \in \mathbb{N} \mid H_U(string(m)) \leq n\}.$$

Chaitin ([39], 80–82, 189) has shown that Σ grows larger than any recursive function, i.e., for every recursive function f, there exists a natural number N, which depends upon f, such that $\Sigma(n) \geq f(n)$, for all $n \geq N$:[32] indeed, there is a constant q such that every program of length n either halts in time less than $\Sigma(n + q)$, or else it never halts.

As $H_U(string(\Sigma(n))) \leq n$, it follows that $U(y_n) = string(\Sigma(n))$, for some string y_n of length less than n. This program y_n takes, however, a huge amount of time to halt: there is a constant c such that for large enough n, $U(y_n)$ takes between $\Sigma(n - c)$ and $\Sigma(n + c)$ units of time to halt. To conclude, the equation (4.1) is compatible with (4.2), but incompatible with (4.3).[33]

4.2 The halting problem revisited

Can the halting problem be solved if one could compute program-size complexity? The answer is **yes** and here is a proof (cf. [43]).

Fix a universal Chaitin computer U and denote H_U simply by H. We have seen that if an n-bit program p halts, then the time t it takes to halt satisfies $H(t) \leq n + q$. So if p has run for time T without halting, and if for all $t \geq T$ one has $H(t) > n + q$, then p will never halt.

Consider the recursively enumerable set of all true upper bounds on H, $Ch = \{(x, k) \in A^* \times \mathbb{N} \mid \{H(x) \leq k\}$. Imagine enumerating this set, and keep track of the running time. Assuming that H is computable, compute $H(x)$ for each n-bit string x. Then enumerate Ch until we get the best possible upper bound on $H(x)$ for all n-bit strings x. Let $\beta(n)$ be defined to be the time it takes to enumerate enough of the set of all true upper bounds on program-size complexity until one obtains the correct value of $H(x)$ for all n-bit strings x. If one is given n and a number greater than $\beta(n)$, one can determine an n-bit bit string x_{max} with maximum possible complexity

$$H(x_{max}) = n + H(string(n)) + O(1).$$

---

[32] The difficulty might be also explained by the fact that Σ grows as fast as the least time necessary for all programs of length less than n that halt on U to stop, [34].

[33] Incidentally, the above discussion shows that, contrary to what Penrose has suggested (see [121], p. 560), there is no incompatibility between strong determinism and computability: it is indeed impossible for a (universal) machine to "learn its own theory".

Thus any number $k \geq \beta(n)$ has

$$n + H(string(n)) - q' < H(x_{max}) \leq H(string(k)) + H(string(n)) + q''$$

and

$$H(k) > n - q' - q''.$$

Thus we can use $\beta(n)$, which is computable from H, to solve the halting problem as follows: an n-bit program p halts iff it halts before time $\beta(n + q + q' + q'')$.

As a bonus we derive the fact that the information-theoretic Busy Beaver function Σ is computable from H: the formula

$$\Sigma(n) = \max\{U(p) \mid |p| \leq n\},$$

proves that Σ is computable relative to the halting problem which, in turn, is computable from H.

4.3 Random strings

Consider the number

one million, one hundred one thousand, one hundred and one.

This number appears to be

the first number not nameable in under ten words.

However, the above expression has only **nine** words, pointing out a naming inconsistency: it is an instance of Berry's paradox.

It follows that the property of nameability[34] is inherently ambiguous and, consequently, too powerful to be freely used.

Of course, the above analysis is rather vague. We can make it more rigorous by using a universal Chaitin computer U. Some programs for U specify positive integers: when we run such a program on U the computation eventually halts and produces the number. In other words, a program for U "specifies" a positive integer in case after running a finite amount of time it prints the number. What we get is the statement:

THE FIRST POSITIVE INTEGER THAT CANNOT BE SPECI-FIED BY A PROGRAM FOR U WITH LESS THAN N BITS.

However, **there is a program for** U, **of length** $\log N + c$, **for calculating the number that supposedly cannot be specified by any program of** N **bits!** And, of course, for large N, $\log N + c$ is much smaller than N.

---

[34] Another famous example refers to the classification of numbers as *interesting* or *dull*. There can be no dull numbers: if they were, the first such number would be *interesting* on account of its dullness.

Suppose that persons A and B give us a sequence of 32 bits each, saying that they were obtained from independent coin flips. If A gives the string $x = 01101000100110101101100110100101$ and B gives the string $y = 00000000000000000000000000000000$, then we would believe A and would not believe B: the string x *seems* to be random, but the string y does not. Why? The strings are extremely different from the point of view of *regularity*: the second string has a maximum regularity which allows us to express it in a very compact way, *only zeros*, while the first one appears to have no shorter definition at all.[35]

Classical probability theory is not sensitive to the above distinction, as strings are all equally probable. Laplace (1749-1827) was, in a sense, aware of the above paradox when he wrote (cf. [101], pp.16-17):

> *In the game of heads and tails, if heads come up a hundred times in a row then this appears to us extraordinary, because after dividing the nearly infinite number of combinations that can arise in a hundred throws into regular sequences, or those in which we observe a rule that is easy to grasp, and into irregular sequences, the latter are incomparably more numerous.*

Non-random strings are strings possessing some kind of regularity, and since the number of all those strings (of a given length) is **small**, the occurrence of such a string is **extraordinary**. The overwhelming majority of strings have hardly any "computable" regularities – they are random. Randomness means the absence of any compression possibility; it corresponds to maximum information content (because after dropping any part of the string, there remains no possibility of recovering it). Borel (1909), Von Mises (1919), Ville (1939), and Church (1940) elaborated on this idea, but a formal model of irregularity was not found until the mid-1960s in the work of Kolmogorov [99] and Chaitin [33]. Currently, it appears that the model due to Chaitin [35], which will be briefly discussed in what follows, is the best adequate model (cf. [37, 41, 24]).

A string is random in case it has maximal program-size complexity when compared with the program-size complexity of all strings of the same length. As for every $n \in \mathbb{N}$, one has:

$$\max_{x \in A^n} H(x) = n + H(string(n)) + O(1),$$

and is led to the following definition: a string $x \in A^*$ is **(Chaitin)** m-**random** (m is a positive integer) if $H(x) \geq \Gamma(|x|) - m$; x is **(Chaitin)** **random** if it is 0-random. Here $\Gamma(n) = \max\{H(x) \mid x \in A^*, |x| = n\}$.

---

[35] The distinction between regular and irregular strings becomes sharper and sharper for longer and longer strings (e.g., it is easier to specify the number

$$10^{10^{10^{10}}}$$

than the first 100 digits of π).

The above definition depends upon the fixed universal computer U; the generality of the approach comes from the Invariance Theorem. Obviously, for every length n and for every $m \geq 0$ there exists an m-random string x of length n.

It is worth noticing that randomness is an asymptotic property: the larger is the difference between $|x|$ and m, the more random is x. There is no sharp dividing line between randomness and pattern, but it was proven that all m-random strings x with $m \leq H(string(|x|))$ have truly random behaviour [24].

A random string cannot be algorithmically compressed. Incompressibility is a non-effective property: no individual string, except finitely many, can be proven random. Under these circumstances it is doubtful that we can exhibit an example of a random string, even though the vast majority of strings are random. However, we can describe a non-effective construction of random strings. We start by asking ourselves: How many strings x of length n have maximal complexity, i.e., $H(x) = \Gamma(|x|)$? Answer: There exists a natural constant $c > 0$ such that

$$\gamma(n) = \#\{x \in A^* \mid |x| = n, H(x) = \Gamma(|x|)\} > Q^{n-c},$$

for all natural n; here Q is the cardinality of the alphabet A.[36]

Now fix a natural base $Q \geq 2$, and write $\gamma(n)$ in base Q. The resulting string (over the alphabet containing the letters $0, 1, \ldots, Q-1$) is itself **random** (cf. [42]).[37]

4.4 From random to regular languages

Recall that $\{string_Q(n) \mid n \geq 0\}$ is the enumeration of all strings over the alphabet A (having $Q \geq 2$ elements) in quasi-lexicographical order. A language $L \subset A^*$ is described by its binary characteristic sequence $1, 1_i = 0$ iff $string_Q(i) \in L$. A language L is **random** if its characteristic function is random (as an infinite, binary sequence). Denote by **RAND** the set of random languages. In what follows we shall present different characterizations of random languages.

We start with a piece of notation. For every sequence $1 = l_1 l_2 \ldots l_n \ldots$ we denote by $1(n) = l_1 l_2 \ldots l_n$, the prefix of length n of 1.

The unbiased discrete probability on $B = \{0, 1\}$ is defined by the function

$$h : 2^A \to [0, 1], h(X) = \frac{\#X}{2},$$

---

[36] How large is c? Out of Q^n strings of length n, at most $Q + Q^2 + \cdots + Q^{n-m-1} = (Q^{n-m} - 1)/(Q - 1)$ can be described by programs of length less than $n - m$. The ratio between $(Q^{n-m} - 1)/(Q - 1)$ and Q^n is less than 10^{-i} as $Q^m \geq 10^i$, irrespective of the value of n. For instance, this happens in case $Q = 2$, $m = 20$, $i = 6$; it says that *less than one in a million among the binary strings of any given length is not* 20-*random*.

[37] This string is random because it represents a large number.

for all subsets $X \subset B$. This uniform measure induces the product measure μ on the set of binary infinite sequences B^ω: for all strings $x \in B^*$, $xB^\omega = \{\mathbf{y} \in B^\omega \mid \mathbf{y}(|x|) = x\}$, and

$$\mu(xB^\omega) = 2^{-|x|}.$$

If $x = x_1 x_2 \ldots x_n \in B^*$ is a string of length n, then $\mu(xB^\omega) = 2^{-n}$ and the expression $\mu(\ldots)$ can be interpreted as "the probability that a sequence $\mathbf{y} = y_1 y_2 \ldots y_n \ldots \in B^\omega$ has the first element $y_1 = x_1$, the second element $y_2 = x_2, \ldots$, the nth element $y_n = x_n$". Independence means that the probability of an event of the form $y_i = x_i$ does not depend upon the probability of the event $y_j = x_j$.

Every open set $G \subset B^\omega$, i.e., a set $G = \cup_{x \in X} xB^\omega$, for some prefix-free subset $X \subset B^*$, is μ measurable and

$$\mu(G) = \sum_{x \in X} 2^{-|x|}.$$

Finally, $S \subset B^\omega$ is a *null set* in case for every real $\varepsilon > 0$ there exists an open set G_ε which contains S and $\mu(G_\varepsilon) < \varepsilon$. For instance, every enumerable subset of B^ω is a null set.

A property P of sequences $\mathbf{x} \in B^\omega$ is *true almost everywhere in the sense of* μ in case the set of sequences not having the property P is a null set. The main example of such a property was discovered by Borel and it is known as the *Law of Large Numbers*. For every sequence $\mathbf{x} = x_1 x_2 \ldots x_m \ldots \in B^\omega$ and natural number $n \geq 1$ put

$$S_n(\mathbf{x}) = x_1 + x_2 + \cdots + x_n.$$

Then,

> The limit of S_n/n, when $n \to \infty$, exists almost everywhere in the sense of μ and has the value $1/2$.

It is clear that a sequence satisfying a property false almost everywhere with respect to μ is very "particular". Accordingly, it is tempting to try to say that

> a sequence $\mathbf{x}$ is "random" iff it satisfies every property true almost everywhere with respect to μ.

Unfortunately, we may define for every sequence $\mathbf{x}$ the property $P_{\mathbf{x}}$ as following:

> $\mathbf{y}$ satisfies $P_{\mathbf{x}}$ iff for every $n \geq 1$ there exists a natural $m \geq n$ such that $x_m \neq y_m$.

Every $P_{\mathbf{x}}$ is an asymptotic property which is true almost everywhere with respect to μ and $\mathbf{x}$ does not have property $P_{\mathbf{x}}$. Accordingly, *no sequence can verify all properties true almost everywhere with respect to μ.* The above definition is vacuous!

However, there is a way (due to Martin-Löf, cf. [108]) to overcome the above difficulty: we consider not *all* asymptotic properties true almost everywhere with respect to μ, but only a *sequence* of such properties. In this context the important question becomes: *What sequences of properties should be considered?* Clearly, the "larger" the chosen sequence of properties is, the "more random" will be the sequences satisfying that sequence of properties. As a constructive selection criterion seems to be quite natural, we will impose the minimal computational restriction on objects: each set of strings will be recursively enumerable, and every convergent process will be regulated by a recursive function.

A **constructively open set** $G \subset B^\omega$ is an open set $G = XB^\omega$ for which $X \subset B^*$ is recursively enumerable. A **constructive sequence of constructively open sets,** for short, **c.s.c.o. sets** is a sequence $(G_m)_{m \geq 1}$ of constructively open sets $G_m = X_m B^\omega$ such that there exists a recursively enumerable set $X \subset B^* \times \mathbb{N}$ with

$$ X_m = \{x \in B^* \mid (x, m) \in X\}, $$

for all natural $m \geq 1$. A **constructively null set** $S \subset B^\omega$ is a set such that there exists a c.s.c.o. sets $(G_m)_{m \geq 1}$ for which

$$ S \subset \bigcap_{m \geq 1} G_m, $$

and

$$ \lim_{m \to \infty} \mu(G_m) = 0, \text{ constructively,} $$

i.e., there exists an increasing, unbounded, recursive function $H : \mathbb{N} \to \mathbb{N}$ such that $\mu(G_m) < 2^{-k}$ whenever $m \geq H(k)$.

It is clear that $\mu(S) = 0$, for every constructive null set, but the converse is not true.

Here are some properties equivalent to randomness:

Theorem 4.3
- (Martin-Löf) A sequence $\mathbf{x} \in B^\omega$ is random iff $\mathbf{x}$ is not contained in any constructive null set.
- (Chaitin-Schnorr) A sequence $\mathbf{x} \in B^\omega$ is random iff there exists a natural $c > 0$ such that

$$ H(\mathbf{x}(n)) \geq n - c, $$

for all natural $n \geq 1$.

- **(Chaitin)** A sequence $\mathbf{x} \in A^\omega$ is random iff

$$\lim_{n \to \infty} (H(\mathbf{x}(n)) - n) = \infty.$$

Random languages have remarkable properties:

Theorem 4.4
- [24] No random language is recursively enumerable; in fact, every random language is immune in the sense that it contains no infinite recursively enumerable language.
- [24] Random languages are closed under finite variation.
- [19, 18] If $L = \{string_Q(i) \mid l_i = 1\}$ is random and $f : \mathbb{N} \to \mathbb{N}$ is recursive and one-one, then the language $\{string_Q(i) \mid l_{f(i)} = 1\}$ is also random.
- [19, 18] Let $L \subset A^\omega$ be a union of constructively closed sets that is closed under finite variation. Then

$$\mu(L) = 1 \text{ iff } X \cap \mathbf{RAND} \neq \emptyset.$$

- [19, 18] Let L be an intersection of constructively open sets that is closed under finite variation. Then

$$\mu(L) = 1 \text{ iff } \mathbf{RAND} \subset L.$$

- [63, 24] Every language is Turing reducible to a random language.

To characterize recursive and regular languages by means of descriptional complexity we need to introduce the *blank-endmarker or Kolmogorov-Chaitin complexity* (see [99, 33]) associated with a universal Turing machine TM: $K(x) = \min\{|y| \mid TM(y) = x\}$.

Recursive languages can be characterized as follows

Theorem 4.5. ([40]) *A language* $L \subset A^*$ *is recursive iff one of the following two equivalent conditions holds true:*

- *there exists a constant* c *(depending upon* L*) such that* $K(l_1 l_2 \cdots l_n) < K(string_Q(n)) + c$, *for all positive integers* n,
- *there exists a constant* c' *(depending upon* L*) such that* $K(l_1 l_2 \cdots l_n) < \log_2 n + c'$, *for all positive integers* n.

It is worth noticing that recursive sequences cannot be characterized by the property "$H(l_1 l_2 \cdots l_n) < H(string_Q(n)) + O(1)$", as shown in [135] (see also [36]).

The above result can be used to describe regular languages. For $L \subset A^*$ and $x \in A^*$ let $\alpha = \alpha_1 \alpha_2 \cdots \alpha_n \cdots$ be the characteristic sequence of the language $L_x = \{y \in A^* \mid xy \in L\}$.

Theorem 4.6. [104] *A language* $L \subset A^*$ *is regular iff one of the following equivalent statements holds true:*

- *There is a constant* c *(depending upon* L*) such that for all* $x \in A^*$ *and all positive integers* n*,* $K(\alpha(n)) \leq K(string_Q(n)) + c$.
- *There is a constant* c' *(depending upon* L*) such that for all* $x \in A^*$ *and all positive integers* n*,* $K(\alpha(n)) \leq \log_Q n + c'$.

Program-size pumping lemmas for regular and context-free languages have been proposed in [104]; however, they are currently superseded by pumping lemmas based on other tools (see for example [149]).

4.5 Trade-offs

In this section we give some examples of trade-offs between program-size and computational complexities.

We start with an example, discussed in [125], of a language having a very low program-size complexity, but a fairly high computational complexity. Let $\alpha_i, i = 0, 1, \ldots$ be an ordering of all regular expressions over the alphabet A. A positive integer i is *saturated* iff the regular language denoted by α_i equals A^*. A real number

$$r = 0.a_0 a_1 \cdots$$

in base Q (the cardinality of A) is now defined by putting $a_i = 1$ iff i is saturated.

It is obvious that r is non-empty, as some indices i are saturated. The language L_r of all prefixes of the sequence $a_0 a_1 \cdots$ has a low program-size complexity; more precisely, there exists a constant k such that $H(a_0 a_1 \cdots a_n) \leq \log_Q n + k$. On the other hand, the computational complexity of the membership problem in L_r is very high: to decide the membership of a string of length i one has to prove that all the i regular languages involved have an empty complement. If instead of r we consider π, then the corresponding language L_π has the same (within an additive constant) program-size complexity, but the computational complexity of the membership in L_r is higher than the membership in L_π.

To an infinite sequence $\mathbf{x}$ we associate also the languages

$$S(\mathbf{x}) = \{u \in A^* \mid \mathbf{x} = vuy, \ v \in A^*, \ \mathbf{y} \in A^\omega\},$$

and

$$P(\mathbf{x}) = \{u \in A^* \mid \mathbf{x} = uy, \ \mathbf{y} \in A^\omega\},$$

that is, $S(\mathbf{x})$ is the set of all finite substrings of $\mathbf{x}$, and $P(\mathbf{x})$ is the set of all finite prefixes of $\mathbf{x}$.

A way to measure the complexity of an infinite sequence $\mathbf{x}$ is to evaluate the complexity of the language $P(\mathbf{x})$. If $\mathbf{x}$ is random, then $S(\mathbf{x}) = A^*$, but the converse relation is obviously false. Following [96], call a sequence **x** *disjunctive* if $S(\mathbf{x}) = A^*$.

Non-recursive disjunctive sequences have been constructed in [96]. Chaitin's Omega Number [39] is Borel normal in any base and, therefore, disjunctive in any base. More generally, every random sequence is Borel normal and, hence, disjunctive (cf. [24]).

Are there recursive disjunctive sequences? A direct application of Rabin's Theorem (see [23]) shows the existence of disjunctive sequences **x** such that $P(\mathbf{x})$ is recursive but arbitrarily complex. A more effective construction of a recursive disjunctive sequence can be obtained by concatenating, in some recursive order, all strings over A. This construction raises two questions: What is the "complexity" of such a sequence? Are there "simpler" ways to produce disjunctive sequences? If we measure the complexity of a sequence **x** by the complexity of the language $P(\mathbf{x})$, then one can prove (see [29]) that the language associated to the sequence consisting of all strings over the binary alphabet arranged in quasi-lexicographical order is context-sensitive and this complexity is the best possible we can obtain (in other terms, no language $P(\mathbf{x})$ can be regular when **x** is disjunctive).

4.6 More about $P =? NP$

We have seen (Section 1.3.2) that the extreme difficulty of the (in)famous problem $P =? NP$ can be explained in terms of oracles [6].

To complete the picture we quote the following two results:

Theorem 4.7
1) [75] There exist two recursive sets A, B with $P(A) \neq NP(A)$ and $P(B) = NP(B)$, but neither result is provable within Zermelo-Fraenkel set theory augmented with the Axiom of Choice.
2) [11] If A is a random oracle, then $P(A) \neq NP(A)$, i.e., with probability one $P(A) \neq NP(A)$.

Finally, consider the exponential complexity classes

$$E = DTIME\left(2^{\text{linear}}\right), \text{ and } E_2 = DTIME\left(2^{\text{polynomial}}\right).$$

There are several reasons for considering these classes ([106, 107]):

1) Both classes E, E_2 have rich internal structures.
2) E_2 is the smallest deterministic time complexity class known to contain NP and $PSPACE$.
3) $P \subset E \subset E_2, E \neq E_2$, and E contains many NP-complete problems.
4) Both classes E, E_2 have been proven to contain intractable problems.

In view of the property 2) there may be well a natural "notion of smallness" for subsets of E_2 such that P is a small subset of E_2, but NP is not. Similarly, it may be the case that P is a small subset of E, but that $NP \cap E$ is not! In the language of constructive measure theory smallness can be translated by "measure zero" (with respect to the induced spaces E or E_2). One can prove that indeed P has constructive measure zero in E and E_2, [106]. This motivates Lutz [107] to adopt the following quantitative hypothesis:

The set NP does not have measure zero.

This is a strong hypothesis, as it implies $P \neq NP$. It is consistent with Zimand's topological analysis [150] (with respect to a natural, constructive topology, if $NP - P$ is non-empty, then it is a *second Baire category* set, while NP-complete sets form a *first category* class) and appears to have more explanatory power than traditional, qualitative hypotheses.

5. Parallelism

5.1 Parallel computation thesis and alternation

Achievements in hardware technologies have made possible the construction of parallel computers involving several thousand processors capable of co-operating to solve one concrete computing task. While the time complexity has been measured as the number of elementary operations needed to compute the output in all sequential computing models this is not more true for the time complexity of parallel machines simultaneously executing lots of operations. The number of processors used by a parallel machine is a new complexity measure (computational resource) considered as a function of the input size. Since sequential time is the amount of work which has to be done we observe that

$$(\text{number of processors}) * (\text{parallel time}) \geq \text{sequential time},$$

for any parallel algorithm solving a problem. If the equality holds we say that the parallel algorithm exhibits an optimal speed-up, because it does not need to execute more operations than the best deterministic algorithm. We learn from the non-equality one important fact: We cannot hope that parallelism helps to compute intractable computing problems. If the sequential time of a problem L is exponential, then each parallel algorithms for L has an exponential number of processors or it works in exponential parallel time. Both are unrealistic and we conclude that the main contribution of parallelism is to speed up sequential computations for the problems in P. This is of crucial importance for designing "real-time" parallel algorithms.

As in the sequential case, to study parallel complexity measures one needs a formal computing model. Unfortunately, we have a lot of parallel computing models and none is generally accepted. Moreover, the differences between there models are essential. This is because we have not only to arrange the cooperation between input, output, memory, and the operating unit as for sequential computing models, but we additionally have to arrange the cooperation (information exchange) between many processors working in parallel. The study of distinct models of parallel computations has lead to an invariant characterizing "reasonable" parallel computing models. This invariant is called **Parallel Computation Thesis** [69] and it says that, for any computing problem L,

"the parallel time of L is polynomially related to
the sequential space of L".

Thus, each parallel computing model fulfilling the Parallel Computation Thesis is considered to be "suitable" for measuring of parallel computation resources.

In what follows we consider the alternating Turing machine as a parallel computing model and we show that it fulfills the Parallel Computation Thesis. Note that alternating machines cannot be really used to model parallel computations in practice because they consider nondeterministic processors working in parallel. But they provide several nice new characterizations of sequential complexity classes obtained by typical approaches of formal language theory.

An **alternating TM M** is a natural extension of the nondeterministic TM introduced in [45]. The states of M are divided into 4 disjoint subsets of **accepting** states, **rejecting** states, **existential** states and **universal** states. A computation tree $Tr_M(x)$ of M on an input x is the same tree as by a nondeterministic TM. The difference is only in the definition of acceptance. A computation of a nondeterministic TM corresponds to a path of the computation tree. A **computation of the alternating TM M on x** is any subtree T of $Tr_M(x)$ having the following properties:

(i)   the root of T is the root of $Tr_M(x)$,
(ii) if an inner node v of T is labeled by a universal configuration (state), then T must contain all sons of v in $Tr_M(x)$,
(iii) if an inner node v of T is labeled by an existential configuration, then T contains exactly one of the sons of v in $Tr_M(x)$,
(iv) every leaf of T is labeled either by an accepting configuration or by a rejecting configuration.

An **accepting computation** of M on x is any computation T whose all leaves are labeled by accepting configurations. A **word x is accepted** by the alternating TM M $(x \in L(M))$ iff there exists an accepting computation of M on x.

We observe that existential states of an alternating TM has the same meaning as states of nondeterministic TMs. One has to choose one of the possible actions, and one accepts if at least one of these possibilities leads to acceptance. A step from a universal state (configuration) corresponds to a parallel branching of the machine into a number of copies continuing to work in parallel. Here, one requires that all branches lead to acceptance.

Let, for any function $f : \mathbb{N} \to \mathbb{N}$, $ASPACE(f(n))$ and $ATIME(f(n))$ denote the alternating time and space complexity classes respectively. We define

$$ALOG = ASPACE(\log_2 n), AP = \bigcup_{k \in \mathbb{N}} ATIME(n^k),$$

and

$$APSPACE = \bigcup_{k \in \mathbb{N}} ASPACE(n^k).$$

Alternation has brought new characterizations of fundamental complexity classes. Some of the most important ones follow.

Theorem 5.1. [45] *For any space-constructible $s : \mathbb{N} \to \mathbb{N}$, $s(n) \geq n$,*

$$NSPACE(s(n)) \subseteq \bigcup_{c>0} ATIME(c(s(n))^2).$$

Idea of proof. The technique is similar to Savitch's Theorem. To check whether an $s(n)$ space-bounded nondeterministic TM M can achieve an accepting configuration C in $2^{s'(n)}$ steps for a suitable function $s'(n) = O(s(n))$ from an initial configuration C_0 on an input x, the alternating TM A guesses nondeterministically a configuration C' and universally branches into two copies continuing to work in parallel. One copy checks whether C' is reachable from C_0 in $2^{s'(n)-1}$ steps and the other one checks whether C is reachable from C' in $2^{s'(n)-1}$ steps.   □

We do not know whether $SPACE(s(n))$ is equal to $NSPACE(s(n))$ or not. Since all deterministic classes are closed under complementation people have hoped to prove $SPACE(s(n))$ is not equal to $NSPACE(s(n))$ by proving that $NSPACE(s(n))$ is not closed under complementation. But Immerman [94] and Szelepcsényi [138] have proved that this idea cannot work because all classes $NSPACE(s(n))$ for $s(n) \geq \log n$ are closed under complementation too.

Theorem 5.2. [45] *For any function $t, t(n) \geq n$,*

$$ATIME(t(n)) \subseteq DSPACE(t(n)).$$

Idea of proof. Let M be an alternating TM, and let $Tr_M(x)$ is the computation tree of M on x. To check whether $Tr_M(x)$ contains an accepting computation one assigns the value 1 (0) to the accepting (rejecting) leaves, the operator disjunction to the existential nodes, and the operator conjunction to universal nodes. Then one looks at $Tr_M(x)$ as a Boolean circuit with inputs on leaves and the output on the root. Obviously, M accepts x iff the output value of this circuit is 1. A deterministic TM B can traverse $Tr_M(x)$ and calculate the output of the circuit in post order. To do it, at any point of the computation B has to store the visiting configuration (node) and a string representing the position of the node in $Tr_M(x)$. Both can be done in $O(t(n))$ space. □

Theorem 5.3. [45] *For any function* $s : \mathbb{N} \to \mathbb{N}, s(n) \geq \log_2 n,$

$$ASPACE(s(n)) = \bigcup_{c>0} DTIME(c^{s(n)}).$$

The above results show that alternation shifts by exactly one level the fundamental hierarchy of deterministic complexity classes because $ALOG = P, AP = PSPACE, APSPACE = EXPTIME$, etc. Another interesting result, proven in [97], is that the class of languages accepted by two-way alternating finite automata is exactly $PSPACE$.

One can observe that the copies of an alternating TM working in parallel in some computation do not have any possibility to communicate (exchange information). Since communication is one of the main ingredients of parallel computations, synchronized alternating TM ($SATM$) has been introduced in [85] as a generalization of alternating TMs. We give a brief informal description of this idea. (The formal definitions can be found in [90].) An $SATM$ M is an alternating TM with an additional finite synchronization alphabet. An internal state of M can be either an usual internal state or a pair (internal state, synchronizing symbol). The latter is called a **synchronizing state**. The synchronizing states are used in a computation as follows. Each time one of the machine copies, working in parallel in a computation, enters a synchronizing state it must wait until all other machines working in parallel enter an accepting state or a synchronizing state with the same synchronizing symbol. When this happens all machines are allowed to move from the synchronizing states (to continue to work). In what follows the usual notation $SATIME(f(n))$ and $SASPACE(f(n))$ is used for the synchronized alternating complexity classes. Analogously $SALOG$ and SAP denote synchronized alternating logspace and synchronized alternating polynomial time respectively. The main results proven in [90] are the following:

Theorem 5.4. [90] *For any space-constructible function* $s : \mathbb{N} \to \mathbb{N},$

$$\bigcup_{c>0} NSPACE(nc^{s(n)}) = SASPACE(s(n)).$$

Corollary 5.1. [90, 133] *For any space-constructible function* $s(n) \geq \log_2 n$

$$SASPACE(s(n)) = \bigcup_{c>0} DSPACE(c^{s(n)})$$

$$= \bigcup_{c>0} ATIME(c^{s(n)}) = \bigcup_{c>0} SATIME(c^{s(n)}).$$

Corollary 5.2. [90, 133] $NSPACE(n)$ *is exactly the class of languages recognized by synchronized alternating finite automata.*

We observe that the equality

$$SASPACE(s(n)) = \bigcup_{c>0} SATIME(c^{s(n)})$$

implies that synchronized alternating machines are able to use the space in an "optimal way". It seems that deterministic, nondeterministic and alternating computing devices do not have this property because if they would have this property then some fundamental complexity hierarchies would collapse. More precisely, if alternating machines would have this property, then $P = NP = PSPACE$. If nondeterministic (deterministic) machines would have this property, then $NLOG = P = NP$ $(DLOG = NLOG = P)$.

In [90] a new characterization of $PSPACE$ by $SALOG$ and by synchronized alternating multihead automata is given. This extends the well-known characterization of fundamental complexity classes $DLOG \subseteq NLOG \subseteq P$ by deterministic, nondeterministic, and alternating finite automata respectively. Since [93] shows that NP can be characterized by synchronized alternating multihead automata with polynomial number of synchronizations we get the characterization of the hierarchy $P \subseteq NP \subseteq PSPACE$ by synchronized alternating multihead automata without synchronization, with polynomial synchronization and with full (unbounded) synchronization, respectively.

5.2 Limits to parallel computation and P-completeness

In the previous section we have explained that parallelism is used to speed up computations for problems in P, and not to address intractable problems. A very natural question arises: Are the parallel algorithms able to essentially speed up the time complexity of any problem in P? We do not believe that it is possible, but we are not able to prove it from the same reason why we are not able to prove $NP - P \neq \emptyset$. We conjecture that there exist feasible problems which are inherently sequential, i.e., which do not allow any high parallel execution because of hard sequential dependence of the operation order. To give more formal arguments we first define the class of **feasible highly parallel problems** as the class of problems allowing very high degree of parallelization.

Definition 5.1. [122] *For any positive integer* i *let*

$NC^i = \{L \mid L$ *can be accepted by uniform Boolean circuits with size*
$$n^{O(1)} \text{ and depth } O((\log_2 n)^i)\}.$$
$NC = \{L \mid L$ *is decidable in parallel time* $(\log_2 n)^{O(1)}$

$$\text{using } n^{O(1)} \text{processors}\} = \bigcup_{i=1}^{\infty} NC^i.$$

Now, to support the strong conjecture that $P - NC \neq \emptyset$ we use the same approach as we have used to argue that P should be a proper subset of NP. Let ANC be a class of all parallel algorithms working in parallel time $(\log_2 n)^{O(1)}$ with $n^{O(1)}$ processors.

Definition 5.2. *We say that a language* $L_1 \subseteq \Sigma_1^*$ *is* **NC-reducible** *to a language* $L_2 \subseteq \Sigma_2^*$ *if there exists a parallel algorithm* $A \in ANC$ *which for any input* $x \in \Sigma_1^*$ *computes an* $A(x) \in \Sigma_2^*$ *such that*

$$x \in L_1 \Longleftrightarrow A(x) \in L_2.$$

A language L *is* *P-complete under* *NC-reducibility if*

(i) $L \in P$, *and*
(ii) *every language in* P *is* *NC-reducible to* L.

We note that if an NC-complete language would be in NC, then $NC = P$.

Again, as for NP-completeness, we have many P-complete problems, and for none of them do we know a highly parallel solution. So, there is plenty of experience saying that P-complete problems according to NC-reducibility do not allow feasible highly parallel solutions (i.e., they are inherently sequential). Another reason to believe $NC \neq P$ is that we do not know any fast general simulation of sequential machines by parallel ones. The best known parallel simulations reduce sequential time T to parallel time $T/\log_2 T$ or $\sqrt{T}$, depending on the parallel model. Furthermore, an exponential number of processors is needed to achieve even these modest speed ups. Thus P-completeness is an instrument helping to study the border between problems having a tractable highly parallel solution and problems which do not have efficient highly parallel solution. More about the classification of computing problems according to the class NC can be found in the excellent book [71] devoted to this topic only.

5.3 Communication in parallel and distributive computing

In many parallel and distributive computations the main running time is devoted to the communication between processes. To built parallel architectures whose communication structure has high communication facilities is one of the central tasks of parallel computing. There are many studies in this direction dealing with the efficiency of the realization of basic communication tasks (like broadcast, gossip [91], routing [102]) in distinct communication structures as well as with the ability to effective simulate the communication facilities of several different interconnection networks (communication structures) on one network candidating for our parallel architecture [112]. Results and methods used in this area are primary connected with discrete mathematics and graph theory [102, 112] and so we do not want to give more details here. We omit discussion of typical parallel models of formal language theory like systolic arrays, Lindenmayer systems, and other kinds of parallel rewriting too, because they are not in any main research streams in the complexity of parallel computing.

We briefly discuss the **parallel communicating grammar systems (PCGSs)** introduced in [119]. Each PCGS can be considered as a directed graph whose nodes are simple regular grammars. If the graph (communication structure) contains a directed edge (G_1, G_2), then the grammar G_2 may ask the grammar G_1 for the submission of the word generated by G_2 (for more details and formal definitions consult [118, 89]). For this model it was shown that some classes of communication structures are absolutely more powerful than other graph classes [105, 88, 89, 118], i.e., that there are languages generated by one communication structure but not by any communication structure from any other graph class. These results are interesting for the theoretical study of communicational aspects in computing because no similar results were achieved for other parallel computing models. Note that it is not realistic to obtain an absolute comparison of interconnection networks from the following reasons. If nodes of networks are standard sequential computers, then already one node can compute anything computable. If every node is a simple processor (finite automaton) and communication structure is a finite graph, then the whole network can be considered as a finite automaton. If we have simple processors and we allow an unbounded growth of the communication structure during the computation, then one can simulate Turing machines with one-dimensional arrays which are the simplest communication structure at all.

Another really important research area, partially influenced by the standard formal language methods, is the study of abstract communication complexity of languages. The communication complexity of a language [1, 148, 117] is the necessary and sufficient number of bits exchanged between two computers in order to decide about the acceptance of the input word whose input bits are distributed to the two computers in a balanced way. As program-size complexity, the communication complexity of computing problems has numerous applications for different computing models. It can be applied to get lower bounds on different complexity measures of Boolean circuits, VLSI circuits, interconnection networks, and many other not primarily parallel computing models (Turing machines, for instance) [86]. Another important fact is that one can prove exponential gaps between determinism, nondeterminism, and Monte Carlo probabilism for communication complexity. Moreover, deterministic and Las Vegas communication complexity are polynomially related. To prove similar results for time complexity is exactly one of the central problemS of theoretical computer science.

Acknowledgement

We are deeply indebted to all authors of referenced works. Indeed, the material of this chapter represents little more than our belated understanding and organization of their original work. Cezar Câmpeanu, Greg Chaitin, Ivana Černá, Jeremy Gibbons, Dana Pardubská, Anna Slobodová, and Karl Svozil have read and made valuable suggestions: we express to all our gratitude.

References

[1]   H. Abelson, Lower bounds on information transfer in distributed computations. In: *Proc. 29th Annual IEEE FOCS*, IEEE 1978, 151–158.

[2]   S. I. Adian, W. W. Boone, G. Higman (eds.), *Word problems II: The Oxford Book*, North-Holland, New York, 1980.

[3]   L. M. Adleman, M. A. Huang, Recognizing primes in random polynomial time, *Technical Report*, Department of Computer Science, Washington, State University, 1988.

[4]   D. Z. Albert, On quantum-mechanical automata, *Phys. Lett.* 98A (1983), 249–252.

[5]   K. Baker, *Minimalism: Art of Circumstance*, Abbeville Press, New York, 1988.

[6]   T. Baker, J. Gill, R. Solovay, Relativizations of the problem $P = ?NP$ question, *SIAM J. Comput.* 4 (1975), 431–442.

[7]   J. L. Balcazar, J. Diaz, J. Gabarro, *Structural Complexity I*, Springer-Verlag, New York, 1988.

[8]   J. L. Balcazar, J. Diaz, J. Gabarro, *Structural Complexity II*, Springer-Verlag, New York, 1990.

[9]   P. Benioff, Quantum mechanical models of Turing machines, *J. Stat. Phys.* 29 (1982), 515–546.

[10] P. Benioff, Quantum mechanical hamiltonian models of computers, *Annals New York Academy of Sciences* 480 (1986), 475–486.

[11] C. H. Bennett, J. Gill, Relative to a random oracle A, $P^A \neq NP^A \neq co - NP^A$, with probability one, *SIAM J. Comput.* 10 (1981), 96–113.

[12] C. H. Bennett, Certainty from uncertainty, *Nature* 371 (1994), 694–696.

[13] C. H. Bennett, E. Bernstein, G. Brassard, U. V. Vazirani, Strength and weaknesses of quantum computing, *Preprint*, 1995, 9pp.

[14] P. Benacerraf, Tasks, super-tasks, and the modern eleatics, *The Journal of Philosophy* 59 (1962), 765–784.

[15] E. Bernstein and U. Vazirani, Quantum complexity theory. In: *Proc. 25th ACM Symp. on Theory of Computation*, 1993, 11–20.

[16] A. Berthiaume and G. Brassard, The quantum challenge to structural complexity theory. In: *Proc. 7th IEEE Conf. on Structure in Complexity Theory*, 1992, 132–137.

[17] M. Blum, A machine-independent theory of the complexity of recursive functions, *J. Assoc. Comput. Mach.* 14 (1967), 322–336.

[18] R. V. Book, The complexity of languages reducible to algorithmically random languages, *SIAM J. Comput.* 23(1995), 1275–1282.

[19] R. V. Book, J. Lutz, K. Wagner, On complexity classes and algorithmically random languages, *Proc. STACS-92*, Lecture Notes Comp. Sci. 577, Springer-Verlag, Berlin, 1992, 319–328.

[20] D. P. Bovet, P. Crescenzi, *Introduction to the Theory of Complexity*, Prentice Hall, 1984.

[21] D. S. Bridges, *Computability – A Mathematical Sketchbook*, Springer-Verlag, Berlin, 1994.

[22] D. S. Bridges, C. Calude, On recursive bounds for the exceptional values in speed-up, *Theoret. Comput. Sci.* 132 (1994), 387–394.

[23] C. Calude, *Theories of Computational Complexity*, North-Holland, Amsterdam, 1988.

[24] C. Calude, *Information and Randomness – An Algorithmic Perspective*, Springer-Verlag, Berlin, 1994.

[25] C. Calude, D. I. Campbell, K. Svozil, D. Ştefănescu, Strong determinism vs. computability. In: *Proceedings of the International Symposium "The Foundational Debate"*, Vienna Circle Institute Yearbook, 3, 1995, Kluwer, Dordrecht, 115–131.

[26] C. Calude, S. Marcus, Gh. Păun, The universal grammar as a hypothetical brain, *Rev. Roumaine Ling.* 27 (1979), 479–489.

[27] C. Calude, Gh. Păun, Global syntax and semantics for recursively enumerable languages, *Fund. Inform.* 4 (1981), 245–254.

[28] C. Calude, Gh. Păun, On the adequacy of a grammatical model of the brain, *Rev. Roumaine Ling.* 27 (1982), 343–351.

[29] C. Calude, S. Yu, Language-theoretic complexity of disjunctive sequences, *Technical Report No 119, 1995*, Department of Computer Science, The University of Auckland, New Zealand, 8 pp.

[30] C. Calude, M. Zimand, A relation between correctness and randomness in the computation of probabilistic algorithms, *Internat. J. Comput. Math.* 16 (1984), 47–53.

[31] C. Calude, M. Zimand, Effective category and measure in abstract complexity theory – extended abstract, *Proceedings FCT'95*, Lectures Notes in Computer Science 965, Springer-Verlag, Berlin, 1995, 156–171.

[32] V. Černý, Quantum computers and intractable (NP-complete) computing problems, *Phys. Rev.* A 48 (1993), 116–119.

[33] G. J. Chaitin, On the length of programs for computing finite binary sequences, *J. Assoc. Comput. Mach.* 13 (1966), 547–569.

[34] G. J. Chaitin, Information-theoretic limitations of formal systems, *J. Assoc. Comput. Mach.* 21 (1974), 403–424.

[35] G. J. Chaitin, A theory of program size formally identical to information theory, *J. Assoc. Comput. Mach.* 22 (1975), 329–340.

[36] G. J. Chaitin, Algorithmic information theory, *IBM J. Res. Develop.* 21 (1977), 350–359, 496.

[37] G. J. Chaitin, *Algorithmic Information Theory*, Cambridge University Press, Cambridge, 1987 (third printing 1990).

[38] G. J. Chaitin, Computing the Busy Beaver function. In: Cover, T. M. and Gopinath, B. (eds.), *Open Problems in Communication and Computation*, Springer-Verlag, Berlin, 1987, 108–112.

[39] G. J. Chaitin, *Information, Randomness and Incompleteness, Papers on Algorithmic Information Theory*, World Scientific, Singapore, 1987 (second edition 1990).

[40] G. J. Chaitin, Information-theoretic characterizations of recursive infinite strings, *Theoret. Comput. Sci.* 2 (1976), 45–48.

[41] G. J. Chaitin, *Information-Theoretic Incompleteness*, World Scientific, Singapore, 1992.

[42] G. J. Chaitin, On the number of N-bit strings with maximum complexity, *Applied Mathematics and Computation* 59 (1993), 97–100.

[43] G. J. Chaitin, A. Arslanov, C. Calude. Program-size complexity computes the halting problem, *EATCS Bull.*, 57 (1995) 198–200.

[44] G. J. Chaitin, J. T. Schwartz, A note on Monte-Carlo primality tests and algorithmic information theory, *Comm. Pure Appl. Math.* 31 (1978), 521–527.

[45] A. K. Chandra, D. C. Kozen, L. J. Stockmeyer, Alternation, *J. Assoc. Comput. Mach.* 28 (1981), 114–133.

[46] A. Cobham, The intrinsic computational difficulty of functions. In: *Proc. Congress for Logic, Mathematics, and Philosophy of Science* 1964, 24–30.

[47] S. A. Cook, The complexity of theorem proving procedure. In: *Proc. 3-rd Annual ACM STOC*, 1971, 151–158.

[48] S. A. Cook, An observation on time-storage trade off. In: *Proc. 5-th Annual ACM STOC*, 1973, 29–33.

[49] S. A. Cook, Deterministic CFL's are accepted simultaneously in polynomial time and log squared space. In: *Proc. ACM STOC'79*, 338–345.

[50] M. Davis, Unsolvable problems. In: Barwise, J. (ed.), *Handbook of Mathematical Logic*, North-Holland, Amsterdam, 1976, 568–594.

[51] P. Davies, *The Mind of God, Science and the Search for Ultimate Meaning*, Penguin Books, London, 1992.

[52] D. Deutsch, Quantum theory, the Church–Turing principle and the universal quantum computer, *Proceedings of the Royal Society of London* A 400 (1985), 97–117.

[53] D. Deutsch, Quantum computational networks, *Proceedings of the Royal Society of London* A 425 (1989), 73–90.

[54] D. Deutsch and R. Jozsa, Rapid solution of problems by quantum computation, *Proceedings of the Royal Society of London* A 439 (1992), 553-558

[55] R. L. Devaney, *An Introduction to Chaotic Dynamical Systems*, Addison-Wessley, 1989.

[56] J. Earman and J. D. Norton, Forever is a day: supertasks in Pitowsky and Malament–Hogarth spacetimes, *Philosophy of Science* 60 (1993), 22–42.

[57] U. Eco, *L'Oeuvre ouverte*, Editions du Seuil, Paris, 1965.

[58] G. M. Edelman, *Bright Air, Brilliant Fire – On the Matter of Mind*, Basic Books, 1992.

[59] S. Feferman, J. Dawson, Jr., S. C. Kleene, G. H. Moore, R. M. Solovay, J. van Heijenoort (eds.), *Kurt Gödel Collected Works*, Volume I, Oxford University Press, New York, 1986.

[60] S. Feferman, J. Dawson, Jr., S. C. Kleene, G. H. Moore, R. M. Solovay, J. van Heijenoort (eds.), *Kurt Gödel Collected Works*, Volume II, Oxford University Press, New York, 1990.

[61] R. P. Feynman, Simulating physics with computers, *International Journal of Theoretical Physics* 21 (1982), 467–488.

[62] R. P. Feynman, Quantum Mechanical Computers, *Opt. News* 11 (1985), 11–20.

[63] P. Gács, Every sequence is reducible to a random one, *Inform. and Control* 70 (1986), 186-192.

[64] M. Garey, D. Johnson, *Computers and Intractability: A Guide to the Theory of NP-Completeness*, W. H. Freeman, New York, 1979.

[65] M. Gardner, *Logic Machines and Diagrams*, University of Chicago Press, Chicago, 1958, 144 (second printing, Harvester Press, 1983).

[66] M. Gardner, *Fractal Music, Hypercards, and More . . .*, W. H. Freeman, New York, 1992, 133.

[67] V. Geffert, Bridging across the $\log(n)$ space frontier. In: *Proc. MFCS'95*, Lect. Notes Comp. Sci. 969, Springer-Verlag, 1995, 50–65.

[68] J. Gill, Computational complexity of probabilistic Turing machines, *SIAM J. Computing* 6 (1977), 675–695.

[69] L. M. Goldschlager, A universal interconnection pattern for parallel computers, *J. Assoc. Comput. Mach.* 29 (1982), 1073–1086.

[70] A. Goswami, R. E. Reed, M. Goswami, *The Self-Aware Universe – How Consciouness Creates the Material World*, G. P. Putnam's Sons, New York, 1993.

[71] R. Greenlaw, H. J. Hoover, W. L. Ruzzo, *Limits to Parallel Computations*, Oxford University Press, Oxford, 1995.

[72] A. Grünbaum, *Modern Science and Zeno's paradoxes*, Allen and Unwin, London, 1968 (second edition).

[73] A. Grünbaum, *Philosophical Problems of Space of Time*, D. Reidel, Dordrecht, 1973. (second, enlarged edition)

[74] B. Grümbaum, G. C. Shephard, *Tilings and Patterns*, W. H. Freeman, New York, 1987.

[75] J. Hartmanis, J. E. Hopcroft, Independence results in computer science, *SIGACT News* 8 (1976), 13-24.

[76] J. Hartmanis, J. E. Hopcroft, An overview of the theory of computational complexity, *J. Assoc. Comput. Mach.* 18 (1971), 444–475.

[77] J. Hartmanis, P. M. Lewis II, R. E. Stearns, Hierarchies of memory limited computations. In: *Proc. 6th Annual IEEE Symp. on Switching Circuit Theory and Logical Design*, 1965, 179–190.

[78] J. Hartmanis, R. E. Stearns, On the computational complexity of algorithms, *Trans. Amer. Math. Soc.* 117 (1965), 285–306.

[79] D. Havel, *Algorithmics: The Spirit of Computing*, Addison-Wesley, 1987.

[80] F. C. Hennie, R. E. Stearns, Two-tape simulation of multitape Turing machines, *J. Assoc. Comput. Mach.* 13 (1966), 533–546.

[81] M. Hogarth, Non-Turing computers and non-Turing computability, *PSA 1994* 1 (1994), 126-138.

[82] J. E. Hopcroft, J. W. Paul, L. Valiant, On time versus space, *J. Assoc. Comput. Mach.* 24 (1977), 332–337.

[83] J. E. Hopcroft, J. D. Ullman, Relations between time and tape complexities, *J. Assoc. Comput. Mach.* 15 (1968), 414–427.

[84] J. E. Hopcroft, J. D. Ullman, *Introduction to Automata Theory, Languages and Computation*, Addison-Wesley, 1979.

[85] J. Hromkovič, How to organize communication among parallel processes in alternating computations, *Unpublished manuscript*, Comenius University, Bratislava, January 1986.

[86] J. Hromkovič, *Communication Complexity and Parallel Computing*, EATCS Texts in Theoretical Computer Science, in preparation.

[87] J. Hromkovič, O. H. Ibarra, N. Q. Trân, Γ, NP and $PSPACE$ characterizations by synchronized alternating finite automata with different communication protocols, *Unpublished manuscript*, 1994.

[88] J. Hromkovič, J. Kari, L. Kari, Some hierarchies for the communication complexity measures of cooperating grammar systems, *Theoretical Computer Science* 127 (1994), 123–147.

[89] J. Hromkovič, J. Kari, L. Kari, D. Pardubská, Two lower bounds on distributive generation of languages. In: *Proc. 19th MFCS'94*, Lect. Notes Comp. Sci. 841, Springer-Verlag, Berlin, 1994, 423–432.

[90] J. Hromkovič, J. Karhumäki, B. Rovan, and A. Slobodová, On the power of synchronization in parallel computations, *Disc. Appl. Math.* 32 (1991), 156–182.

[91]   J. Hromkovič, R. Klasing, B. Monien, R. Peine, Dissemination of information in interconnection networks (broadcasting & gossiping). In: Hsu, F., Du, D.-Z. eds., *Combinatorial Network Theory*, Science Press & AMS 1995, to appear.

[92]   J. Hromkovič, B. Rovan, A. Slobodová, Deterministic versus nondeterministic space in terms of synchronized alternating machines, *Theoret. Comp. Sci.* 132 (1994), 319–336.

[93]   O. H. Ibarra, N. Q. Trân, On communication-bounded synchronized alternating finite automata, *Acta Informatica* 31 (1994), 315–327.

[94]   N. Immerman, Nondeterministic space is closed under complementation, *SIAM J. Comput.* 17 (1988), 935–938.

[95]   D. S. Johnson, A catalog of complexity classes. In: van Leeuwen, J. (ed.), *Handbook of Theoretical Computer Science*, Vol. A, Elsevier, Amsterdam, 1990, 67–161.

[96]   H. Jürgensen, G. Thierrin, Some structural properties of ω-languages, *13th Nat. School with Internat. Participation "Applications of Mathematics in Technology"*, Sofia, 1988, 56–63.

[97]   K. N. King, Alternating multihead finite automata. In: *Proc. 8-th ICALP'81*, Lect. Notes Comp. Sci. 115, Springer-Verlag, Berlin, 1981, 506–520.

[98]   L. G. Khachiyan, A polynomial algorithm in linear programming, *Soviet Mathematics Doklad* 20 (1979), 191–194.

[99]   A. N. Kolmogorov, Three approaches for defining the concept of "information quantity", *Problems Inform. Transmission* 1 (1965), 3–11.

[100]   K.-J. Lange, Complexity and Structure in Formal Language Theory. *Unpublished manuscript*, University of Tübingen, Germany.

[101]   P. S. Laplace, *A Philosophical Essay on Probability Theories*, Dover, New York, 1951.

[102]   F. T. Leighton, *Introduction to Parallel Algorithms and Architectures: Array · Trees · Hypercubes*, Morgan Kaufmann Publishers, San Mateo, California, 1992.

[103]   L. A. Levin, Randomness conservation inequalities: information and independence in mathematical theories, *Problems Inform. Transmission* 10 (1974), 206–210.

[104]   M. Li, P. M. Vitányi, A new approach to formal language theory by Kolmogorov complexity, *SIAM J. Comput.* 24 (1995), 398–410.

[105]   M. Lukáč, *About Two Communication Structures of PCGS*, Master Thesis, Dept. of Computer Science, Comenius University, Bratislava, 1992.

[106]   J. H. Lutz, Almost everywhere high nonuniform complexity, *J. Comput. System Sci.* 44 (1992), 220–258.

[107]   J. H. Lutz, The quantitative structure of exponential time, *Proceedings of the Eighth Annual Structure in Complexity Theory Conference*, (San Diego, CA, May 18–21, 1993), IEEE Computer Society Press, 1993, 158–175.

[108]   P. Martin-Löf, The definition of random sequences, *Inform. and Control* 9 (1966), 602–619.

[109]   Yu. V. Matiyasevich, *Hilbert's Tenth Problem*, MIT Press, Cambridge, Massachusetts, London, 1993.

[110]   M. Mendès France, A. Hénaut, Art, therefore entropy, *Leonardo* 27 (1994), 219–221.

[111]   G. L. Miller, Riemann's hypothesis and tests for primality, *J. Comput. Syst. Sci.* 13 (1976), 300–317.

[112]   B. Monien, I. H. Sudborough, Embedding one interconnection network in another, *Computing Suppl.* 7 (1990), 257–282.

[113]   C. Moore, Real-valued, continuous-time computers: a model of analog computation, Part I, *Manuscript*, 1995, 15pp.

[114] P. Odifreddi, *Classical Recursion Theory*, North-Holland, Amsterdam, Vol.1, 1989.

[115] H. R. Pagels, *The Dreams of Reason*, Bantam Books, New York, 1989.

[116] C. H. Papadimitriou, *Computational Complexity*, Addison-Wesley, New York, 1994.

[117] Ch. Papadimitriou, M. Sipser, Communication complexity, *J. Comp. Syst. Sci.* 28 (1984), 260–269.

[118] D. Pardubská, *On the Power of Communication Structure for Distributive Generation of Languages*, Ph.D. Thesis, Comenius University, Bratislava 1994.

[119] Gh. Păun, L. Sântean, Parallel communicating grammar systems: The regular case, *Ann. Univ. Buc. Ser. Mat.-Inform.* 37 (1989), 55–63.

[120] R. Penrose, *The Emperor's New Mind. Concerning Computers, Minds, and the Laws of Physics*, Oxford University Press, Oxford, 1989.

[121] R. Penrose, *Shadows of the Mind. A Search for the Missing Science of Consciousness*, Oxford University Press, Oxford, 1994.

[122] N. Pipinger, On simultaneous resource bounds (preliminary version). In: *Proc. 20th IEEE Symp. FOCS*, IEEE, New York 1979, 307–311.

[123] I. Pitowsky, The physical Church–Turing thesis and physical complexity theory, *Iyyun, A Jerusalem Philosophical Quarterly* 39 (1990), 81–99.

[124] M. O. Rabin, Probabilistic algorithms. In: Traub, J., ed., *Algorithms and Complexity: New Directions and Recent Results*, Academic Press, New York, 1976, 21–40.

[125] G. Rozenberg, A. Salomaa, *Cornerstones of Undecidability*, Prentice Hall, 1994.

[126] R. Rucker, *Infinity and the Mind*, Bantam Books, New York, 1983.

[127] R. Rucker, *Mind Tools*, Houghton Mifflin, Boston, 1987.

[128] A. Salomaa, *Computation and Automata*, Cambridge University Press, Cambridge, 1985.

[129] W. J. Savitch, Relationships between nondeterministic and deterministic tape complexities, *J. Comp. Syst. Sciences* 4 (1970), 177–192.

[130] J. R. Searle, *The Rediscovery of the Mind*, MIT Press, Cambridge, Mass. (third printing 1992).

[131] P. W. Shor, Algorithms for quantum computation: discrete logarithms and factoring. In: *Proc. 35th Annual IEEE FOCS*, IEEE, 1995 (in press).

[132] H. T. Siegelmann, Computation beyond the Turing limit, *Science* 268, 28 April (1995), 545–548.

[133] A. Slobodová, On the power of communication in alternating machines. In: *Proc. 13th MFCS'88*, Lect. Notes Comp. Sci. 324, Springer-Verlag, Berlin, 1988, 518–528.

[134] R. M. Smullyan, *Diagonalization and Self-Reference*, Clarendon Press, Oxford, 1994.

[135] R. M. Solovay, *Draft of a paper (or series of papers) on Chaitin's work ... done for the most part during the period of Sept.–Dec. 1974*, unpublished manuscript, IBM Thomas J. Watson Research Center, Yorktown Heights, New York, May 1975, 215pp.

[136] R. Solovay, V. Strassen, A fast Monte Carlo test for primality, *SIAM J. Computing* 6 (1977), 84–85.

[137] E. Strickland, *Minimalism – Origins*, Indiana University Press, Bloomington, 1993.

[138] R. Szelepcsényi, The method of forced enumeration for nondeterministic automata, *Acta Informatica* 26 (1988), 279–284.

[139] A. Szepietowski, *Turing Machines with Sublogarithmic Space*, Lect. Notes Comp. Sci. 843, Springer-Verlag, Berlin, 1994.

[140] K. Svozil, *Randomness and Undecidability in Physics*, World Scientific, Singapore, 1993.

[141] K. Svozil, On the computational power of physical systems, undecidability, the consistency of phenomena and the practical uses of paradoxes. In: Greenberger, D. M., Zeilinger, A. (eds.), *Fundamental Problems in Quantum Theory: A Conference Held in Honor of Professor John A. Wheeler, Annals of the New York Academy of Sciences* 755 (1995), 834–842.

[142] K. Svozil, Quantum computation and complexity theory, I, *EATCS Bull.* 55 (1995), 170–207.

[143] K. Svozil, Quantum computation and complexity theory, II, *EATCS Bull.* 56 (1995), 116–136.

[144] A. M. Turing, On computable numbers, with an application to the Entscheidungsproblem, *Proc. London Math. Soc. Ser. 2* 42 (1936), 230–265.

[145] J. F. Thomson, Tasks and super-tasks, *Analysis* 15 (1954), 1–13.

[146] H. Weyl, *Philosophy of Mathematics and Natural Science*, Princeton University Press, Princeton, 1949.

[147] I. Xenakis, Musique formelle, *La Revue Musicale* 253/4 (1963), 10.

[148] A. C. Yao, Some complexity questions related to distributed computing. In: *Proc. 11th Annual ACM STOC*, ACM 1981, 308–311.

[149] S. Yu, A pumping lemma for deterministic context-free languages, *Inform. Process. Lett.* 31 (1989), 47–51.

[150] M. Zimand, If not empty, $NP - P$ is topologically large, *Theoret. Comput. Sci.* 119 (1993), 293–310.

[151] P. van Emde Boas, Machine models and simulations. In: van Leeuwen, J. (ed.), *Handbook of Theoretical Computer Science*, Vol. A, Elsevier, Amsterdam, 1990, 525–632.

[152] D. J. Velleman, *How to Prove It. A Structural Approach*, Cambridge University Press, Cambridge, 1994.

[153] H. Wang, On 'computabilism' and physicalism: some subproblems. In: Cornwell, J., (ed.), *Nature's Imagination*, Oxford University Press, Oxford, 1995, 161–189.

Parsing of Context-Free Languages

Klaas Sikkel and Anton Nijholt

Summary. Parsing is the process of assigning structure to sentences. The structure is obtained from the grammatical description of the language. Both in Computer Science and in Computational Linguistics, context-free grammars and associated parsing algorithms are among the most useful tools. Numerous parsing algorithms have been developed. Special subclasses of the context-free grammars have been introduced in order to allow and induce efficient parsing algorithms. Special super-classes of the context-free grammars have been introduced in order to allow use of variants of efficient parsing methods that had been developed for context-free grammars. At first sight many parsing algorithms seem to be different, but nevertheless related. Some unifying approaches have been attempted in the past, but none survived the changing field. This chapter introduces a unifying approach at a level between grammars and algorithms, introducing so-called parsing schemata. In the parsing schemata framework the essentials of different parsing algorithms can be compared and it can be shown how to derive an algorithm from another one. The insight that is obtained this way also allows the derivation of new algorithms and it allows less tedious observations about correctness than usual. The framework can also be applied to grammar formalisms beyond the context-free grammars.

1. Introduction

In computer science, grammars are human-constructed formalisms that are meant to define languages. These can be programming languages or, in theoretical computer science formal languages. This description is often partial. It is not unusual to see a formal description of the syntactic structure of a language, while the semantic part remains ill-defined. Finding the syntactic structure of a program (which is a sentence in the language) is part of the compilation process of a program. The construction of this structure is called parsing. The result of the parsing process is a hierarchical account of the elements that make up the program. This account makes it possible to assign semantics to a program.

Formal syntactic descriptions of languages were first given by the linguist Noam Chomsky. Because the descriptions were formal, the languages were also formal: sequences of symbols that satisfied descriptions based on finite state automata or regular grammars, context-free grammars, or context-sensitive grammars. In Chomsky's view, these descriptions were the starting point for descriptions of the syntax of natural, human spoken, languages.

Moreover, these descriptions would allow one to assign meaning to sentences. Interestingly, at about the same time Chomsky introduced different classes of grammars and languages, a committee defining a programming language (ALGOL) introduced a programming language description formalism called Backus – Naur Form (BNF) which turned out to be equivalent to one of Chomsky's grammar classes, the so-called context-free grammars.

In Chomsky's view, human language grammars were not human-constructed formalisms. The rules of the formalism, or, more generally, the principles that determine the rules, are supposed to be innate. This view led to a distinction between competence and performance in human language use. Each language user has a language competence that allows him to construct all kinds of sentences using the rules of a grammar. Constructing sentences can be compared with using rules to compute a multiplication or a division in arithmetic. Language users can construct sentences using rules of syntax. Due to environmental circumstances in normal man-to-man communication, these rules are not always obeyed. Performance differs from competence.

It is much easier, however, to do research on self-chosen rules of sentence construction and analysis than to do research on actual language behaviour. For this obvious reason, grammar formalisms and their parsing methods have drawn so much attention by computer scientists and computational linguists. It should be admitted, on the other hand, that nowadays natural and programming language processing systems can be built on the basis of these formalisms. Whether or not formalisms that are used for natural language processing meet certain linguistic principles in some way or other, or even some principles of human language innateness, is not the main concern of those doing research and development in this area.

1.1 Parsing algorithms

Parsing algorithms have been defined for all kinds of language descriptions. After the introduction of the well-known Chomsky hierarchy in the late 1950s and early 1960s, we see a common interest of computer scientists and computational linguists in parsing methods for context-free languages. The quest for efficient parsing methods led to polynomial-time algorithms for general context-free grammars in the middle and late 1960s. Among them, the so-called Cocke – Younger – Kasami and the Earley parsing algorithms. In computer science, however, these formalisms were thought to be unnecessarily general for describing the syntactic properties of programming languages, and therefore to be unnecessarily inefficient. Linear-time algorithms like LL and LR were introduced. These are sufficiently general for dealing with the syntactic backbone of programming languages. Interest in general context-free methods diminished, or was left to theoretical computer scientists. In computational linguistics there were other reasons to become critical of the context-free grammar formalism. Its descriptional adequacy, that is, its ability to cover linguistic generalities in a natural way, was considered to be too weak. It was also doubted whether it provides sufficient generative capacity. The LL and LR approaches favoured in computer science were clearly much less suitable, because these do not allow representation of syntactic ambiguities.

It is remarkable that in the late seventies and early eighties we see a growing interest in LR-like methods and context-free grammars in computational linguistics and a growing interest in general context-free grammar descriptions in computer science. How can this be explained? In computational linguistics, first of all, the so-called "determinism hypothesis" attracted a lot of attention. The idea is that, in general, people do not "backtrack" while analysing a sentence. Backtracking becomes necessary only when a started analysis cannot be continued at some point in the sentence. Mitch Marcus introduced an LR-like "wait and see" stack formalism in order to parse sentences "deterministically". Reviewing the literature from that period, one sees lots of misconceptions and confusion among researchers. Apparently these are partly due to lack of knowledge about formal parsing methods such as, for example, Earley's method and how issues like "backtracking", "determinism", and "efficiency" relate to these algorithms. Since then, however, knowledge of formal methods has become more widespread. This can also be illustrated with the introduction of formalisms like Lexical Functional Grammar (LFG), Generalized Phrase Structure Grammar (GPSG), Head-Driven Phrase Structure Grammar (HPSG), Unification Formalisms, Definite Clause Grammars and Tree Adjoining Grammars (TAGs) in the early 1980s. It led to a new discussion on the question whether the generative capacity of context-free formalisms would suffice to describe the syntax of natural languages, it led to a systematic comparison of grammar formalisms, yielding the weakly context-sensitive languages as a newly discovered class for which adequate generative capacity was claimed [JVW91], and it led to many less efficient, but nevertheless polynomial variants of general context-free parsing algorithms.

The formalisms mentioned above are certainly much more general than a pure context-free formalism. However, their backbone is context-free or the way the formalisms are defined and used bear very much resemblance to the context-free paradigm.

1.2 Parsing technology

We have not yet mentioned one of the main influences that caused researchers in computational linguistics and natural language processing to shift their attention to existing formal parsing methods and possible extensions of these methods. That influence was the increasing demand of society, military, and funding organisations to produce research results that could be used to build tools and systems for practical natural language processing applications. Applications like speech understanding systems, natural language interfaces to information systems, machine translation of texts, information retrieval, help systems for complex software and machinery, knowledge extraction from documents, text image processing, and so on. The availability of a comprehensive cognitive and linguistic theory does not seem to be a precondition for applications in the area of natural language understanding. Many applications do not require this comprehensive theory. Moreover, many applications can be built using research results that are not influenced in any way by cognitive, psycho-linguistic or linguistic principles. Generalized LR parsing, introduced in the mid-1980s by Masaru Tomita, is such a method that was introduced as a simple, straightforward and efficient parser for general context-free languages and grammars. Due to its straightforwardness, like the deterministic LR method, it has attracted a lot of attention and it has been used in many natural language processing research projects and in some applications.

In computer science, it was stated above, more and more attention has been devoted to general context-free parsing methods. The just mentioned generalized LR method, for example, has been used in grammar, parser and compiler development environments. In general, software engineering environments may offer their users syntax-dependent tools. The compiler construction level is only one level, and a rather low one, where descriptions based on formal grammar play a role. Furthermore, computer science is a growing science in the sense that borders between so-called "pure" computer science and several application areas are disappearing. Grammars and parsing methods play a role in pattern recognition, they have been used to describe and analyse command and action languages (human interaction with a computer system through key presses, cursor movements, etc.), to describe screen layouts, etc. Human factors have become important in computer science. Increasing the accessibility of computer systems through the use of speech and natural language in the man-machine interface is an aim worth pursuing. It is obvious that computer scientists and computational linguists will meet each other here and that they can learn from each other's methods to deal with languages.

Finally, we would like to mention another influence that caused computer scientists to go back to general context-free parsing methods. Parallelism is the keyword here. The introduction of new types of machine architectures and the possibility to implement algorithms on single chips have led to new research on existing parsing algorithms. Some research has been purely theoretical, in the sense that all kinds of efficiency limits were explored. Some research has been practical, in the sense that all kinds of extensions and variations of existing parsing algorithms have been investigated in order to make them suitable for parallel implementations and in the sense that these implementations have been realized, analysed and evaluated.

1.3 About this chapter

Parsing schemata are introduced in this chapter and used to bring some order in the field of context-free language parsing. In order to compare the essentials of parsing algorithms we want to abstract as much as possible from implementation issues, including the data, control and communication structures. The general idea is that all kinds of different parsers are "item" based, in the sense that they start with an initial set of items (constructed from the sentence), compute intermediate sets of items and deliver a final set of items. Items can have different interpretations. In this context it suffices to consider an item as a set of constraints on a (partial or complete) parse tree. This is, for example, a possible interpretation of an item in an LR or Earley parsing algorithm. Recognition or production of items can be interpreted as logical deduction from a set of hypotheses (initial items) to a set of final items (representing completed parse trees) by applying deduction steps. Underlying the parsers we are familiar with are different "deduction" systems using different items and different deduction steps. Relations between parsing algorithms can be found by defining operations on items and deduction steps. A parsing schema is the formal model, on a level of abstraction between grammar and algorithm, in which these ideas are expressed. Having these parsing schemata and being able, as we claim, to use them to understand the relations between parsing algorithms, it becomes also possible to port improvements and optimizations for one algorithm to related algorithms.

In the next section, parsing schemata are introduced by means of some informal examples. These examples are the standard CYK algorithm and the Earley algorithm. A formal introduction to parsing schemata follows in Section 3. Here we also introduce the notion of correctness of a parsing schema. That is, we need a way to say that a parse tree satisfies the constraints expressed in the items. Proving correctness on the level of parsing schemata is a less tedious task than on the level of algorithms since all details about data, control and communication structures are not present.

Section 4 and Section 5 are concerned with generalization and filtering, respectively. Generalizations and filters are relations between parsing schemata. Relations between different parsers can be uncovered when relations between their underlying parsing schemata have been established.

Generalization takes a more fine-grained look on the parsing process. It leads to more items and more steps. Generalization can be decomposed into several "primitive" relationsships. These primitive generalizations and their combinations allow, for example, the derivation of a simple version of the Earley parser from the CYK parser.

The aim of filtering is to reduce the number of steps and items. The different basic kinds of filtering that are introduced allow, for example, the derivation of the canonical Earley parser from the simple (bottom-up) Earley parser that was obtained as a generalization of the CYK parser. In a similar way, the well known and efficient Graham – Harrison – Ruzzo parser can be filtered from the Earley parser using basic filtering techniques on the underlying parsing schemata. Some observations can be made about the conse-

quences of using these techniques for run-time and compile-time optimization
and consequences for parallel implementations.

After having formalized and illustrated the theoretical concepts we can, in
Section 6, show more elaborate examples of how filtering and generalization
can be used to relate parsing schemata underlying different parsers. Rytter's
parallel parsing algorithm and Left-Corner parsing are among the algorithms
that will be related to previously mentioned algorithms through comparisons
of the underlying parsing schemata.

Section 7 surveys some other well-known approaches to the parsing problem and shows how they relate to the parsing schemata framework that is
introduced here. In particular, the relation with LR-like methods is exam-
ined. In Section 8 we briefly review some other grammar formalisms in computational linguistics and discuss how they relate to the parsing schemata
framework. Conclusions are summarized in the final section.

2. An informal introduction

We introduce the general idea of a parsing schema by means of a few informal
examples. A more rigorous treatment follows in Section 3. A comprehensive
discussion of parsing schemata will appear in [Sik97].

The following conventions apply throughout this chapter:

A *context-free grammar* is a 4-tuple $G = (N, \Sigma, P, S)$, with N a set of
nonterminal symbols, Σ a set of terminal symbols, P a finite set of productions, and $S \in N$ the start symbol. Furthermore, $N \cap \Sigma = \emptyset$. We write V for
$N \cup \Sigma$.

We write $A, B, \ldots \in N$ for nonterminals; $a, b, \ldots \in \Sigma$ for terminals;
$X, Y, \ldots \in V$ for arbitrary variables; $\alpha, \beta, \ldots \in V^*$ for strings of arbitrary
variables; ε for the empty string. The letters $i, j, \ldots$ denote nonnegative integers.

We write $A \rightarrow \alpha$ for a production (A, α) in P. The relation $\Rightarrow$ on $V^* \times V^*$ is
defined by $\alpha \Rightarrow \beta$ if there are $\alpha_1, \alpha_2, A, \gamma$ such that $\alpha = \alpha_1 A \alpha_2$, $\beta = \alpha_1 \gamma \alpha_2$
and $A \rightarrow \gamma \in P$.

The class of context-free grammars is denoted by $\mathcal{CFG}$. A subclass of $\mathcal{CFG}$
is $\mathcal{CNF}$, the class of grammars in Chomsky Normal Form. If $G \in \mathcal{CNF}$ then
P contains productions of the form $A \rightarrow BC$ and $A \rightarrow a$ only.

A very simple parsing algorithm is the so-called CYK algorithm [Kas65,
You67], named after Cocke, Younger, and Kasami. It is restricted to grammars in Chomsky Normal Form.

Assume that we have some grammar $G \in \mathcal{CNF}$ and a string $a_1 \ldots a_n$ to
be parsed. The CYK algorithm recognizes *items* $[A, i, j]$ that satisfy $A \Rightarrow^*$
$a_{i+1} \ldots a_j$.

The canonical way to implement this is to use a triangular matrix T with
cells $T_{i,j}$ for all applicable value pairs of i and j. Recognition of an item

$[A, i, j]$ is denoted by adding A to $T_{i,j}$. If we have $a = a_j$ and $A \to a \in P$ then A can be added to entry $T_{j-1,j}$. If we have $B \in T_{i,k}$, $C \in T_{k,j}$ and $A \to BC \in P$ then A can be added to $T_{i,j}$. The CYK algorithm gives an obvious control structure to make sure that all items are recognized that can be recognized.

It is worth noting that the output of the algorithm is *not* a parse tree, or a collection of parse trees. The output of the CYK algorithm (abstracting from its canonical data structure) is a *set of items*

$$\{[A, i, j] \mid A \Rightarrow^* a_{i+1} \ldots a_j\}.$$

The string is correct if and only if $[S, 0, n]$ is in this set. Moreover, if the string is correct, a parse forest or a particular (e.g. leftmost) parse can be constructed fairly easy from the items in this set. If we have $[S, 0, n]$ then there must be B, C, and k such that $S \to BC \in P$ and $[B, 0, k]$ and $[C, k, n]$ have been recognized as well. So, in a strict sense, CYK is not a parser but a recognizer enhanced with information that facilitates parse tree construction. It is common practice to call this a parser as well, and most parsers discussed in the remainder of this chapter will be of the same nature.

The way in which the CYK algorithm recognizes items for a given grammar $G \in \mathcal{CNF}$ and string $a_1 \ldots a_n$ can be denoted by a logical deduction system, called a *parsing system*.

Example 2.1. (*CYK*)
Firstly, we define a domain of items

$$\mathcal{I}_{\text{CYK}} = \{[A, i, j] \mid A \in N \wedge 0 \le i < j\}.$$

One could restrict $\mathcal{I}$ to items with $j \le n$, of course, but there are some advantages in choosing the domain of items independent of the given sentence. Secondly, we need a set of so-called *hypotheses*[1]

$$H = \{[a, i-1, i] \mid a = a_i \wedge 1 \le i \le n\}$$

that represent the string.
Thirdly, we need inference rules. We specify an inference rule by a *set of deduction steps* that covers all instances of inferences[2]. A set of inference rules, therefore, can be denoted by the union of corresponding sets of deduction steps. For CYK we define:

$$D^{(1)} = \{[a, i-1, i] \vdash [A, i-1, i] \mid A \to a \in P\},$$

---

[1] Whether the hypotheses are included in the domain of items or not does not really matter. It will turn out te be more convenient to define a separate set of hypotheses.

[2] This way of specifying rules has been chosen because it allows a certain flexibility. For example, it allows specification of *conditional rules*, to be applied only in certain circumstances, simply by restricting the set of deduction steps.

$$D^{(2)} = \{[B,i,j],[C,j,k] \vdash [A,i,k] \mid A \rightarrow BC \in P\},$$

$$D_{\text{CYK}} = D^{(1)} \cup D^{(2)}.$$

As with the domain $\mathcal{I}$, we have not bothered to restrict the deduction steps to items with $j \leq n$. The parsing system $\mathbb{P}_{\text{CYK}}$ for G and $a_1 \ldots a_n$ is defined by the triple $\langle \mathcal{I}, H, D \rangle$.

A *parsing schema* **CYK** is a generalization of $\mathbb{P}_{\text{CYK}}$ to arbitrary strings and arbitrary grammars in $\mathcal{CNF}$. One can see a parsing schema as a function that yields a parsing system for a given grammar and a given string over the alphabet of that grammar.

The CYK algorithm has the disadvantage that it is restricted to grammars in Chomsky Normal Form. A similar algorithm for arbitrary context-free grammars has been discovered by Earley [Ear68, Ear70]. Different variants of Earley's algorithm exist. First we investigate the one that is closest to CYK, the *bottom-up* Earley parser.

Example 2.2. (*bottom-up Earley*)
An Earley item has the form $[A \rightarrow \alpha \bullet \beta, i, j]$, with $A \rightarrow \alpha\beta \in P$. The bottom-up Earley parser recognizes the item set

$$\{[A \rightarrow \alpha \bullet \beta, i, j] \mid \alpha \Rightarrow^* a_{i+1} \ldots a_j\} \quad \text{for } G \text{ and } a_1 \ldots a_n.$$

A recognized item denotes partial recognition of a production. If $\beta = \varepsilon$, we have recognized a full production – and hence the left-hand side A, corresponding to $[A, i, j]$ in the CYK case. Partially recognized productions can be expanded by "moving the dot rightwards", i.e., recognizing the symbol behind the dot. How to organize this and store the results does not concern us here. We only specify the domain of items, the hypotheses and the deduction steps. For some grammar G and string $a_1 \ldots a_n$ we specify a parsing system $\mathbb{P}_{\text{buE}}$ by[3]

$$\mathcal{I}_{\text{buE}} = \{[A \rightarrow \alpha \bullet \beta, i, j] \mid A \rightarrow \alpha\beta \in P \wedge 0 \leq i \leq j\};$$

$$H_{\text{buE}} = \{[a, i-1, i] \mid a = a_i \wedge 1 \leq i \leq n\};$$

$$D^{\text{Init}} = \{\vdash [A \rightarrow \bullet \gamma, i, i]\},$$

$$D^{\text{Scan}} = \{[A \rightarrow \alpha \bullet a\beta, i, j], [a, j, j+1] \vdash [A \rightarrow \alpha a \bullet \beta, i, j+1]\},$$

$$D^{\text{Compl}} = \{[A \rightarrow \alpha \bullet B\beta, i, j], [B \rightarrow \gamma \bullet, j, k] \vdash [A \rightarrow \alpha B \bullet \beta, i, k]\},$$

$$D_{\text{buE}} = D^{\text{Init}} \cup D^{\text{Scan}} \cup D^{\text{Compl}}.$$

---

[3] From the usual set notation $\{\ldots \mid \ldots\}$ we omit the second part if there are no further constraints on the elements that comprise the set. It should be evident (and it will be formally stated in Section 3.1) that only items are used that relate to productions of the grammar G.

Deduction steps D^{Init} are needed to start the deduction of further valid items, hence these have no antecedents. In the definition of D^{Init} there is no need to state explicitly that $A{\rightarrow}\gamma \in P$ is required, as the deduction steps are only meaningful for items drawn from from $\mathcal{I}$ and H.
D^{Scan} and D^{Compl} conform to the *scan* and *complete* steps of Earley's algorithm. In Fig. 2.1 it is sketched how the *complete* step produces an item representing a larger partial parse from two known partial parses.

Earley's original algorithm is more restrictive in the items it recognizes. It makes use of top-down filtering. That is, the recognition of a production is started only if there is a need to do so. Only if we have an item $[A{\rightarrow}\alpha{\bullet}B\beta, i, j]$ there is a need to start recognizing a nonterminal B that produces $a_{j+1}\ldots a_k$ for some k. Top-down filtering reduces the number of recognized items, but also reduces the possibilities for parallel processing. Earley's algorithm is essentially left-to-right. Initial items start at position 0 and a parser has to work its way rightwards, unlike the bottom-up case where one can start recognizing items at all positions in the sentence in parallel.

Example 2.3. (*canonical Earley*)
The parsing system $\mathbb{P}_{\text{Earley}}$ for a given context-free grammar G and string $a_1\ldots a_n$ is defined by $\mathcal{I}$ and H as in $\mathbb{P}_{\text{buE}}$ (cf. Example 2.2) and by D_{Earley} as follows:

$$D^{\text{Init}} = \{\vdash [S{\rightarrow}{\bullet}\gamma, 0, 0]\},$$

$$D^{\text{Pred}} = \{[A{\rightarrow}\alpha{\bullet}B\beta, i, j] \vdash [B{\rightarrow}{\bullet}\gamma, j, j]\},$$

$$D^{\text{Scan}} = \{[A{\rightarrow}\alpha{\bullet}a\beta, i, j], [a, j, j+1] \vdash [A{\rightarrow}\alpha a{\bullet}\beta, i, j+1]\},$$

$$D^{\text{Compl}} = \{[A{\rightarrow}\alpha{\bullet}B\beta, i, j], [B{\rightarrow}\gamma{\bullet}, j, k] \vdash [A{\rightarrow}\alpha B{\bullet}\beta, i, k]\},$$

$$D_{\text{Earley}} = D^{\text{Init}} \cup D^{\text{Scan}} \cup D^{\text{Compl}} \cup D^{\text{Pred}}.$$

The Earley parsing system for G and $a_1\ldots a_n$ yields the following set of recognized items:

$$\{[A{\rightarrow}\alpha{\bullet}\beta, i, j] \mid \alpha \Rightarrow^* a_{i+1}\ldots a_j \wedge S \Rightarrow^* a_1\ldots a_i A\gamma \text{ for some } \gamma\}.$$

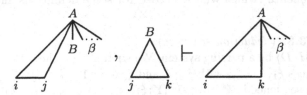

Fig. 2.1. The *complete* step

3. Parsing schemata

Parsing systems and parsing schemata are formally introduced in Sections 3.1 and 3.2, respectively. Section 3.3 discusses the nature of items and introduces a concept of parsing schema correctness.

3.1 Parsing systems

Definition 3.1. (*parsing system*)
A parsing system $\mathbb{P}$ for some grammar G and string $a_1 \ldots a_n$ is a triple $\mathbb{P} = \langle \mathcal{I}, H, D \rangle$, in which

- $\mathcal{I}$ is a set of items[4], called the *domain* or the *item set* of $\mathbb{P}$;
- H is a finite set of items called the *hypotheses* of $\mathbb{P}$;
- $D \subseteq \wp_{fin}(H \cup \mathcal{I}) \times \mathcal{I}$ is a set of deduction steps.

Note that H need not be a subset of $\mathcal{I}$. $\wp_{fin}$ in the above definition denotes the powerset restricted to finite sets. As a more convenient notation for deduction steps, we write $\eta_1, \ldots, \eta_k \vdash \xi$ rather than $(\{\eta_1, \ldots, \eta_k\}, \xi)$. Furthermore if we have $Y = \{\eta_1, \ldots, \eta_k\}$, we may also write $Y \vdash \xi$ as an abbreviation for $\eta_1, \ldots, \eta_k \vdash \xi$.

To be formally correct, however, we make a distinction between the set of deduction steps D and the inference relation $\vdash$ on $\wp_{fin}(H \cup \mathcal{I}) \times \mathcal{I}$. We want the inference relation to have the following conventional property:

if $\eta_1, \ldots, \eta_k \vdash x$ holds, then also $\eta_1, \ldots, \eta_k, \zeta \vdash x$ for any ζ.

Therefore we define $\vdash$ as the closure of D under addition of antecedents to an inference:

Definition 3.2. (*inference relation* $\vdash$)
Let $\mathbb{P} = \langle \mathcal{I}, H, D \rangle$ be a parsing system. The relation $\vdash \subseteq \wp_{fin}(H \cup \mathcal{I}) \times \mathcal{I}$ is defined by

$Y \vdash \xi$ if $(Y', \xi) \in D$ for some $Y' \subseteq Y$.

Before we define the transitive closure of $\vdash$ we introduce the notion of a *deduction sequence* (which will be needed for some definitions and proofs in Section 4.2).

Definition 3.3. (*deduction sequence*)
Let $\mathbb{P} = \langle \mathcal{I}, H, D \rangle$ be a parsing system. We write $\mathcal{I}^+$ for the set of non-empty, finite sequences $\xi_1, \ldots, \xi_j$, with $j \geq 1$ and $\xi_i \in \mathcal{I}$ $(1 \leq i \leq j)$.
A deduction sequence in $\mathbb{P}$ is a pair $(Y; \xi_1, \ldots, \xi_j) \in \wp_{fin}(H \cup \mathcal{I}) \times \mathcal{I}^+$, such that $Y \cup \{\xi_1, \ldots, \xi_{i-1}\} \vdash \xi_i$ for $1 \leq i \leq j$.

---

[4] Here we treat 'item' as an undefined basic concept. A discussion about the nature of items follows in Section 3.3.

As a practical informal notation we write $Y \vdash \xi_1 \vdash \ldots \vdash \xi_j$ for a deduction sequence $(Y; \xi_1, \ldots, \xi_j)$.

Definition 3.4. (Δ)
The *set of deduction sequences* $\Delta \subseteq \wp_{fin}(H \cup \mathcal{I}) \times \mathcal{I}^+$ for a parsing system $\mathbb{P} = \langle \mathcal{I}, H, D \rangle$ is defined by

$$\Delta = \{(Y; \xi_1, \ldots, \xi_j) \in \wp_{fin}(H \cup \mathcal{I}) \times \mathcal{I}^+ \mid Y \vdash \xi_1 \vdash \ldots \vdash \xi_j\}.$$

Definition 3.5. $(\vdash^*)$
For a parsing system $\mathbb{P} = \langle \mathcal{I}, H, D \rangle$ we define the relation $\vdash^*$ on $\wp_{fin}(H \cup \mathcal{I}) \times \mathcal{I}$ by

$$Y \vdash^* \xi \quad \text{if} \quad \xi \in Y \quad \text{or} \quad Y \vdash \ldots \vdash \xi.$$

Definition 3.6. (*valid items*)
For a parsing system $\mathbb{P} = \langle \mathcal{I}, H, D \rangle$ the set of valid items is defined by

$$\mathcal{V}(\mathbb{P}) = \{\xi \in \mathcal{I} \mid H \vdash^* \xi\}.$$

We do not make a distinction between semantic validity (usually denoted $\models \xi$) and syntactic provability (i.e., $H \vdash^* \xi$).

3.2 Parsing schemata

A parsing system has been defined for a fixed grammar and string. In two steps we will extend this to a parsing schema for arbitrary grammmars and strings.

Definition 3.7. (*uninstantiated parsing system*)
An uninstantiated parsing system for a grammar G is a triple $\langle \mathcal{I}, \mathcal{H}, D \rangle$ with $\mathcal{H}$ a function that assigns a set of hypotheses to each string $a_1 \ldots a_n \in \Sigma^*$, such that $\langle \mathcal{I}, \mathcal{H}(a_1 \ldots a_n), D \rangle$ is a parsing system.

A function $\mathcal{H}$ that will be used throughout the remainder of this chapter (unless specifically stated otherwise) is

$$\mathcal{H}(a_1 \ldots a_n) = \{[a, i - 1, i] \mid a = a_i \wedge 1 \le i \le n\}.$$

In the sequel we will omit the hypotheses H from the specification of a parsing system when the default $\mathcal{H}(a_1 \ldots a_n)$ applies.

Definition 3.8. (*parsing schema*)
A parsing schema for some (sub)class of context-free grammars $\mathcal{CG} \subseteq \mathcal{CFG}$ is a function that assigns an uninstantiated parsing system to every grammar $G \in \mathcal{CG}$.

Schema 3.9. (**CYK**)
The parsing schema **CYK** is defined for any $G \in \mathcal{CNF}$ and for any $a_1 \ldots a_n \in \Sigma^*$ by $\mathbf{CYK}(G)(a_1 \ldots a_n) = \mathbb{P}_{\text{CYK}}$ as in Example 2.1.

Schema 3.10. (buE)
The parsing schema **buE** is defined for any $G \in C\mathcal{F}G$ and for any $a_1 \ldots a_n \in \Sigma^*$ by $\mathbf{buE}(G)(a_1 \ldots a_n) = \mathbb{P}_{\text{buE}}$ as in Example 2.2.

Schema 3.11. (Earley)
The parsing schema **Earley** is defined for any $G \in C\mathcal{F}G$ and for any $a_1 \ldots a_n \in \Sigma^*$ by $\mathbf{Earley}(G)(a_1 \ldots a_n) = \mathbb{P}_{\text{Earley}}$ as in Example 2.3.

3.3 Correctness of parsing schemata

In order to define a notion of correctness, some understanding of the nature of items is needed. We have seen two kinds of items so far, there are other parsing algorithms that involve different kinds of items. What, exactly *is* an item?

An item lists a set of constraints on a (partial or complete) parse tree. Recognition of an Earley item $[A{\rightarrow}\alpha{\bullet}\beta, i, j]$ means: There is *some* tree that has a root labelled A with children labelled $\alpha\beta$ (concatenated from left to right). Moreover, the nodes labelled α are the roots of sub-trees that yield $a_{i+1} \ldots a_j$ whereas the nodes labelled β are leaves, cf. Fig. 3.1.

One way to interpret an item is to identify it with a *set of trees*, viz., all trees that satisfy the constraints stated in the item. This approach is taken in [Sik93a]. Pursuing this line of thought, an item set is defined by a congruence relation on a set of trees with respect to the deduction relation.

A rather simpler approach is to regard an item as a partial specification of a tree. We assume that there is some general item specification language and that all items used in practical algorithms are (efficient notations for) specific instances of this specification language. We will not further formalize this, because in all practical cases it is abundantly clear what is meant by the various types of items.

Before we define correctness, there are two regularity properties on item sets that have to be stated explicitly.

Firstly, we have tacitly assumed that there is a clear separation between *final items*, denoting completed parse trees, and *intermediate items*, denoting partial, not yet completed trees.

It is possible – but admittedly rather artifical – to contruct *mixed items* that denote a combination of both types. Consider, for example, a grammar

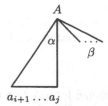

Fig. 3.1. A partially specified tree

in Chomsky Normal Form that has productions $A{\to}SC$ and $A{\to}BC$, with S and B not occuring anywhere else in the right-hand side of a production. For the recognition of A, therefore, it is irrelevant whether $[S, i, j]$ or $[B, i, j]$ has been recognized. So we could replace these two items by a single item $[(S, B), i, j]$. But then we have a problem with the item $[(S, B), 0, n]$. If this item is recognized, it is unclear whether it denotes the existence of a parse tree.

Secondly, we assume that for each parse tree of a sentence, this parse tree conforms to the partial specification of some item in $\mathcal{I}$.

Definition 3.12. (*semiregularity*)[5]
A parsing system $\mathbb{P} = \langle \mathcal{I}, H, D \rangle$ for a grammar G and string $a_1 \ldots a_n$ is called semiregular if $\mathcal{I}$ does not contain mixed items and each parse tree of $a_1 \ldots a_n$ conforms to the specification of some item in $\mathcal{I}$.
A parsing schema $\mathbf{P}$ for a class of grammars $\mathcal{CG}$ is semiregular if $\mathbf{P}(G)(a_1 \ldots a_n)$ is semiregular for all $G \in \mathcal{CG}$ and all $a_1 \ldots a_n \in \Sigma^*$.

Definition 3.13. (*correct final items*)
We write $\mathcal{F}(\mathbb{P}) \subseteq \mathcal{I}$ for the set of the final items of a parsing system $\mathbb{P}$ for a grammar G and a string $a_1 \ldots a_n$.
A final item is correct if there is a parse tree for $a_1 \ldots a_n$ that conforms to the specification expressed by this item. We write $\mathcal{C}(\mathbb{P}) \subseteq \mathcal{F}(\mathbb{P})$ for the set of correct final items of $\mathbb{P}$.

Example 3.14. (*final and correct final items*)

- $\mathcal{F}(\mathbb{P}_{\text{CYK}}) = \{[S, 0, n]\};$
- $\mathcal{C}(\mathbb{P}_{\text{CYK}}) = \{[S, 0, n]\}$ if $a_1 \ldots a_n \in L(G)$,
  $\mathcal{C}(\mathbb{P}_{\text{CYK}}) = \emptyset$ if $a_1 \ldots a_n \notin L(G)$;
- $\mathcal{F}(\mathbb{P}_{\text{buE}}) = \mathcal{F}(\mathbb{P}_{\text{Earley}}) = \{[S{\to}\alpha\bullet, 0, n] \mid S{\to}\alpha \in P\};$
- $\mathcal{C}(\mathbb{P}_{\text{buE}}) = \mathcal{C}(\mathbb{P}_{\text{Earley}}) = \{[S{\to}\alpha\bullet, 0, n] \mid \alpha \Rightarrow^* a_1 \ldots a_n\}.$

Definition 3.15. (*correctness of a parsing schema*)
A semiregular parsing system $\mathbb{P}$ is *sound* if $\mathcal{F}(\mathbb{P}) \cap \mathcal{V}(\mathbb{P}) \subseteq \mathcal{C}(\mathbb{P})$, i.e., all valid final items are correct.
A semiregular parsing system $\mathbb{P}$ is *complete* if $\mathcal{F}(\mathbb{P}) \cap \mathcal{V}(\mathbb{P}) \supseteq \mathcal{C}(\mathbb{P})$, i.e., all correct final items are valid.
A semiregular parsing system is *correct* if $\mathcal{F}(\mathbb{P}) \cap \mathcal{V}(\mathbb{P}) = \mathcal{C}(\mathbb{P})$, i.e., it is sound and complete.
A semiregular parsing schema $\mathbf{P}$ is sound/complete/correct for a class of grammars $\mathcal{CG}$ if $\mathbf{P}(G)(a_1 \ldots a_n)$ is sound/complete/correct for all $G \in \mathcal{CG}$ and $a_1 \ldots a_n \in \Sigma^*$.

---

[5] The notion *regularity* was introduced in [Sik93a] for parsing systems and schemata that do not contain inconsistent specifications, viz. the empty set of items. We do not need the regularity property in this context.

CYK, **buE**, and **Earley** are correct semiregular parsing schemata (and so are the other schemata that will be proposed in the remainder of this chapter). This is well-known from the literature and we will not further explore the issue of how to prove the correctness of a parsing schema.

4. Generalization

Various kinds of relations between parsing algorithms can be formally established by defining relations between their underlying parsing schemata. In this section we will look at *generalization* of a schema, that can be obtained by *refinement* into a more detailed parsing schema and/or *extension* to a larger class of grammars.

Adding detail to a schema means more (refined) items, more deduction steps, hence more work to parse a sentence. This is useful if it leads to *qualitative* improvements in the parsing algorithm. The canonical example (that we spell out first as an illustration) is that the bottom-up Earley parser is a generalization of the CYK parser.

4.1 Some examples

More precisely, but still informally, we distinguish the following basic kinds of generalizations:

- A parsing schema P_2 is an *item refinement* of a schema P_1 if a single item in P_1 is broken down into multiple items in P_2 (and the set of deduction steps adapted accordingly);
- A parsing schema P_2 is a *step refinement* of a schema P_1 if a single deduction step in P_1 is decomposed into a sequence of deduction steps in P_2 (and new items are introduced, when needed, to store the refined intermediate results);
- A schema P_2 is called an *extension* of a schema P_1 if it is defined for a larger class of grammars.

A relation is called a *refinement* if it is a step refinement, an item refinement or a combination of these. A relation is called a *generalization* if it is a refinement, an extension or a combination of these. We write $\overset{ir}{\Longrightarrow}$ for item refinement, $\overset{sr}{\Longrightarrow}$ for step refinement, $\overset{ref}{\Longrightarrow}$ for refinement, $\overset{ext}{\Longrightarrow}$ for extension, and $\overset{gen}{\Longrightarrow}$ for generalization.

Example 4.1. (CYK $\overset{gen}{\Longrightarrow}$ buE)
In order to generalize **CYK** into **buE** we introduce two intermediate parsing systems **CYK′** and **ECYK**, such that

$$\text{CYK} \overset{ir}{\Longrightarrow} \text{CYK}' \overset{sr}{\Longrightarrow} \text{ECYK} \overset{ext}{\Longrightarrow} \text{buE}.$$

The only thing we change in **CYK'** is that CYK items $[A, i, j]$ are replaced by completed Earley items $[A \to \alpha\bullet, i, j]$. We define **CYK'** by specifying a parsing system $\mathbb{P}_{\text{CYK'}}$ for an arbitrary grammar in $\mathcal{CNF}$ as follows:

$$\mathcal{I}_{\text{CYK'}} = \{[A \to \alpha\bullet, i, j] \mid A \to \alpha \in P \wedge 0 \le i \le j\};$$

$$D^{(1)} = \{[a, j-1, j] \vdash [A \to a\bullet, j-1, j]\},$$

$$D^{(2)} = \{[B \to \beta\bullet, i, j], [C \to \gamma\bullet, j, k] \vdash [A \to BC\bullet, i, k]\},$$

$$D_{\text{CYK'}} = D^{(1)} \cup D^{(2)}.$$

Note that a single CYK item $[A, i, j]$ may correspond to multiple Earley items $[A \to \alpha\bullet, i, j]$, $[A \to \beta\bullet, i, j]$, etc., if there are different productions with left-hand side A. That is why this is an item refinement, not merely a change of notation.

In the next step, we refine a single CYK deduction step

$$[B \to \beta\bullet, i, j], [C \to \gamma\bullet, j, k] \vdash [A \to BC\bullet, i, k]$$

into a sequence of deduction steps

$$\vdash [A \to \bullet BC, i, i],$$
$$[A \to \bullet BC, i, i], [B \to \beta\bullet, i, j] \vdash [A \to B\bullet C, i, j],$$
$$[A \to B\bullet C, i, j], [C \to \gamma\bullet, j, k] \vdash [A \to BC\bullet, i, k].$$

This is encorporated in the parsing schema **ECYK**, defined by a parsing system $\mathbb{P}_{\text{ECYK}}$ for an arbitrary grammar in $\mathcal{CNF}$:

$$\mathcal{I}_{\text{ECYK}} = \{[A \to \alpha\bullet\beta, i, j] \mid A \to \alpha\beta \in P \wedge 0 \le i \le j\};$$

$$D^{\text{Init}} = \{\vdash [A \to \bullet\alpha, j, j]\},$$

$$D^{\text{Scan}} = \{[A \to \alpha\bullet a\beta, i, j], [a, j, j+1] \vdash [A \to \alpha a\bullet\beta, i, j+1]\},$$

$$D^{\text{Compl}} = \{[A \to \alpha\bullet B\beta, i, j], [B \to \gamma\bullet, j, k] \vdash [A \to \alpha B\bullet\beta, i, k]\},$$

$$D_{\text{ECYK}} = D^{\text{Init}} \cup D^{\text{Scan}} \cup D^{\text{Compl}}.$$

ECYK is identical to **buE**, cf. Example 2.2, except for the fact that **ECYK** is defined only for grammars in $\mathcal{CNF}$. Hence, obviously, **buE** is an extension of **ECYK**.

4.2 Formalization

We will now formalize the concepts that have been informally introduced and illustrated above. In the sequel we write $\mathbb{P}_i$ for a parsing system $\mathbb{P}_i = \langle \mathcal{I}_i, H, D_i \rangle$; we write $\vdash_i$ and $\vdash_i^*$ for an inference relation and its closure, based on D_i, cf. Section 3.1.

For the definition of item refinement we make use of an *item mapping* $f : \mathcal{I}_2 \rightarrow \mathcal{I}_1$ that maps items of $\mathbb{P}_2$ to items of $\mathbb{P}_1$. The function f can be extended to cover sets of items in the usual way: For $Y \subseteq \mathcal{I}_2$ we define

$$f(Y) = \{\xi \in \mathcal{I}_1 \mid \exists \eta \in Y : f(\eta) = \xi\}$$

Moreover, we extend[6] f to a function $f : \mathcal{I}_2 \cup H \rightarrow \mathcal{I}_1 \cup H$ by letting $f(h) = h$ for $h \in H$. Then we can apply f to deduction steps by letting

$$f(\eta_1, \ldots, \eta_k \vdash \xi) = f(\eta_1), \ldots, f(\eta_k) \vdash f(\xi).$$

In the same fashion we can extend f to deduction sequences[7], to sets of deduction steps and to sets of deduction sequences. The equation

$$\Delta_1 = f(\Delta_2)$$

is a clear and concise notation for: $Y_1 \vdash_1 x_1 \vdash_1 \ldots \vdash_1 x_j$ if and only if there are $Y_2 \in \wp_{fin}(H_2 \cup \mathcal{I}_2)$ and $x'_2, \ldots, x'_j \in \mathcal{I}_2$ with $f(Y_2) = Y_1$ and $f(x_i) = x'_i$ for $1 \leq i \leq j$, such that $Y_2 \vdash_2 x'_1 \vdash_1 \ldots \vdash_1 x'_j$.

Definition 4.2. (*item refinement*)

The relation $\mathbb{P}_1 \xrightarrow{\text{ir}} \mathbb{P}_2$ holds between parsing systems $\mathbb{P}_1$ and $\mathbb{P}_2$ if there is an item mapping $f : \mathcal{I}_2 \rightarrow \mathcal{I}_1$ such that

(*i*) $\mathcal{I}_1 = f(\mathcal{I}_2)$,
(*ii*) $\Delta_1 = f(\Delta_2)$,

Let $\mathbf{P}_1$ and $\mathbf{P}_2$ be parsing schemata for some class of grammars $\mathcal{CG}$. The relation $\mathbf{P}_1 \xrightarrow{\text{ir}} \mathbf{P}_2$ holds if $\mathbf{P}_1(G)(a_1 \ldots a_n) \xrightarrow{\text{ir}} \mathbf{P}_2(G)(a_1 \ldots a_n)$ for all $G \in \mathcal{CG}$ and for all $a_1 \ldots a_n$.

The first condition in Definition 4.2 states that no items are 'lost' in the refinement; the second condition ensures that deduction sequences are carried over into the refined system. Deduction sequences are needed in the definition of refinement, in order to guarantee the transitivity of generalization. A weaker condition

(*ii*)′ $\vdash_1^* = \vdash_2^*$,

which might seem sufficient, will not do. An example of a relation that satisfies (*i*) and (*ii*)′ but is not a refinement is given by

$$\mathbb{P}_1 = \langle \{\xi, \eta, \zeta\}, \{h\}, \{h \vdash \xi, \ h \vdash \eta, \ \xi \vdash \zeta, \ \eta \vdash \zeta\}\rangle,$$

$$\mathbb{P}_2 = \langle \{\xi_1, \xi_2, \eta, \zeta\}, \{h\}, \{h \vdash \xi_1, \ h \vdash \eta, \ \xi_2 \vdash \zeta, \ \eta \vdash \zeta\}\rangle,$$

with $f(\xi_i) = \xi$, f is the identity function otherwise.

---

[6] Assuming that $H \cap \mathcal{I} = \emptyset$; otherwise we demand that f restricted to $H \cap \mathcal{I}$ is the identity function.

[7] Note, however, that the image of a deduction sequence is not necessarily a deduction sequence (cf. Definition 3.3), hence $f : \Delta_2 \rightarrow \Delta_1$ and $f : \wp(\Delta_2) \rightarrow \wp(\Delta_1)$ are partial functions.

Definition 4.3. (*step refinement*)

The relation $\mathbb{P}_1 \xrightarrow{\text{sr}} \mathbb{P}_2$ holds between parsing systems $\mathbb{P}_1$ and $\mathbb{P}_2$ if

(*i*) $\mathcal{I}_1 \subseteq \mathcal{I}_2$,

(*ii*) $\vdash_1^* \subseteq \vdash_2^*$.

Let $\mathbf{P}_1$ and $\mathbf{P}_2$ be parsing schemata for some class of grammars $\mathcal{CG}$. The relation $\mathbf{P}_1 \xrightarrow{\text{sr}} \mathbf{P}_2$ holds if $\mathbf{P}_1(G)(a_1 \ldots a_n) \xrightarrow{\text{sr}} \mathbf{P}_2(G)(a_1 \ldots a_n)$ for all $G \in \mathcal{CG}$ and for all $a_1 \ldots a_n$.

A sufficient condition[8] for (*ii*) in Definition 4.3 is $D_1 \subseteq \vdash_2^*$, that is, a single deduction step in $\mathbb{P}_1$ is emulated by a sequence of deduction steps in $\mathbb{P}_2$. Furthermore, the domain of $\mathbb{P}_2$ may contain items that did not exist in $\mathbb{P}_1$.

Definition 4.4. (*refinement*)

Let $\mathbf{P}_1$ and $\mathbf{P}_2$ be parsing schemata for a class of grammars $\mathcal{CG}$. The relation $\mathbf{P}_1 \xrightarrow{\text{ref}} \mathbf{P}_2$ holds if there is a parsing schema $\mathbf{P}'$ such that $\mathbf{P}_1 \xrightarrow{\text{ir}} \mathbf{P}' \xrightarrow{\text{sr}} \mathbf{P}_2$.

Definition 4.5. (*extension*)

Let $\mathbf{P}_1$ be a parsing schema for a class of grammars $\mathcal{CG}_1$, $\mathbf{P}_2$ a parsing schema for a class of grammars $\mathcal{CG}_2$ and $\mathcal{CG}_1 \subseteq \mathcal{CG}_2$. The relation $\mathbf{P}_1 \xrightarrow{\text{ext}} \mathbf{P}_2$ holds if $\mathbf{P}_1(G) = \mathbf{P}_2(G)$ for all $G \in \mathcal{CG}_1$.

Definition 4.6. (*generalization*)

Let $\mathbf{P}_1$ be a parsing schema for a class of grammars $\mathcal{CG}_1$, $\mathbf{P}_2$ a parsing schema for a class of grammars $\mathcal{CG}_2$ and $\mathcal{CG}_1 \subseteq \mathcal{CG}_2$. The relation $\mathbf{P}_1 \xrightarrow{\text{gen}} \mathbf{P}_2$ holds if there is parsing schema $\mathbf{P}'$ such that $\mathbf{P}_1 \xrightarrow{\text{ref}} \mathbf{P}' \xrightarrow{\text{ext}} \mathbf{P}_2$.

4.3 Properties of generalization

Proposition 4.7.

Each of the relations $\xrightarrow{\text{ir}}$, $\xrightarrow{\text{sr}}$, $\xrightarrow{\text{ext}}$ is transitive and reflexive.   □

If xR_1y implies xR_2y for relations R_1, R_2, we write $R_1 \subseteq R_2$ (the set inclusion is in the Cartesian product of the domain of the relations).

Corollary 4.8.

(*a*) $\xrightarrow{\text{ir}} \subseteq \xrightarrow{\text{ref}}$;

(*b*) $\xrightarrow{\text{sr}} \subseteq \xrightarrow{\text{ref}}$;

(*c*) $\xrightarrow{\text{ref}} \subseteq \xrightarrow{\text{gen}}$;

(*d*) $\xrightarrow{\text{ext}} \subseteq \xrightarrow{\text{gen}}$.

---

[8] We write $\vdash_1^*$ just for symmetry, because all other relations defined in this section and the next one display the same kind of symmetry.

Next, we will establish the transitivity of $\overset{\text{ref}}{\Longrightarrow}$ and $\overset{\text{gen}}{\Longrightarrow}$. The former is not entirely trivial.

Lemma 4.9. (*refinement lemma*)

Let $\mathbb{P}_1$, $\mathbb{P}_2$, $\mathbb{P}_3$ be parsing systems such that $\mathbb{P}_1 \overset{\text{sr}}{\longrightarrow} \mathbb{P}_2 \overset{\text{ir}}{\longrightarrow} \mathbb{P}_3$. Then there is a parsing system $\mathbb{P}_4$ such that $\mathbb{P}_1 \overset{\text{ir}}{\longrightarrow} \mathbb{P}_4 \overset{\text{sr}}{\longrightarrow} \mathbb{P}_3$.
Let $\mathbf{P}_1$, $\mathbf{P}_2$, $\mathbf{P}_3$ be parsing schemata for some class of grammars CG, such that $\mathbf{P}_1 \overset{\text{sr}}{\Longrightarrow} \mathbf{P}_2 \overset{\text{ir}}{\Longrightarrow} \mathbf{P}_3$. Then there is a parsing schema $\mathbf{P}_4$ for CG such that $\mathbf{P}_1 \overset{\text{ir}}{\Longrightarrow} \mathbf{P}_4 \overset{\text{sr}}{\Longrightarrow} \mathbf{P}_3$.

Proof. It suffices to prove the lemma for parsing systems, the generalization to parsing schemata is trivial. Let $f : \mathcal{I}_3 \rightarrow \mathcal{I}_2$ be the item mapping from $\mathbb{P}_3$ to $\mathbb{P}_2$. Then we define $\mathbb{P}_4$ by

$$\mathcal{I}_4 = \{x \in \mathcal{I}_3 \mid f(x) \in \mathcal{I}_1\},$$
$$D_4 = \{(Y,x) \in \wp_{fin}(H \cup \mathcal{I}_4) \times \mathcal{I}_4 \mid f((Y,x)) \in D_1 \wedge Y \vdash_3^* x\}.$$

We now have to show that $\mathbb{P}_4$ is a parsing schema, that $\mathbb{P}_1 \overset{\text{ir}}{\longrightarrow} \mathbb{P}_4$ holds, and that $\mathbb{P}_4 \overset{\text{sr}}{\longrightarrow} \mathbb{P}_2$ holds. We prove $\mathbb{P}_1 \overset{\text{ir}}{\longrightarrow} \mathbb{P}_4$, as an exemplary case (the only one that needs some care in spelling out the details) and omit the other parts.
From the definition of $\mathbb{P}_4$ it is clear that $\mathcal{I}_1 = f(\mathcal{I}_4)$, hence it remains to be shown that $\Delta_1 = f(\Delta_4)$.

(*i*) $f(\Delta_4) \subseteq \Delta_1$.
This follows from the definition of $\mathbb{P}_4$ with induction on the length of deduction sequences.

(*ii*) $\Delta_1 \subseteq f(\Delta_4)$.
We use an ad-hoc notation $Y \vdash^* x_1 \vdash^* \ldots \vdash^* x_j \in \Delta$, meaning that there are (possibly empty) sequences $z_{i,1}, \ldots, z_{i,m_i}$ for $1 \leq i \leq j$ such that $Y \vdash z_{1,1} \vdash \ldots \vdash z_{1,m_1} \vdash x_1 \vdash \ldots \vdash z_{j,1} \vdash \ldots \vdash z_{j,m_j} \vdash x_j \in \Delta$.

Assume now that $Y \vdash_1 x_1 \vdash_1 \ldots \vdash_1 x_j \in \Delta_1$. Because $\mathbb{P}_1 \overset{\text{sr}}{\longrightarrow} \mathbb{P}_2$ it must hold that $Y \vdash_2^* x_1 \vdash_2^* \ldots \vdash_2^* x_j \in \Delta_2$. Moreover, from $\mathbb{P}_2 \overset{\text{ir}}{\longrightarrow} \mathbb{P}_3$ we obtain $Y' \in \wp_{fin}(H \cup \mathcal{I}_3) \times \mathcal{I}_3$ with $f(Y') = Y$ and $x_1', \ldots, x_j'$ with $f(x_1') = x_1, \ldots, f(x_j') = x_j$, such that $Y' \vdash_3^* x_1' \vdash_3^* \ldots \vdash_3^* x_j' \in \Delta_3$. From the definition of $\mathbb{P}_4$ it follows that $Y' \vdash_4 x_1' \vdash_4 \ldots \vdash_4 x_j' \in \Delta_4$. Thus we have shown $Y \vdash_1 x_1 \vdash_1 \ldots \vdash_1 x_j \in f(\Delta_4)$ which proves (*ii*). $\square$

Lemma 4.10.

Let $\mathbf{P}_1$ be a parsing schema for a class of grammar CG_1 and $\mathbf{P}_2$, $\mathbf{P}_3$ be parsing schemata for a class of grammar CG_2, such that $\mathbf{P}_1 \overset{\text{ext}}{\Longrightarrow} \mathbf{P}_2 \overset{\text{ref}}{\Longrightarrow} \mathbf{P}_3$. Then there is a parsing schema $\mathbf{P}_4$ such that $\mathbf{P}_1 \overset{\text{ref}}{\Longrightarrow} \mathbf{P}_4 \overset{\text{ext}}{\Longrightarrow} \mathbf{P}_3$.

Proof. Define $\mathbf{P}_4(G) = \mathbf{P}_2(G)$ for $G \in \mathcal{CG}_1$.   □

Theorem 4.11. The relation $\overset{\text{gen}}{\Longrightarrow}$ is transitive and reflexive.

Proof. Straightforward from Lemmata 4.9 and 4.10.   □

Correctness of parsing schemata is, in general, not preserved by generalization. A useful partial result is that completeness is preserved by $\overset{\text{sr}}{\Longrightarrow}$, that is, if $\mathbf{P}_1$ is a complete semiregular parsing schema and $\mathbf{P}_1 \overset{\text{sr}}{\Longrightarrow} \mathbf{P}_2$, then $\mathbf{P}_2$ is a complete semiregular parsing schema.

5. Filtering

Generalization increases the number of steps that have to be performed, but the more fine-grained look on the parsing process may allow qualitative improvements. Filtering is, in a way, the reverse. The purpose is to obtain quantitative improvements in parsing algorithms, by decreasing the number of items and deduction steps. It is often possible to argue that some kinds of items need not be recognized, because they cannot contribute to a valid parse. Discarding those items from the parsing schema means less work for the algorithm that implements the schema, but sometimes a more complicated description of the schema.

We distinguish three kinds of filtering:

- *static filtering*: redundant parts of a parsing schema are simply discarded;
- *dynamic filtering*: the validity of some items can be made dependent on the validity of other items, hence context information can be taken into account;
- *step contraction*: sequences of deduction steps are replaced by single deduction steps.

The theoretical framework is simple and elegant. As in the previous section we assume that a parsing system $\mathbb{P}_i$ is defined as $\langle \mathcal{I}_i, H, D_i \rangle$, with inference relations $\vdash_i$ and $\vdash_i^*$ on $\mathbb{P}_i$ according to Section 3.1.

5.1 Static filtering

Static filtering can be demonstrated by means of a very simple example (a more exciting example will follow in Section 6.2).

A nonterminal $A \in N$ is called *reduced* if

(i) there are $v, w \in \Sigma^*$ such that $S \Rightarrow^* vAw$,
(ii) there is some $w \in \Sigma^*$ such that $A \Rightarrow^* w$.

A grammar is called reduced if all its nonterminals are reduced. Let $G \in \mathcal{CFG}$ be an arbitrary context-free grammar. We can define a reduced grammar G' by

$$N' = \{A \in N \mid A \text{ is reduced}\}$$
$$P' = \{A \to \alpha \in P \mid A \in N' \land \alpha \in (N' \cup \Sigma)^*\}$$
$$G' = (N', \Sigma, P', S).$$

If G is reduced, then $G = G'$. Furthermore, it is clear that G and G' yield the same parse trees for any sentence.

Example 5.1. (*reduced* **buE**)
Let $\mathbb{P}_{\text{buE}}$ be a parsing system for some grammar G and string $a_1 \ldots a_n$. We define a parsing system $\mathbb{P}_{\text{buE}'}$ by

$$\mathcal{I}_{\text{buE}'} = \{[A \to \alpha \bullet \beta, i, j] \mid A \to \alpha\beta \in P'\}$$

and $D_{\text{buE}'}$ as in Example 2.2. Because $D \subseteq \wp_{fin}(H \cup \mathcal{I}) \times \mathcal{I}$ by definition, only deduction steps remain that contain non-reduced nonterminals wherever applicable.
A parsing schema **buE**$'$ is defined for all $G \in \mathcal{CFG}$ and $a_1 \ldots a_n \in \Sigma^*$ by **buE**$'(G)(a_1 \ldots a_n) = \mathbb{P}_{\text{buE}'}$, as usual.

Definition 5.2. (*static filtering*)
The relation $\mathbb{P}_1 \xrightarrow{\text{sf}} \mathbb{P}_2$ holds if

(*i*) $\mathcal{I}_1 \supseteq \mathcal{I}_2$
(*ii*) $D_1 \supseteq D_2$.

Let $\mathbf{P}_1$ and $\mathbf{P}_2$ be parsing schemata for some class of grammars $\mathcal{CG}$.

The relation $\mathbf{P}_1 \xRightarrow{\text{sf}} \mathbf{P}_2$ holds if $\mathbf{P}_1(G)(a_1 \ldots a_n) \xrightarrow{\text{sf}} \mathbf{P}_2(G)(a_1 \ldots a_n)$ for all $G \in \mathcal{CG}$ and for all $a_1 \ldots a_n$.

5.2 Dynamic filtering

The purpose of dynamic filtering is to take context information into account. If some type of constituent can only occur directly after another type of constituent, we may defer recognizing the former until we have established the latter. The technique to do this is to add antecedents to deduction steps. If we decided that an item ξ is to be valid only if some other item ζ is also valid, we simply replace deduction steps $\eta_1, \ldots, \eta_k \vdash \xi$ by deduction steps $\eta_1, \ldots, \eta_k, \zeta \vdash \xi$.

Definition 5.3. (*dynamic filtering*)
The relation $\mathbb{P}_1 \xrightarrow{\text{df}} \mathbb{P}_2$ holds if

(*i*) $\mathcal{I}_1 \supseteq \mathcal{I}_2$
(*ii*) $\vdash_1 \supseteq \vdash_2$.

Let $\mathbf{P}_1$ and $\mathbf{P}_2$ be parsing schemata for some class of grammars $\mathcal{CG}$.

The relation $\mathbf{P}_1 \xRightarrow{\text{df}} \mathbf{P}_2$ holds if $\mathbf{P}_1(G)(a_1 \ldots a_n) \xrightarrow{\text{df}} \mathbf{P}_2(G)(a_1 \ldots a_n)$ for all $G \in \mathcal{CG}$ and for all $a_1 \ldots a_n$.

Proposition 5.4. buE $\overset{\text{sf}}{\Longrightarrow}$ Earley.

Proof. We consider the parsing systems $\mathbb{P}_{\text{buE}}$ and $\mathbb{P}_{\text{Earley}}$ for some given G and $a_1 \ldots a_n$. By the definition of the two parsing systems (cf. Examples 2.3 and 2.2), $\mathcal{I}_{\text{buE}} = \mathcal{I}_{\text{Earley}}$ holds. In order to prove $\vdash_{\text{buE}} \supseteq \vdash_{\text{Earley}}$, it suffices to show that $\vdash_{\text{buE}} \supseteq D_{\text{Earley}}$. For each deduction step $\mathbb{P}_{\text{Earley}}$ we show that it is an inference in $\mathbb{P}_{\text{buE}}$.

Every *init*, *scan*, and *complete* step in $\mathbb{P}_{\text{Earley}}$ also exists in $\mathbb{P}_{\text{buE}}$. Only the Earley *predict* steps have to be accounted for. Let $[A \rightarrow \alpha \bullet \beta, i, j] \vdash_{\text{Earley}} [B \rightarrow \bullet \gamma, j, j]$ be such a *predict* step. Then $\mathbb{P}_{\text{buE}}$ contains an *init* step $\vdash_{\text{buE}} [B \rightarrow \bullet \gamma, j, j]$, hence, by definition of the inference relation $\vdash$, it holds that $[A \rightarrow \alpha \bullet \beta, i, j] \vdash_{\text{buE}} [B \rightarrow \bullet \gamma, j, j]$. $\qquad \square$

Another example of dynamic filtering is the application of *look-ahead*. Recognition of an item does not need to take place if the next symbol(s) in the string cannot logically follow, given the context of the item. For the sake of convenience, we augment the grammar with an *end-of-sentence marker* \$ and a new start symbol S'. Assuming $\{S', \$\} \cap V = \emptyset$, we define

$$N' = N \cup \{S'\},$$

$$\Sigma' = \Sigma \cup \{\$\},$$

$$P' = P \cup \{S' \rightarrow S\$\},$$

$$G' = (N', \Sigma', P', S').$$

There is only a single final item: $[S' \rightarrow S \bullet \$, 0, n]$. Furthermore, we define the function FOLLOW $: N \rightarrow \wp(\Sigma')$ by

$$\text{FOLLOW}(A) = \{a \mid \exists \alpha, \beta : S' \Rightarrow^* \alpha A a \beta\}.$$

Schema 5.5. (E(1))
The parsing schema **E(1)** is defined by a parsing system $\mathbb{P}_{\text{E}(1)}$ for any $G \in \mathcal{CFG}$ and for any $a_1 \ldots a_n \in \Sigma^*$ by

$$\mathcal{I}^{\text{Compl}} = \{[A \rightarrow \alpha \bullet \beta, i, j] \mid A \rightarrow \alpha\beta \in P' \wedge 0 \le i \le j\};$$

$$H = \{[a, i-1, i] \mid a = a_i \wedge 1 \le i \le n\} \cup \{[\$, n, n+1]\};$$

$$D^{\text{Init}} = \{\vdash [S' \rightarrow \bullet S\$, 0, 0]\},$$

$$D^{\text{Pred}} = \{[A \rightarrow \alpha \bullet B\beta, i, j] \vdash [B \rightarrow \bullet \gamma, j, j]\},$$

$$D^{\text{Scan}} = \{[A \rightarrow \alpha \bullet a\beta, i, j], [a, j, j+1] \vdash [A \rightarrow \alpha a \bullet \beta, j, j+1]\},$$

$$
\begin{aligned}
D^{\text{Compl}} = \{&[A \rightarrow \alpha \bullet B\beta, h, i], [B \rightarrow \gamma \bullet, i, j], [a, j, j+1] \\
&\qquad \vdash [A \rightarrow \alpha B \bullet \beta, h, j] \mid a \in \text{FOLLOW}(B)\},
\end{aligned}
$$

$$D_{\text{E}(1)} = D^{\text{Init}} \cup D^{\text{Pred}} \cup D^{\text{Scan}} \cup D^{\text{Compl}}.$$

The astute reader may wonder why the look-ahead is restricted to $a \in$ FOLLOW(B) and not extended to, for example, $a \in$ FIRST(β FOLLOW(A)). A similar filter, moreover, could be applied to the *scan* steps. This schema incorporates the look-ahead that is used in the construction of an SLR(1) parsing table. It can be shown that an SLR(1) parser is an implementation of $\mathbf{E(1)}$[9]. More examples of dynamic filtering will follow in Sections 6.2

A few important remarks must be made about static and dynamic filtering.

Firstly, dynamic filtering reduces the number of valid items, but at the same time reduces the possibilities for parallel processing. The bottom-up Earley parser has been introduced as a non-filtered version of Earley's algorithm, specifically because it can be carried out in parallel in a straightforward manner.

Secondly, the two types of filters refer to different optimization techniques in parser implementation. Static (i.e., *compile-time*) optimization can take the specific grammar structure into account, but is necessarily unrelated to the sentence. Dynamic optimization is *run-time*, and hence can take into account those parts of the sentence that have been analysed already. It is exactly this difference that is expressed on a higher level of abstraction.

Note that every static filter is also a dynamic filter. This means that any static optimization could also be done run-time, rather than compile-time (but the former is generally less efficient).

5.3 Step contraction

The last and most powerful type of filtering is step contraction. This is the inverse of step refinement, cf. Definition 4.3.

Definition 5.6. (*step contraction*)
The relation $\mathbb{P}_1 \xrightarrow{\text{SC}} \mathbb{P}_2$ holds if

(i) $\mathcal{I}_1 \supseteq \mathcal{I}_2$
(ii) $\vdash_1^* \supseteq \vdash_2^*$.

Let $\mathbf{P}_1$ and $\mathbf{P}_2$ be parsing schemata for some class of grammars $\mathcal{CG}$.
The relation $\mathbf{P}_1 \xRightarrow{\text{SC}} \mathbf{P}_2$ holds if $\mathbf{P}_1(G)(a_1 \ldots a_n) \xrightarrow{\text{SC}} \mathbf{P}_2(G)(a_1 \ldots a_n)$ for all $G \in \mathcal{CG}$ and for all $a_1 \ldots a_n$.

As a realistic example we give the parsing schema that underlies the improved Earley algorithm of Graham, Harrison and Ruzzo [GHR80]. This is a combination of two different step contractions:

– *nullable symbols* (i.e., symbols that can be rewritten to the empty string) can be skipped when the dot is worked rightwards through a production;

---

[9] Note, however, that a deterministic SLR(1) parser is defined only for a suitably small subclass of context-free grammars. See also Section 7.

- *chain derivations* (i.e., derivations of the form $A \Rightarrow^+ B$) are reduced to single steps.

Schema 5.7. (GHR)

The parsing schema **GHR** is defined by a parsing system $\mathbb{P}_{\text{GHR}}$ for any $G \in \mathcal{CFG}$ and for any $a_1 \ldots a_n \in \Sigma^*$ by

$$\mathcal{I}_{\text{GHR}} = \{[A{\rightarrow}\alpha{\bullet}\beta, i, j] \mid A{\rightarrow}\alpha\beta \in P \wedge 0 \leq i \leq j\};$$

$$D^{\text{Init}} = \{\vdash [S{\rightarrow}\beta{\bullet}\gamma, 0, 0] \mid \beta \Rightarrow^* \varepsilon\},$$

$$D^{\text{Scan}} = \{[A{\rightarrow}\alpha{\bullet}a\beta\gamma, i, j], [a, j, j+1] \vdash [A{\rightarrow}\alpha a\beta{\bullet}\gamma, i, j+1] \\ \mid \beta \Rightarrow^* \varepsilon\},$$

$$D^{\text{C1}} = \{[A{\rightarrow}\alpha{\bullet}B\beta\gamma, i, j], [B{\rightarrow}\delta{\bullet}, j, k] \vdash [A{\rightarrow}\alpha B\beta{\bullet}\gamma, i, k] \\ \mid i < j < k \wedge \beta \Rightarrow^* \varepsilon\},$$

$$D^{\text{C2}} = \{[A{\rightarrow}\alpha{\bullet}B\beta\gamma, i, i], [C{\rightarrow}\delta{\bullet}, i, j] \vdash [A{\rightarrow}\alpha B\beta{\bullet}\gamma, i, j] \\ \mid i < j \wedge B \Rightarrow^* C \wedge \beta \Rightarrow^* \varepsilon\},$$

$$D^{\text{Pred}} = \{[A{\rightarrow}\alpha{\bullet}B\beta, i, j] \vdash [C{\rightarrow}\alpha'{\bullet}\beta', j, j] \mid B \Rightarrow^* C\gamma \wedge \alpha' \Rightarrow^* \varepsilon\},$$

$$D_{\text{GHR}} = D^{\text{Init}} \cup D^{\text{Scan}} \cup D^{\text{C1}} \cup D^{\text{C2}} \cup D^{\text{Pred}}.$$

Proposition 5.8. Earley $\overset{\text{sc}}{\Longrightarrow}$ GHR.   □

Other examples of step refinement follow in Section 6.1.

5.4 Properties of filtering relations

Unlike similar properties of refinement and generalization, the following are trivial.

Proposition 5.9. $\overset{\text{sf}}{\Longrightarrow} \subseteq \overset{\text{df}}{\Longrightarrow} \subseteq \overset{\text{sc}}{\Longrightarrow}$.   □

Proposition 5.10. $\overset{\text{sf}}{\Longrightarrow}$, $\overset{\text{df}}{\Longrightarrow}$, and $\overset{\text{sc}}{\Longrightarrow}$ are transitive and reflexive.   □

Proposition 5.11. $\overset{\text{sf}}{\Longrightarrow}$, $\overset{\text{df}}{\Longrightarrow}$, and $\overset{\text{sc}}{\Longrightarrow}$ preserve soundness.   □

6. Some larger examples

The emphasis in the previous sections was on formalizing the theoretical concepts. In Sections 6.1–6.3 we present some nontrivial examples of parsing schemata and relations between them. In Section 6.4 we review the value of these exercises.

6.1 Left-corner parsing

As a more elaborate example of how filters can be used to relate parsing schemata to one another, we will precisely establish the relation between Earley parsing and Left-Corner parsing. We define a parsing schema **LC** that underlies the (generalized) Left-Corner algorithm that is known from the literature, cf. [M&a83, Ned93] (as opposed to *deterministic* LC parsing [RL70]). Along the way will show that **Earley** $\overset{sc}{\Longrightarrow}$ **LC**. As a conceptual aid, we will first consider a "bottom-up Left-Corner" parser that is a rather trivial step contraction of bottom-up Earley.

Consider an item of the form $[A{\rightarrow}B{\bullet}\beta, i, j]$ in $\mathbb{P}_{\text{buE}}$. The item is valid if some $[B{\rightarrow}\gamma{\bullet}, i, j]$ is valid, because $[A{\rightarrow}{\bullet}B\beta, i, i]$ is valid by definition. So we can contract the sequence of deduction steps

$$\vdash \quad [A{\rightarrow}{\bullet}B\beta, i, i],$$
$$[A{\rightarrow}{\bullet}B\beta, i, i], [B{\rightarrow}\gamma{\bullet}, i, j] \quad \vdash \quad [A{\rightarrow}B{\bullet}\beta, i, j].$$

to a single deduction step

$$[B{\rightarrow}\gamma{\bullet}, i, j] \vdash [A{\rightarrow}B{\bullet}\beta, i, j].$$

A similar argument applies to items of the form $[A{\rightarrow}a{\bullet}\beta, i, j]$ and the appropriate *scan* step. The (bottom-up) *left-corner* step is illustrated in Fig. 6.1. This is incorporated in the following *bottom-up left-corner* parsing schema.

Schema 6.1. (buLC)
The parsing schema **buLC** is defined by a parsing system $\mathbb{P}_{\text{buLC}}$ for any $G \in \mathcal{CFG}$ and for any $a_1 \ldots a_n \in \Sigma^*$ by

$$\mathcal{I}^{(1)} = \{[A{\rightarrow}X\alpha{\bullet}\beta, i, j] \mid A{\rightarrow}X\alpha\beta \in P \wedge 0 \leq i \leq j\},$$

$$\mathcal{I}^{(2)} = \{[A{\rightarrow}{\bullet}, j, j] \mid A{\rightarrow}\varepsilon \in P \wedge j \geq 0\},$$

$$\mathcal{I}_{\text{buLC}} = \mathcal{I}^{(1)} \cup \mathcal{I}^{(2)};$$

$$D^\varepsilon = \{\vdash [A{\rightarrow}{\bullet}, j, j]\},$$

$$D^{\text{LC}(a)} = \{[a, j-1, j] \vdash [B{\rightarrow}a{\bullet}\beta, j-1, j]\},$$

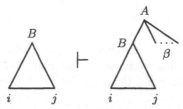

Fig. 6.1. The (bottom-up) *left-corner* step

$$D^{\mathrm{LC}(A)} = \{[A\to\alpha\bullet, i, j] \vdash [B\to A\bullet\beta, i, j]\},$$

$$D^{\mathrm{Scan}} = \{[A\to\alpha\bullet a\beta, i, j], [a, j, j+1] \vdash [A\to\alpha a\bullet\beta, i, j+1]\},$$

$$D^{\mathrm{Compl}} = \{[A\to\alpha\bullet B\beta, i, j], [B\to\gamma\bullet, j, k] \vdash [A\to\alpha B\bullet\beta, i, k]\},$$

$$D_{\mathrm{buLC}} = D^\varepsilon \cup D^{\mathrm{LC}(a)} \cup D^{\mathrm{LC}(A)} \cup D^{\mathrm{Scan}} \cup D^{\mathrm{Compl}}.$$

Proposition 6.2. $\mathbf{buE} \overset{\mathrm{sc}}{\Longrightarrow} \mathbf{buLC}$. □

Things get more interesting –and rather more complicated – if we apply the same transformation to **Earley**, rather than **buE**. It is *not* the case that $[A\to\bullet B\beta, i, i]$ is always valid. Therefore, the replacement of $[A\to\bullet B\beta, i, i]$, $[B\to\gamma\bullet, i, j] \vdash [A\to B\bullet\beta, i, j]$ by a deduction $[B\to\gamma\bullet, i, j] \vdash [A\to B\bullet\beta, i, j]$ should be allowed *only in those cases* where $[A\to\bullet B\beta, i, i]$ is actually valid. Under which conditions is this the case?

The item $[A\to\bullet B\beta, i, i]$ is predicted by **Earley** only if there is some valid item of the form $[C\to\alpha\bullet A\delta, h, i]$. But if, by chance, $\alpha = \varepsilon$, then this is one of the very items that we seek to eliminate. In that case we continue the search for an item that licences the validity of $[C\to\bullet A\delta, i, i]$. This search can end in two ways: either we find some item with the dot not in leftmost position, or (only in case $i = 0$) we may move all the way up to $[S\to\bullet\gamma, 0, 0]$.

Definition 6.3. (*left-corner relation*)
The *left corner* of a non-empty production is the leftmost right-hand side symbol (i.e., production $A\to X\alpha$ has left corner X); the left corner of an empty production is ε.
The relation $>_\ell$ on $N \times (N \cup \Sigma \cup \{\varepsilon\})$ is defined by

$A >_\ell U$ if there is $p = A\to\alpha \in P$ with U the left corner of p.

The transitive and reflexive closure of $>_\ell$ is denoted $>_\ell^*$.

We can now proceed to define a schema **LC** for a left-corner parser. Clearly, $[A\to\bullet B\beta, i, i]$ will be recognized by the Earley algorithm if there is some valid item $[C\to\alpha\bullet E\delta, h, i]$ with $E >_\ell^* A$. Moreover, there is such an item with $\alpha \neq \varepsilon$, unless, perhaps, $i = 0$ and $C = S$. For this exceptional case we retain items $[S\to\bullet\gamma, 0, 0]$ as usual. The discarded *complete* steps are replaced by *left-corner* steps as follows:

$$[C\to\alpha\bullet E\delta, h, i], [B\to\gamma\bullet, i, j] \vdash [A\to B\bullet\beta, i, j] \text{ only if } E >_\ell^* A,$$

see Fig. 6.2; similarly for consequents of the form $[A\to a\bullet\beta, j-1, j]$.

The general idea of a *left-corner* step involves slightly different details for nonterminal, terminal, and empty left corners. Thus we obtain the following parsing schema.

Schema 6.4. (LC)
The parsing schema **LC** is defined by a parsing system $\mathbb{P}_{\mathrm{LC}}$ for any $G \in \mathcal{CFG}$ and for any $a_1 \ldots a_n \in \Sigma^*$ by

$$\mathcal{I}^{(1)} = \{[A\to X\alpha\bullet\beta, i, j] \mid A\to X\alpha\beta \in P \wedge 0 \le i \le j\},$$

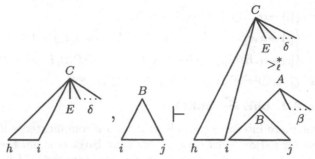

Fig. 6.2. The (*predictive*) *left-corner* step

$$\mathcal{I}^{(2)} = \{[A\rightarrow\bullet, j, j] \mid A\rightarrow\varepsilon \in P \wedge j \geq 0\},$$

$$\mathcal{I}^{(3)} = \{[S\rightarrow\bullet\gamma, 0, 0] \mid S\rightarrow\gamma \in P\},$$

$$\mathcal{I}_{\mathrm{LC}} = \mathcal{I}^{(1)} \cup \mathcal{I}^{(2)} \cup \mathcal{I}^{(3)};$$

$$D^{\mathrm{Init}} = \{\vdash [S\rightarrow\bullet\gamma, 0, 0]\},$$

$$D^{\mathrm{LC}(A)} = \{[C\rightarrow\gamma\bullet E\delta, h, i], [A\rightarrow\alpha\bullet, i, j] \vdash [B\rightarrow A\bullet\beta, i, j] \mid E >_{\ell}^{*} B\},$$

$$D^{\mathrm{LC}(a)} = \{[C\rightarrow\gamma\bullet E\delta, h, i], [a, i, i+1] \vdash [B\rightarrow a\bullet\beta, i, i+1] \mid E >_{\ell}^{*} B\},$$

$$D^{\mathrm{LC}(\varepsilon)} = \{[C\rightarrow\gamma\bullet E\delta, h, i] \vdash [B\rightarrow\bullet, i, i] \mid E >_{\ell}^{*} B\},$$

$$D^{\mathrm{Scan}} = \{[A\rightarrow\alpha\bullet a\beta, i, j], [a, j, j+1] \vdash [A\rightarrow\alpha a\bullet\beta, i, j+1]\},$$

$$D^{\mathrm{Compl}} = \{[A\rightarrow\alpha\bullet B\beta, i, j], [B\rightarrow\gamma\bullet, j, k] \vdash [A\rightarrow\alpha B\bullet\beta, i, k]\},$$

$$D_{\mathrm{LC}} = D^{\mathrm{Init}} \cup D^{\mathrm{LC}(a)} \cup D^{\mathrm{LC}(A)} \cup D^{\mathrm{LC}(\varepsilon)} \cup D^{\mathrm{Scan}} \cup D^{\mathrm{Compl}}.$$

Proposition 6.5. buLC $\overset{\mathrm{df}}{\Longrightarrow}$ LC. □

Proposition 6.6. Earley $\overset{\mathrm{sf}}{\Longrightarrow}$ LC. □

When it comes to implementing the schema **LC**, a practical simplification can be made. In order to apply a *left-corner* step we have to look for *some* item of the form $[C\rightarrow\alpha\bullet E\delta, h, i]$, with arbitrary C, α, β, and h. We can introduce a special *predict item*, denoted $[i, E]$, to indicate that E has been predicted as a feasible constituent at position i. The details are straightforward and need not be spelled out here.

6.2 De Vreught and Honig's algorithm

We define several variants of a parsing schema for an algorithm defined by de Vreught and Honig [VH89, VH91], primarily intended for parallel processing. Rather than working through a production from left to right, as is done in Earley's algorithm, one could start at an arbitrary position in the right-hand side and from there extend the recognized part in both directions. To this end, we use *double-dotted items* of the form $[A\rightarrow\alpha.\beta.\gamma, i, j]$. Recognition of such an item indicates that $\beta \Rightarrow^* a_{i+1} \ldots a_j$, while α and γ still have to be expanded. An item $[A\rightarrow.\alpha.\beta, i, j]$ corresponds to the canonical Earley item.

The algorithm of de Vreught and Honig has two basic steps, called *include* and *concatenate*. The idea of both steps is illustrated in Fig. 6.3. The following schema for our first version of the algorithm should be clear.

Schema 6.7. (dVH1)
The parsing schema **dVH1** is defined by a parsing system $\mathbb{P}_{\text{dVH1}}$ for any $G \in \mathcal{CFG}$ and for any $a_1 \ldots a_n \in \Sigma^*$ by

$$\mathcal{I}_{\text{dVH1}} = \{[A\rightarrow\alpha.\beta.\gamma, i, j] \mid A\rightarrow\alpha\beta\gamma \in P \wedge 0 \le i \le j$$
$$\wedge (\beta \ne \varepsilon \text{ or } \alpha\gamma = \varepsilon)\};$$

$$D^{\text{Init}} = \{[a, j-1, j] \vdash [A\rightarrow\alpha.a.\gamma, j-1, j]\},$$

$$D^{\varepsilon} = \{\vdash [B\rightarrow..., j, j]\},$$

$$D^{\text{Incl}} = \{[B\rightarrow.\beta., i, j] \vdash [A\rightarrow\alpha.B.\gamma, i, j]\},$$

$$D^{\text{Concat}} = \{[A\rightarrow\alpha.\beta_1.\beta_2\gamma, i, j], [A\rightarrow\alpha\beta_1.\beta_2.\gamma, j, k]$$
$$\vdash [A\rightarrow\alpha.\beta_1\beta_2.\gamma, i, k]\},$$

$$D_{\text{dVH1}} = D^{\text{Init}} \cup D^{\varepsilon} \cup D^{\text{Incl}} \cup D^{\text{Concat}}.$$

Next, we observe that D_{dVH1} is redundant, in the following way. An item $[A\rightarrow\alpha.XYZ.\gamma, i, j]$ can be concatenated in two different ways:

$$[A\rightarrow\alpha.X.YZ\gamma, i, k], [A\rightarrow\alpha X.YZ.\gamma, k, j] \vdash [A\rightarrow\alpha.XYZ.\gamma, i, j];$$

$$[A\rightarrow\alpha.XY.Z\gamma, i, l], [A\rightarrow\alpha XY.Z.\gamma, l, j] \vdash [A\rightarrow\alpha.XYZ.\gamma, i, j].$$

Moreover, if $[A\rightarrow\alpha.XYZ.\gamma, i, j]$ is valid, then each of the four antecedents is also valid for some value of k and l. Hence, if we delete the former deduction

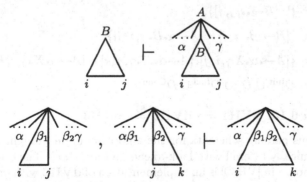

Fig. 6.3. The *include* and *concatenate* steps

step from D, the set of valid items is not affected. For items with more than 3 symbols between the dots, the redundancy in deduction steps increases accordingly.

Schema 6.8. (dVH2)
In the specification of $\mathbb{P}_{\mathrm{dVH1}}$ in Schema 6.7 we replace D^{Concat} by

$$D^{\mathrm{Concat}} = \{[A{\rightarrow}\alpha{\bullet}\beta{\bullet}X\gamma, i, j], [A{\rightarrow}\alpha\beta{\bullet}X{\bullet}\gamma, j, k] \vdash [A{\rightarrow}\alpha{\bullet}\beta X{\bullet}\gamma, i, k]\},$$

and leave $\mathcal{I}$, D^{Init}, D^{ε} and D^{Incl} as in Schema 6.7.

Further optimization of **dVH2** is possible. Observe that items of the form $[A{\rightarrow}\alpha{\bullet}\beta{\bullet}, i, j]$ with $|\alpha| \geq 1$ and $|\beta| \geq 2$ are useless in $\mathbb{P}_{\mathrm{dVH2}}$, in the sense that they do not occur as an antecedent in any derivation step. Hence, these items can be discarded. Similarly, any item of the form $[A{\rightarrow}\alpha{\bullet}\beta{\bullet}\gamma, i, j]$ with $|\alpha| \geq 1$, $|\beta| \geq 2$ and $|\gamma| \geq 1$ can concatenate to the right, but cannot contribute to the recognition of an item of the form $[A{\rightarrow}{\bullet}\beta{\bullet}, i, j]$. Hence the whole set

$$\{[A{\rightarrow}\alpha{\bullet}\beta{\bullet}\gamma, i, j] \mid |\alpha| \geq 1 \wedge |\beta| \geq 2\}$$

can be considered useless; these items can only be used to recognize further items in this set, but none of these items can be used to recognize an item outside this set. Hence we delete this set and discard all deduction steps that have one of these items as antecedent or as consequent.

Schema 6.9. (dVH3)
The parsing schema **dVH3** is defined by a parsing system $\mathbb{P}_{\mathrm{dVH3}}$ for any $G \in \mathcal{CFG}$ and for any $a_1 \ldots a_n \in \Sigma^*$ by

$$
\begin{aligned}
\mathcal{I}^{(1)} &= \{[A{\rightarrow}\alpha{\bullet}X{\bullet}\gamma, i, j] \mid A{\rightarrow}\alpha X\gamma \in P \wedge 0 \leq i \leq j\}, \\
\mathcal{I}^{(2)} &= \{[A{\rightarrow}{\bullet}X\beta{\bullet}\gamma, i, j] \mid A{\rightarrow}X\beta\gamma \in P \wedge 0 \leq i \leq j\}, \\
\mathcal{I}^{(3)} &= \{[A{\rightarrow}{\bullet}{\bullet}, j, j] \mid A{\rightarrow}\varepsilon \in P \wedge j \geq 0\}, \\
\mathcal{I}_{\mathrm{dVH3}} &= \mathcal{I}^{(1)} \cup \mathcal{I}^{(2)} \cup \mathcal{I}^{(3)}; \\
D^{\mathrm{Init}} &= \{[a, j-1, j] \vdash [A{\rightarrow}\alpha{\bullet}a{\bullet}\gamma, j-1, j]\}, \\
D^{\varepsilon} &= \{\vdash [B{\rightarrow}{\bullet}{\bullet}, j, j]\}, \\
D^{\mathrm{Incl}} &= \{[B{\rightarrow}{\bullet}\beta{\bullet}, i, j] \vdash [A{\rightarrow}\alpha{\bullet}B{\bullet}\gamma, i, j]\}, \\
D^{\mathrm{Concat}} &= \{[A{\rightarrow}{\bullet}\alpha{\bullet}X\gamma, i, j], [A{\rightarrow}\alpha{\bullet}X{\bullet}\gamma, j, k] \vdash [A{\rightarrow}{\bullet}\alpha X{\bullet}\gamma, i, k]\}, \\
D_{\mathrm{dVH3}} &= D^{\mathrm{Init}} \cup D^{\varepsilon} \cup D^{\mathrm{Incl}} \cup D^{\mathrm{Concat}}.
\end{aligned}
$$

Proposition 6.10. $\mathbf{dVH1} \overset{\mathrm{sf}}{\Longrightarrow} \mathbf{dVH2} \overset{\mathrm{sf}}{\Longrightarrow} \mathbf{dVH3}$. $\qquad\qquad\square$

The various parsing schemata for the algorithm of de Vreught and Honig can be dynamically filtered with look-ahead and *look-back*. The original algorithm as proposed in [VH89] is an implementation of **dVH2** with one position look-ahead and look-back.

Proposition 6.11. dVH3 $\overset{sc}{\Longrightarrow}$ buLC.

Proof. In order to show that **dVH3** can be filtered to **buLC**, we have to realize that a single-dotted item $[A\rightarrow\alpha\bullet\beta, i, j]$ and a double-dotted item $[A\rightarrow\bullet\alpha\bullet\beta, i, j]$ are merely different notations for the same object. Hence, clearly, $\mathcal{I}_{\text{buLC}} \subset \mathcal{I}_{\text{dVH3}}$.

It remains to be shown that $\vdash^*_{\text{buLC}} \subseteq \vdash^*_{\text{dVH3}}$. To this end it suffices to show that for every deduction step $\eta_1 \ldots, \eta_k \vdash \xi \in D_{\text{buLC}}$ it holds that $\eta_1 \ldots, \eta_k \vdash^*_{\text{dVH3}} \xi$.

An arbitrary deduction step in D^{Compl} in $\mathbb{P}_{\text{buLC}}$

$$[A\rightarrow\bullet\alpha\bullet B\beta, i, j], [B\rightarrow\bullet\gamma\bullet, j, k] \vdash [A\rightarrow\bullet\alpha B\bullet\beta, i, k]$$

is emulated in $\mathbb{P}_{\text{dVH3}}$ by

$$[B\rightarrow\bullet\gamma\bullet, j, k] \vdash [A\rightarrow\alpha\bullet B\bullet\beta, j, k],$$
$$[A\rightarrow\bullet\alpha\bullet B\beta, i, j], [A\rightarrow\alpha\bullet B\bullet\beta, j, k] \vdash [A\rightarrow\bullet\alpha B\bullet\beta, i, k];$$

similarly for D^{Scan}.

The other cases are trivial, hence **dVH3** $\overset{sc}{\Longrightarrow}$ **buLC**. □

By Propositions 6.5 6.10, and 6.11 we have shown that **dVH1** $\overset{sc}{\Longrightarrow}$ **LC**. The conclusion should *not* be, however, that de Vreught and Honig's algorithm is a sub-optimal version of (bottom-up) left-corner parsing. A subtle but decisive change took place in the seemingly harmless static filter **dVH1** $\overset{sf}{\Longrightarrow}$ **dVH2**, where we laid down that the the concatenation of right-hand side elements is from left to right.

A different, more general way to eliminate the redundancy in **dVH1** is to start expanding the right-hand side from the "most interesting" symbol, called the *head* of a production, rather than the leftmost symbol. This leads to so-called *Head Corner* (HC) parsers. A bottom-up head-corner parser that is very similar to the one defined by Satta and Stock[10] [SS89] can be obtained as a step contraction of the Vreught and Honig's algorithm along similar lines; top-down prediction can be added as in [SA96].

A context-free head grammar, in which every production has some right-hand side symbol assigned as head, can be seen as a generalization of a context-free grammar (take the left corner by default if no head has been specified explicitly). A head-corner parsing schema **HC** can be specified that is a generalization of **LC**.

---

[10] If there are right-hand side symbols both to the left and the right of a head, there is a choice in which direction the item should be expanded first. Satta and Stock leave this choice to the parser but block the other step when a choice has been made. This leads to a nondeterministic set of valid items, which is rather undesirable in this framework. The nondeterminism can be removed by prescribing a choice at the level of the parsing schema.

6.3 Rytter's algorithm

Another example of step refinement is provided by Rytter's algorithm [Ryt85, GR88]. This algorithm has theoretical, rather than practical value. It allows parallel recognition in logarithmic time, but requires $O(n^6)$ processors to do so. The algorithm is based on CYK – hence only defined for grammars in Chomsky Normal Form – but can generalized to arbitrary context-free grammars just as **CYK** was generalized to **buE**.

In addition to the conventional CYK items we introduce *Rytter items*, denoted $[A, h, k; B, i, j]$. A Rytter item is recognized if

$$A \Rightarrow^* a_{h+1} \ldots a_i B a_{j+1} \ldots a_k,$$

that is, the part $B \Rightarrow^* a_{i+1} \ldots a_j$ is still missing, cf. Fig. 6.4.

Fig. 6.4. A Rytter item $[A, h, k; B, i, j]$

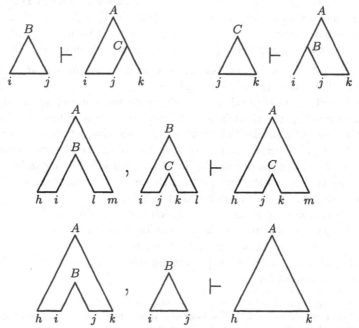

Fig. 6.5. Different types of deduction steps in Rytter's algorithm

Another way to interpret a Rytter item is as a *conditional item*: if $[B, i, j]$ is valid, then $[A, i, j]$ is also valid. The different kinds of deduction steps (excluding the initial CYK steps) are shown in Fig. 6.5.

A CYK deduction step $[B, i, j], [C, j, k] \vdash [A, i, k]$ can be refined into

$$[B, i, j] \quad \vdash \quad [A, i, k; C, j, k]$$
$$[A, i, k; C, j, k], [C, j, k] \quad \vdash \quad [A, i, k].$$

Note that, given $[B, i, j]$ and $[C, j, k]$, we also could have used $[A, i, k; B, i, j]$ as an intermediate conditional item. In fact, unless $i + 1 = j = k - 1$, there are many more ways to recognize $[A, i, k]$, by combining conditional items that have been created at various stages. There is a massive redundancy in the different ways in which a single item can be recognized. It is this redundancy, that guarantees the existence of a balanced recognition tree for each item, which – given enough computing resources – allows for parallel parsing in logarithmic time. For a proof, see [GR88] or [Sik93a, Sik97].

A parsing schema for Rytter's algorithm is defined as follows. The operations associated with the sets of deduction steps $D^{(1)}$, $D^{(2)}$, and $D^{(3)}$, are originally called *activate*, *square*, and *pebble*, respectively. In this context the original names do not make much sense and we rather use numbers.

Schema 6.12. (Rytter)
The parsing schema **Rytter** is defined by a parsing system $\mathbb{P}_{\text{Rytter}}$ for any $G \in \mathcal{CNF}$ and for any $a_1 \ldots a_n \in \Sigma^*$ by

$$\mathcal{I}^{(1)} \quad = \{[A, i, j] \mid A \in N \wedge 0 \le i < j\};$$

$$\mathcal{I}^{(2)} \quad = \{[A, h, k; B, i, j] \mid [A, h, k] \in \mathcal{I} \wedge [B, i, j] \in \mathcal{I}$$
$$\wedge \ h \le i < j \le k \wedge (h \ne i \text{ or } j \ne k)\}$$

$$\mathcal{I}_{\text{Rytter}} \quad = \mathcal{I}^{(1)} \cup \mathcal{I}^{(2)};$$

$$D^{(0)} \quad = \{[a, i - 1, i] \vdash [A, i - 1, i] \mid A \rightarrow a \in P\},$$

$$D^{(1a)} \quad = \{[B, i, j] \vdash [A, i, k; C, j, k] \mid A \rightarrow BC \in P\},$$

$$D^{(1b)} \quad = \{[C, j, k] \vdash [A, i, k; B, i, j] \mid A \rightarrow BC \in P\},$$

$$D^{(2)} \quad = \{[A, h, m; B, i, l], [B, i, l; C, j, k] \vdash [A, h, m; C, j, k]\},$$

$$D^{(3)} \quad = \{[A, h, k; B, i, j], [B, i, j] \vdash [A, h, k]\},$$

$$D_{\text{Rytter}} \quad = D^{(0)} \cup D^{(1a)} \cup D^{(1b)} \cup D^{(2)} \cup D^{(3)}.$$

Between CYK and Rytter's algorithm another algorithm is hiding that is not uninteresting. It addresses the problem of parallel *on-line* parsing: The symbols of a string arrive one by one, as the words of a sentence in spoken natural language. If the processing of each word is finished when the next-word arrives, parsing can be done in real time. For on-line parsing, "gaps" in items are useful in rightmost position, so that the still missing words can be anticipated with a partial syntactic analysis. But the large number of position markers, which accounts for the excessive resources required by Rytter's algorithm, can be reduced.

These ideas underly the definition of the following parsing schema **OCYK** (for *on-line* CYK). In addition to $[A, i, j]$ we introduce items $[A, i, j; B]$ to denote $A \Rightarrow^* a_{i+1} \ldots a_j B$. There is no need to specify the size of the "gap" $[B, j, ?]$.

Schema 6.13. (OCYK)
The parsing schema **OCYK** is defined by a parsing system $\mathbb{P}_{\text{OCYK}}$ for any $G \in \mathcal{CNF}$ and for any $a_1 \ldots a_n \in \Sigma^*$ by

$$\mathcal{I}^{(1)} = \{[A, i, j] \mid A \in N \wedge 0 \leq i < j\},$$

$$\mathcal{I}^{(2)} = \{[A, i, j; B] \mid A, B \in N \wedge 0 \leq i < j\},$$

$$\mathcal{I}_{\text{OCYK}} = \mathcal{I}^{(1)} \cup \mathcal{I}^{(2)};$$

$$D^{(0)} = \{[a, j-1, j] \vdash [A, j-1, j] \mid A \rightarrow a \in P\},$$

$$D^{(1)} = \{[B, i, j] \vdash [A, i, j; C] \mid A \rightarrow BC \in P\},$$

$$D^{(2)} = \{[A, i, j; B], [B, j, k; C] \vdash [A, i, k; C]\},$$

$$D^{(3)} = \{[A, i, j; B], [B, j, k] \vdash [A, i, k]\},$$

$$D_{\text{OCYK}} = D^{(0)} \cup D^{(1)} \cup D^{(2)} \cup D^{(3)}.$$

The schema **OCYK** can be implemented on a parallel random access machine with $O(n^2)$ processors such that only constant time per word is needed, cf. [Sik93b].

Proposition 6.14. CYK $\overset{\text{sr}}{\Longrightarrow}$ OCYK $\overset{\text{sr}}{\Longrightarrow}$ Rytter. □

The schemata **OCYK** and **Rytter** can be generalized from $\mathcal{CNF}$ to $\mathcal{CFG}$ in a way that is similar to the generalization **CYK** $\overset{\text{gen}}{\Longrightarrow}$ **buE** as discussed in Example 4.1.

Another way to obtain a logarithmic-time parallel parsing algorithm with $O(n^6)$ processors is a step refinement of any of the **dVH** schemata, similar to **CYK** $\overset{\text{sr}}{\Longrightarrow}$ **Rytter**. This has been worked out in [VH91].

6.4 Some general remarks

After these extensive examples, some general remarks are due.

We have shown that the parsing schemata framework is able to handle nontrivial algorithms of different flavours. Very general frameworks, offering a good insight at a high level of abstraction, have a risk of becoming unwieldy when confronted with problems that stretch far beyond their canonical example. The fact that we were able to cover these algorithms with sufficient clarity provides circumstantial evidence that the parsing schemata framework makes abstractions that are *right* in some way.

The above examples are taken from the computer science and computational linguistics literature on parsing, but some variants (**dVH3** and **OCYK** in the discussed examples) have been discovered as a *result* of analysing these algorithms by means of parsing schemata. Also, the close relation between Earley parsing and Left-Corner parsing was noticed in [Sik93a] for the first time.

An inevitable weakness of the formalism is a diminished understanding of practical algorithm efficiency. Because of the absence of any kind of data structure it is not immediately clear what the algorithmic complexity of a parsing schema is. Improving the efficiency of parsers, therefore, cannot be done only at the level of schemata. How such an improved schema is to be realized in a more efficient implementation is a matter that has not been (and need not be) addressed in this context.

But in order to find optimizations of algorithms, one must have a very good insight in the characteristic behaviour of an algorithm. It is such an insight that is offered by the parsing schemata framework.

7. From schemata to algorithms

We will not discuss parsing algorithms in detail, but briefly review how some well-known classes of parsing algorithms relate to the framework presented in the preceding sections.

Parsing schemata are a generalization of *chart parsers* [Kay80, Kay82, Win83]. From the view that has been unfolded in the previous sections, we can see a chart parser as the canonical implementation of a parsing schema.

A chart parser employs two data structures: An *agenda*, containing items that will be actively used to search for new items that can be recognized, and a *chart*, storing the items that need no further attention. At each step, one item, say ξ, is taken from the agenda and put on the chart. The chart is searched for all (combinations of) items $\eta_1 \ldots \eta_k$ such that $\xi, \eta_1 \ldots \eta_k \vdash \zeta \in D$. All ζ that are found in the way and were not recognized before are added to the agenda. An Earley chart parser, for example, is initialized with items $[a, i - 1, i]$ on the chart and $[S \rightarrow \bullet \gamma, 0, 0]$ on the agenda.

The control structure of the chart parser guarantees that the final chart, which is reached when the agenda is empty, contains $\mathcal{V}(\mathbb{P})$. An issue that has to be addressed (but not in this context) is how to structure the chart and agenda so as to make the parser efficient. For an overview of chart based approaches to CYK and Earley parsing, see [Nij94].

The chart parsing framework offers a generic way to handle *nondeterminism*. Hence it is not surprising that chart parsers has attracted widespread attention in computational linguistics.

Logical deduction as a basis for the description of chart parsers is due to Pereira and Warren [PW83]. But our framework has a rather different emphasis. While the "Parsing as Deduction" approach is primarily interested in connecting the parsing logic with unification-based grammar formalisms, parsing schemata use deduction merely as a convenient notation for describing the essential traits of arbitrary parsing algorithms. We come back to unification grammars in Section 8.

A different parsing paradigm is provided by the pushdown automaton (PDA). A fundamental theorem in formal language theory states that the class of languages accepted by (nondeterministic) PDA's is equal to the class of languages generated by context-free languages. Many parsing algorithms from the field of compiler construction are based on the PDA paradigm. The canonical example is the family of *LR-parsers*, discovered by Knuth [Knu65] and extended to the more practical *SLR* and *LALR* parsers by DeRemer [DeR69, DeR71]. See [ASU86] for a good introduction and [Nij83] for an extensive bibliography of LR parsing.

While deterministic LR parsers on restricted classes of context-free grammars are particularly efficient, nondeterministic LR parsers (known as *generalized LR (GLR)* parsers) have been introduced to cover wider classes of grammars, in particular for use in computational linguistics. A general method to handle nondeterministic PDA's in an efficient manner has been given by Lang [Lan74]. Generalized LR parsing has attracted more attention in the form of Tomita's algorithm [Tom85], based on a graph-structured stack as the data structure to handle the ambiguities that occur during parsing.

Tomita's algorithm cannot handle certain classes of grammars (cyclic grammars and grammars with hidden left-recursion[11]). Rekers has improved Tomita's algorithm to handle these grammars as well [Rek92]. See also [Nij91] for a historic overview and [Ned94] for some optimizations in generalized LR parsing.

The question arises how PDA-based algorithms like LR relate to parsing schemata. LR parsers use items *compile-time* in the construction of the parsing table. The states of an LR parser are in fact sets dotted productions $A \to \alpha \bullet \beta$. A state comprises those productions that could apply at the current point in the parsing process. One can partially uncompile the various LR-type algorithms and make these dotted rules visible during parsing. It is easy to add position markers to a dotted rule (viz., the position where a particular production was started and the current position). This yields Earley-type

---

[11] A grammar is called hidden left-recursive if $A \Rightarrow^+ \alpha A \beta$ and $\alpha \Rightarrow^+ \varepsilon$. The term has been coined in [NS93].

items. An item is recognized when a state that contains the dotted rule, in combination with the appropriate position markers, is pushed onto the stack.

An LR parser, therefore, implements some underlying parsing schema. Let **LR(0)** denote the parsing schema that underlies a generalized LR(0) parser. Then it holds that **LR(0)** $\stackrel{\mathbf{ext}}{\Longrightarrow}$ **Earley**[12]. In Section 5.2 we defined **E(1)** such that **SLR(1)** $\stackrel{\mathbf{ext}}{\Longrightarrow}$ **E(1)**.

Having uncovered the close relation between the Algorithms of Earley and Tomita, it is possible to apply many optimizations, extensions and variants of one algorithm to the other as well. An interesting example of such cross-fertilization is a parallel bottom-up Tomita parser [SL92, Sik93a, Sik97] which applies the usual "vertical" parallelization of Earley to the graph-structured stack of Tomita (in contrast to the "horizontal" parallelization as in [TN89, NT90, TDL91] that seems more obvious from the LR point of view), which showed reasonable efficiency in a test implementation.

An ambiguous grammar can be supplied with various kinds of disambiguation rules so as to guarantee that only a single valid parse tree remains. Consider, for example, the grammar

$$E \rightarrow E + E \mid E * E \mid a$$

for arithmetic expressions. The string $a + a * a * a$ has five parse trees. But by introducing *operator precedence* and *associativity* one can specify that $[a + [[a * a] * a]]$ is the only valid parse.

Rather than constructing all parse trees and discarding the invalid ones, an efficient parser must apply the disambiguation rules locally during the parsing process. For the above mentioned grammar for arithmetic expressions. it is in fact possible to construct a deterministic LR-type parser by disambiguating the states and transitions in the LR parsing table.

A general formal framework for this type of optimization is given by Visser [KV94, Vis95] who applies disambiguation rules[13] at the level of parsing schemata.

Head-Driven parsing [Kay89] does not proceed through the sentence from left to right, but starts with the most informative parts. This introduces some extra administrative burden on the parser (as it jumps up and down the sentence) but may lead to substantial savings if the semantic information captured in the head excludes possibilities that would have been explored without this information available. Bouma and van Noord [BN93] have done several experiments with head-driven parsing and conclude (unsurprisingly) that the efficiency of the algorithm is critically dependent on the discriminative nature of the information captured in the heads.

---

[12] GLR parsers typically assume grammars to be reduced. Hence, to be formally correct, GLR applies only to a subset of $\mathcal{CFG}$

[13] called *filters* in the cited publications, but not to be confused with the filters in Section 5.

The parsing schema for Head-Corner parsing presented in [SA96] has been extended with typed structures and implemented by Moll [Mol95]. It is used as a parser in an experimental natural language dialogue system [A&a95].

8. Beyond context-free grammars

The parsing schemata framework has been specified for context-free grammars, but it can easily be extended to other grammar formalisms as well.

Unification-based grammars are the predominant class of grammar formalisms in current computational linguistics. Between 1980 and 1990 a variety of different kinds of unification grammar has been introduced. Some of these, like Definite Clause Grammars [PW80], Functional Unification Grammar [Kay85], and PATR-II [Shi86] have been introduced primarily to offer powerful formalisms for grammar description; others like Lexical-Functional Grammar [KB82] and Head-Driven Phrase Structure Grammar [PS87, PS94] with the aim to provide a theory of linguistic phenomena in natural language.

Unification grammars are related to *attribute grammars* [Knu68, Knu71], which are typically used in the field of compiler construction. There are some fundamental differences in the underlying logic, but these cannot be explained satisfactorily in a few lines. The interested reader is referred to [Shi92] and [Car92] for a thorough treatment of unification logics.

One of the properties of context-free grammars that was felt constraining for the description of natural languages is the rigid order of right-hand side elements in a production. In Generalized Phrase Structure Grammar [G&a85], the notion of *ID/LPgrammars* was introduced. There are separate specifications for the set of right-hand side elements of a production (*immediate dominance*) and constraints on the order in which these elements may appear (*linear precedence*). A parsing schema for unification-based ID/LP-grammars is given by Morawietz [Mor95].

Unification grammars treat syntactic and semantic information in a uniform manner. One can reduce the role of syntax and consider syntactic category as a feature like any other. Indeed there seems to be a trend that less and less information is stored in the context-free backbone of a grammar – i.e., the cat feature in a feature structure – because various syntactic properties can be expressed more elegantly by other kinds of feature constraints. A typical example is *subcategorization* of verbs: all verbs have syntactic category *verb*; constraints on the various kinds of objects that a verb can take are denoted in the subcat feature of the particular verb.

Nagata [Nag92] and Maxwell and Kaplan [MK93] have independently pointed out that this is convenient for writing natural language grammars, but that it has repercussions on parsing efficiency. Context-free parsing is much more efficient than feature structure unification. Hence is it not surprising that the experiments reported in [Nag92] and [MK93] show that the efficiency of unification grammar parsing can be increased by retrieving an (implicit) context-free backbone from a unification grammar that covers more than just the cat feature and using this context-free part for syntactic analysis.

9. Conclusions

Parsing schemata provide a general framework for description, analysis and comparison of parsing algorithms, both sequential and parallel. Data structures, control structures and (for parallel algorithms) communication structures are abstracted from. This framework constitutes an intermediate, well-defined level of abstraction between grammars (defining what valid parses are) and parsing algorithms (prescribing how to compute these).

At this high level of abstraction, the essential traits of a particular type of parser stand out more clearly. Moreover, it is possible to clarify exactly the relationships between different parsers that have some fundamental principles common – even though the realizations of these parsers may look radically different. The price to be paid for this improved clarity and insight is the loss of some details that are of practical importance. The notion of algorithm complexity is strongly related to the data structures used to store intermediate results.

The prime strength of parsing schemata, therefore, is in the analysis of algorithms known to exist. Clarifying the basic traits of an algorithm may suggest improvements that have been overlooked so far. Also, showing that different algorithms have closely related parsing schemata improves cross-fertilization of extensions and optimizations to these algorithms.

We have presented many examples of parsing schemata and their usage, in order to show that this framework, rather then being a mere theoretical nicety, constitutes are a valuable contribution to parsing theory.

References

[A&a95] op den Akker R., ter Doest H., Moll M., Nijholt A. (1995): Parsing in Dialogue Systems Using Typed Feature Structures. *4th International Workshop on Parsing Technologies*, Prague, Czech Republic, 10–11.

[ASU86] Aho A.V., Sethi R., Ullman J.D. (1986): *Compilers: Principles, Techniques and Tools*. Addison-Wesley, Reading, Mass.

[BN93]   Bouma G., van Noord G. (1993): Head-driven Parsing for Lexicalist Grammars: Experimental Results. *6th Meeting of the European Chapter of the Association of Computational Linguistics*, Utrecht, 71–80.

[Car92] Carpenter B. (1992): *The Logic of Typed Feature Structures*. Cambridge University Press, Cambridge, UK.

[DeR69]   DeRemer F.L. (1969): Practical Translators for LR(k) Languages. Ph.D.
          Thesis, MIT, Cambridge, Mass.
[DeR71]   DeRemer F.L. (1971): Simple LR(k) grammars. *Communications of the
          ACM* **14**, 94–102.
[Ear68]   Earley J. (1986): An Efficient Context-Free Parsing Algorithm. Ph.D.
          Thesis, Carnegie-Mellon University, Pittsburgh, Pa.
[Ear70]   Earley J. (1970): An Efficient Context-Free Parsing Algorithm. *Commu-
          nications of the ACM* **13**, 94–102.
[G&a85]   Gazdar G., Klein E., Pullum G.K., Sag I.A. (1985): *Generalized Phrase
          Structure Grammar.* Harvard University Press, Cambridge, Mass.
[GHR80]   Graham S.L., Harrison M.A., Ruzzo W.L. (1980): An Improved Context-
          Free Recognizer. *ACM Transactions on Programming Languages and Sys-
          tems* **2**, 415–462.
[GR88]   Gibbons A., Rytter W. (1988): *Efficient Parallel Algorithms.* Cambridge
          University Press, Cambridge, UK.
[JVW91]   Joshi A., Vijay-Shanker K., Weir D., (1991): The Convergence of Mildly
          Context-Sensitive Grammar Formalisms. In: Sells P., Shieber S.M., Wa-
          sow T. (Eds), *Foundational Issues in Natural Language Processing*, MIT
          Press, Cambridge, Mass., 31–81.
[Kas65]   Kasami T. (1965): An Efficient Recognition and Syntax Analysis Algo-
          rithm for Context-Free Languages. Scientific Report AFCLR-65-758, Air
          Force Cambridge Research Laboratory, Bedford, Mass.
[Kay80]   Kay M. (1980): Algorithm Schemata and Data Structures in Syntactic
          Processing. Report CSL-80-12, Xerox PARC, Palo Alto, Ca.
[Kay82]   Kay M. (1982): Algorithm Schemata and Data Structures in Syntactic
          Processing. In: Grosz B.J., Sparck Jones, K., Webber B.L. (Eds), *Read-
          ings in Natural Language Processing*, Morgan Kaufmann, Los Altos, Ca.
[Kay85]   Kay M. (1985): Parsing in Functional Unification Grammar. In: D.R.
          Dowty, L. Karttunen, and A. Zwicky (Eds.), *Natural Language Parsing*,
          Cambridge University Press, Cambridge, UK, 251–278.
[Kay89]   Kay M. (1989): Head Driven Parsing. *1st International Workshop on
          Parsing Technologies*, Pittsburgh, Pa., 52–62.
[KB82]   Kaplan R.M., Bresnan J. (1982): Lexical-Functional Grammar: a formal
          system for grammatical representation. In: J. Bresnan (Ed.), *The Mental
          Representation of Grammatical Relations*, MIT Press, Cambridge, Mass.,
          173–281.
[Knu65]   Knuth D.E. (1965): On the Translation of Languages from Left to Right.
          Information and Control **8**, 607–639.
[Knu68]   Knuth D.E. (1968): Semantics of Context-Free Languages. *Mathematical
          Systems Theory* **2**, 127–145.
[Knu71]   Knuth D.E. (1971): Semantics of Context-Free Languages, Correction.
          Mathematical Systems Theory **5**, 95–96.
[KV94]   Klint P., Visser E. (1994): Using Filters for the Disambiguation of
          Context-free Grammars. Proc. ASMICS Workshop on Parsing Theory,
          Milan, October 1994, Report 126-94, Dept. of Computer Science, Univer-
          sity of Milan, Italy.
[Lan74]   Lang B. (1974): Deterministic Techniques for Efficient Non-Deterministic
          Parsers. *2nd Colloquium on Automata, Languages and Programming*, Lec-
          ture Notes in Computer Science 14, Springer-Verlag, Berlin, 255–269.
[M&a83]   Matsumoto Y., Tanaka H., Hirakawa H., Miyoshi H., Yasukawa H. (1983):
          BUP: a bottom-up parser embedded in Prolog. *New Generation Comput-
          ing* **1**, 145–158.

[MK93]   Maxwell J.T., Kaplan R.M. (1993): The Interface between Phrasal and Functional Constraints. *Computational Linguistics* **19**, 571–590.

[Mol95]   Moll M. (1995): Head-Corner Parsing using Typed Feature Structures. M.Sc. Thesis, University of Twente, Dept. of Computer Science, Enschede, the Netherlands.

[Mor95]   Morawietz F. (1995): A Unification-based ID/LP Parsing Schema. *4th International Workshop on Parsing Technologies*, Prague, Czech Republic, 162–173.

[Nag92]   Nagata M. (1992): An Empirical Study on Rule Granularity and Unification Interleaving Toward an Efficient Unification-Based Parsing System. *14th International Conference on Computational Linguistics*, Nantes, France, 177–183.

[Ned93]   Nederhof M.J. (1993): Generalized Left-Corner Parsing. *6th Meeting of the European Association of Computational Linguistics*, Utrecht, the Netherlands, 305–314.

[Ned94]   Nederhof M.J. (1994): Linguistic Parsing and Program Transformations. Ph.D. Thesis, University of Nijmegen, the Netherlands.

[Nij83]   Nijholt A. (1983): Deterministic Top-Down and Bottom-Up Parsing: Historical Notes and Bibliographies. Mathematisch Centrum, Amsterdam, the Netherlands.

[Nij91]   Nijholt A. (1991): (Generalized) LR parsing: From Knuth to Tomita. In: *Tomita's Algorithm: Extensions and Applications*. Proceedings Twente Workshop on Language Technology 1 (TWLT1), University of Twente, Dept. of Computer Science, 1–8.

[Nij94]   Nijholt A. (1994): Parallel approaches to context-free language parsing. In: Hahn U., Adriaens G. (Eds.), *Parallel Natural Language Processing*. Ablex Publishing Corporation, Norwood, New Jersey.

[NS93]   Nederhof M.J., Sarbo J.J. (1993): Increasing the Applicability of LR Parsing. *3rd International Workshop on Parsing Technologies* Tilburg and Durbuy, Netherlands/Belgium, 187–201.

[NT90]   Numazaki H., Tananaka H. (1990): A New Parallel Algorithm for Generalized LR Parsing, *13th International Conference on Computational Linguistics*, Helsinki, Vol. 2, 304–310.

[PS87]   Pollard C., Sag I.A. (1987): *An Information-Based Syntax and Semantics, Vol. 1: Fundamentals*. CSLI Lecture Notes 13, Center for the Study of Language and Information, Stanford University, Stanford, Ca.

[PS94]   Pollard C., Sag I.A. (1994): *Head-Driven Phrase Structure Grammar*, University of Chicago Press, Chicago, Ill.

[PW80]   Pereira F.C.N., Warren D.H.D. (1980): Definite Clause Grammars for Language Analysis – A Survey of the Formalism and a Comparison with Augmented transition Networks. *Artificial Intelligence* **13**, 231–278.

[PW83]   Pereira F.C.N., Warren, D.H.D. (1983): Parsing as Deduction. *21th Annual Conference of the Association of Computational Linguistics*, Cambridge, Mass., 137–144.

[Rek92]   Rekers J. (1992): Parser Generation for Interactive Environments. Ph.D. Thesis, University of Amsterdam.

[RL70]   Rosenkrantz D.J., Lewis P.M. (1970): Deterministic Left Corner Parsing. *11th Annual Symposium on Switching and Automata Theory*, 139–152.

[Ryt85]   Rytter W. (1985): On the recognition of context-free languages. *5th Symposium on Fundamentals of Computation Theory*, Lecture Notes in Computer Science 208, Springer-Verlag, 315–322.

[SA96]   Sikkel K., op den Akker R. (1996): Predictive Head-Corner Chart Parsing. In: Bunt H., Tomita M. (Eds), *Recent Advances in Parsing Technology*, Kluwer, Boston, Mass., 1996, 171–184.

[Shi86]   Shieber S.M. (1986): *An Introduction to Unification-Based Approaches to Grammar*. CSLI Lecture Notes 4, Center for the Study of Language and Information, Stanford University, Stanford, Ca.

[Shi92]   Shieber S.M. (1992): *Constraint-Based Grammar Formalisms: Parsing and Type Inference for Natural and Computer Languages*. The MIT Press, Cambridge, Mass.

[Sik93a]   Sikkel K. (1993): Parsing schemata. Ph.D. Thesis, University of Twente, Enschede, the Netherlands.

[Sik93b]   Sikkel K. (1993): On-line Parsing in Constant Time per Word. *Theoretical Computer Science* **120**, 303–310.

[Sik97]   Sikkel K.: *Parsing schemata – a framework for specification and analysis of parsing algorithms*. Texts in Theoretical Computer Science - An EATCS Series. Springer-Verlag, Berlin (in preparation).

[SL92]   Sikkel K., Lankhorst M. (1992): A Parallel Bottom-Up Tomita Parser. *1. Konferenz Verarbeitung natürlicher Sprache*, Nürnberg, Germany, 238–247.

[SS89]   Satta G., Stock O. (1989): Head-Driven Bidirectional Parsing: A Tabular Method. *1st International Workshop on Parsing Technologies*, Pittsburgh, Pa., 43–51.

[TDL91]   Thompson H.S., Dixon M., Lamping J. (1991): Compose-Reduce Parsing. *29th Annual Meeting of the Association of Computational Linguistics*, Berkeley, Ca., 87–97,

[TN89]   Tanaka H., Numazaki H. (1989): Parallel Generalized LR Parsing based on Logic Programming. *1st International Workshop on Parsing Technologies*, Pittsburgh, Pa., 329–338.

[Tom85]   Tomita M. (1985): *Efficient Parsing for Natural Language*. Kluwer Academic Publishers, Boston, Mass., 1985.

[VH89]   de Vreught J.P.M., Honig H.J. (1989): A Tabular Bottom-Up Recognizer. Report 90-31, Dept. of Applied Mathematics and Informatics, Delft University of Technology, Delft, the Netherlands.

[VH91]   de Vreught J.P.M., Honig H.J. (1991): Slow and fast parallel recognition. *2nd International Workshop on Parsing Technologies*, Cancun, Mexico, 127–135.

[Vis95]   Visser E. (1995): A Case Study in Optimizing Parsing Schemata by Disambiguation Filters. Report P9507, Dept. of Computer Science, University of Amsterdam, the Netherlands.

[Win83]   Winograd T. (1983): *Language as a Cognitive Process. Vol. I: Syntax*. Addison-Wesley, Reading, Mass.

[You67]   Younger D.H. (1967): Recognition of context-free languages in time n^3, *Information and Control* **10**, 189–208.

Grammars with Controlled Derivations

Jürgen Dassow, Gheorghe Păun, and Arto Salomaa

1. Introduction and notations

In [13], N. Chomsky says that "the main problem of immediate relevance to the theory of language is that of determining where in the hierarchy of devices the grammars of natural languages lie." Formulated in other terms, the question is "where are the natural languages placed in the Chomsky hierarchy?" The debate started in 1959 and is not yet settled. Various arguments over English [4], Mohawk [92], Swiss German [112], Bambara [16], Chinese [95], etc., were given, refuted, rehabilitated – see pro and con arguments as well as further bibliographical information in [38] and [70]. The main difficulty is not a mathematical one but a linguistic one: what is English, what is a natural language, can we separate the syntax and the morphology from semantics or pragmatics ? Whatever is or will be the position with respect to these questions, the linguists seem to agree (see again [70]) that "all" natural languages contain *constructions* which cannot be described by context-free grammars. Three basic such features of natural languages are:

- *reduplication*, leading to languages of the form $\{xx \mid x \in V^*\}$,
- *multiple agreements*, modeled by languages of the form $\{a^n b^n c^n \mid n \geq 1\}$, $\{a^n b^n c^n d^n \mid n \geq 1\}$, etc.,
- *crossed agreements*, as modeled by $\{a^n b^m c^n d^m \mid n, m \geq 1\}$.

All these languages are well known as examples of (context-sensitive) non-context-free languages.

Of course, Chomsky's question, although formulated with respect to natural languages, stands also for other classes of languages. This is especially the case for programming languages, where the compiler construction essentially depends on the type of the language. In this area, the things are much simpler, the programming languages being much more precisely defined than natural languages. Floyd's proof [34] that *Algol 60* is not context-free seems to be unanimously accepted and, moreover, it can be directly extended for all programming languages which request identifier declaration, label definition, or other similar long-distance agreements.

Besides natural and programming languages, many other areas where languages can be identified provide examples of non-context-free constructions.

Seven such areas are briefly discussed in Section 0.4 of [27], concluding with the remark that "the world seems to be non-context-free..."

However, in the Chomsky hierarchy beyond the context-free family there is the context-sensitive one. This is large enough. "Almost any language one can think of is context-sensitive", [55], page 224, because "the only known proofs that certain languages are not context-sensitive are ultimately based on diagonalization". Also the linguists agree, in general, that context-sensitive grammars are sufficient for describing the (constructions appearing in) natural languages. But they are also convinced that context-sensitive grammars are "too much"; their explanatory power is reduced just because their generative power is so strong. See [70] for a detailed discussion on this point. The same holds for artificial languages, programming languages included. For instance, it is wellknown that the family of context-sensitive languages has some bad properties, e.g., the known algorithms for the membership problem have exponential complexity and the emptiness problem is undecidable. Furthermore, by the possible structure of context-sensitive (or monotone) productions it is very hard to determine the generated language.

Therefore it is of interest to use context-free grammars and to add some mechanisms which extract some subset (from the generated language) in order to cover some aspects of natural and/or programming languages (and/or other fields of interest). The aim is to get a "mild" subfamily of context-sensitive languages (the term "mild" is used in mathematical linguistics with a precise meaning: semilinear and parsable in polynomial time), with as many context-free-like properties as possible, but able to cover the non-context-free features of the languages we work with.

In this chapter we shall present some of such mechanisms and study some properties of the associated families of languages. For the proofs of most of the results covered by [27] we refer the reader to that monograph.

Throughout this chapter we assume that the reader is familiar with the basic concepts and results of the theory of the grammars of the Chomsky hierarchy, Lindenmayer systems, and their associated languages and language families (see Chapters 4 and 5 of this *Handbook* or [39], [51], [98], [55], [108], [101]).

We settle the following notations.

λ denotes the empty word. $|w|$ and $|w|_M$ give the length of a word w and the number of occurrences of letters of the set M in the word w, respectively. If M consists of a single symbol a, we also write $|w|_a$.

We specify a Chomsky grammar by $G = (N, T, P, S)$, where N is the set of nonterminals, T is the set of terminals, $N \cap T = \emptyset$, P is the set of productions and $S \in N$ is the axiom. Moreover, we set $V_G = N \cup T$. A production is written in the form $p = \alpha \rightarrow \beta$.

By CF, CS, and RE we denote the families of context-free, context-sensitive (or equivalently monotone), and arbitrary phrase-structure (or equivalently Type-0) grammars, respectively.

An extended tabled Lindenmayer system, abbreviated as ET0L system, is specified by $G = (V, T, P, w)$ where V is an alphabet, T is a subset of V (called the set of terminals), P is a finite set of finite substitutions from V to V^*, and $w \in V^+$. Such a system is called deterministic, abbreviated as EDT0L system, if the substitutions of P are morphisms.

By $ET0L$ and $EDT0L$ we denote the families of ET0L and deterministic ET0L systems, respectively.

For a family X of grammars or systems, we denote the associated family of languages by $\mathcal{L}(X)$.

2. Some types of controlled derivations and their power

The idea behind all control mechanisms is the following one: Given a context-free grammar we restrict by the control the application of rules such that some derivations are avoided which are possible in the usual context-free derivation process. Therefore the set of words generated by the restrictions is a subset of the context-free language generated in the usual way. We shall see that these generated subsets can be non-context-free languages. Therefore these mechanisms are more powerful than the context-free grammars.

2.1 Prescribed sequences

In this section we consider some mechanisms which define sequences of productions, and we take only such words in the generated language which can be obtained by a derivation where the productions are applied according to defined sequences.

First we define the allowed sequences as a language over the set of productions.

Definition 2.1. *([42])*
i) *A regularly controlled (context-free) grammar with appearance checking is a 6-tuple*

$$G = (N, T, P, S, R, F),$$

where
– N, T, P and S are specified as in a context-free grammar,
– R is a regular set over P, and
– F is a subset of P.

ii) *For a rule* $p = A \to w \in P$ *and* $x, y \in V_G^*$, *we define the application*

$$x \Longrightarrow_p^{ac} y$$

of a production p *in* appearance checking *mode by*

$$x = x_1 A x_2 \text{ and } y = x_1 w x_2$$

or

$$x = y, \quad A \text{ does not appear in } x, \quad \text{and } p \in F.$$

iii) *The* language $L(G)$ *generated by* G *with appearance checking* consists of *all words* $w \in T^*$ *such that there is a derivation*

$$S \Longrightarrow_{p_1}^{ac} w_1 \Longrightarrow_{p_2}^{ac} w_2 \Longrightarrow_{p_3}^{ac} \ldots \Longrightarrow_{p_n}^{ac} w_n = w$$

with

$$p_1 p_2 p_3 \ldots p_n \in R.$$

iv) *We say that* G *is a* regularly controlled grammar without appearance checking *if and only if* $F = \emptyset$.

Note that the modes with and without appearance checking coincide if the production is not contained in F and if the left-hand nonterminal occurs in the sentential form. In these cases we also write $x \Longrightarrow_p y$ instead of $x \Longrightarrow_p^{ac} y$, i.e., we use the standard notation.

Obviously, if $R = P^*$, then there is no restriction with respect to the sequences of productions, i.e., the grammar works as a context-free grammar.

Example 2.1. Let

$$G_1 = (\{S, A, X\}, \{a\}, \{p_1, p_2, p_3, p_4, p_5\}, S, (p_1^* p_2 p_3^* p_4)^* p_5^*, \{p_2, p_4\})$$

with

$$p_1 = S \to AA, \quad p_2 = S \to X, \quad p_3 = A \to S, \quad p_4 = A \to X, \quad p_5 = S \to a$$

be a regularly controlled grammar with appearance checking. Let us consider a word S^n ($n = 1$ holds for the axiom) and a derivation according to the sequence $p_1^k p_2 p_3^l p_4$. First we have to replace k occurrences of S by AA. If $k < n$, then the derived word contains at least one S which has to be replaced by X according to the application of p_2. Now we cannot derive a word over $\{a\}$ since there is no production which replaces X. Hence $k = n$ has to hold. Thus the only derivation which can be terminated is

$$S^n \underbrace{\Longrightarrow_{p_1} \cdots \Longrightarrow_{p_1}}_{n \text{ times}} A^{2n} \Longrightarrow_{p_2}^{ac} A^{2n} \underbrace{\Longrightarrow_{p_3} \cdots \Longrightarrow_{p_3}}_{2n \text{ times}} S^{2n} \Longrightarrow_{p_4}^{ac} S^{2n}$$

(by the same argumentation as above, $l = 2n$ has to hold), i.e. the only derivation of the considered type (which can be terminated) doubles the number of S. (The applications of p_2 and p_4 in the appearance checking mode ensure that all occurrences of S and A are replaced by p_1 and p_3, respectively.)

By the definition, such a derivation can be performed an arbitrary number of steps. Hence according to a sequence of $(p_1^* p_2 p_3^* p_4)^*$ we obtain a word S^{2^m} for certain $m \geq 1$. Finally the production p_5 replaces the nonterminal S by the terminal letter a, and in order to obtain a word over the termninal alphabet this production has to be applied until all occurrences of S are replaced.

Therefore we get

$$L(G_1) = \{a^{2^m} \mid m \geq 1\}$$

which is a non-context-free language.   □

Example 2.2. We consider the regularly controlled grammar

$$G_2 = (\{S, A, B\}, \{a, b, c\}, \{p_1, p_2, p_3, p_4, p_5\}, S, \{p_1(p_2 p_3)^n p_4 p_5 \mid n \geq 0\}, \emptyset),$$

without appearance checking, where

$$p_1 = S \to AB, \ p_2 = A \to aAb, \ p_3 = B \to cB, \ p_4 = A \to ab, \ p_5 = B \to c.$$

By the definition, any correct derivation has to start with an application of p_1 which gives $S \Longrightarrow_{p_1} AB$. We now consider a word $a^n A b^n c^n B$ with $n \geq 0$ (the word with $n = 0$ is obtained by the application of p_1). Then we get

$$a^n A b^n c^n B \Longrightarrow_{p_2} a^{n+1} A b^{n+1} c^n B \Longrightarrow_{p_3} a^{n+1} A b^{n+1} c^{n+1} B$$

and

$$a^n A b^n c^n B \Longrightarrow_{p_4} a^{n+1} b^{n+1} c^n B \Longrightarrow_{p_5} a^{n+1} b^{n+1} c^{n+1}.$$

Hence by a derivation according to the sequence $p_1(p_2 p_3)^n p_4 p_5$, $n \geq 0$, we obtain the word $a^{n+1} b^{n+1} c^{n+1}$, and thus

$$L(G_2) = \{a^n b^n c^n \mid n \geq 1\}$$

which is a well-known non-context-free language, too.   □

By $\lambda r C_{ac}$ we denote the family of regularly controlled grammars with appearance checking (and with arbitrary context-free productions). If we restrict to regularly controlled grammars without appearance checking we omit the lower index ac, and if we restrict to grammars without erasing rules we omit the λ.

The following theorem presents the hierarchy of the corresponding language families and their relation to the language families of the Chomsky hierarchy.

Theorem 2.1.

i) $\mathcal{L}(CF) \subset \mathcal{L}(rC) \subset \mathcal{L}(rC_{ac}) \subset \mathcal{L}(CS)$,

ii) $\mathcal{L}(CF) \subset \mathcal{L}(rC) \subset \mathcal{L}(\lambda rC) \subset \mathcal{L}(\lambda rC_{ac}) = \mathcal{L}(RE)$,

iii) The family $\mathcal{L}(\lambda rC)$ is incomparable with $\mathcal{L}(CS)$ and $\mathcal{L}(rC_{ac})$.

Remarks on the proof. $\mathcal{L}(rC_{ac}) \subsetneq \mathcal{L}(CS)$. Let $G = (N, T, P, S, R, F)$ be a regularly controlled grammar (with appearance checking). Further let $\mathcal{A} = (Z, P, z_0, Q, \delta)$ be a deterministic finite automaton (with the set Z of states, the initial state z_0, the set Q of accepting states, the input set P (of productions of G) and the transition function $\delta : Z \times X \to Z$) which accepts R.

For any $a \in T$, let X_a be an additional letter. We set

$$\overline{N} = N \cup \{X_a \mid a \in T\}.$$

For $w \in V_G^*$, let $w_T \in (\overline{N})^*$ be the word which is obtained from w by replacing any $a \in T$ by X_a.

We construct the monotone grammar $G' = (N', T', P', S)$ where

$$
\begin{aligned}
N' &= \overline{N} \cup \{S', X\} \cup \{(A, z) \mid A \in \overline{N}, z \in Z\} \\
&\quad \cup \{(A, z, p, t) \mid A \in \overline{N}, z \in Z, p \in P, t \in \{0, 1\}\}, \\
T' &= T \cup \{\#\},
\end{aligned}
$$

(X is an additional symbol which blocks the termination, $\#$ is a marker) and P' consists of all rules of the following forms:

$$S' \to \#(S, z_0, p, 0)\#, \quad \text{for } p \in P,$$
$$\#(A, z) \to \#(A, z, p, 0), \quad \text{for } p \in P, A \in \overline{N}$$

(we try to apply the production p),

$(A, z, p, t)B \to A(B, z, p, t)$, for $A, B \in \overline{N}, z \in Z, p \in P, t \in \{0, 1\}$,

$(A, z, p, 0) \to (A, z, p, 1)$, for $z \in Z, A \in N, p = A \to w \in P, w \in V_G^+$,

$(A, z, p, 1) \to (x_T, \delta(z, p))y_T$, for $z \in Z, A \in N, p = A \to xy \in P, x \in N \cup T, y \in V_G^*$

(we move the nonterminals with four components to the right searching for a nonterminal to which p can be applied; if we find the nonterminal of the left side of p, we change the fourth component to 1, i.e., we remember that p is applicable to the sentential form; we simulate the application of p and change the state according to the work of $\mathcal{A}$),

$$(A, z, p, 1)\# \to X\#, \text{ for } A \in N, z \in Z, p = B \to w \in P, A \neq B,$$

$$(A, z, p, 0)\# \to X\#, \text{ for } A \in N, z \in Z, p = B \to w \in P - F, A \neq B,$$

$$(A, z, p, 0)\# \to (A, \delta(z, p))\#, \text{ for } A \in N, z \in Z, p = B \to w \in F, A \neq B$$

(we reach the right marker; if p is applicable, i.e., $t = 1$, and has not been applied we block the derivation; if p is not applicable, we simulate the application of p in the appearance checking mode),

$$B(A, z) \rightarrow (B, z)A, \quad \text{for } A, B \in \overline{N}, z \in Z$$

(we move the nonterminal with two components to the left),

$$\#(X_a, z) \rightarrow \#a, \quad \text{for } a \in T, z \in Q,$$
$$X_a \rightarrow a, \quad \text{for } a \in T$$

(after the application of a production $\#(X_a, z) \rightarrow \#a$, none of the above productions is applicable to the derived sentential form, i.e., we have finished the derivation; if we apply productions $X_a \rightarrow a$ earlier in the process, the terminals a block the derivation or only occur at places which are not involved in the further derivation). Note that any sentential form – besides those over $\overline{N} \cup T$ – contains a letter $(A, \delta(z_0, p_1 p_2 \ldots p_k))$ or $(A, \delta(z_0, p_1 p_2 \ldots p_k), p, t)$ where $p_1, p_2, \ldots, p_k$ is the sequence of productions which we have simulated. Moreover, we can only finish the derivation with a terminal word if and only if the sequence of simulated productions belongs to R). By these explanations,

$$L(G') = \#L(G)\#$$

and then by the closure properties of $\mathcal{L}(CS)$, $L(G) \in \mathcal{L}(CS)$ follows easily.

We omit the necessary modifications in order to prove $\mathcal{L}(\lambda r C_{ac}) \subseteq \mathcal{L}(RE)$.

For the proof of $\mathcal{L}(RE) \subseteq \mathcal{L}(\lambda r C)$, we refer to [27], Theorem 1.2.5, or [108], Theorem 5.1 (in combination with Theorem 2.2 or Theorem 2.4 below).

All the remaining inclusions follow by definition. Furthermore, the above examples prove that the family of context-free languages is strictly included in those generated by regularly controlled grammars. The strictness of the inclusions obtained by adding the appearance checking is implied by Example 1 and the following fact which we give without proof.

Fact. *All languages of $\mathcal{L}(\lambda r C)$ over a unary alphabet are regular.*

(The proof of this fact is given in [53] using deep results of the theory of Petri nets and a relation between matrix grammars and Petri nets; furthermore, one has to use Theorem 2.2 below.)

For a proof of the properness of the inclusions if erasing rules are added and of the strict inclusion of $\mathcal{L}(r C_{ac})$ in $\mathcal{L}(CS)$ we refer to [54] (see the correction in [29]), [99] and [27].

In [11] it is also proven that $\mathcal{L}(\lambda r C) - \mathcal{L}(CS) \neq \emptyset$. A proof of the fact that appearance checking increases the power of regular control grammars can be also found in [30]. □

Definition 2.2. *([1])*
i) A matrix grammar with appearance checking *is a quintuple*

$$G = (N, T, M, S, F),$$

where

- N, T and S are specified as in a context-free grammar,
- $M = \{m_1, m_2, \ldots, m_n\}$, $n \geq 1$, is a finite set of sequences $m_i = (p_{i_1}, \ldots, p_{i_{k(i)}})$, $k(i) \geq 1$, $1 \leq i \leq n$, where any p_{i_j}, $1 \leq i \leq n$, $1 \leq j \leq k(i)$, is a context-free production,
- F is a subset of all productions occurring in the elements of M, i.e. $F \subseteq \{p_{i_j} \mid 1 \leq i \leq n, 1 \leq j \leq k(i)\}$.

ii) For m_i, $1 \leq i \leq n$, and $x, y \in V_G^*$, we define $x \Longrightarrow_{m_i} y$ by

$$x = x_0 \Longrightarrow_{p_{i_1}}^{ac} x_1 \Longrightarrow_{p_{i_2}}^{ac} x_2 \Longrightarrow_{p_{i_3}}^{ac} \cdots \Longrightarrow_{p_{i_{k(i)}}}^{ac} x_{k(i)} = y,$$

where the application of productions in the appearance checking mode is defined as in Definition 1. ii).

iii) The language $L(G)$ generated by G (with appearance checking) is defined as the set of all words $w \in T^*$ such that there is a derivation

$$S \Longrightarrow_{m_{j_1}} y_1 \Longrightarrow_{m_{j_2}} y_2 \Longrightarrow_{m_{j_3}} \cdots \Longrightarrow_{m_{j_s}} w,$$

for some $s \geq 1$, $1 \leq j_i \leq n$, $1 \leq i \leq s$.

iv) We say that M is a matrix grammar without appearance checking if and only if $F = \emptyset$.

The elements of M are called matrices.

Example 2.3. Let

$$G_3 = (\{S, A, A', C, D, D', D'', X\}, \{a, b, c, d\},$$

$$\{m_1, \ldots, m_{10}\}, S, \{A \to X, A' \to X\}),$$

with

$$m_1 = (S \to ASC), \ m_2 = (S \to SC), \ m_3 = (S \to ADC),$$
$$m_4 = (D \to D, A \to A'b), \ m_5 = (D \to D', A \to X, C \to c),$$
$$m_6 = (D' \to D', A' \to A), \ m_7 = (D' \to D, A' \to X),$$
$$m_8 = (D' \to D'', A' \to X), \ m_9 = (D'' \to D'', A \to a),$$
$$m_{10} = (D'' \to d, A \to X),$$

be a matrix grammar with appearance checking.

Any derivation has to start with applications of the matrices m_1, m_2, m_3 which yields $A^n D C^m$ with $1 \leq n \leq m$.

Without loss of generality we consider a word of the form $(Ab^k)^n Dc^k C^{m-k}$ (by the first phase of a derivation we obtain such a word with $k = 0$). Now we can apply m_4 or m_5. In the latter case the generated sentential form contains the letter X which blocks the derivation. Thus we have to apply m_4 until no A occurs in the sentential form, i.e.

$$(Ab^k)^n Dc^k C^{m-k} \underbrace{\Longrightarrow_{m_4} \cdots \Longrightarrow_{m_4}}_{n \ times} (A'b^{k+1})^n Dc^k C^{m-k}.$$

Now the only applicable matrix is m_5 which (without loss of generality) gives

$$(A'b^{k+1})^n Dc^k C^{m-k} \Longrightarrow_{m_5} (A'b^{k+1})^n D'c^{k+1} C^{m-k-1}.$$

By analogous arguments we now have to replace all occurrences of A' using m_6 and then we have to apply m_7 or m_8. Thus we obtain

$$(A'b^{k+1})^n D'c^{k+1} C^{m-k-1} \qquad \underbrace{\Longrightarrow_{m_6} \cdots \Longrightarrow_{m_6}}_{n \ times} (Ab^{k+1})^n D'c^{k+1} C^{m-k-1}$$

$$\Longrightarrow_{m_7} (Ab^{k+1})^n Dc^{k+1} C^{m-k-1}$$

or

$$(A'b^{k+1})^n D'c^{k+1} C^{m-k-1} \qquad \underbrace{\Longrightarrow_{m_6} \cdots \Longrightarrow_{m_6}}_{n \ times} (Ab^{k+1})^n D'c^{k+1} C^{m-k-1}$$

$$\Longrightarrow_{m_8} (Ab^{k+1})^n D''c^{k+1} C^{m-k-1}.$$

In the former case we get a word of the same structure, and we can iterate the procedure. In the latter case we have to apply m_9 as long as A occurs in the sentential form, and then we finish the derivation by an application of m_{10}. Thus we obtain

$$(Ab^{k+1})^n D''c^{k+1} C^{m-k-1} \qquad \underbrace{\Longrightarrow_{m_9} \cdots \Longrightarrow_{m_9}}_{n \ times} (ab^{k+1})^n D''c^{k+1} C^{m-k-1}$$

$$\Longrightarrow_{m_{10}} (ab^{k+1})^n dc^{k+1} C^{m-k-1} = v.$$

Obviously, if $m = k + 1$, then a terminal word is derived; otherwise, the derivation is blocked since no matrix can be applied to v.

By these arguments it is easy to see that

$$L(G_3) = \{(ab^m)^n dc^m \mid 1 \le n \le m\}.$$

We note that this language cannot be generated by an ET0L system (if we cancel all occurrences of d and c by a morphism we obtain a well-known non-ET0L-language, see [101] or Chapter 5 of this *Handbook*).   □

Example 2.4. We consider the matrix grammar

$$G_4 = (\{S, A, B\}, \{a, b\}, \{m_1, m_2, m_3, m_4, m_5\}, S, \emptyset),$$

without appearance checking, where

$$\begin{aligned} m_1 &= (S \to AB), & m_2 &= (A \to aA, B \to aB), \\ m_3 &= (A \to bA, B \to bB), & m_4 &= (A \to a, B \to a), \\ m_5 &= (A \to b, B \to b). \end{aligned}$$

Any derivation has to start with the application of the matrix $(S \to AB)$ yielding AB.

We now consider a word $wAwB$, $w \in \{a,b\}^*$ (after the first derivation step we obtain such a word with $w = \lambda$). Then we can apply any of the matrices m_2, m_3, m_4, m_5 and obtain

$$wAwB \Longrightarrow_{m_2} waAwaB, \quad wAwB \Longrightarrow_{m_3} wbAwbB,$$
$$wAwB \Longrightarrow_{m_4} wawa, \quad wAwB \Longrightarrow_{m_5} wbwb.$$

Thus we obtain a word of the same structure, again, or a word over the terminal set. Hence

$$L(G_4) = \{ww \mid w \in \{a,b\}^+\}.$$

Note that this is the non-context-free language which models the reduplication phenomenon in natural (and artificial) languages.   □

Obviously, if every matrix contains exactly one production, the matrix grammar works as a context-free grammar.

By λM_{ac} we denote the family of matrix grammars with appearance checking and with erasing rules. As for regularly controlled grammars we omit the index ac and/or the λ, if we restrict to matrix grammars without appearance checking and/or without erasing rules.

Theorem 2.2. $\mathcal{L}(M) = \mathcal{L}(rC)$, $\mathcal{L}(\lambda M) = \mathcal{L}(\lambda rC)$, $\mathcal{L}(M_{ac}) = \mathcal{L}(rC_{ac})$, $\mathcal{L}(\lambda M_{ac}) = \mathcal{L}(\lambda rC_{ac})$.

Remarks on the proof. We only prove the second equality (the modification for the non-erasing case uses the closure of the families under derivatives with letters and instead of an erasing in the finally applied matrix we introduce by this matrix the additional letter (for this method, see the proof of Theorem 2.8 below); the necessary changes for the case with appearance checking are obvious).

$\mathcal{L}(\lambda M) \subseteq \mathcal{L}(\lambda rC)$. It is easy to see that the control by matrices corresponds to a control by the regular set M^*.

$\mathcal{L}(\lambda rC) \subseteq \mathcal{L}(\lambda M)$. Let $G = (N, T, P, S, R)$ be a regularly controlled grammar (without appearance checking), and let $\mathcal{A} = (Z, P, z_0, Q, \delta)$ be a deterministic finite automaton (with the set Z of states, the initial state z_0, the set Q of accepting states, the input set P of productions of G and the transition function $\delta : Z \times P \to Z$) which accepts R. Without loss of generality we assume that $N \cap Z = \emptyset$. Then we construct the matrix grammar

$$G' = (N \cup Z \cup \{S'\}, T, \{m\} \cup \{m_z \mid z \in Q\} \cup \{m_p \mid p \in P\}, S', \emptyset),$$

with

$$m = (S' \to Sz_0),$$
$$m_p = (p, z \to \delta(z, p)), \quad \text{for} \quad p \in P, z \in Z,$$
$$m_z = (z \to \lambda), \quad \text{for} \quad z \in Q.$$

By the construction, any derivation in G' has the form

$$S' \Longrightarrow_m Sz_0 \Longrightarrow_{m_{p_1}} w_1\delta(z_0, p_1) \Longrightarrow_{m_{p_2}} w_2\delta(z_0, p_1p_2)$$
$$\Longrightarrow_{m_{p_3}} \ldots \Longrightarrow_{m_{p_n}} w_n\delta(z_0, p_1p_2 \ldots p_n) \Longrightarrow_{m_z} w_n, \qquad (*)$$

where

$$S \Longrightarrow_{p_1} w_1 \Longrightarrow_{p_2} w_2 \Longrightarrow_{p_3} \ldots \Longrightarrow_{p_n} = w_n \qquad (**)$$

holds. Moreover, the last step of the derivation $(*)$ is possible iff $\delta(z_0, p_1p_2 \ldots p_n)$ is in Q, i.e. iff $p_1p_2 \ldots p_n \in R$. This condition proves that $(**)$ is a correct derivation in G. Hence $L(G') \subseteq L(G)$.

On the other hand it is easy to see that, for any derivation of the form $(**)$ in G, there is a derivation of the form $(*)$ in G' which proves the converse inclusion, $L(G) \subseteq L(G')$. $\qquad \square$

We now introduce a slight modification of the control set used in the case of matrix grammars (see the first part of the proof of Theorem 2.2).

Definition 2.3. *([14])*
i) *An (unordered) vector grammar is a quadruple $G = (V, T, M, S)$ where N, T, M and S are defined as for matrix grammars.*
ii) *The language $L(G)$ generated by G is defined as the set of all words $w \in T^*$ such that there is a derivation*

$$S \Longrightarrow_{p_1} w_1 \Longrightarrow_{p_2} w_2 \Longrightarrow_{p_3} \ldots \Longrightarrow_{p_n} w,$$

where $p_1p_2 \ldots p_n$ is a permutation of some element of M^*.

Example 2.5. We consider the grammar G_4 given in Example 4 and use the notation $m_1 = (p_1)$ and $m_i = (p_{i1}, p_{i2})$, $2 \le i \le 5$.

If we regard G_4 as a matrix grammar, then the only derivation of $abaaba$ is

$$S \Longrightarrow_{p_1} AB \Longrightarrow_{p_{21}} aAB \Longrightarrow_{p_{22}} aAaB \Longrightarrow_{p_{31}} abAaB \Longrightarrow_{p_{32}} abAabB$$
$$\Longrightarrow_{p_{41}} abaabB \Longrightarrow_{p_{42}} abaaba.$$

If we consider G_4 as an (unordered) vector grammar, this derivation is only one of 20 different derivations yielding $abaaba$. We only present one further such derivation:

$$S \Longrightarrow_{p_1} AB \Longrightarrow_{p_{22}} AaB \Longrightarrow_{p_{32}} AabB \Longrightarrow_{p_{21}} aAabB \Longrightarrow_{p_{31}} abAabB$$
$$\Longrightarrow_{p_{41}} abaaB \Longrightarrow_{p_{42}} abaaba.$$

Moreover, the (unordered) vector grammar also allows derivations which are forbidden in the matrix grammar, e.g.,

$$S \Longrightarrow_{p_1} AB \quad \Longrightarrow_{p_{31}} \quad bAB \Longrightarrow_{p_{21}} baAB \Longrightarrow_{p_{41}} baaAB \Longrightarrow_{p_{22}} baaaB$$
$$\Longrightarrow_{p_{32}} \quad baaabB \Longrightarrow_{p_{42}} baaaba.$$

It is easy to see that the (unordered) vector grammar G_4 generates

$$\{wxw'x \mid x \in \{a,b\}, w \in \{a,b\}^*, w' \in Perm(\{w\})\},$$

where $Perm(R)$ contains all words which can be obtained by permutations from the words in R. □

Note that the control language used by an (unordered) vector grammar can be a non-context-free language. This can be seen by our example. Obviously, the productions p_{21} and p_{22} have to be used the same number of times, and the same holds for p_{31} and p_{32}. Thus

$$Perm(M^*) \cap p_1 p_{21}^* p_{31}^* p_{22}^* p_{32}^* p_{41} p_{42} = \{p_1 p_{21}^n p_{31}^m p_{22}^n p_{32}^m p_{41} p_{42} \mid n, m \geq 0\},$$

which is not a context-free language.

By λuV and uV we denote the families of (unordered) vector grammars with and without erasing rules, respectively.

Without proof we give the following relations for the associated families of languages (for a proof see [27], [86], [118]).

Theorem 2.3. $\mathcal{L}(CF) \subset \mathcal{L}(uV) = \mathcal{L}(\lambda uV) \subset \mathcal{L}(M)$. □

Matrix grammars and (unordered) vector grammars can be interpreted as grammars with special control sets. In [22] further families of languages are studied where the control languages are special regular sets (e.g., accepted by special types of finite automata or with special algebraic properties).

Definition 2.4. *([99])*
i) *A programmed grammar is a quadruple $G = (N,T,P,S)$, where*
    – N, T and S are specified as in a context-free grammar and
    – P is a finite set of triples $r = (p, \sigma, \varphi)$ where p is a context-free production and σ and φ are subsets of P.
ii) *The language $L(G)$ generated by G is defined as the set of all words $w \in T^*$ such that there is a derivation*

$$S = w_0 \Longrightarrow_{r_1} w_1 \Longrightarrow_{r_2} w_2 \Longrightarrow_{r_3} \ldots \Longrightarrow_{r_k} w_k = w,$$

   $k \geq 1$, and there is $r_{k+1} \in P$ such that, for $r_i = (A_i \rightarrow v_i, \sigma_i, \varphi_i)$, $1 \leq i \leq k$, one of the following conditions holds:

$$w_{i-1} = w'_{i-1} A_i w''_{i-1}, \; w_i = w'_{i-1} v_i w''_{i-1}, \text{ for some } w'_{i-1}, w''_{i-1} \in V_G^* \text{ and}$$
$$r_{i+1} \in \sigma_i,$$

or

$$A_i \text{ does not occur in } w_{i-1}, \ w_{i-1} = w_i \text{ and } r_{i+1} \in \varphi_i.$$

iii) If $r = (p, \sigma, \emptyset)$ holds for each $r \in P$, then we say that G is a programmed grammar without appearance checking. Otherwise G is a programmed grammar with appearance checking.

If $r = (p, \sigma, \varphi)$, then σ and φ are called the *success field* and the *failure field* of r, respectively.

Example 2.6. Let

$$G_5 = (\{S, S'\}, \{a\}, \{r_1, r_2, r_3\}, S)$$

be a programmed grammar with

$$r_1 = (S \rightarrow S'S', \{r_1\}, \{r_2\}), \ r_2 = (S' \rightarrow S, \{r_2\}, \{r_1, r_3\}),$$

$$r_3 = (S \rightarrow a, \{r_3\}, \emptyset).$$

We consider a word S^{2^n} for some $n \geq 0$ and assume that r_1 and r_3 are applicable (this situation holds for the axiom with $n = 0$).

If we apply r_1, then we have to do this as long as S appears in the sentential form, since the success field only contains r_1, i.e., r_1 has to be applied 2^n times. Since any application introduces two occurrences of S' for one occurrence of S, we obtain $(S')^{2^{n+1}}$. Then we have to continue with r_2 as long as S' occurs in the sentential form, which gives $S^{2^{n+1}}$. This word has the same structure and r_1 and r_3 are applicable, again.

If we apply r_3 to S^{2^n}, then we obtain a^{2^n}.

Hence G_5 generates the non-context-free language

$$L(G_5) = \{a^{2^n} \mid n \geq 0\}. \qquad \square$$

Example 2.7. We consider the programmed grammar

$$G_6 = (\{S, A, A', B, B'\}, \{a, b\}, \{r_1, r_2, \ldots, r_9\}, S),$$

without appearance checking, where

$$r_1 = (S \rightarrow AB, \{r_2, r_6\}, \emptyset), \qquad r_2 = (A \rightarrow aA', \{r_4\}, \emptyset),$$
$$r_3 = (A' \rightarrow aA, \{r_5\}, \emptyset), \qquad r_4 = (B \rightarrow B'B', \{r_4, r_3, r_7\}, \emptyset),$$
$$r_5 = (B' \rightarrow BB, \{r_5, r_2, r_6\}, \emptyset), \quad r_6 = (A \rightarrow a, \{r_8, r_9\}, \emptyset),$$
$$r_7 = (A' \rightarrow a, \{r_8, r_9\}, \emptyset), \qquad r_8 = (B \rightarrow b, \{r_8, r_9\}, \emptyset),$$
$$r_9 = (B' \rightarrow b, \{r_8, r_9\}, \emptyset).$$

We prove that G_6 generates

$$L(G_6) = \{a^n b^m \mid 1 \le n \le m \le 2^n\}$$

(note that $L(G_6)$ is not a semilinear language).

This can be seen by a combination of the following facts.

We consider two types of situations: The first type is given by words w such that

$$w = a^{n-1} A w', \quad |w'|_B \ge 1, \quad n \le |w'|_B + |w'|_{B'} \le 2^n$$

and at most r_2, r_5, and r_6 are applicable to w (such a situation with $n = 1$ is obtained after the application of r_1 which has to be applied in the first step of any derivation in G). The second type is given by words v such that

$$v = a^{n-1} A' v', \quad |v'|_{B'} \ge 1, \quad n \le |v'|_B + |v'|_{B'} \le 2^n$$

and at most r_3, r_4, and r_7 can be applied to v.

If we apply r_6 (i.e., we substitute A by a) to a word of the first type, we have to continue with r_8 or r_9 until all nonterminals B and B' are replaced by b. Thus we get a terminal word $a^n b^m$ with $n \le m \le 2^n$ which has the desired form. The same holds if we apply r_7 to a word of second type.

If we apply r_2 to a word w of the first type, we have to continue with some, say k, $1 \le k \le |w'|_B$, applications of r_4. This yields

$$w \Longrightarrow_{r_2} a^n A' w' \underbrace{\Longrightarrow_{r_4} \cdots \Longrightarrow_{r_4}}_{k \ times} a^n A' w'',$$

where w'' contains at least one occurrence of B', at least $n + k$ letters (any application of r_4 increases the length by 1), and at most 2^{n+1} letters (the upper bound is reached if $w' = B^{2^n}$ and $k = 2^n$). Moreover, besides r_4 the only applicable rules are r_3 and r_7. Therefore we have derived a word of the second type.

Furthermore, by analogous arguments the application of r_3 to a word of the second type and the continuation with r_5 (as long as we want) give a word of the first type.   □

In order to describe the relations between the productions $r = (p_r, \sigma_r, \varphi_r)$ of a programmed grammar one can use the following graph with two types of edges, say, *green* and *red*: the set of nodes consists of the context-free rules p_r, $r \in P$; the first type of edges (green) connects p_r with all first components of σ_r and any of these edges is labelled by p_r; the second type of edges (red) connects p_r with all first components of φ_r and any of these edges is labelled by p_r, again. The appearance checking controls which type of edges is used. Now the sequence of context-free rules which have to be applied in succession in a programmed grammar corresponds to a path in the associated graph. Hence in programmed grammars without appearance checking we use sequences prescribed by a graph with green edges only.

Clearly, if $\sigma_r = P$ and $\varphi_r = \emptyset$ holds for every $r \in P$, then a programmed grammar works as a context-free grammar.

By λP_{ac} we denote the family of programmed grammars with appearance checking and with erasing rules. As for matrix grammars we omit the lower index ac and/or the letter λ, if we consider only programmed grammars without appearance checking and/or without erasing productions.

Theorem 2.4. $\mathcal{L}(P) = \mathcal{L}(rC)$, $\mathcal{L}(\lambda P) = \mathcal{L}(\lambda rC)$, $\mathcal{L}(P_{ac}) = \mathcal{L}(rC_{ac})$, $\mathcal{L}(\lambda P_{ac}) = \mathcal{L}(\lambda rC_{ac})$.

Sketch of the proof. We only prove the second relation.

$\mathcal{L}(\lambda P) \subseteq \mathcal{L}(\lambda rC)$. Obviously, the language is not changed if we add the new axiom S' and the new production $S' \to S$ with an empty failure field and a success field consisting of all rules with the left side S.

The graph associated with this programmed grammar according to the above description can be interpreted as a graphical interpretation of a finite automaton. We choose the production $S' \to S$ as the initial state and the set of all states as the set of accepting states. Then it is easy to see that the set R accepted by this automaton controls the derivations as in the programmed grammar.

$\mathcal{L}(\lambda rC) \subseteq \mathcal{L}(\lambda P)$. By Theorem 2.2, it is sufficient to show that programmed grammars can simulate matrix grammars. In order to do this we associate with a matrix $(p_1, p_2, \ldots, p_k)$ the programmed rules

$$r_i = (p_i, \{r_{i+1}\}, \emptyset), \text{ for } 1 \leq i \leq k-1, \quad \text{and} \quad r_k = (p_k, M_1, \emptyset),$$

where M_1 consists of $(X \to \lambda, \emptyset, \emptyset)$ (where X is an additional nonterminal) and all programmed productions associated with a starting rule in a matrix. Adding the new axiom S' and the production $(S' \to SX, M_1, \emptyset)$ we ensure that we start with the first rule of a matrix. By $(X \to \lambda, \emptyset, \emptyset)$ we ensure that we cannot terminate inside a matrix. Now the equality of the generated languages follows easily.   □

2.2 Control by context conditions

In this section we consider some mechanisms where the control works as follows: a production can only be applied to a sentential form if the form satisfies some conditions. For example, by such a condition we can require that certain symbols have to occur in the sentential form.

By such a control the sequence of applicable productions is not described in advance; it is controlled by the generated sequence of sentential forms.

Definition 2.5. *([36])*
i)   A conditional grammar is a quadruple $G = (N, T, P, S)$, where
     – N, T and S are specified as in a context-free grammar, and

– P is a finite set of pairs $r = (p, R)$, where p is a context-free production and R is a regular set over V_G.

ii) For $x, y \in V_G^*$, we say that x directly derives y, written as $x \Longrightarrow y$, iff there is a pair $r = (A \to w, R) \in P$ such that $x = x'Ax''$ and $y = x'wx''$ for some $x', x'' \in V_G^*$ and $x \in R$. (We also use the notation $x \Longrightarrow_r y$ or $x \Longrightarrow_p y$ if we want to specify the pair or the production which is applied.)

iii) The language L(G) generated by G is defined as

$$L(G) = \{w \in T^* \mid S \Longrightarrow^* w\}.$$

By definition, we can apply the pair $r = (p, R)$ (and thus the production p) to x iff x is contained in the regular set associated with p.

If the regular set associated with every production is the set of all words over V_G, then the condition for its application is always satisfied. Therefore such a conditional grammar works as a context-free grammar.

Example 2.8. Let

$$G_7 = (\{S, S'\}, \{a\}, \{r_1, r_2, r_3\}, S)$$

be a conditional grammar with

$$r_1 = (S \to S'S', (S')^*S^+), \quad r_2 = (S' \to S, S^*(S')^+), \quad r_3 = (S \to a, a^*S^+).$$

We consider a word S^{2^n} for some $n \geq 0$ (the axiom satisfies this condition). By the definition we can apply only r_1 and r_3.

Assume that we apply r_3. Then we obtain $S^s a S^{2^n - s - 1}$. If $s \geq 1$, then the derivation is blocked, since the derived sentential form is not contained in any of the regular sets associated with r_1, r_2 or r_3. Hence $s = 0$ has to hold, i.e., $a S^{2^n - 1}$ is generated. Now we can apply only r_3, and we obtain $a^2 S^{2^n - 2}$ by analogous arguments. This process has to be iterated until we derive the terminal word a^{2^n}.

Assume now that we apply r_1 to S^{2^n}. Then we generate $S'S'S^{2^n - 1}$, since by arguments as above we have to replace the first S in order to ensure that the derivation is not blocked. Therefore we get the derivation

$$S^{2^n} \Longrightarrow_{r_1} (S')^2 S^{2^n - 1} \Longrightarrow_{r_1} \cdots \Longrightarrow_{r_1} (S')^{2k} S^{2^n - k} \Longrightarrow_{r_1} \cdots \Longrightarrow_{r_1} (S')^{2^{n+1}}.$$

Now we can only apply r_2 and obtain the derivation

$$(S')^{2^n} \Longrightarrow_{r_2} S(S')^{2^{n+1} - 1} \Longrightarrow_{r_2} \cdots \Longrightarrow_{r_2} S^t S^{2^{n+1} - t} \Longrightarrow_{r_2} \cdots \Longrightarrow_{r_2} S^{2^{n+1}},$$

i.e. we have doubled the number of S.

Combining these facts, we obtain

$$L(G_7) = \{a^{2^n} \mid n \geq 0\}.$$

This example shows that we can generate non-context-free languages by conditional grammars. $\qquad\square$

By λC and C we denote the family of all conditional grammars with and without erasing productions, respectively.

Theorem 2.5. $\mathcal{L}(\lambda C) = \mathcal{L}(RE)$ and $\mathcal{L}(C) = \mathcal{L}(CS)$.

Proof. We only prove the second equality; the first one can be shown analogously.

$\mathcal{L}(C) \subseteq \mathcal{L}(CS)$. Let $G = (N, T, P, S)$ be a conditional grammar. For any $r = (p, R_p) \in P$, we also write $p \in P$.

For any $a \in T$, let X_a be an additional letter. We set

$$N' = N \cup \{X_a \mid a \in T\}.$$

For $w \in V_G^*$, let $w_T \in (N')^*$ be the word which is obtained from w by replacing any $a \in T$ by X_a. Moreover, for $p \in P$, let $A_p = (Z_p, N', z_{0p}, F_p, \delta_p)$ be the deterministic finite automaton which accepts $\{w_T \mid w \in R_p\}$.

We now construct the monotone grammar

$$G' = (\overline{N}, T \cup \{\#\}, P', \overline{S}),$$

where

$$
\begin{aligned}
\overline{N} \;=\; & N' \cup \{A \mid A \in N'\} \cup \{\overline{S}, \#', \#''\} \\
& \cup \{(A, p, z) \mid A \in N', p \in P, z \in Z_p\} \cup \{(A, p) \mid A \in N', p \in P\}
\end{aligned}
$$

and P' is the set of all rules of the forms

(1) $\overline{S} \to \#'S\#''$,
(2) $\#' \to (\#', p, z_{0p})$, $p \in P$,
(3) $(A, p, z)B \to A(B, p, z')$, $A \in N' \cup \{\#'\}$, $B \in N'$, $p \in P$, $z' = \delta_p(z, B)$,
(4) $A, p, z)\#'' \to (A, p)\#''$, $A \in N'$, $p \in P$, $z \in F_p$,
(5) $B(A, p) \to (B, p)A$, $A, B \in N'$, $p \in P$,
(6) $(A, p) \to w_T$, $p = A \to w$,
(7) $\#' \to \#$, $\#'' \to \#$, $X_a \to a$, $a \in T$.

Let $r = (p, R_p)$ with $p = A \to w$. Then the derivation $y = y_1 A y_2 \Longrightarrow_r y_1 w y_2$ with $y \in R_p$ in G is simulated in G' in the following way: We start from the word $\#' y_T \#''$. First we apply the rule of type (2) associated with p yielding $(\#', p, z_{0p}) y_T \#''$. Then we shift the symbol of the form (B, p, z) by productions of type (3) from left to right and obtain $\#' y' (X, p, \delta(z_{0p}, y_T)) \#''$ where $y_T = y'X$. If $\delta(z_{0p}, y_T) \notin F_p$, then the derivation is blocked. Otherwise, i.e. $y \in R_p$, we continue with the production of type (4) associated with p which gives $\#' y'(X, p)\#''$. Then we shift the letter of the form (B, p) by rules of type (5) from right to left until we derive $\#'(y_1)_T (A, p)(y_2)_T \#''$ and apply the production of type (6) generating the desired word $\#'(y_1)_T w_T (y_2)_T \#''$.

The production of type (1) ensures the correct start of the derivations, and the productions of type (7) terminate the derivations.

By these considerations it is easy to see that $\#L(G)\# \subseteq L(G')$. $L(G') \subseteq \#L(G)\#$ follows because, essentially, only derivations of the form described above terminate. Thus $L(G') = \#L(G)\#$, and then $L(G) \in \mathcal{L}(CS)$ follows by the closure properties of the family of context-sensitive languages.

$\mathcal{L}(CS) \subseteq \mathcal{L}(C)$. Let $G = (N, T, P, S)$ be a context-sensitive grammar. We can assume that G is in the Kuroda normal form, i.e., all productions are of the form $A \to w$ or $AB \to CD$, where $A, B, C, D \in N$ and $w \in V_G^+$ (we omit the special considerations which are necessary if $\lambda \in L(G)$). Without loss of generality we may assume that $A \neq B$ for each rule $AB \to CD$ in P (if $AA \to CD \in P$, then we replace it by $A \to A'$, $AA' \to CD$, for a new symbol, A', associated with this rule). We now construct the conditional grammar

$$G' = (N \cup \{A_p \mid A \in N, p \in P\}, T, P', S),$$

where P' contains the production

$$(A \to w, \ V_G^*),$$

for every rule $A \to w \in P$ and the productions

$$(A \to A_p \ V_G^* ABV_G^*),$$
$$(B \to B_p, \ V_G^* A_p BV_G^*),$$
$$(A_p \to C, \ V_G^* A_p B_p V_G^*),$$
$$(B_p \to D, \ V_G^* CB_p V_G^*),$$

for $AB \to CD \in P$. Now it can easily be seen that G' can simulate the derivations in G and that only simulations of derivations in G are possible in G' which implies $L(G') = L(G)$. $\qquad\square$

Definition 2.6. *([60], [90])*
i)   A semi-conditional grammar *is a quadruple* $G = (N, T, P, S)$, *where*
    – N, T *and* S *are specified as in a context-free grammar, and*
    – P *is a finite set of triples* $r = (p, R, Q)$, *where* p *is a context-free production and* R *and* Q *are disjoint finite sets of words over* V_G.
ii) *For* $x, y \in V_G^*$, *we say that* x *directly derives* y, *written as* $x \Longrightarrow y$, *iff there is a triple* $r = (A \to w, R, Q) \in P$ *such that* $x = x'Ax''$ *and* $y = x'wx''$ *for some* $x', x'' \in V_G^*$, *every word of* R *is a subword of* x, *and no word of* Q *is a subword of* x. *(Again, we use the notation* $x \Longrightarrow_r y$ *or* $x \Longrightarrow_p y$ *if we want to specify the pair or the production which is applied.)*
iii) *The language* L(G) *generated by* G *is defined as*

$$L(G) = \{w \in T^* \mid S \Longrightarrow^* w\}.$$

R *and* Q *are called the* permitted context *and* forbidden context *associated with* r *(or* p*), respectively.*

The condition that x contains every subword of a finite set $R = \{w_1, w_2, \ldots, w_n\}$ and no subword of Q is equivalent with the requirements $x \in \bigcap_{i=1}^{n} V_G^* \{w_i\} V_G^*$ (if R is the empty set, i.e. no context is required, one has to use $x \in V_G^*$) and $x \notin V_G^* Q V_G^*$. Hence p is applicable to x iff $x \in (\bigcap_{i=1}^{n} V_G^* \{w_i\} V_G^*) - V_G^* Q V_G^*)$. Therefore semi-conditional grammars are a special case of conditional grammars.

Definition 2.7. *([120]) A random context grammar is a semi-conditional grammar where the permitted and forbidden contexts of all productions are subsets of the set of nonterminals.*

By definition, random context grammars are special cases of semi-conditional grammars and therefore of conditional grammars, too.

Example 2.9. We first show that we can construct a semi-conditional grammar G_8 which allows the same derivations as the conditional grammar G_7 of Example 8. To this end we consider the semi-conditional grammar

$$G_8 = (\{S, S'\}, \{a\}, \{r_1, r_2, r_3\}, S),$$

with

$$r_1 = (S \rightarrow S'S', \emptyset, \{SS', a\}), \quad r_2 = (S' \rightarrow S, \emptyset, \{S'S, a\}),$$

$$r_3 = (S \rightarrow a, \emptyset, \{Sa, S'\}).$$

Since the terminal a is contained in the forbidden context of the production r_1 we can apply r_1 only to words over $\{S, S'\}$. Moreover, since SS' is a forbidden subword for the application of r_1 and the application of r_1 requires the existence of at least one S in the sentential form, we can apply r_1 only to words in $(S')^* S^+$ which is the regular language associated with r_1 in G_7.

Analogously, we can see that the context conditions of r_2 and r_3 in G_8 describe exactly the regular sets associated with r_2 and r_3 in G_7.

Hence we can apply a production to a sentential form in G_8 if and only if this production can also be applied to the sentential form in G_7. Thus both grammars generate the same language, i.e.,

$$L(G_8) = \{a^{2^n} \mid n \geq 0\}.$$

We now show that this language can also be generated by a random context grammar with only forbidden contexts. In order to do this we consider the random context grammar

$$G_9 = (\{S, S', S'', A\}, \{a\}, \{r_1, r_2, r_3, r_4, r_5\}, S),$$

with

$$r_1 = (S \rightarrow S'S', \emptyset, \{S'', A\}), \quad r_2 = (S' \rightarrow S'', \emptyset, \{S\}),$$
$$r_3 = (S'' \rightarrow S, \emptyset, \{S'\}), \quad\quad r_4 = (S \rightarrow A, \emptyset, \{S'\}),$$
$$r_5 = (A \rightarrow a, \emptyset, \{S\}).$$

Let us consider S^{2^n} with $n \geq 0$ (the axiom satisfies this condition with $n = 0$). The only applicable rules are r_1 and r_4.

We first discuss the application of r_1. We obtain $S^k S' S' S^{2^n-k}$. If $n = 0$, then $S'S' = (S')^{2^{n+1}}$. If $n \geq 1$, then v contains S and S' which are the forbidden letters for the application of r_2 and r_4. Thus the only applicable production is r_1, again. Moreover, this situation is preserved until any occurrence of S is replaced by $S'S'$. Hence we derive $(S')^{2^{n+1}}$. Now we have to continue with r_2, and as above we can show that we have to apply r_2 as long as S' is present in the sentential form. Therefore we obtain $(S'')^{2^{n+1}}$. Analogously, we have to continue with r_3 until $S^{2^{n+1}}$ is generated. By this derivation we have doubled the number of S. Furthermore, we can apply r_1 and r_4, again. Thus we have reproduced the situation which we started in.

If we apply r_4 to S^{2^n}, we obtain A (if $n = 0$) or a word containing S and A. In the latter case r_1 and r_5 are not applicable, i.e., r_4 is the only applicable rule, and because this situation holds until every S is replaced by A, we generate A^{2^n}. Now we have to finish the derivation by applications of r_5 which yields a^{2^n}.

Combining these facts it is easy to see that

$$L(G_9) = \{a^{2^n} \mid n \geq 0\},$$

too. $\square$

By the grammars of Example 9 we have shown that the language generated by the conditional grammar of Example 8 can also be obtained by a conditional grammar where the regular sets associated to the productions are of the special form $V_G^* - V_G^* Q V_G^*$ where Q contains only words consisting of at most 2 letters or only nonterminals.

By λsC and sC we denote the families of semi-conditional grammars with and without erasing rules, respectively. By λRC_{ac} we denote the family of random context grammars with erasing productions. If we restrict to non-erasing productions and/or to random context grammars where all forbidden contexts are empty, we omit the λ and/or the index ac. λfRC and fRC denote the families of of random context grammars with and without erasing rules, respectively, where all permitted contexts are empty.

Theorem 2.6. $\mathcal{L}(\lambda sC) = \mathcal{L}(RE)$ and $\mathcal{L}(sC) = \mathcal{L}(CS)$.

Proof. We only prove the second equality.
The inclusion $\mathcal{L}(sC) \subseteq \mathcal{L}(CS)$ follows from Theorem 2.5 and the above mentioned fact that productions of a semi-conditional grammar can be interpreted as productions of a conditional grammar with special regular sets.

If we consider the construction of the conditional grammar in the second part of the proof of Theorem 2.5, we see that the regular sets associated with the productions correspond to the requirement that the permitted context is empty or a set of the form $X_1 X_2$ with nonterminals X_1 and X_2 and the forbidden contexts are subsets of the set of nonterminals. Therefore that proof shows that any context-sensitive language can be generated by a semiconditional grammar.   □

Theorem 2.7.
i) $\mathcal{L}(CF) \subset \mathcal{L}(RC) \subset \mathcal{L}(RC_{ac}) = \mathcal{L}(M_{ac}) \subset \mathcal{L}(\lambda RC_{ac}).$
ii) $\mathcal{L}(RC) \subseteq \mathcal{L}(\lambda RC) \subset \mathcal{L}(\lambda RC_{ac}) = \mathcal{L}(\lambda M_{ac}).$
iii) $\mathcal{L}(RC) \subseteq \mathcal{L}(M).$

Sketch of a proof. We only show that random contexts can be simulated by matrices; for proofs of the other inclusions we refer to [27]; the strictness of the inclusion $\mathcal{L}(CF) \subset \mathcal{L}(RC)$ follows from examples above and the strictness of the other inclusions follows by the Fact mentioned in the proof of Theorem 2.1.

With a production

$$p = (A \to w, \{A_1, A_2, \ldots, A_r\}, \{B_1, B_2, \ldots, B_s\})$$

of a random context grammar we associate the matrix

$$m_p = (A_1 \to A_1, \ldots, A_r \to A_r, B_1 \to X, \ldots, B_s \to X, A \to w),$$

where X is an additional letter. Moreover the set F of productions which can be omitted (if the left side does not occur in the sentential form) is given by all productions of the form $B_i \to X$. Obviously, the application of a matrix m_p checks whether all symbols $A_1, A_2, \ldots, A_r$ are present in the sentential form (if one of these symbols does not occur, then the matrix is not applicable) and whether all symbols $B_1, B_2, \ldots, B_s$ are absent (if a rule $B_i \to X$ is applicable, then we generate an occurrence of X which cannot be replaced) before the production $A \to w$ is applied. Therefore the random context grammar G and the matrix grammar G' with

$$G = (N, T, P, S) \text{ and } G' = (N \cup \{X\}, T, \{m_p \mid p \in P\}, S, \{B \to X \mid B \in N\})$$

generate the same language.   □

Semi-conditional and random context grammars are special cases of conditional grammars because there special regular sets are used as conditions. Further families of languages defined by other special regular sets as conditions are investigated in [24]. On the other hand, for conditional grammars one can impose the restriction to have the same regular language associated to all productions and the generated family of languages is the same as in the general case, see [27], [63].

Definition 2.8. *([36])*
i)   An ordered grammar is a quadruple $G = (N, T, P, S)$, where
  – N, T and S are specified as in a context-free grammar and
  – P is a finite (partially) ordered set of context-free productions.
ii) For $x, y \in V_G^*$, we say that x directly derives y, written as $x \Longrightarrow y$, if
  and only if there is a production $p = A \to w \in P$ such that $x = x'Ax''$,
  $y = x'wx''$ and there is no production $q = B \to v \in P$ such that $p \prec q$
  and B occurs in x. If we want to specify the applied production we write
  $x \Longrightarrow_p y$.
iii) The language $L(G)$ generated by G is defined as

$$L(G) = \{w \in T^* \mid S \Longrightarrow^* w\}.$$

By point ii) of the definition we have to apply a production which is maximal in the set of applicable productions.

Example 2.10. We consider the ordered grammar

$$G_{10} = (\{S, S', S'', A, Z\}, \{a\}, P, S),$$

where the productions and the (partial) order of P can be seen from the following diagram:

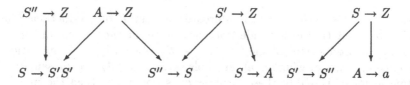

Since P does not contain a production with left side Z, the derivation is blocked if a production of the form $X \to Z$ is applied, $X \in \{A, S, S', S''\}$. Hence the only applicable productions are those of the form $X \to w$ with $w \neq Z$. By the definition of the order,

– $S \to S'S'$ is only applicable if the sentential form does not contain S'' and
  A,
– $S'' \to S$ is only applicable if the sentential form does not contain S' and A,
– $S' \to S''$ is only applicable if the sentential form does not contain S,
– $S \to A$ is only applicable if the sentential form does not contain S',
– $A \to a$ is only applicable if the sentential form does not contain S.

This describes exactly the forbidden contexts of the rules in the random context grammar G_9. Thus the order allows exactly the same derivations in G_{10} as the contexts in G_9. Therefore both grammars generate the same language, i.e.,

$$L(G_{10}) = \{a^{2^n} \mid n \geq 0\}.$$
□

By the definition of an ordered grammar, there is no restriction by contexts; however, by the preceding example it can be seen that forbidden contexts can be simulated by orders. This construction can be generalized and proves that random context grammars with empty permitted contexts are a special case of ordered grammars. Therefore we have introduced ordered grammars in this section.

By λO and O we denote the families of ordered grammars with and without erasing rules.

Theorem 2.8.
i) $\mathcal{L}(O) = \mathcal{L}(fRC) \subseteq \mathcal{L}(\lambda O) = \mathcal{L}(\lambda fRC) \subset \mathcal{L}(RE)$.
ii) $\mathcal{L}(ET0L) \subset \mathcal{L}(O) \subset \mathcal{L}(rC_{ac})$.

Sketch of the proof. As already mentioned above, the inclusions $\mathcal{L}(fRC) \subseteq \mathcal{L}(O)$ and $\mathcal{L}(\lambda fRC) \subseteq \mathcal{L}(\lambda O)$ follow by a generalization of the method presented in Example 10.

For a proof of the converse inclusions we refer to [27], Theorem 2.3.4.

$\mathcal{L}(ET0L) \subseteq \mathcal{L}(O)$. Let $L \in \mathcal{L}(ET0L)$. Then

$$L = \bigcup_{a \in T} aL_a, \quad \text{where} \quad L_a = \{w \mid aw \in L\}, \text{ for } a \in T.$$

By the closure properties of $\mathcal{L}(ET0L)$, for any $a \in T$, L_a is an ET0L language. Thus $L_a = L(G_a)$ for some ET0L system $G_a = (V, T, \{h_1, h_2, \ldots, h_k\}, w)$ without erasing rules.

Let $V' = \{x' \mid x \in V\}$, and let h be the morphism from V^* to $(V')^*$ given by $h(x) = x'$, for $x \in V$.

We construct the ordered grammar $G' = (N, T, P, S)$, where

$$N = V' \cup \{S', Z, A\} \cup \{(x, t) \mid x \in V \cup \{A\}, 1 \leq t \leq k\}$$

and all rules of P and their order are given by

(1) $S' \to (A, t)h(w), \quad 1 \leq t \leq k$
(2) $x' \to (x, t) \prec (A, s) \to Z, \quad x \in V, 1 \leq t \leq k, 1 \leq s \leq k, s \neq t,$
    $x' \to (x, t) \prec A \to Z, \quad x \in V, 1 \leq t \leq k,$
(3) $(x, t) \to y \prec z' \to Z, \quad x \in V, 1 \leq t \leq k, y \in h(h_t(x)), z \in V$
    $(x, t) \to y \prec A \to Z, \quad x \in V, 1 \leq t \leq k, y \in h(h_t(x)),$
(4) $(A, t) \to (A, s) \prec (x, t) \to Z, \quad 1 \leq t \leq k, 1 \leq s \leq k, x \in V,$
(5) $(A, t) \to A \prec (x, t) \to Z, \quad 1 \leq t \leq k, x \in V$
    $(A, t) \to A \prec a' \to Z, \quad 1 \leq t \leq k, a \in V - T,$
(6) $a' \to a \prec (A, t) \to Z, \quad a \in T, 1 \leq t \leq k,$
    $A \to a \prec x' \to Z, \quad x \in V.$

We consider a word

$$v = (A, t)h(x_1 x_2 \ldots x_n) = (A, t)x_1' x_2' \ldots x_n'$$

with $x_i \in V$, for $1 \leq i \leq n$ (such a word is produced by any application of a rule of type (1) to the axiom).

If we apply a production according to (4), we generate a word of the same form, and we can start the process, again.

If we apply a production according to (2), we have to continue with such productions until we obtain $(A, t)(x_1, t)(x_2, t) \ldots (x_n, t)$. Then we have to use productions according to (3) and derive

$$v' = (A, t)h(h_t(x_1))h(h_t(x_2)) \ldots h(h_t(x_n)) = (A, t)h(h_t(x_1 x_2 \ldots x_n)),$$

i.e. we have simulated one derivation step in the ETOL system G_a. Now we can start this process, again, because v' has the same form as v.

If we apply a production according to (5), then $x_1 x_2 \ldots x_n$ only contains letters of T, and we have to replace any primed letter by its non-primed version according to (6), and finally we apply $A \rightarrow a$.

Hence $L(G') = aL(G_a) = aL_a$ holds. Using the classical construction in order to generate the union of languages, we can prove that L as a union of such sets can also be generated.

The random context grammar $G = (\{S, A, A', B, B', C, D\}, \{a, b, c\}, P, S)$ with

$$P = \{(S \rightarrow ASB, \emptyset, \emptyset), (S \rightarrow AB, \emptyset, \emptyset), (A \rightarrow A'b, \emptyset, \{S, B', C, D\}),$$

$$(B \rightarrow C, \emptyset, \{A, C\}), (B \rightarrow B', \emptyset, \{A\}), (A' \rightarrow A, \emptyset, \{B\}),$$

$$(B' \rightarrow B, \emptyset, \{A'\}), (C \rightarrow c, \emptyset, \{A'\}), (A \rightarrow D, \emptyset, \{B\}), (D \rightarrow a, \emptyset, \{A\})\}$$

generates $\{(ab^m)^n c^n \mid m \geq n \geq 1\}$ which is not in $\mathcal{L}(ETOL)$ (see [101] and the remark at the end of Example 3.)

$\mathcal{L}(fRC) \subseteq \mathcal{L}(RC_{ac})$ and $\mathcal{L}(fRC) \subseteq \mathcal{L}(RE)$ follow by definition and Theorem 2.7. For a proof of the strictnesses we refer to [29] and [30]. □

2.3 Grammars with partial parallelism

In a context-free grammar a derivation step consists of the substitution of one nonterminal according to a production; in a Lindenmayer system a derivation step consists in the parallel substitution of all letters according to productions, where in the deterministic case we have to use the same production for all occurrences of a symbol. In this section we shall present some type of grammars where the mode of substitution is intermediate between the purely sequential context-free and parallel one.

We start with Indian parallel grammars where a derivation step consists in the substitution of all occurrences of one nonterminal according to the same production. Formally, this gives the following definition.

Definition 2.9. *([114])*

i) *An* Indian parallel grammar *is a quadruple* $G = (N, T, P, S)$, *where* N, T, P, *and* S *are specified as in a context-free grammar.*

ii) *For* $x, y \in V_G^*$, *we say that* x *directly derives* y, *written as* $x \Longrightarrow y$, *if and only if the following conditions hold:*

- $x = x_1 A x_2 A \ldots x_n A x_{n+1}$, *with* $n \geq 0$, $A \in N$, *and* $x_i \in (V_G - \{A\})^*$, *for* $1 \leq i \leq n + 1$,

- $y = x_1 w x_2 w \ldots x_n w x_{n+1}$,

- $A \to w \in P$.

If we want to specify the applied production $p = A \to w$, *we use the notation* $x \Longrightarrow_p y$.

iii) *The language* $L(G)$ *generated by* G *is defined as*

$$L(G) = \{w \in T^* \mid S \Longrightarrow^* w\}.$$

Example 2.11. We consider the Indian parallel grammar

$$G_{11} = (\{S, A\}, \{a, b\}, \{S \to AA, A \to aA, A \to bA, A \to a, A \to b\}, S).$$

Consider a word $wAwA$ (by the first step of any derivation we obtain AA which has this form). Using the four different productions with left side A we get the following derivations:

$$wAwA \Longrightarrow waAwaA, \quad wAwA \Longrightarrow wbAwbA,$$
$$wAwA \Longrightarrow wawa, \quad wAwA \Longrightarrow wbwb.$$

It is easy to see that

$$L(G_{11}) = \{ww \mid w \in \{a, b\}^*\}. \qquad \square$$

By λIP and IP we denote the families of Indian parallel grammars with and without erasing rules, respectively.

Theorem 2.9.

i) $\mathcal{L}(\lambda IP) = \mathcal{L}(IP)$.

ii) *The families* $\mathcal{L}(CF)$ *and* $\mathcal{L}(IP)$ *are incomparable.*

iii) $\mathcal{L}(IP) \subset \mathcal{L}(EDT0L)$.

Sketch of a proof.

i) This can be shown by the standard elimination of erasing rules (e.g, used for context-free grammars, see [55]).

ii) Example 11 shows that there are non-context-free languages which can be generated by Indian parallel grammars. Without proof we mention that the (context-free) Dyck language cannot be generated by an Indian parallel grammar (see [115] for a proof). (Note that there are languages which are contained in the intersection, e.g. all regular languages are in $\mathcal{L}(CF) \cap \mathcal{L}(IP)$.)

iii) $\mathcal{L}(IP) \subseteq \mathcal{L}(EDT0L)$. Let $G = (N, T, P, S)$ be an Indian parallel grammar. Then we construct the EDT0L system

$$G' = (N \cup T, T, \{h_p \mid p \in P\}, S),$$

where, for $p = A \to w \in P$, h_p is defined by

$$h_p(A) = w \quad \text{and} \quad h_p(x) = x, \text{ for } x \in (N \cup T) - \{A\}.$$

Obviously, $v_1 \Longrightarrow_p v_2$ holds in G if and only if $h_p(v_1) = v_2$. Thus $L(G') = L(G)$.

The inclusion is proper since $L = \{a^n b^n c^n \mid n \geq 1\}$ is not in $\mathcal{L}(IP)$ (see [27], Example 2.4.3 or [119]) and it is easy to construct an EDT0L system which generates L. □

Note that the situation differs from that for the other control mechanisms given in this chapter which define proper extensions of the family of context-free languages.

The derivation in an Indian parallel grammar is characterized by a parallel application of one rule to all occurrences of one letter. We now introduce a type of grammars where a derivation step consists in a parallel application of a certain number of productions (which are not necessarily different).

Definition 2.10. *([61], [43])*
i) *A k-grammar is a quadruple $G = (N, T, P, S)$, where N, T, P and S are specified as in a context-free grammar and $k \geq 1$.*
ii) *For $x, y \in V_G^*$, we say that x directly derives y if and only if*

$$x = S \quad \text{and} \quad S \to y \in P$$

or

$$x = x_1 A_1 x_2 A_2 \ldots x_k A_k x_{k+1}, \text{ for certain } x_1, x_2, \ldots, x_{k+1} \in V_G^*,$$
$$y = x_1 w_1 x_2 w_2 \ldots x_k w_k x_{k+1},$$
$$A_1 \to w_1, A_2 \to w_2, \ldots, A_k \to w_k \in P.$$

iii) *The language $L_k(G)$ generated by G is defined as the set*

$$L_k(G) = \{w \in T^* \mid S \Longrightarrow^* w\}.$$

By definition, besides the first step of a derivation (which is one in the context-free mode) we have to replace in parallel k occurrences of nonterminals according to productions of the grammar. Thus a derivation is blocked if – besides the axiom – the sentential form contains less than k nonterminals.

Example 2.12. Let

$$G_{12} = (\{S, A, B, C\}, \{a, b, c\}, P, S)$$

with

$$P = \{S \to ABC, A \to aA, B \to bB, C \to cC, A \to a, B \to b, C \to c\}.$$

First we consider G_{12} as a 2-grammar. Then the following derivation is a typical one:

$$S \Longrightarrow ABC \Longrightarrow aAbBC \Longrightarrow aaAbBcC \Longrightarrow aaaAbbcC \Longrightarrow aaaabbcc.$$

Obviously, by any step of derivation in the mode of the 2-grammar G_{12} – besides the first one – we generate two additional occurrences of terminals. Therefore

$$L_2(G_{12}) = \{a^n b^m c^r \mid n + m + r = 2t, \ t \geq 1\}.$$

Now we consider G_{12} as a 3-grammar. All possible derivations are described in the following diagram

$$
\begin{array}{ccccccccc}
S & \Longrightarrow & ABC & \Longrightarrow & aAbBcC & \Longrightarrow & \cdots & \Longrightarrow & a^n Ab^n Bc^n C & \Longrightarrow & \cdots \\
& & \Downarrow & & \Downarrow & & & & \Downarrow & & \\
& & abc & & a^2 b^2 c^2 & & & & a^{n+1} b^{n+1} c^{n+1} & &
\end{array}
$$

and thus

$$L_3(G_{12}) = \{a^n b^n c^n \mid n \geq 1\}.$$

If we consider G_{12} as a k-grammar with $k \geq 4$, then we obtain

$$L_k(G_{12}) = \emptyset,$$

since the first step of derivation yields ABC and then the derivation is blocked because the sentential form does not contain k occurrences of nonterminals.

Note that $L_2(G_{12})$ and $L_k(G_{12})$, for $k \geq 4$, are regular languages, whereas $L_3(G_{12})$ is a non-context-free language. $\qquad \square$

By kG and λkG we denote the families of k' grammars without and with erasing rules, respectively, with $k \leq k'$.

Theorem 2.10. *For $k \geq 1$, $\mathcal{L}(kG) \subset \mathcal{L}((k+1)G)$, $\mathcal{L}(\lambda kG) \subset \mathcal{L}(\lambda(k+1)G)$.*

Proof. The inclusions are obvious. For the proof of the strictness we refer to [109] and [110], where it is proven that

$$\{a_1^n a_2^n \ldots a_{2k+1}^n \mid n \geq 1\} \in \mathcal{L}((k+1)G) - \mathcal{L}(\lambda kG),$$

for $k \geq 1$. $\qquad \square$

Theorem 2.11. *For $k \geq 1$, $\mathcal{L}(kG) \subseteq \mathcal{L}(M)$ and $\mathcal{L}(\lambda kG) \subseteq \mathcal{L}(\lambda M)$.*

Proof. Let $G = (N, T, P, S)$ be a k-grammar. Assume that k is minimal. We construct the matrix grammar

$$G' = (N \cup \{A' \mid A \in N\} \cup \{S_0\}, T, M, S_0),$$

where M consists of all matrices

$$(S_0 \to w), \quad \text{for } S \to w \in P,$$

and all matrices

$$(A_1 \to A'_1, A_2 \to A'_2, \ldots, A_k \to A'_k, A'_1 \to w_1, A'_2 \to w_2, \ldots, A'_k \to w_k),$$

where $A_i \to w_i$, $1 \leq i \leq k$ are k not necessarily different productions of P. Obviously, $v_1 \Longrightarrow v_2$ holds in G if and only if there is a matrix $m \in M$ such that $v_1 \Longrightarrow_m v_2$ holds in G'. This implies $L(G') = L(G)$. $\qquad \square$

In a k-grammar we have to apply in parallel k productions, however, the choice of productions is not restricted. The following type of grammars also works with a parallel application of a certain number of productions but there are restrictions on the possible combinations of productions and on the places where the productions can be applied.

Definition 2.11. *([46])*
i)   A scattered context grammar *is a quadruple $G = (N, T, P, S)$, where*
   – N, T and S are specified as in a context-free grammar, and
   – P is a finite set of matrices

$$(A_1 \to w_1, A_2 \to w_2, \ldots, A_k \to w_k),$$

   where $k \geq 1$, $A_i \in N$, and $w_i \in V_G^*$, for $1 \leq i \leq k$ (the number k depends on the matrix and can differ from matrix to matrix).
ii) For $x, y \in V_G^*$, we say that x directly derives y, written as $x \Longrightarrow y$, if and only if the following conditions are satisfied:

$$x = x_1 A_1 x_2 A_2 \ldots x_k A_k x_{k+1}, \quad \text{for some } x_i \in V_G^*, 1 \leq i \leq k+1,$$
$$y = x_1 w_1 x_2 w_2 \ldots x_k w_k x_{k+1},$$
$$(A_1 \to w_1, A_2 \to w_2, \ldots, A_k \to w_k) \in P.$$

iii) The language $L(G)$ generated by G is defined as

$$L(G) = \{w \in T^* \mid S \Longrightarrow^* w\}.$$

Example 2.13. Let

$$G_{13} = (\{S, A, B, C\}, \{a, b, c\}, P, S)$$

be a scattered context grammar with

$$P = \{(S \to AAA), (A \to aA, A \to bA, A \to cA), (A \to a, A \to b, A \to c)\}.$$

Then any derivation in G_{13} has the form

$$S \implies AAA \implies aAbAcA \implies a^2Ab^2Ac^2A \implies \ldots$$
$$\implies a^nAb^nAc^nA \implies a^{n+1}b^{n+1}c^{n+1},$$

which proves

$$L(G_{13}) = \{a^nb^nc^n \mid n \geq 1\}. \qquad \square$$

We now define a slight modification of scattered context grammars by restricting the possible combination of context-free rules but the order of the nonterminals in the sentential form can differ from the order of the context-free rules in a matrix. Formally this type of grammars is given in the following definition.

Definition 2.12. *([80])*
i) *An* unordered scattered context grammar *is a quadruple* $G = (N, T, P, S)$, *where* N, T, P *and* S *are specified as in a scattered context grammar.*
ii) *For* $x, y \in V_G^*$, *we say that* x *directly derives* y, *written as* $x \implies y$, *if and only if the following conditions are satisfied:*

$$x = x_1A_{i_1}x_2A_{i_2}\ldots x_kA_{i_k}x_{k+1}, \text{ for some } x_i \in V_G^*, 1 \leq i \leq k+1,$$
$$y = x_1w_{i_1}x_2w_{i_2}\ldots x_kw_{i_k}x_{k+1},$$
$$(A_1 \to w_1, A_2 \to w_2, \ldots, A_k \to w_k) \in P,$$
$$(i_1, i_2, \ldots, i_k) \text{ is a permutation of } (1, 2, \ldots, k).$$

iii) *The language* $L(G)$ *generated by* G *is defined as in the Definition 11 using the derivation of part ii) of this definition.*

Example 2.14. We consider the grammar G_{13} of Example 13, again; however, we consider it as an unordered scattered context grammar. Then

$$S \implies AAA \implies aAcAbA \implies abAcaAbcA \implies abacabbcc$$

is a derivation in the unordered scattered context grammar G_{13} which is not allowed in the scattered context grammar G_{13}. It is easy to see that the language generated by the unordered scattered context grammar G_{13} is

$$\{x_1x_2\ldots x_ny_1y_2\ldots y_nz_1z_2\ldots z_n \mid n \geq 1, \{x_i, y_i, z_i\} = \{a, b, c\}, 1 \leq i \leq n\}.$$
$$\square$$

By SC (uSC) and λSC (λuSC) we denote the families of (unordered) scattered context grammars without and with erasing productions.

Theorem 2.12.
i) $\mathcal{L}(M) = \mathcal{L}(uSC) \subset \mathcal{L}(SC) \subseteq \mathcal{L}(CS)$.
ii) $\mathcal{L}(\lambda M) = \mathcal{L}(\lambda uSC) \subset \mathcal{L}(\lambda SC) = \mathcal{L}(RE)$.

Proof. $\mathcal{L}(\lambda uSC) \subseteq \mathcal{L}(\lambda M)$ can be shown by a modification of the proof of Theorem 2.11.

$\mathcal{L}(\lambda M) \subseteq \mathcal{L}(\lambda uSC)$. Let $L \in \mathcal{L}(\lambda M)$. Then $L = L(G)$ for some matrix grammar $G = (N, T, M, S, \emptyset)$, where $N = N_1 \cup N_2 \cup \{S\}$ for some sets N_1 and N_2 with $N_1 \cap N_2 = \emptyset$ and all matrices of M have one of the forms $(S \to AB)$ or $(B \to \lambda)$ or $(A \to w, B \to B')$, with $A \in N_1$ and $B, B' \in N_2$ (for a proof of this fact, see the proof of Theorem 2.2). Obviously, the application of matrices of such forms does not differ for matrix grammars and unordered scattered context grammars.

$\mathcal{L}(\lambda uSC) \subseteq \mathcal{L}(\lambda SC)$. Let $G = (N, T, P, S)$ be an unordered scattered context grammar. We construct the scattered context grammar

$$G' = (N, T, \bigcup_{p \in P} H_p, S),$$

where, for $p = (A_1 \to w_1, A_2 \to w_2, \ldots, A_k \to w_k) \in P$,

$$H_p = \{(A_{i_1} \to w_{i_1}, A_{i_2} \to w_{i_2}, \ldots, A_{i_k} \to w_{i_k}) \mid$$
$$(i_1, i_2, \ldots, i_k) \text{ is a permutation of } (1, 2, \ldots k)\}.$$

Then $v_1 \Longrightarrow_p v_2$ in G if and only if there is an element $p' \in H_p$ such that $v_1 \Longrightarrow_{p'} v_2$. This implies $L(G') = L(G)$.

For a proof of the strictness of this inclusion, we refer to [30].

The modifications of the preceding proofs for the non-erasing grammars are left to the reader.

$\mathcal{L}(SC) \subseteq \mathcal{L}(CS)$. Using constructions as in the proofs of Theorem 2.1 and Theorem 2.5 we move special symbols (remembering the matrix and the rule of the matrix which we try to apply) from left to right and apply the productions in the given order.

$\mathcal{L}(\lambda SC) \subseteq \mathcal{L}(RE)$ can be shown analogously.

For the proof of $\mathcal{L}(RE) \subseteq \mathcal{L}(\lambda SC)$ we refer to [45], [77] and [75]. $\square$

We now define a special type of scattered context grammars.

Definition 2.13. *([57], [64])*
i)   A k-simple matrix grammar is a $(k + 3)$-tuple

$$G = (N_1, N_2, \ldots, N_k, T, M, S),$$

where
– $N_1, N_2, \ldots, N_k$ and T are pairwise disjoint alphabets (the sets N_i, $1 \leq i \leq k$, are the sets of nonterminals; T is the set of terminals),
– $S \notin T \cup N_1 \cup N_2 \cup \ldots N_k$ (the axiom),
– M is a finite set of matrices which have one of the following forms:

$$(S \to w), \ w \in T^*,$$
$$(S \to A_1 A_2 \ldots A_k), \ A_i \in N_i, 1 \le i \le k,$$
$$(A_1 \to w_1, A_2 \to w_2, \ldots, A_k \to w_k),$$
$$A_i \in N_i, w_i \in (N_i \cup T)^*, |w_i|_{N_i} = |w_j|_{N_j},$$
$$1 \le i \le k, 1 \le j \le k, i \ne j.$$

ii) For $x, y \in (\{S\} \cup T \cup N_1 \cup N_2 \cup \ldots \cup N_k)^*$, we say that x directly derives y, written as $x \Longrightarrow y$ if and only if either

$$x = S \quad and \quad (S \to y) \in M$$

or

$$x = x_1 A_1 z_1 x_2 A_2 z_2 \ldots x_k A_k z_k, \ x_i \in T^*, z_i \in (T \cup N_i)^*, 1 \le i \le k,$$
$$y = x_1 w_1 z_1 x_2 w_2 z_2 \ldots x_k w_k z_k,$$
$$(A_1 \to w_1, A_2 \to w_2, \ldots, A_k \to w_k) \in M.$$

iii) The language $L(G)$ generated by G is defined as

$$L(G) = \{w \in T^* \mid S \Longrightarrow^* w\}.$$

By definition, the matrices have to be applied as in scattered context grammars, however, the rule $A_i \to w_i$ has to be applied to the leftmost occurrence of a letter of N_i.

Moreover, since the nonterminals on the left sides of the rules in a production are in disjoint sets – besides the leftmost restriction – the application of a matrix in the mode of an unordered scattered context grammar does not differ from the application of a matrix in the mode of a scattered context grammar.

Furthermore, by the disjointness of the set of nonterminals, no rule $A_i \to w_i$ can be applied to a nonterminal introduced by the application of a rule $A_j \to w_j$, $j < i$. Thus – besides the leftmost restriction – the application of a matrix in the mode of a matrix grammar does also not differ from the application of a matrix in the mode of a scattered context grammar.

By these remarks – besides the leftmost restriction – simple matrix grammars are special cases of scattered context grammars, unordered scattered context grammars and matrix grammars (without appearance checking).

If $k = 1$, then we obtain the usual context-free grammars with leftmost restriction in the derivation (which leads to no change with respect to the generated language and language family).

Example 2.15. We consider the 2-simple matrix grammar

$$G_{14} = (\{A\}, \{B\}, \{a, b\}, M, S),$$

with

$$M = \{(S \to AB), (A \to aA, B \to aB), (A \to bA, B \to bB),$$
$$(A \to a, B \to a), (A \to b, B \to b)\},$$

and the 3-simple matrix grammar

$$G_{15} = (\{A\}, \{B\}, \{C\}, \{a, b, c\}, \{(S \to ABC),$$
$$(A \to aA, B \to bB, C \to cC), (A \to a, B \to b, C \to c)\}, S).$$

It is easy to see that the derivations in G_{14} and G_{15} are exactly the same derivations as those in the matrix grammar G_4 (Example 4) and the 3-grammar G_{12} (Example 12). Thus

$$L(G_{14}) = \{ww \mid w \in \{a, b\}^+\} \quad \text{and} \quad L(G_{15}) = \{a^n b^n c^n \mid n \geq 1\}. \qquad \square$$

By λkSM and kSM we denote the families of k'-simple matrix grammars with and without erasing rules, respectively, with $k \leq k'$. Moreover, we define λSM and SM as the union of all families λkSM and kSM, respectively, where the union is taken over all integers.

Theorem 2.13.
i) For $k \geq 2$, $\mathcal{L}(kSM) \subset \mathcal{L}(\lambda kSM)$, $\mathcal{L}(kSM) \subset \mathcal{L}(SM)$ and $\mathcal{L}(\lambda kSM) \subset \mathcal{L}(\lambda SM)$.
ii) For $k \geq 1$, $\mathcal{L}(kSM) \subset \mathcal{L}((k+1)SM)$, $\mathcal{L}(\lambda kSM) \subset \mathcal{L}(\lambda(k+1)SM)$.
iii) $\mathcal{L}(SM) \subset \mathcal{L}(\lambda SM) \subset \mathcal{L}(CS)$.

Proof. We only prove ii) and refer to [27] for the other statements.

The inclusions are obvious by the definitions. The proof of the strictness will be given in some steps.

Let V be an alphabet and $n \geq 0$ an integer. We set

$$[V, n] = V \times \{1, 2, \dots, n\}$$

and define the mappings $\tau_n : [V, k]^* \to (V^*)^k$ by

$$\tau_k(\lambda) = (\lambda, \lambda, \dots, \lambda),$$

$$\tau_k((a, i)x) = (x_1, x_2, \dots, x_{i-1}, ax_i, x_{i+1}, x_{i+2} \dots, x_k),$$

for $a \in V, 1 \leq i \leq k, x \in [V, k]^*, \tau_n(x) = (x_1, x_2, \dots, x_i, \dots, x_k)$,

and $f : (V^*)^k \to V^*$ and $p_i : (V^*)^k \to V^*, 1 \leq i \leq k$, by

$$f(x_1, x_2, \dots, x_k) = x_1 x_2 \dots x_k \quad \text{and} \quad p_i(x_1, x_2, \dots, x_k) = x_i,$$

respectively.

Assertion 2.1. *For a language* $L \subseteq V^*$ *in* $\mathcal{L}(\lambda kSM)$, *there is a context-free language* L' *such that* $L' \subseteq [V, n]^*$ *and* $f(\tau_n(L')) = L$.

Proof of Assertion 1. Let $G = (N_1, N_2, \ldots, N_k, T, M, S)$ be a k-simple matrix grammar. We construct the context-free grammar

$$G' = (\{S\} \cup (N_1 \times N_2 \times \ldots \times N_k), [T, k], P, S),$$

where P consists of all the following rules:

$$S \to \alpha, \text{ for } (S \to w) \in M, w \in T^*, \alpha \in \tau_k^{-1}(\lambda, \lambda, \ldots, \lambda, w),$$

$$S \to (A_1, A_2, \ldots, A_k), \text{ for } (S \to A_1 A_2 \ldots A_k) \in M, A_i \in N_i, 1 \leq i \leq k,$$

$$(A_1, A_2, \ldots, A_k) \to \alpha, \text{ for } (A_1 \to w_1, A_2 \to w_2, \ldots, A_k \to w_k) \in M,$$

$$A_i \in N_i, w_i \in T^*, 1 \leq i \leq n, \alpha \in \tau_k^{-1}(w_1, w_2, \ldots, w_k),$$

$$(A_1, A_2, \ldots, A_k) \to \alpha_1(A_{11}, A_{21}, \ldots, A_{k1}) \ldots \alpha_t(A_{1t}, A_{2t}, \ldots, A_{kt}) \ldots \alpha_{t+1}, \text{for}$$

$$(A_1 \to x_{11} A_{11} x_{12} A_{12} \ldots x_{1t} A_{1t} y_1, \ldots, A_k \to x_{k1} A_{k1} x_{k2} A_{k2} \ldots x_{kt} A_{kt} y_k) \in M,$$

$$A_{ij} \in N_i, x_{ij} \in T^*, y_i \in T^*, \alpha_j \in \tau_k^{-1}(x_{1j}, x_{2j}, \ldots, x_{kj}),$$

$$\alpha_{t+1} \in \tau_k^{-1}(y_1, y_2, \ldots, y_k), 1 \leq i \leq k, 1 \leq j \leq t, t \geq 1.$$

The equality $L(G) = f(\tau_k(L(G')))$ can easily be shown.

Assertion 2.2. *Let $L \in T^*$ be an infinite language in $\mathcal{L}(\lambda k SM)$. Then there exist integers p and q such that each $z \in L$ with $|z| > p$ can be written as*

$$z = u_1 v_1 w_1 x_1 u_2 v_2 w_2 x_2 \ldots u_k v_k w_k x_k u_{k+1}$$

such that $|v_1 x_1 v_2 x_2 \ldots v_k x_k| > 0$, $|u_1 w_1 x_1 u_2 w_2 x_2 \ldots u_k w_k x_k| \leq q$, and

$$u_1 v_1^n w_1 x_1^n u_2 v_2^n w_2 x_2^n \ldots u_k v_k^n w_k x_k^n u_{k+1} \in L, \quad \text{for } n \geq 0.$$

Proof of Assertion 2. We transfer the pumping lemma for context-free languages to k-simple matrix languages using the mappings f and τ_k and Asssertion 1:

$$\tau_k(uv^n w x^n y) = \tau_k(u)(\tau_k(v))^n \tau_k(w)(\tau_k(x))^n \tau_k(y) =$$

$$= (u_1, \ldots, u_k)(v_1, \ldots, v_k)^n (w_1, \ldots, w_k)(x_1, \ldots, x_k)^n (y_1, \ldots, y_k) =$$

$$= (u_1 v_1^n w_1 x_1^n y_1, u_2 v_2^n w_2 x_2^n y_n, \ldots, u_k v_k^n w_k x_k^n y_k)$$

and so on.

The language

$$L_k = \{a_1^n a_2^n \ldots a_{2k+1}^n \mid n \geq 1\}$$

is generated by the $(k+1)$-simple matrix grammar

$$G = (\{A_1\}, \{A_2\}, \ldots, \{A_{k+1}\}, \{a_1, a_2, \ldots, a_{2k+1}\}, M, S),$$

with

$$M = \{(S \to A_1 A_2 \ldots A_{k+1}),$$

$$(A_1 \to a_1 a_2, \ldots, A_k \to a_{2k-1} a_{2k}, A_{k+1} \to a_{2k+1}),$$

$$(A_1 \to a_1 A_1 a_2, \ldots, A_k \to a_{2k-1} A_t a_{2k}, A_{k+1} \to a_{2k+1} A_{k+1})\}.$$

Thus $L_k \in \mathcal{L}((k+1)SM)$. Using Assertion 2 we can prove that $L_k \notin \mathcal{L}(\lambda k SM)$. $\square$

2.4 Indexed grammars

The constructs CALL-BY-VALUE and CALL-BY-NAME of programming languages are the basis of some generalizations of context-free grammars. We present here one of the grammatical models, already briefly discussed in Chapter 4 of this *Handbook*.

Definition 2.14. *([2])*
i) An indexed grammar *is a quintuple*

$$G = (N, T, I, P, S),$$

where
- *N, T and S are specified as in a context-free grammar,*
- *I is a finite set of finite sets of productions of the form $A \to w$, with $A \in N$ and $w \in V_G^*$, and*
- *P is a finite set of productions of the form $A \to \alpha$, with $A \in N$ and $\alpha \in (NI^* \cup T)^*$.*

ii) For $x, y \in (NI^* \cup T)^*$, we say that x directly derives y, written as $x \Longrightarrow y$, if either

$$x = x_1 A \beta x_2, \quad \text{for some} \quad x_1, x_2 \in (NI^* \cup T)^*, A \in N, \beta \in I^*,$$
$$A \to X_1 \beta_1 X_2 \beta_2 \ldots X_k \beta_k \in P,$$
$$y = x_1 X_1 \gamma_1 X_2 \gamma_2 \ldots X_k \gamma_k x_2,$$

with $\gamma_j = \beta_j \beta$, for $X_j \in N$, and $\gamma_j = \lambda$, for $X_j \in T, 1 \le j \le k$,

or

$$x = x_1 A i \beta x_2, \quad \text{for some} \quad x_1, x_2 \in (NI^* \cup T)^*, A \in N, i \in I, \beta \in I^*,$$
$$A \to X_1 X_2 \ldots X_k \in i,$$
$$y = x_1 X_1 \gamma_1 X_2 \gamma_2 \ldots X_k \gamma_k x_2$$

with $\gamma_j = \beta$, for $X_j \in N$, and $\gamma_j = \lambda$, for $X_j \in T, 1 \le j \le k$.

iii) The language $L(G)$ generated by G is defined as

$$L(G) = \{w \in T^* \mid S \Longrightarrow^* w\}.$$

The control is done by means of the elements of I, the so-called indexes. The application of rules of P gives sentential forms where the nonterminals are followed by a sequence of indexes (in a certain sense, this sequence describes the history of the nonterminal), and an index can only be erased by productions contained in the index (where the erasing is done in the reversed order of the appearance).

If the set of indexes is empty, then we obtain a context-free grammar.

Example 2.16. Let

$$G = (\{S, A, B\}, \{a, b, c\}, \{f, g\}, P, S)$$

be an indexed grammar, where

$$f = \{B \to b\}, \ g = \{B \to bB\},$$
$$P = \{S \to aAfc, A \to aAgc, A \to B\}.$$

The productions of the indexes f and g can only be used if B occurs in the sentential form, and its application does not derive additional occurrences of S and A. On the other hand, P contains no production with B on its left side. Therefore each derivation can be divided into two phases: the first one uses only productions of P, the second one consists of applications of the rules in the indexes.

The first phase is shown in the following diagram.

$$
S \ \Longrightarrow \ aAfc \ \Longrightarrow \ a^2 agfc^2 \ \Longrightarrow \ \cdots \ \Longrightarrow \ a^n Ag^{n-1}fc^n \ \Longrightarrow \ \cdots
$$
$$
\qquad\quad \Downarrow \qquad\qquad \Downarrow \qquad\qquad\qquad\qquad \Downarrow
$$
$$
\qquad\quad aBfc \qquad\ a^2 Bgfc^2 \qquad\qquad\qquad a^n Bg^{n-1}fc^n
$$

The only possibility to rewrite $a^n Bg^{n-1}fc^n$ by $B \to bB \in g$ and $B \to b \in f$ is

$$a^n Bg^{n-1}fc^n \Longrightarrow a^n bBg^{n-2}fc^n \Longrightarrow \cdots$$
$$\Longrightarrow a^n b^{n-2}Bgfc^n \Longrightarrow a^n b^{n-1}Bfc^n \Longrightarrow a^n b^n c^n.$$

Hence

$$L(G) = \{a^n b^n c^n \mid n \geq 1\}. \qquad\qquad \Box$$

By I and λI we denote the families of indexed grammars without and with erasing rules, respectively.

Without proof we mention the following result on indexed grammars.

Theorem 2.14. $\mathcal{L}(ET0L) \subseteq \mathcal{L}(I) = \mathcal{L}(\lambda I) \subset \mathcal{L}(CS).$

2.5 Hierarchies of families with controlled derivations

In the preceding sections we have defined some types of grammars with controlled derivations. By $\mathcal{G}$ we denote the set of the associated families of grammars excluding the families associated with k-grammars and simple matrix grammars, i.e.

$$
\begin{aligned}
\mathcal{G} \ = \ & \{rC, \lambda rC, rC_{ac}, \lambda rC_{ac}, M, \lambda M, M_{ac}, \lambda M_{ac}, uV, \lambda uV, P, \lambda P, \\
& P_{ac}, \lambda P_{ac}, C, \lambda C, sC, \lambda sC, RC, \lambda RC, RC_{ac}, \lambda RC_{ac}, fRC, \lambda fRC, \\
& O, \lambda O, IP, \lambda IP, SC, \lambda SC, uSC, \lambda uSC, I, \lambda I\}.
\end{aligned}
$$

We summarize the results obtained in the preceding sections for the language families associated with the elements of $\mathcal{G}$.

Theorem 2.15.

i) The following equalities are valid:

$$
\begin{aligned}
\mathcal{L}(RE) &= \mathcal{L}(\lambda M_{ac}) = \mathcal{L}(\lambda r C_{ac}) = \mathcal{L}(\lambda P_{ac}) = \mathcal{L}(\lambda R C_{ac}) \\
&= \mathcal{L}(\lambda C) = \mathcal{L}(\lambda s C) = \mathcal{L}(\lambda S C), \\
\mathcal{L}(CS) &= \mathcal{L}(C) = \mathcal{L}(sC), \\
\mathcal{L}(\lambda M) &= \mathcal{L}(\lambda r C) = \mathcal{L}(\lambda P) = \mathcal{L}(\lambda u S C), \\
\mathcal{L}(M_{ac}) &= \mathcal{L}(r C_{ac}) = \mathcal{L}(P_{ac}) = \mathcal{L}(R C_{ac}), \\
\mathcal{L}(M) &= \mathcal{L}(r C) = \mathcal{L}(P) = \mathcal{L}(u S C), \\
\mathcal{L}(uV) &= \mathcal{L}(\lambda u V), \\
\mathcal{L}(\lambda O) &= \mathcal{L}(\lambda f R C), \\
\mathcal{L}(O) &= \mathcal{L}(f R C), \\
\mathcal{L}(IP) &= \mathcal{L}(\lambda IP), \\
\mathcal{L}(I) &= \mathcal{L}(\lambda I).
\end{aligned}
$$

ii) .The diagram in Figure 1 holds: if two families are connected by a line (an arrow), then the upper family includes (includes properly) the lower family; if two families are not connected then they are not necessarily incomparable.

iii) Let X and Y be two families of $\mathcal{G}$ such that $\mathcal{L}(X) \subseteq \mathcal{L}(Y)$. Then there is an algorithm which, for any grammar $G \in X$, produces a grammar $G' \in Y$ such that $L(G') = L(G)$.

Let X be a type of grammars, and let $k \geq 1$ be an integer. We define $\mathcal{L}_k(X)$ as the set of all languages L such that there is a grammar $G \in X$ such that $L = L(G)$ and, for any word $w \in L$, there is a derivation

$$S \Longrightarrow w_1 \Longrightarrow w_2 \Longrightarrow \ldots \Longrightarrow w_n = w$$

in G with

$$|w_i|_N \leq k, \quad \text{for } 1 \leq i \leq n,$$

where N and S are the set of nonterminals and the axiom of G, respectively. Further we denote

$$\mathcal{L}_{fin}(X) = \bigcup_{k \geq 1} \mathcal{L}_k(X).$$

If $L \in \mathcal{L}_k(X)$ or $L \in \mathcal{L}_{fin}(X)$, we say that L is of *index k* or of *finite index*, respectively.

Theorem 2.16.

i) Each of the following families of languages coincides with $\mathcal{L}_{fin}(M)$:
$\mathcal{L}_{fin}(\lambda M)$, $\mathcal{L}_{fin}(M_{ac})$, $\mathcal{L}_{fin}(\lambda M_{ac})$, $\mathcal{L}_{fin}(rC)$, $\mathcal{L}_{fin}(\lambda rC)$,
$\mathcal{L}_{fin}(rC_{ac})$, $\mathcal{L}_{fin}(\lambda rC_{ac})$, $\mathcal{L}_{fin}(P)$, $\mathcal{L}_{fin}(\lambda P)$, $\mathcal{L}_{fin}(P_{ac})$, $\mathcal{L}_{fin}(\lambda P_{ac})$,
$\mathcal{L}_{fin}(RC_{ac})$, $\mathcal{L}_{fin}(\lambda RC_{ac})$, $\mathcal{L}_{fin}(O)$, $\mathcal{L}_{fin}(\lambda O)$, $\mathcal{L}_{fin}(SC)$, $\mathcal{L}_{fin}(\lambda SC)$.

ii) $\mathcal{L}_{fin}(M) \subseteq \mathcal{L}_{fin}(C)$.

iii) $\mathcal{L}_{fin}(RC) \subseteq \mathcal{L}_{fin}(\lambda RC) \subseteq \mathcal{L}_{fin}(M)$.

iv) $\mathcal{L}_{fin}(CF) \subset \mathcal{L}_{fin}(IP) \subset \mathcal{L}_{fin}(M) \subset M$.

v) $\mathcal{L}_{fin}(M)$ *and* $\mathcal{L}(CF)$ *are incomparable.*

vi) For any $k \geq 1$, *all relations given in i) - iv) remain valid if we consider* $\mathcal{L}_k(X)$ *instead of* $\mathcal{L}_{fin}(X)$; *v) remains valid if we put* $k \geq 2$ *instead of* *fin.*

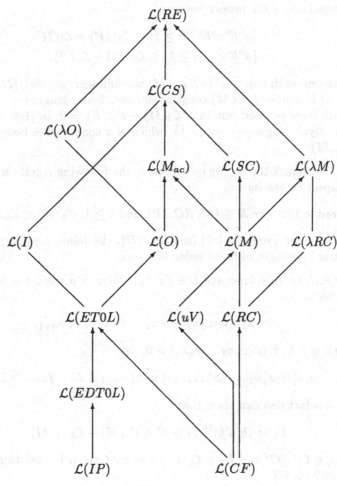

Fig. 1

Remarks on the proof. For the proof of i), ii) and iii) we refer to [84], [104], [105] and [27] where, in addition, the equality of $\mathcal{L}_{fin}(M)$ to $\mathcal{L}_{fin}(ET0L)$ and $\mathcal{L}_{fin}(EDT0L)$ (for finite index definition for Lindenmayer systems, see also [101]) has been proven. In some cases the simulations given in Sections 2.1–2.3 can be used.

iv) Since any sentential form contains a bounded number of nonterminals one can ensure that, for any nonterminal A, there is at most one occurrence of A in a sentential form. Then context-free and Indian parallel derivations coincide. Thus the first inclusion holds.

The second inclusion follows by adding special nonterminals which remember the number of occurrences of any nonterminal. By these nonterminals we can control that a matrix replaces all occurrences of some nonterminal according to the same rule and changes the additional nonterminal.

The inclusions are proper because

$$\{a^n b^n a^n b^n \mid n \geq 1\} \in \mathcal{L}_2(IP) - \mathcal{L}(CF),$$
$$\{a^n b^n c^n \mid n \geq 1\} \in \mathcal{L}_2(M) - \mathcal{L}(IP).$$

Moreover, each language in $\mathcal{L}_{fin}(M)$ is semilinear (see [88], [27]), whereas, in view of Example 7, $\mathcal{L}(M)$ contains non-semilinear languages.

v) We have pointed out that $\mathcal{L}_2(M) - \mathcal{L}(CF) \neq \emptyset$; in [106] it is proved that the Dyck language over $\{a, b\}$, which is a context-free language, is not in $\mathcal{L}_{fin}(M)$. □

By Theorem 2.16, vi), we only present the following result on hierarchies with respect to the index.

Theorem 2.17. *For $X \in \{M, RC, IP\}$ and $k \geq 1$, $\mathcal{L}_k(X) \subset \mathcal{L}_{k+1}(X)$.*

Remarks on the proof. In [87] (see also [27]) the following pumping lemma for matrix languages of finite index is shown:

For each infinite language $L \in \mathcal{L}_k(M)$, there is a word $z \in L$ which can be written as

$$z = u_1 v_1 w_1 x_1 u_2 v_2 w_2 x_2 \ldots u_n v_n w_n x_n u_{n+1},$$

such that $n \leq k$, $|v_1 x_1 v_2 x_2 \ldots v_n x_n| > 0$, and

$$u_1 v_1^i w_1 x_1^i u_2 v_2^i w_2 x_2^i \ldots u_n v_n^i w_n x_n^i u_{n+1} \in L, \quad for \ i \geq 1.$$

Using this fact one can show that

$$L_k = \{b(a^i b)^{2k} \mid i \geq 0\} \in \mathcal{L}_k(M) - \mathcal{L}_{k-1}(M).$$

Since $L_k \in \mathcal{L}_k(RC)$ and $L_k \in \mathcal{L}_k(IP)$ also hold, the other relations follow by Theorem 2.16, vi). □

For further types of grammars with controlled derivations, e.g. time-variant grammars ([107], vector grammars ([14]), valence grammars ([86]), grammars controlled by graphs ([123]), tree controlled grammars ([15]), Russian parallel grammars ([69]), k-limited L systems ([121], [122]), walk grammars ([91]), we refer to [27].

3. Basic properties

3.1 Operations on language families

In this section we present some results on the behaviour of some language families introduced in the preceding sections with respect to algebraic, set-theoretic and other operations. This is a subject well investigated in the theory of formal languages. Moreover, results of this area can be used in other fields, too.

In this section we consider the following operations (where we assume that the reader is familiar with the definitions):

- the classical set-theoretic operations of union, intersection and complement,
- the algebraic operations of concatenation (or product) and Kleene-closures,
- morphisms, their variations and generalizations and inverses of them,
- intersection by regular sets (a basic operation in the theory of abstract families of languages, see [40], [5]),
- quotient by a regular set and quotient by a letter.

We summarize the results in the following table. If the family $\mathcal{L}$ is closed (not closed) under the operation τ, then we write a "+" ("−") at the meet of the row associated with $\mathcal{L}$ and the column associated with τ; "?" denotes that it is not known whether or not $\mathcal{L}$ is closed under τ.

Theorem 3.1. *Table 1 holds.*

We omit the proofs. Most of them can be given by modifications of the methods used for the corresponding proofs in case of the families of the Chomsky hierarchy; details can be also found in [27]. Not presented in [27] are the non-closure of $\mathcal{L}(M)$ under gsm mappings and quotient by regular sets and the non-closure of $\mathcal{L}(M_{ac})$ under quotient by regular sets.

They can be proved as follows. Take a matrix grammar without appearance checking $G = (N, T, M, S, \emptyset)$ (the modifications for the appearance checking case are left to the reader). Without loss of generality, we may assume that the only S-rule appears in a matrix $(S \to A)$, for some symbol $A \in N$ (if necessary, we add such a matrix to M, for S a new axiom and A the former axiom of G). We construct the matrix grammar

$$G' = (N', T \cup \{c, d\}, M', S', \emptyset),$$

where

$$N' = N \cup \{S', X\} \cup \{\overline{A} \mid A \in N\},$$

c, d are new symbols, and M' is constructed as follows.

Table 1. Closure properties[1]

operation	SC	I	M_{ac}	λM	M	uV	λO	O	λRC	RC	IP
union	+	+	+	+	+	+	+	+	+	+	+
intersection	+	-	?	-	-	-	-	-	-	-	-
complementation	?	-	?	-	-	-	-	-	-	-	-
intersection by											
regular sets	+	+	+	+	+	+	+	+	+	+	?
concatenation	+	+	+	+	+	+	+	+	+	+	+
Kleene-closure	+	+	+	?	-	-	+	+	+	+	?
λ-free morphisms	+	+	+	+	+	+	+	+	+	+	+
morphisms	-	+	-	+	-	+	+	?	+	?	+
inverse morphisms	+	+	+	+	+	?	+	+	+	+	?
λ-free gsm-mappings	+	+	+	+	+	?	+	+	+	+	?
gsm-mappings	-	+	-	+	-	?	+	+	+	?	?
quotient by											
regular sets	-	+	-	+	-	?	+	+	+	?	?
quotient by letters	+	+	+	+	+	+	+	+	+	+	?

For a string $w \in (N \cup T)^*$, we define the set of pairs

$$\rho(w) = \{(\rho_1(w), \rho_2(w)) \mid w = w_1 B_1 w_2 B_2 \ldots w_k B_k w_{k+1}, k \geq 0,$$
$$w_i \in (N \cup T)^*, 1 \leq i \leq k+1, B_i \in N, 1 \leq i \leq k,$$
$$\rho_1(w) = w_1 w_2 \ldots w_{k+1}, \rho_1(w) \neq \lambda,$$
$$\rho_2(w) = \overline{B}_1 \ldots \overline{B}_k \}.$$

(Note that $(w, \lambda) \in \rho(w)$.)

For each matrix $(A_1 \rightarrow w_1, \ldots, A_r \rightarrow w_r) \in M, r \geq 1$, proceed as follows:

– if there is $i, 1 \leq i \leq r$, such that $A_i \rightarrow \lambda$, then replace this rule with $\overline{A}_i \rightarrow c$;
– if there is $i, 1 \leq i \leq r$, such that $A_i \rightarrow w_i$ has $w_i = B_i \in N$, then, nondeterministically, either leave the rule unchanged, or replace it with $\overline{A}_i \rightarrow \overline{B}_i$;
– each rule $A_i \rightarrow w_i$ which is not of the previous types is replaced by

$$A_i \rightarrow \rho_1(w_i), X \rightarrow X \rho_2(w_i),$$

for some $(\rho_1(w_i), \rho_2(w_i)) \in \rho(w_i)$.

Then, M' consists of all matrices obtained as above, plus

$$\{(S' \rightarrow SX), (X \rightarrow d)\}.$$

From the construction of M', one can easily see that the λ-rules in M are simulated in M' by rules introducing occurrences of the symbol c to the right-hand of X (in turn, X is eventually replaced by d). Therefore, $L(G') \subseteq L(G)dc^*$ and for each $w \in L(G)$ there is $i \geq 0$ such that

---

[1] Recently, H. Fernau and K. Reinhardt have proved that $\mathcal{L}(uV)$ is closed under inverse morphisms, λ-free gsm mappings and quotients by regular sets; therefore, the question marks on the uV column should be replaced by +.

$wdc^i \in L(G')$. Consequently, the closure of $\mathcal{L}(M)$ under right quotient with regular languages or under gsm mappings would imply $\mathcal{L}(\lambda M) \subseteq \mathcal{L}(M)$, a contradiction.

The non-closure of $\mathcal{L}(O), \mathcal{L}(\lambda O)$ under intersection and complement as well as the closure under gsm mappings are proven in [31]. $\qquad\qquad$ $\square$

3.2 Decision problems

In this subsection we study the status of decidability of some basic problems of the theory of formal languages, and in the case of decidability we present some results on the complexity of the problem.

First we mention that the equivalence problem is undecidable for all families of grammars defined in Section 2 because this problem is already undecidable for the family of linear languages which is a proper subset for all families we are interested in. An analogous statement holds for other problems which are undecidable for linear or context-free grammars.

We now consider a problem which is of special interest for such families of grammars, which give proper extensions of the family of context-free languages and is undecidable for all such families.

Theorem 3.2. *Let X be a family of grammars defined in Section 2. Then the problem*

> **Instance:** *grammar $G \in X$*
> **Answer:** *"Yes" if and only if G generates a context-free language.*

is undecidable.

Proof. Any family X considered in this statement satisfies the following properties:

- Any linear language is contained in $\mathcal{L}(X)$, and $\mathcal{L}(X)$ contains a non-context-free language.
- $\mathcal{L}(X)$ is closed under union and concatenation.

We now consider an arbitrary linear language $L_1 \subseteq V_1^*$ given by a linear grammar G_1 and a fixed language $L_2 \in \mathcal{L}(X) - \mathcal{L}(CF)$, $L_2 \subseteq V_2^*$, where $V_1 \cap V_2 = \emptyset$. By the (effective) closure under union and concatenation,

$$L = L_1 V_2^* \cup V_1^* L_2 \in \mathcal{L}(X)$$

and a grammar $G \in X$ can be effectively constructed such that $L = L(G)$.

If $L_1 = V_1^*$, then $L = V_1^* V_2^*$ is a regular (and hence a context-free) language.

If $L_1 \subset V_1^*$, then there is a word $w \in V_1^*$ such that $w \notin L_1$. Assume that $L \in \mathcal{L}(CF)$. Then, by the closure of $\mathcal{L}(CF)$ under intersection by regular sets, the language $L \cap \{w\}V_2^* = \{w\}L_2$ is context-free. Further, by the closure of

$\mathcal{L}(CF)$ under quotients by regular sets, L_2 is context-free. This contradicts the choice of L_2. Thus L is not context-free.

We have shown that L is context-free if and only if $L_1 = V_1^*$ holds. By the undecidability of the universality problem for linear languages, the statement of the theorem follows.   □

The situation with respect to the membership problem, the emptiness problem and the finiteness problem can differ from family to family, especially if we take into consideration complexity aspects.

Table 2 presents the results for the families given in Section 2 which do not generate language families of the Chomsky hierarchy. If a problem P is undecidable for a family X, we write a "–" at the meet of the row associated with X and the column associated with P. Analogously, "+" denotes that the problem is decidable and, moreover, all known algorithms are at least exponential in time. By "?" we denote that it is not known whether or not the problem is decidable. Furthermore, we mention the complexity class in which the problem is contained or give some remark on the complexity.

Theorem 3.3. *Table 2 holds.*

Table 2. Decidability properties

family	membership	emptiness	finiteness
I	NP-complete [100]	+ [2]	?
SC	+	– (A)	– (A)
λO	+, NP-hard (B)	?	?
O	+ [7], NP-hard (B)	+ [7]	?
M_{ac}	+ , NP-hard	– (A)	– (A)
λM	+ (C)	+ (C), NP-hard	+ [53] , NP-hard
M	+	+ , NP-hard	+ , NP-hard
λuV	$\in$ LOGCFL [111]	+ , NP-hard	+ , NP-hard
uV	$\in$ LOGCFL	+ , NP-hard (D)	+ , NP-hard (D)
RC	+	+ , NP-hard (D)	+ , NP-hard (D)
IP	$\in$ P (E)	$\in$ LIN-SPACE (E)	$\in$ LIN-SPACE (E)
kG	$\in$ P [45]	+	+
kSM	$\in$ DTIME(n^3) (F)	$\in$ DTIME(n^2) (F)	$\in$ DTIME(n^2) (F)

Remarks on the proof. We do not give a complete proof of the theorem. We only establish some statements to which we refer by a Latin letter in brackets. In the other cases we refer to a paper in the references. All those statements without some hint follow by the inclusions mentioned in Section 2.5.

(A). We give the proof only for matrix grammars with appearance checking and without erasing productions. The proof for scattered context grammars can be given analogously.

First we show the undecidability of the emptiness problem. Obviously, it is undecidable whether or not an arbitrarily given (type-0) grammar G generates the empty set. We first construct a matrix grammar G' with appearance checking and with erasing rules such that $L(G') = L(G)$. Then we construct the matrix grammar G'' with appearance checking, which is obtained from G' by replacing any production $A \to \lambda$ by $A \to c$ where c is an additional terminal letter. Then $L(G'')$ is empty if and only if $L(G')$ is empty, which holds if and only if $L(G)$ is empty.

If we construct G''' from G'' by adding the nonterminal S' and the matrices $(S' \to cS')$ and $(S' \to S)$, then $L(G''')$ is finite if and only if $L(G''')$ is empty if and only if $L(G'')$ is empty.

(B). The statement follows from the NP-completeness of the membership for ET0L systems (see [65]).

(C). We first prove the decidability of the emptiness problem by a reduction to the reachability problem for vector addition systems.

An n-dimensional *vector addition system* is a couple (x_0, K) where $x_0 \in \mathbf{N}^n$ and K is a finite subset of $\mathbf{Z}^n$. A vector $y \in \mathbf{N}^n$ is called reachable within the system (x_0, K) if and only if there are vectors $v_1, v_2, \ldots, v_t \in K$, $t \geq 1$, such that

$$x_0 + \sum_{i=1}^{j} v_i \in \mathbf{N}^n, \text{ for } 1 \leq j \leq t \quad \text{and} \quad x_0 + \sum_{i=1}^{t} v_i = y,$$

i.e. we can reach the vector y from x_0 by addition of some vectors of K such that all intermediate vectors only have non-negative components.

We shall use that the *reachability problem*:

Instance: an n-dimensional vector addition system (x_0, K) and a vector $y \in \mathbf{N}^n$

Answer: "Yes" if and only if y is reachable within (x_0, K)

is decidable (see [62], [72]). We note that the reachability problem requires exponential space (see [68]).

Let $G = (N, T, M, S, \emptyset)$ be a matrix grammar without appearance checking. Without loss of generality (see the proof of Theorem 2.2) we can assume that $N = N_1 \cup N_2$ and any matrix of M has one of the following forms:

(1) $(B \to w)$,

(2) $(B_1 \to w_1, B_2 \to w_2)$,   with $B_1 \in N_1, B_2 \in N_2$.

Let

$$N = \{A_1, A_2, \ldots, A_n\}.$$

For any matrix m of form (1), let

$$v(m) = (|w|_{A_1} - |B|_{A_1}, |w|_{A_2} - |B|_{A_2}, \ldots, |w|_{A_n} - |B|_{A_n}),$$

and for any matrix m of form (2), let

$$v(m) = (|w_1 w_2|_{A_1} - |B_1 B_2|_{A_1}, \ldots, |w_1 w_2|_{A_n} - |B_1 B_2|_{A_n}).$$

Further, for any word w, we set

$$p(w) = (|w|_{A_1}, |w|_{A_2}, \ldots, |w|_{A_n}).$$

One can easily see that $L(G)$ is non-empty if and only if the vector $(0, 0, \ldots, 0)$ is reachable in the vector addition system

$$(p(S), \{v(m) \mid m \in M\}).$$

First we recall that by the results mentioned in the last section, for any matrix grammar G without appearance checking and any word w, we can construct effectively a matrix grammar G' without appearance checking such that $L(G') = L(G) \cap \{w\}$. Moreover, for a matrix grammar G without appearance checking and a word w, $w \in L(G)$ if and only if $L(G) \cap \{w\} \neq \emptyset$. Thus $w \in L(G)$ holds if and only if $L(G')$ is not empty.

(D). We give a reduction of the 3-*partition problem* given by

Instance: a multiset $\{t_1, t_2, \ldots, t_{3m}\}$ of integers and an integer t
Answer: "Yes",

iff there is partition $\{Q_1, Q_2, \ldots, Q_m\}$ of $\{t_1, t_2, \ldots, t_{3m}\}$
such that $card(Q_i) = 3$ and $\sum_{s \in Q_i} s = t$, for $1 \leq i \leq m$.

which is **NP**-complete (see [37]), to the emptiness problem for unordered vector grammars and random context grammars. Let

$$U = \{(i, j, k) \mid 1 \leq i < j < k \leq 3m, t_i + t_j + t_k = t\}.$$

The reduction is done by the the (unordered) vector grammar

$$G = (\{S, A_1, A_2, \ldots, A_{3m}\}, \{a_1, a_2, \ldots, a_{3m}\}, M, S),$$

where

$$M = \{(S \to A_1 A_2 \ldots A_{3m})\} \cup \{(A_i \to a_i, A_j \to a_j, A_k \to a_k) \mid (i, j, k) \in U\}$$

and the random context grammar

$$G' = (N, \{a_1, a_2, \ldots, a_{3m}\}, P, S),$$

where

$$N = \{S\} \cup \{A_i \mid 1 \leq i \leq 3m\} \cup \{(i, u) \mid 1 \leq i \leq 3m, u \in U\}$$

and P consists of all rules of the form

$$(S \to A_1 A_2 \ldots A_{3m}, \emptyset, \emptyset),$$
$$(A_i \to (i, (i, j, k)), \emptyset, \emptyset), 1 \leq i \leq 3m, (i, j, k) \in U,$$
$$(A_j \to (j, (i, j, k)), \{(i, (i, j, k))\}, \emptyset), 1 \leq j \leq 3m, (i, j, k) \in U,$$
$$(A_k \to (k, (i, j, k)), \{(i, (i, j, k)), (j, (i, j, k))\}, \emptyset), 1 \leq k \leq 3m, (i, j, k) \in U,$$
$$((i, (i, j, k)) \to a_i, \{(j, (i, j, k)), (k, (i, j, k))\}, \emptyset), 1 \leq i \leq 3m, (i, j, k) \in U,$$
$$((j, (i, j, k)) \to a_j, \{(k, (i, j, k))\}, \emptyset), 1 \leq j \leq 3m, (i, j, k) \in U,$$
$$((k, (i, j, k)) \to a_k, \emptyset, \emptyset), 1 \leq j \leq 3m, (i, j, k) \in U.$$

Obviously, the sizes of G and G' are equal to $O(m^3)$ and $O(m^4)$, respectively, and hence they can be constructed in polynomial time. Moreover, if the given instance of the 3-partition problem has a solution, then

$$L(G) = L(G') = \{a_1 a_2 \ldots a_{3m}\};$$

and if the instance has no solution, then G and G' generate the empty set.

The modification for the finiteness problem is analogous to that given in **(A)**.

(E). By the proof of Theorem 2.9, for any Indian parallel grammar G, we can construct in polynomial time an EDT0L system G' which generates $L(G)$. The presented results are the (best) upper bounds known for the corresponding problems for EDT0L systems (see [51], [58], [59]).

(F). Given a simple matrix grammar G, we construct the context-free grammar G' according to the proof of Assertion 1 in the proof of Theorem 2.13. Then, for a word w, $w \in L(G)$ holds if and only if there is a $w' \in (f \circ \tau_k)^{-1}(w))$ such that $w' \in L(G')$. Thus we can decide whether or not $w \in L(G)$ in the following way: First we determine all elements of $(f \circ \tau_k)^{-1}(w)$ and this can be done in time $O(|w|^k)$; then, for any $w' \in (f \circ \tau_k)^{-1}(w)$, we check whether or not $w' \in L(G')$, which can be done in time $O(|w|^3)$, because $|w'| = |w|$. Therefore we can decide in time $O(|w|^{k+3})$ whether or not $w \in L(G)$.

Obviously, $L(G)$ is empty (finite) if and only if $L(G')$ is empty (finite). However, the construction of G' presented in the proof of Theorem 2.13 increases the size of the grammar considerably (the size of G' is of the order of the k-th power of the size of G). However, the statement on emptiness and finiteness remain true if with any production of G we associate only one production in G', i.e. instead using all elements of τ_k^{-1} we choose only one element. This modification ensures that G and G' have the same size. Thus the complexity of the emptiness and finiteness problems for simple matrix grammars are of the same order as that of the same problem for context-free grammars. □

3.3 Descriptional complexity

Formal grammars represent a finite description of (possibly) infinite languages. There the problem whether or not the description is efficient arises naturally. In this section we study the number of nonterminals as a parameter for the quality of the description. For context-free grammars and languages, this parameter has been studied intensively (see [47], [48]). We show that, in a certain sense, matrix and programmed grammar give a more concise description than random context grammars, and moreover, all these three types of grammars present more efficient descriptions than context-free grammars.

We start with the formal definition.

Definition 3.1.

i) For a grammar G, by $Var(G)$ we denote the cardinality of its set of non-terminals.

ii) Let X be a family of grammars. For a language $L \in \mathcal{L}(X)$, we define

$$Var_X(L) = \min\{Var(G) \mid G \in X, \ L(G) = L\}.$$

Theorem 3.4. *For any recursively enumerable language L,*

i) $Var_{\lambda P_{ac}}(L) - 2 \leq Var_{\lambda M_{ac}}(L) \leq Var_{\lambda P_{ac}}(L)$,

ii) $Var_{\lambda M_{ac}}(L) \leq Var_{\lambda RC_{ac}}(L) + 1$ and $Var_{\lambda P_{ac}}(L) \leq Var_{\lambda RC_{ac}}(L) + 2$.

Proof.

i) Let G be a matrix grammar such that $L = L(G)$ and $Var_{\lambda M_{ac}}(L) = Var(G)$. Then we construct the programmed grammar G' with $L(G') = L(G)$ as in the proof of Theorem 2.4. Thus

$$Var_{\lambda P_{ac}}(L) \leq Var(G') = Var(G) + 2 = Var_{\lambda M_{ac}}(L) + 2,$$

which proves the first relation.

Conversely, let $H = (N, T, \{p_1, p_2, \ldots, p_r\}, S)$ be a programmed grammar such that $L = L(G)$ and $Var_{\lambda P_{ac}}(L) = Var(H)$. For the i-th production $p_i = (A_i \to w_i, U_i, W_i)$, $1 \leq i \leq r$, we put

$$M_i = \{(\underbrace{C \to \lambda, C \to \lambda, \ldots, C \to \lambda}_{r \ times}, C \to C^{r+1}, A \to wC^j) \mid p_j \in U_i\} \cup$$

$$\cup \{(\underbrace{C \to \lambda, \ldots, C \to \lambda}_{r \ times}, C \to C^{r+1}, A \to C^{r+1}, B \to BC^j) \mid p_j \in W_i, B \in N\} \cup$$

$$\cup \{(\underbrace{C \to \lambda, C \to \lambda, \ldots, C \to \lambda}_{r \ times}, C \to C^{r+1}, A \to w\}$$

(the first group of matrices simulates the application of $A \to w$ in the presence of A; the second one simulates the application in the absence of A; in both cases we have to continue with an "application" of p_j; the third group simulates a terminating step of the derivation). Then we construct the matrix grammar

$$H' \ = \ (N \cup \{C\}, T, \{(C \to C^{r+1}, S \to C^k S) \mid 1 \leq k \leq r\} \cup$$
$$\bigcup_{i=1}^{r} M_i, S, \{C \to C^{r+1}, A \to C^{r+1}\}).$$

It is easy to see that $L(H') = L(H)$ holds, hence

$$Var_{\lambda M_{ac}}(L) \leq Var(H') = Var(H) + 1 = Var_{\lambda P_{ac}}(L) + 1$$

which proves the second relation.

ii) The first relation follows by the simulation of random context grammars by matrix grammars presented in the proof of Theorem 2.7.

The second relation follows by a modification of this construction.   □

By Theorem 3.4, the measures $Var_{\lambda M_{ac}}$ and $Var_{\lambda P_{ac}}$ differ only by a fixed constant. This situation does not hold with respect to random context grammars and programmed or matrix grammars as can be seen from the following theorem.

Theorem 3.5.

i) For any recursively enumerable language L,

$$Var_{\lambda M_{ac}}(L) \leq 6 \quad and \quad Var_{\lambda P_{ac}}(L) \leq 8.$$

ii) There is a sequence of recursively enumerable languages L_n, $n \geq 1$, such that

$$f(n) \leq Var_{\lambda RC_{ac}}(L_n) \leq [\log_2 n] + 3, \quad for \ n \geq 1$$

where f is an unbounded function from $\mathbf{N}$ into $\mathbf{N}$.

iii) For any integer $n \geq 1$, there is a language $L_n \in \mathcal{L}(IP)$ such that

$$Var_{IP}(L_n) = n.$$

For a proof we refer to [89], [17], [96], and [27], Section 4.

We mention that Theorem 3.5 states that the families of languages which can be generated by a bounded number of nonterminals form an infinite hierarchy, if we use random context grammars or Indian parallel grammars, whereas the hierarchy has at most 6 or 8 levels, if we use matrix or programmed grammars with appearance checking and erasing productions. It is an open question whether or not these bounds are tight. Some results on the position of the families of linear languages, metalinear languages and languages of finite index in these hierarchies can be found in [25] (see also [27], Section 4).

Further, we observe that, for context-free languages, there holds a statement analogous to Theorem 3.5 iii) (see [48]).

Hitherto we have only considered the relation between different types of control. Obviously, by the definitions the description of a context-free language by grammars with control (besides the Indian parallel grammars) is at least as efficient as the description by a context-free grammar (we only have to interpret the context-free grammar as a grammar with control). However, there are languages where the control allows much more efficient descriptions. Note that this already holds for control mechanisms which only generate subfamilies of $\mathcal{L}(CS)$.

Theorem 3.6. *There is a sequence of context-free languages L_n, $n \geq 1$, such that*

$$Var_{CF}(L_n) = n, \quad Var_M(L_n) \leq 4, \quad Var_P(L_n) = 1, \quad Var_{RC_{ac}}(L_n) \leq 8.$$

A proof can be given by the techniques presented in [47], [17], and [27], Section 4.

Finally we mention that analogous statements can be given for the number of productions. For details we refer to [25] and [27], Section 4.

4. Further topics

In the preceding sections we have only discussed grammars where the productions are context-free. Obviously, one can study grammars with control mechanism and right-linear or context-sensitive or monotone productions. One can show that in most cases the families of languages defined by such grammars with control are the families of regular or context-sensitive languages, again, i.e., there is no increase of the generative power by adding the control (see [27]). A notable exception are the valence grammars, see [86], [81], which can generate non-context-free languages using regular rules.

Pure grammars (see [71]) form the sequential counterpart of non-extended Lindenmayer systems. They are characterized by the following features: there is no distinction between terminals and nonterminals (i.e., any generated sentential form belongs to the language and there can exist productions for all letters), a sequential process of derivation is used (i.e., only one letter is replaced in one step of derivation) and a finite set of start words.

Pure versions of grammars with controlled derivations have been defined and investigated in [20] and [26]. We note that the hierarchy of the associated families of pure languages differs essentially from that given in Section 2.5. Besides the inclusions which directly follow from the definitions, different control mechanisms may lead to incomparable families of pure languages.

For a grammar, we define the degree of nondeterminism as the maximal number of different right-hand side of productions with the same letter on the left-hand side. In case of usual grammars (with and without control), one can prove that any language can be generated by a grammar whose degree of nondeterminism is at most 2 (see [6]). On the contrary, for pure versions with respect to the degree of nondeterminism we obtain infinite hierarchies for grammars with and without control (see [6]).

The use of a control is very natural for Lindenmayer systems by the biological motivation of these systems. For example: the change from a growing phase to a flowering phase or the change from light (day) to darkness (night) or the change of the seasons can be modelled by a control mechanism which ensures a certain sequence of tables in the derivation (this can be done as in the case of regularly controlled grammars); certain steps in the development of an organism are only possible, if certain cells are present or absent in it, and this can be modelled by the use of context conditions for the application of the tables.

The hierarchy of languages obtained by controlled Lindenmayer systems also differs from that in Section 2.5. If we consider non-extended tabled systems, programs are more powerful than matrices, and both these mechanisms generate language families which are incomparable with that obtained by random contexts. In the case of extended tabled systems, programs, matrices, and regular control languages do not increase the generative power whereas random contexts do. For details we refer to [41], [3], [83], [97], [116], [102], [103], [23], [27].

In [18], [19], and [21] deterministic tabled systems with some mechanism of control are presented such that the sequence or language equivalence is decidable.

Usually, grammars are considered as generating devices, however, one can consider an accepting mode, too, i.e. w belongs to a language *accepted* by a grammar G if there is a sequence $w_0, w_1, \ldots, w_n$ of words such that

$$w = w_0 \models w_1 \models w_2 \models \ldots \models w_{n-1} \models w_n = S,$$

where $v \models v'$ if and only if $v' \implies v$ in G. Obviously, for grammars of the Chomsky hierarchy, the accepting and generating mode define the same language.

In [7] and [8] it has been shown that, for grammars with controlled derivations, the situation can be different. For example, for programmed grammars with appearance checking and without erasing rules, the accepting mode is more powerful than the generating mode.

If $\mathcal{L}(X) \subseteq \mathcal{L}(Y)$ holds for two families X and Y of grammars, then $\mathcal{A}(X) \subseteq \mathcal{A}(Y)$ also holds for the associated families of languages obtained in the accepting mode. However, the hierarchy of language families accepted by grammars with control differs from that presented in Section 2.5 (see [7] and [8]).

All results and comments presented hitherto concern string generating devices. Some concepts of control which have been considered above are studied for graph and array grammars, too. We refer to [12], [28], [49], [117], [50], [67], [82], [35], [33].

Acknowledgement

The work of the second author has been supported by the Academy of Finland, Project 11281. Useful comments by Henning Fernau and Lucian Ilie are gratefully acknowledged.

References

[1]   S. Abraham, Some questions of phrase-structure grammars, *Comput. Linguistics*, 4 (1965), 61–70.

[2]   A. V. Aho, Indexed grammars – an extension of context-free grammars, *Journal of the ACM*, 15 (1968), 647–671.

[3]   P. R. J. Asveld, Controlled iteration grammars and full hyper-AFLs, *Inform. Control*. 34 (1976), 248–269.

[4]   Y. Bar-Hillel, E. Shamir, Finite-state languages: formal representations and adequacy problems, *Bull. of Res. Council of Israel*, 8F (1960), 155–166.

[5]   J. Berstel, *Transductions and Context-Free Languages*, Teubner, Stuttgart, 1979.

[6]   H. Bordihn, Pure languages and the degree of nondeterminism, *J. Inform. Process. Cybern.*, 28 (1992), 231–240.

[7]   H. Bordihn, H. Fernau, Accepting grammars with regulation, *Internat. J. Comp. Math.*, 53 (1994), 1–18.

[8]   H. Bordihn, H. Fernau, Accepting grammars and systems: an overview, *Proc. of Development in Language Theory Conf.*, Magdeburg, 1995, 199–208.

[9]   H. Bordihn, H. Fernau, Remarks on accepting parallel systems, *Intern. J. Computer Math.*, 56 (1995), 51–67.

[10] F. J. Brandenburg, On context-free grammars with regular control set, in : *Proc. 7th Conf. Automata and Formal Languages*, Salgotarjan, 1993.

[11] F. J. Brandenburg, Erasing in context-free control languages. Petri nets languages make the difference, manuscript, 1994.

[12] H. Bunke, Programmed graph grammars, in: *Graph Grammars and their Application to Computer Science and Biology, Lecture Notes in Computer Science* 73, Springer-Verlag, Berlin, 1982, 89–112.

[13] N. Chomsky, On certain formal properties of grammars, *Inform. Control*, 1 (1959), 91–112.

[14] A. B. Cremers, O. Mayer, On matrix languages, *Inform. Control*, 23 (1973), 86–96.

[15] K. Culik II, H. A. Maurer, Tree controlled grammars, *Computing*, 19 (1977), 129–139.

[16] C. Culy, The complexity of the vocabulary of Bambara, *Linguistic Inquiry*, 8 (1985), 345–351.

[17] J. Dassow, Remarks on the complexity of regulated rewriting, *Fundamenta Informaticae*, 7 (1984), 83–103.

[18] J. Dassow, On deterministic random context L systems in: *Proc. 2nd Frege Conf.*, Akademie-Verlag, Berlin, 1984, 309–313.

[19] J. Dassow, On compound Lindenmayer systems, in: *The Book of L* (G. Rozenberg, A. Salomaa, eds.), Springer-Verlag, Berlin, 1986, 75–86.

[20] J. Dassow, Pure grammars with regulated rewriting, *Rev. Roum. Math. Pures Appl.*, 31 (1988), 657–666.

[21] J. Dassow, On the sequence equivalence problems for 0L systems with controlled derivations, *Computers and Artificial Intelligence*, 7 (1988), 97–105.

[22] J. Dassow, Subregularly controlled derivations: context-free case, *Rostock Math. Colloq.*, 34 (1988), 61–70.

[23] J. Dassow, U. Fest, On regulated L systems, *Rostock Math. Kolloq.*, 25 (1984), 99–118.

[24] J. Dassow, H. Hornig, Conditional grammars with subregular conditions, in: *Proc. 2nd Internat. Conf. Words, Languages and Combinatorics*, Kyoto, World Scientific, Singapore, 1994, 71–86.

[25] J. Dassow, Gh. Păun, Further remarks on the complexity of regulated rewriting, *Kybernetika*, 21 (1985), 213–227.

[26] J. Dassow, Gh. Păun, Further remarks on pure grammars with regulated rewriting, *Rev. Roum. Math. Pures Appl.*, 31 (1986), 855–864.

[27] J. Dassow, Gh. Păun, *Regulated Rewriting in Formal Language Theory*, Akademie-Verlag, Berlin, and Springer-Verlag, Berlin, 1989.

[28] H. Ehrig, A. Habel, Graph grammars with application conditions, in: *The Book of L* (G. Rozenberg, A. Salomaa, eds.), Springer-Verlag, Berlin, 1986, 87–100.

[29] H. Fernau, On grammar and language families, *Fundamenta Informaticae*, 25 (1996), 17–34.

[30] H. Fernau, A predicate for separating language classes, *Bulletin EATCS*, 56 (1995), 96–97.

[31] H. Fernau, Closure properties of ordered languages, *Bulletin EATCS*, 58 (1996), 159–162.

[32] H. Fernau, On unconditional transfer, *Proc. 21th MFCS Conf.*, Cracow, Lecture Notes in Computer Science, 1113, Springer-Verlag, Berlin, 1996, 348–359.

[33] R. Freund, Control mechanisms on #-context-free array grammars, in: *Mathematical Aspects of Natural and Formal Languages* (Gh. Păun, ed.), World Scientific, Singapore, 1995, 97–136.

[34] R. W. Floyd, On the non-existence of a phrase-structure grammar for Algol 60, *Comm. of the ACM*, 5 (1962), 483–484.

[35] R. Freund, Gh. Păun, One-dimensional matrix grammars, *J. Inform. Process. Cybern.*, 29 (1993), 357–374.

[36] I. Fris, Grammars with a partial ordering of the rules, *Inform. Control*, 12 (1968), 415–425.

[37] M. R. Garey, D. J. Johnson, *Computers and Intractability. A Guide to the Theory of NP-completeness*, Freeman and Co., San Francisco, 1979.

[38] G. Gazdar, G. K. Pullum, Computationally relevant properties of natural languages and their grammars, *New Generation Computing*, 3 (1985), 273–306.

[39] S. Ginsburg, *The Mathematical Theory of Context-Free Languages*, McGraw-Hill, New York, 1966.

[40] S. Ginsburg, *Algebraic and Automata-Theoretic Properties of Formal Languages*, North-Holland, Amsterdam, 1975.

[41] S. Ginsburg, G. Rozenberg, T0L schemes and control sets, *Inform. Control*, 27 (1974), 109–125.

[42] S. Ginsburg, E. H. Spanier, Control sets on grammars, *Math. Syst. Th.*, 2 (1968), 159–177.

[43] J. Gonczarowski, E. Shamir, Pattern selector grammars and several parsing algorithms in the context-free style, *J. Comp. Syst. Sci.*, 30 (1985), 249–273.

[44] J. Gonczarowski, M. K. Warmuth, Applications of scheduling theory to formal language theory, *Theor. Comp. Sci.*, 37 (1985), 217–243.

[45] J. Gonczarowski, M. K. Warmuth, Scattered and context-sensitive rewriting, *Acta Informatica*, 20 (1989), 391–411.

[46] S. Greibach, J. E. Hopcroft, Scattered context grammars, *J. Comp. Syst. Sci.*, 3 (1969), 233–247.

[47] J. Gruska, On a classification of context-free languages, *Kybernetika*, 13 (1967), 22–29.

[48] J. Gruska, The descriptional complexity of context-free languages, in: *Proc. 2nd MFCS Conf.*, 1973, 71–83.

[49] A. Habel, *Hyperedge Replacement: Grammars and Languages*, Lecture Notes in Computer Science 643, Springer-Verlag, Berlin, 1992.

[50] A. Habel, R. Heckel, G. Taentzer, Graph grammars with negative application conditions, *Fundamenta Informaticae*, to appear.
[51] M. A. Harrison, *Introduction to Formal Language Theory*, Addison-Wesley, Reading, 1978.
[52] T. Harju, A polynomial recognition algorithm for the EDT0L languages, *Elektron. Informationsverarb. Kybern.*, 13 (1977), 169–177.
[53] D. Hauschildt, M. Jantzen, Petri net algorithms in the theory of matrix grammars, *Acta Informatica*, 31 (1994), 719–728.
[54] F. Hinz, J. Dassow, An undecidability result for regular languages and its application to regulated rewriting, *Bulletin EATCS*, 38 (1989), 168–173.
[55] J. E. Hopcroft, J. D. Ullman, *Introduction to Automata Theory, Languages, and Computation*, Addison-Wesley, Reading, 1979.
[56] G. Hotz, G. Pitsch, On parsing coupled context-free languages, *Theor. Comp. Sci.*, 161 (1996), 205–233.
[57] O. H. Ibarra, Simple matrix languages, *Inform. Control*, 17 (1970), 359–394.
[58] N. D. Jones, S. Skyum, Recognition of deterministic ET0L languages in logarithmic space, *Inform. Control*, 35 (1977), 177–181.
[59] N. D. Jones, S. Skyum, Upper bounds on the complexity of some problems concerning L systems, *Techn. rep. DAIMI PB-69*, University of Aarhus, Dept. of Comp. Sci., 1977.
[60] J. Kelemen, Conditional grammars: motivations, definitions and some properties, in: *Proc. Conf. Automata Languages and Math. Sciences*, Salgotarjan, 1984, 110–123.
[61] H. C. M. Kleijn, G. Rozenberg, Context-free like restrictions on selective rewriting, *Theor. Comp. Sci.*, 16 (1981), 237–269.
[62] S. R. Kosaraju, Decidability of reachability in vector addition systems, in: *Proc. 14th Symp. Theory of Computing*, 1982, 267–281.
[63] J. Kral, A note on grammars with regular restrictions, *Kybernetika*, 9 (1973), 159–161.
[64] W. Kuich, H. A. Maurer, Tuple languages, in: *Proc. Intern. Comp. Symp.*, Bonn, 1970, 881–891.
[65] J. van Leeuwen, The membership question for ET0L systems is polynomially complete, *Inform. Proc. Letters*, 3 (1975), 138–143.
[66] J. van Leeuwen (ed.), *Handbook of Theoretical Computer Science*, Elsevier, Amsterdam, 1994.
[67] I. Litovski, Y. Metivier, Computing with graph rewriting systems with priorities, *Theor. Comp. Sci.*, 115 (1993), 191–224.
[68] R. J. Lipton, The reachability problem requires exponential space, *Research rep. 62*, Yale University, Dept. of Comp. Sci., 1976.
[69] M. K. Levitina, On some grammars with global productions, *NTI*, Ser. 2, No. 3, (1972), 32–36
[70] A. Manaster Ramer, Uses and misuses of mathematics in linguistics, in: *Proc. Xth Congress on Natural and Formal Languages*, Sevilla, 1994.
[71] H. A. Maurer, A. Salomaa, D. Wood, Pure grammars, *Inform. Control*, 42 (1979), 103–141.
[72] E. W. Mayr, An algorithm for the general Petri net reachability problem, in: *Proc. 13th Symp. Theory of Computing*, 1981, 238–246.
[73] A. Meduna, Generalized forbidding grammars, *Intern. J. Computer Math.*, 36 (1990), 31–38.
[74] A. Meduna, Global context-conditional grammars, *J. Inform. Process. Cybern.*, 27 (1991), 159–165.

[75] A. Meduna, A trivial method of characterizing the family of recursively enumerable languages by scattered context grammars, *Bulletin EATCS*, 56 (1995), 104–106.

[76] A. Meduna, Matrix grammars under leftmost and rightmost restriction, in: *Mathematical Linguistics and Related Topics* (Gh. Păun, ed.), Ed. Academiei, Bucharest, 1995, 243–257.

[77] A. Meduna, Syntactic complexity of scattered context grammars, *Acta Informatica*, 32 (1995), 285–298.

[78] A. Meduna, A. Gopalaratnam, On semi-conditional grammars with productions having either forbidding or permitting conditions, *Acta Cybernetica*, 14 (1994), 307–323.

[79] A. Meduna, M. Sarek, C. Crooks, Syntactic complexity of regulated rewriting, *Kybernetika*, 30 (1994), 177–186.

[80] D. Milgram, A. Rosenfeld, A note on scattered context grammars, *Inform. Proc. Letters*, 1 (1971), 47–50.

[81] V. Mitrana, Valence grammars on a free generated group, *Bulletin EATCS*, 47 (1992), 174–179.

[82] U. Montanari, Separable graphs, planar graphs and web grammars, *Inform. Control*, 16 (1970), 243–267.

[83] M. Nielsen, E0L systems with control devices, *Acta Informatica*, 4 (1975), 373–386.

[84] Gh. Păun, On the family of finite index matrix languages, *J. Comp. Syst. Sci.*, 18 (1979), 267–280.

[85] Gh. Păun, Some further remarks on the family of finite index matrix languages, *RAIRO, Informatique theorique*, 13 (1979), 289–297.

[86] Gh. Păun, A new generative device: valence grammars, *Rev. Roum. Math. Pures Appl.*, 25 (1980), 911–924.

[87] Gh. Păun, An infinite hierarchy of matrix languages, *Stud. Cerc. Mat.*, 32 (1980), 697–707.

[88] Gh. Păun, *Matrix Grammars* (in Romanian), The Scientific and Encyclopaedic Publ. House, Bucharest, 1981.

[89] Gh. Păun, Six nonterminals are enough for generating each r.e. language by a matrix grammar, *Intern. J. Comp. Math.*, 15 (1984), 23–37.

[90] Gh. Păun, A variant of random context grammars: semi-conditional grammars, *Theor. Comp. Sci.*, 41 (1985), 1–17.

[91] Gh. Păun, Some recent restriction in the derivation of context-free grammars, in: *Proc. 4th IMYCS, Lecture Notes in Computer Science* 281, Springer-Verlag, Berlin, 1986, 96–108.

[92] P. M. Postal, Limitations of phrase structure grammars, in: *The Structure of Language: Readings in the Philosophy of Languages* (J. A. Fodor, J. J. Katz, eds.), Prentice-Hall, Englewood Cliffs, N. J., 1964, 137–151.

[93] O. Rambow, Imposing vertical context conditions on derivations, in: *Proc. 2nd Conf. Developments in Language Theory*, Magdeburg, 1995.

[94] O. Rambow, G. Satta, A rewriting system for natural language syntax that is non-local and mildly context-sensitive, in: *Current Issues in Mathematical Linguistics* (C. Martin-Vide, ed.), North-Holland, Amsterdam, 1994, 121–130.

[95] D. Radzinski, Unbounded syntactic copying in Mandarin Chinese, *Linguistics and Philosophy*, 13 (1990), 113–127.

[96] B. Reichel, A remark on some classifications of Indian parallel languages, in: *Machines, Languages and Complexity, Lecture Notes in Computer Science* 381, Springer-Verlag, Berlin, 1989, 55–63.

[97] P. J. A. Reusch, Regulierte T0L Systeme, *Techn. rep. No. 81*, Gesellsch. f. Math. u. Datenverarbeitung, Bonn, 1974.

[98] G. E. Revesz, *Introduction to Formal Language Theory*, McGraw-Hill, New York, 1983.

[99] D. J. Rosenkrantz, Programmed grammars and classes of formal languages, *Journal of the ACM*, 16 (1969), 107–131.

[100] W. C. Rounds, Complexity of recognition in intermediate-level languages, in: *Proc. 14th IEEE Symp. Switching and Automata Theory*, 1973, 145–158.

[101] G. Rozenberg, A. Salomaa, *The Mathematical Theory of L Systems*, Academic Press, New York, 1980.

[102] G. Rozenberg, S. H. von Solms, Some aspects of random context grammars, *Informatik-Fachberichte*, 5 (1976), 63–75.

[103] G. Rozenberg, S. H. von Solms, Priorities on context conditions in rewriting systems, *Inform. Sci.*, 14 (1978), 15–51.

[104] G. Rozenberg, D. Vermeir, On the effect of the finite index restriction on several families of grammars I, *Inform. Control*, 39 (1978), 284–302.

[105] G. Rozenberg, D. Vermeir, On the effect of the finite index restriction on several families of grammars II, *Found. Control Engineering*, 3 (1978), 125–142.

[106] B. Rozoy, The Dyck language $D_1'^*$ is not generated by any matrix grammar of finite index, *Inform. Control*, 74 (1987), 64–89.

[107] A. Salomaa, Periodically time-variant context-free grammars, *Inform. Control*, 17 (1970), 294–211.

[108] A. Salomaa, *Formal Languages*, Academic Press, New York, 1973.

[109] K. Salomaa, Hierarchy of k-context-free languages I, *Intern. J. Comp. Math.*, 26 (1989), 69–90.

[110] K. Salomaa, Hierarchy of k-context-free languages II, *Intern. J. Comp. Math.*, 26 (1989), 193–205.

[111] G. Satta, The membership problem for unordered vector languages, in: *Proc. 2nd Conf. Developments in Language Theory*, Magdeburg, 1995, 267–275.

[112] S. M. Shieber, Evidence against the context-freeness of natural languages, *Linguistics and Philosophy*, 8 (1985), 333–343.

[113] E. Shamir, C. Beeri, Checking stacks and context-free programmed grammars accept P-complete languages, *Lecture Notes Computer Science*, 14 Springer-Verlag Berlin, 1974, 27–33.

[114] R. Siromoney, K. Krithivasan, Parallel context-free grammar, *Inform. Control*, 24 (1974), 155–162.

[115] S. Skyum, Parallel context-free languages, *Inform. Control*, 26 (1974), 280–285.

[116] S. H. von Solms, Some notes on ET0L languages, *Intern. J. Comp. Math.*, 5 (1976), 285–296.

[117] S. H. von Solms, Node-label controlled graph grammars with context conditions, *Intern. J. Comp. Math.*, 15 (1984), 39–49.

[118] R. Stiebe, *Kantengrammatiken*, Dissertation, Magdeburg Univ., 1996.

[119] E. D. Stotskij, Remarks on a paper by M. K. Levitina, *NTI*, Ser. 2, No. 4 (1972), 40–45.

[120] A. P. J. van der Walt, Random context grammars, in: *Proc. Symp. on Formal Languages*, 1970.

[121] D. Wötjen, k-limited OL systems and languages, *J. Inform. Process. Cybern.*, 24 (1988), 267–288.

[122] D. Wötjen, On k-uniformly-limited T0L systems and languages, *J. Inform. Process. Cybern.*, 26 (1990), 229–238.

[123] D. Wood, Bicolored digraph grammar systems, *RAIRO, Rech. Oper.*, R-1 (1973), 45–50.

Grammar Systems

Jürgen Dassow, Gheorghe Păun, and Grzegorz Rozenberg

1. Introduction

In classic formal language and automata theory, grammars and automata were modeling classic computing devices. Such devices were "centralized" – the computation was accomplished by one "central" agent. Hence in classic formal language theory a language is generated by *one* grammar or recognized by *one* automaton. In modern computer science distributed computation plays major role. Analyzing such computations in computer networks, distributed data bases, etc., leads to notions such as distribution, parallelism, concurrency, and communication. The theory of grammar systems was developed as a grammatical model for distributed computation where these notions could be defined and analyzed.

A grammar system is a *set* of grammars, working together, according to a specified protocol, to generate *one* language. There are many reasons to consider such a generative mechanism: to model distribution, to increase the generative power, to decrease the (descriptional) complexity,... The crucial element here is the protocol of cooperation. The theory of grammar systems may be seen as the "grammatical" theory of cooperation protocols. The central problems are the "functioning" of systems under specific protocols, and the influence of various protocols on various properties of systems considered.

One can distinguish two basic classes of grammar systems: *sequential* and *parallel*.

A *cooperating distributed* (CD) grammar system is sequential. Here all component grammars have a common sentential form. Initially, this is a common axiom. At each moment only one grammar is active – it rewrites the current sentential form. The matters such as which component grammar can become active at a given moment, and when an active grammar becomes inactive leaving the current sentential form to the other component grammars, are determined by the cooperation protocol. Examples of stop conditions (becoming inactive) are: the active component has to work for exactly k steps, at least k steps, at most k steps, or the maximal number of steps (a step means an application of a rewriting rule). Many other stop conditions are considered in the literature – we will consider some of them in this chapter. The language of terminal strings generated in this way is the language generated by the system.

In a *parallel communicating* (PC) grammar system, each component has its own sentential form. Within each time unit (there is a clock common to all component grammars) each component applies a rule, rewriting its own sentential form. The key feature of a PC grammar system is its "communication through queries" mechanism. Special (query) symbols are provided with each symbol pointing to a component of the system. When a component i introduces (generates) the query symbol Q_j, then the current sentential form of the component j will be sent to the component i, replacing all the occurrences of Q_j. One component of the system is designated as the *master*, and the language it generates is the language generated by the system. Several variants of the communication mechanism can be considered. They are determined by things such as the shape of the communication graph, or the action a component has to perform after communicating. We will consider many variants in this chapter.

In the architecture of a CD grammar system one can recognize the structure of a blackboard model, as used in problem-solving area [99]. The common sentential form is the blackboard (the common data structure containing the current state of the problem to be solved), the component grammars are the knowledge sources (agents, processors, procedures, etc.) contributing to solving the problem by changing the contents of the blackboard according to their competences, the protocol of cooperation encodes the control of the knowledge sources (for instance, the sequence in which the knowledge sources can contribute to the solution).

This was the explicit motivation of [15], the paper where the study of CD grammar systems in the form we consider here was initiated. (In the late 1980s other papers discussed the connection between the blackboard model and cooperating grammar systems; see, e.g., [13], [21].) The explicit notion of a "cooperating grammar system" has been introduced in [82], with motivations related to two-level grammars. A somewhat similar idea appears in [1], where one considers compound and serial grammars; this latter notion is a particular type of CD grammar systems. The main aim of [1] was to increase the power of (matrix) grammars; a similar motivation is behind the modular grammars of [4] which considers time varying grammars as the basic model.

PC grammar systems were introduced in [124], as a grammatical model of *parallelism* in the broad sense.

In [17], a modification of the blackboard model has been proposed, leading to problem-solving architectures similar to the structure of PC grammar systems. In this model

- each agent (knowledge source, processor, grammar) has its own "notebook" containing the description of a particular subproblem of a given problem which the whole group of agents has to solve,

– each agent operates only on its own "notebook" and one distinguished agent operates the "blackboard"; this distinguished agent has the description of the whole problem (in particular this agent decides when the problem is solved),
– the agents communicate on request the contents of their "notebooks".

This model is called in [17] *the classroom model* of problem solving. The classroom leader (the "master") operates the blackboard, the pupils have particular problems to solve, using their own notebooks, they may communicate with each other either in each step of the problem solving, or only when the leader requests this. The global problem is solved through such a cooperation on the blackboard.

Information about the early development of CD and PC grammar systems is given in [32] and [131], respectively. The first chapter of [17] discusses in some detail many relationships of CD and PC grammar systems to issues related to distribution, cooperation, parallelism in artificial intelligence, cognitive psychology, robotics, complex systems study, etc. Almost all results on grammar systems published until the middle of 1992 are presented in [17] (hence we will often refer to [17]).

There is a number of important topics concerning grammar systems that are not discussed in [17]. Among such topics are the applications of grammar systems to the study of natural languages, to the modeling of ecosystem-like societies of interrelated agents "living" in a common environment (see, e.g., [24] and the contributions to [111]), and to DNA computing, [20], [35].

2. Formal language prerequisites

For an alphabet V, we denote by V^* the free monoid generated by V under the operation of concatenation; the empty string is denoted by λ and $V^+ = V^* - \{\lambda\}$. The length of $x \in V^*$ is denoted by $|x|$. If $x \in V^*$ and $U \subseteq V$, then $|x|_U$ is the number of occurrences in x of symbols from U (the length of the string obtained by erasing from x all symbols in $V - U$).

A Chomsky grammar is denoted by $G = (N, T, S, P)$, where N is the nonterminal alphabet, T is the terminal alphabet, $S \in N$ is the axiom, and P is the set of rewriting rules (written in the form $u \to v$, $u, v \in (N \cup T)^*$, $|u|_N \geq 1$). The direct derivation step is defined by

$$x \Longrightarrow y \text{ iff } x = x_1 u x_2 \text{ and } y = x_1 v x_2 \text{ for some } u \to v \in P.$$

Denoting by $\Longrightarrow^*$ the reflexive and transitive closure of the relation $\Longrightarrow$, the language generated by G is defined by

$$L(G) = \{x \in T^* \mid S \Longrightarrow^* x\}.$$

We may also write $x \Longrightarrow_P y$ and $\Longrightarrow_P^*$ if P is not clear from the context of considerations.

We denote by *REG, LIN, CF, CS,* and *RE* the families of regular, linear, context-free, context-sensitive, and recursively enumerable languages, respectively. By *MAT* we denote the family of languages generated by matrix grammars with λ-free context-free rules and without appearance checking. We use subscript *ac* and superscript λ to denote that either appearance checking or λ-rules or both are allowed; in this way we get MAT_{ac}, MAT^{λ} and MAT^{λ}_{ac}, respectively. We use *ET0L* and *EDT0L* to denote the families of languages generated by ET0L systems and by EDT0L systems, respectively.

Regulated rewriting is treated in depth in [41], and [129] is a good reference for L systems. As a general reference for formal language theory we use [130]. Of course, other chapters of this handbook can be also useful as a reference for various topics from formal language theory.

3. CD grammar systems

3.1 Definitions

Definition 3.1. *A CD grammar system of degree* $n, n \geq 1$, *is a construct*

$$\Gamma = (N, T, S, P_1, \ldots, P_n),$$

where N, T *are disjoint alphabets,* $S \in N$, *and* $P_1, \ldots, P_n$ *are finite sets of rewriting rules over* $N \cup T$.

The elements of N *are nonterminals, those of* T *are terminals;* $P_1, \ldots, P_n$ *are called components of the system.*

Returning to the blackboard systems motivation, the components correspond to the agents solving the problem on the blackboard; any rule represents an action of the agent which results in a possible change on the blackboard. The axiom S is the formal counterpart of the problem on the blackboard in the beginning. The alphabet T contains the letters which correspond to such knowledge pieces which are accepted as solutions/part of solutions. The nonterminals represent "questions" to be answered; those introduced by a component and rewritten by another one can be interpreted as questions formulated by the one component and answered by the other (thus, the components can communicate through messages inserted in the current state of the solution, as encoded by the sentential form which is the contents of the blackboard).

If we want to specify explicitly *grammars* as components of a CD grammar system as above, then we can write Γ in the form $\Gamma = (N, T, S, G_1, \ldots, G_n)$, where $G_i = (N, T, S, P_i), 1 \leq i \leq n$.

Definition 3.2. *Let* $\Gamma = (N, T, S, P_1, \ldots, P_n)$ *be a CD grammar system.*
1. *For each* $i \in \{1, \ldots, n\}$, *the terminating derivation by the i-th component, denoted* $\Longrightarrow^t_{P_i}$, *is defined by*

$$x \Longrightarrow^t_{P_i} y \text{ iff } x \Longrightarrow^*_{P_i} y \text{ and there is no } z \in V^* \text{ with } y \Longrightarrow_{P_i} z.$$

2. *For each* $i \in \{1, \ldots, n\}$, *the* k-*steps derivation by the* i-*th component, denoted* $\Longrightarrow_{P_i}^{=k}$, *is defined by*

$$x \Longrightarrow_{P_i}^{=k} y \quad \text{iff} \quad \text{there are } x_1, \ldots, x_{k+1} \in (N \cup T)^* \text{ such that}$$
$$x = x_1, y = x_{k+1}, \text{ and, for each } 1 \leq j \leq k,$$
$$x_j \Longrightarrow_{P_i} x_{j+1}.$$

3. *For each* $i \in \{1, \ldots, n\}$, *the at most* k-*steps derivation by the* i-*th component, denoted* $\Longrightarrow_{P_i}^{\leq k}$, *is defined by*

$$x \Longrightarrow_{P_i}^{\leq k} y \text{ iff } x \Longrightarrow_{P_i}^{=k'} y \text{ for some } k' \leq k.$$

4. *For each* $i \in \{1, \ldots, n\}$, *the at least* k-*steps derivation by the* i-*th component, denoted* $\Longrightarrow_{P_i}^{\geq k}$, *is defined by*

$$x \Longrightarrow_{P_i}^{\geq k} y \text{ iff } x \Longrightarrow_{P_i}^{=k'} y \text{ for some } k' \geq k.$$

The normal $*$-mode of derivation, $\Longrightarrow_{P_i}^*$, describes the case where the agent works at the blackboard as long as it wants. The t-mode of derivation corresponds to the strategy where an agent *has* to contribute to the solving process on the blackboard as long as it can do it (maximal use of its competence). The $= k$ mode of derivation corresponds to k direct derivation steps in succession using rules of the i-th component, and this corresponds to k actions on the blackboard by one of the agents. Then, the $\leq k$-derivation mode corresponds to a time limitation, since the agent can perform at most k changes, and the $\geq k$-derivation mode requires that the agent performs at least k actions, thus this derivation mode asks for a certain minimal competence of the agent.

Let $D = \{*, t\} \cup \{\leq k, = k, \geq k \mid k \geq 1\}$.

Definition 3.3. *The language generated by a CD grammar system* $\Gamma = (N, T, S, P_1, \ldots, P_n)$ *in the derivation mode* $f \in D$ *is*

$$L_f(\Gamma) = \{w \in T^* \mid S \Longrightarrow_{P_{i_1}}^f w_1 \Longrightarrow_{P_{i_2}}^f \ldots \Longrightarrow_{P_{i_m}}^f w_m = w,$$
$$m \geq 1, 1 \leq i_j \leq n, 1 \leq j \leq m\}.$$

In the above several languages are associated with Γ, using the *stop conditions* in D. A component P_i may start working (gets enabled) on a sentential form w whenever w contains an occurrence of the left-hand side of a production from P_i; which of the enabled components "gets" the current sentential form is decided by the nondeterministic choice. One can consider also various *start conditions*. For instance, a component is enabled on a sentential form only if certain "random context conditions" are satisfied, or when a more general predicate is true for the sentential form, or if an external control (for example, a graph specifying the sequence of components enabling) decides so. These, and many other variants are discussed in [17]. Here we confine ourselves to the basic model and to some extensions that do not appear in [17].

3.2 Examples

Consider the following CD grammar systems:

$\Gamma_1 = (\{S, A, A', B, B'\}, \{a, b, c\}, S, P_1, P_2),$

$\qquad P_1 = \{S \rightarrow S, S \rightarrow AB, A' \rightarrow A, B' \rightarrow B\},$

$\qquad P_2 = \{A \rightarrow aA'b, B \rightarrow cB', A \rightarrow ab, B \rightarrow c\}.$

$\Gamma_2 = (\{S, A\}, \{a\}, S, P_1, P_2, P_3),$

$\qquad P_1 = \{S \rightarrow AA\},$

$\qquad P_2 = \{A \rightarrow S\},$

$\qquad P_3 = \{A \rightarrow a\},$

$\Gamma_3 = (\{S, A_1, \ldots, A_k, A'_1, \ldots, A'_k\}, \{a, b\}, S, P_1, P_2),$

$\qquad P_1 = \{S \rightarrow S, S \rightarrow A_1 b A_2 b \ldots b A_k b\} \cup \{A'_i \rightarrow A_i \mid 1 \leq i \leq k\},$

$\qquad P_2 = \{A_i \rightarrow aA'_i a, A_i \rightarrow aba \mid 1 \leq i \leq k\}, \ k \geq 1,$

$\Gamma_4 = (\{S, S'\} \cup \{A_i, A'_i, A''_i \mid 1 \leq i \leq k\}, \{a, b\}, S, P_0, P_1, \ldots, P_{3k}),$

$\qquad P_0 = \{S \rightarrow S', S' \rightarrow A_1 b A_2 b \ldots b A_k b\},$

$\qquad P_1 = \{A_1 \rightarrow A_1, A_1 \rightarrow aA'_1 a\},$

$\qquad P_{i+1} = \{A'_i \rightarrow A''_i, A_{i+1} \rightarrow aA'_{i+1} a\}, \ 1 \leq i \leq k - 1,$

$\qquad P_{k+1} = \{A'_k \rightarrow A''_k, A''_1 \rightarrow A'_1\},$

$\qquad P_{k+i+1} = \{A'_i \rightarrow A_i, A''_{i+1} \rightarrow A'_{i+1}\}, \ 1 \leq i \leq k - 2,$

$\qquad P_{2k} = \{A'_{k-1} \rightarrow A_{k-1}, A''_k \rightarrow A_k\},$

$\qquad P_{2k+i} = \{A_i \rightarrow A_i, A_i \rightarrow aba\}, \ 1 \leq i \leq k, k \geq 2,$

$\Gamma_5 = (\{S, A, A'\}, \{a, b\}, S, P_1, P_2, P_3),$

$\qquad P_1 = \{S \rightarrow S, S \rightarrow AA, A' \rightarrow A\},$

$\qquad P_2 = \{A \rightarrow aA', A \rightarrow a\},$

$\qquad P_3 = \{A \rightarrow bA', A \rightarrow b\}.$

The reader may verify that:

$\qquad L_f(\Gamma_1) = \{a^n b^n c^m \mid m, n \geq 1\}, \ f \in \{= 1, \geq 1, *, t\} \cup \{\leq k \mid k \geq 1\},$

$\qquad L_{=2}(\Gamma_1) = L_{\geq 2}(\Gamma_1) = \{a^n b^n c^n \mid n \geq 1\},$

$\qquad L_{=k}(\Gamma_1) = L_{\geq k}(\Gamma_1) = \emptyset, \ k \geq 3,$

$\qquad L_t(\Gamma_2) = \{a^{2^n} \mid n \geq 1\},$

$\qquad L_{=k}(\Gamma_3) = L_{\geq k}(\Gamma_3) = \{(a^n b)^{2k} \mid n \geq 1\},$

$\qquad L_{=2}(\Gamma_4) = L_{\geq 2}(\Gamma_4) = \{(a^n b)^{2k} \mid n \geq 1\},$

$\qquad L_{=2}(\Gamma_5) = L_{\geq 2}(\Gamma_5) = \{ww \mid w \in \{a, b\}^+\}.$

Let us consider in some detail the way that Γ_1 works in the mode $= 2$.

We have to start from S. Only P_1 can be used. Applying twice the rule $S \rightarrow S$ changes nothing, hence eventually we shall perform the step

$$S \Longrightarrow_{P_1} S \Longrightarrow_{P_1} AB.$$

From now on, S will never appear again. Only P_2 can be applied to AB. If we use the nonterminal rules, we get

$$AB \Longrightarrow_{P_2} aA'bB \Longrightarrow_{P_2} aA'bcB'.$$

In general, from a string of the form $a^i Ab^j c^k B$ (initially we have $i = j = k = 0$), we can obtain in this way

$$a^i Ab^j c^k B \Longrightarrow_{P_2}^{=2} a^{i+1} A'b^{j+1} c^{k+1} B'.$$

To such a string we have to apply P_1 again and then we get

$$a^{i+1} A'b^{j+1} c^{k+1} B' \Longrightarrow_{P_1}^{=2} a^{i+1} Ab^{j+1} c^{k+1} B.$$

This is the only possibility of using P_1. However, P_2 can be applied to a string $a^i Ab^j c^k B$ in the $= 2$ mode also using only one nonterminal rule (replacing either A or B by A' or B', respectively), and one terminal rule (removing the remaining symbol A or B). To a string containing only one nonterminal (which is different from S), none of the two components can be applied. Consequently, we have to use, in turn, the first component and the nonterminal rules of the second one, and we have to finish the derivation by using the terminal rules of P_2. This means that all derivations in Γ_1 in the mode $= 2$ are of the form

$$S \Longrightarrow_{P_1}^{=2} AB \Longrightarrow_{P_2}^{=2} aA'bcB' \Longrightarrow_{P_1}^{=2} aAbcB \Longrightarrow_{P_2}^{=2} \ldots$$
$$\Longrightarrow_{P_2}^{=2} a^n A'b^n c^n B' \Longrightarrow_{P_1}^{=2} a^n Ab^n c^n B \Longrightarrow_{P_2}^{=2} a^{n+1} b^{n+1} c^{n+1},$$

for some $n \geq 0$. Hence, $L_{=2}(\Gamma_1) = \{a^n b^n c^n \mid n \geq 1\}$.

The following observations are easy to verify: (1) the mode ≥ 2 is identical in this case with the mode $= 2$, (2) for $f \in \{= 1, \geq 1, *, t\} \cup \{\leq k \mid k \geq 1\}$ every component, when enabled, may use only one rule, which implies that $L_f(\Gamma_1) = L(G)$ for $G = (N, T, S, P_1 \cup P_2)$, and (3) for $= k, \geq k$ with $k \geq 3$ the only possible derivations are of the form $S \Longrightarrow_{P_1}^* S \Longrightarrow \ldots$, never leading to a terminal string.

The maximal mode of using the components of Γ_2 means that at every step all current nonterminals are rewritten by the unique rule of each component, hence the possible derivations are of the form

$$S \Longrightarrow_{P_1}^t A^2 \Longrightarrow_{P_2}^t S^2 \Longrightarrow_{P_1}^t A^4 \Longrightarrow_{P_2}^t S^4 \Longrightarrow_{P_1}^t \ldots \Longrightarrow_{P_1}^t A^{2^n} \Longrightarrow_{P_3}^t a^{2^n}.$$

The reader may wish to examine the working of the systems Γ_3, Γ_4. They are quite instructive when comparing the complexity of the generated languages with the complexity of the systems (where the size is the number of components), or with derivation modes (where one considers the number of rewritings done at each step). Through such examples one also observes that no usual pumping lemmas, with a fixed number of pumped positions, can be found for CD grammar systems working in the modes $t, = k, \geq k$, for $k \geq 2$. This makes the task of finding examples outside various classes of languages considerably more difficult (such counterexamples are needed, e.g., when establishing negative closure properties).

Note that well-known examples of non-context-free languages, such as $\{a^n b^n c^n \mid n \geq 1\}$, $\{ww \mid w \in \{a,b\}^+\}$ and $\{a^n b^m a^n b^m \mid n, m \geq 1\}$ can be easily generated by CD grammar systems. Moreover, they can be generated in a "context-free-like" manner, with a clear correspondence between the used nonterminals and the "blocks" they generate in the strings of a language considered. This is convenient when interpreting the CD grammar systems involved from the point of view of motivations coming from various features of natural and artificial languages (e.g., multiple agreements, duplication, and long distance crossed agreements).

3.3 On the generative capacity

It is easily seen that for CD grammar systems working in any of the modes discussed above and having regular, linear, context-sensitive, or type-0 components, the generative power does not change, i.e., they generate the families of regular, linear, context-sensitive, or recursively enumerable languages, respectively. Therefore, CD grammar systems with such components are of no further interest to us as far as the generative power is concerned. For this reason from now on we consider only CD systems with context-free rules.

We denote by $CD_n(f)$ the family of languages generated by λ-free (context-free) CD grammar systems of degree at most $n, n \geq 1$, working in the derivation mode $f \in D$. When the number of components is not limited, we replace n by ∞. The union of all families $CD_\infty(= k)$ and the union of all families $CD_\infty(\geq k), k \geq 1$, are denoted by $CD_\infty(=)$ and $CD_\infty(\geq)$, respectively. When the λ-rules are allowed, we add the superscript λ obtaining $CD_n^\lambda(f), CD_\infty^\lambda(f)$, etc.

The following theorem summarizes the basic results concerning the generative power of CD grammar systems.

Theorem 3.1.
(i) $CD_\infty(f) = CF$, for all $f \in \{= 1, \geq 1, *\} \cup \{\leq k \mid k \geq 1\}$.
(ii) $CF = CD_1(f) \subset CD_2(f) \subseteq CD_r(f) \subseteq CD_\infty(f) \subseteq MAT$, for all $f \in \{= k, \geq k \mid k \geq 2\}, r \geq 3$.
(iii) $CD_r(= k) \subseteq CD_r(= sk)$, for all $k, r, s \geq 1$.
(iv) $CD_r(\geq k) \subseteq CD_r(\geq k + 1)$, for all $r, k \geq 1$.
(v) $CD_\infty(\geq) \subseteq CD_\infty(=)$.
(vi) $CF = CD_1(t) = CD_2(t) \subset CD_3(t) = CD_\infty(t) = ET0L$.
(vii) Except for the inclusion $CD_\infty(f) \subseteq MAT$ (which must be replaced with $CD_f^\lambda(f) \subseteq MAT^\lambda$), all the previous relations are true also for CD grammar systems with λ-rules.

Two surprising relations are those in (vi): $CD_2(t) \subseteq CF$ and $CD_\infty(t) \subseteq CD_3(t)$ (the hierarchy on the number of components collapses in the case of the t mode of derivation). Proofs can be found in [15], [4], [17]. We present here the proof of the second inclusion above, in a form similar to that in [4] (in [15] one starts from ET0L systems, thus obtaining $ET0L \subseteq CD_3(t)$; this relation is not noticed in [4]).

Take a CD grammar system $\Gamma = (N, T, S, P_1, \ldots, P_n)$ and construct the system $\Gamma' = (N', T, [S, 1], P_1', P_2', P_3')$, with

$$N' = \{[A, i] \mid A \in N, 0 \le i \le n\},$$
$$P_1' = \{[A, i] \to [w, i] \mid A \to w \in P_i, 1 \le i \le n\},$$
$$P_2' = \{[A, i] \to [A, i+1] \mid A \in N, 1 \le i < n, i \text{ odd}\}$$
$$\cup \{[A, n] \to [A, 0] \mid A \in N, \text{ if } n \text{ is odd}\},$$
$$P_3' = \{[A, i] \to [A, i+1] \mid A \in N, 0 \le i < n, i \text{ even}\}$$
$$\cup \{[A, n] \to [A, 1] \mid A \in N, \text{ if } n \text{ is even}\},$$

where $[w, i]$ denotes the string obtained by replacing each nonterminal $A \in N$ appearing in w by $[A, i]$ and leaving the terminals unchanged.

Then $L_t(\Gamma) = L_t(\Gamma')$. This can be easily seen: the rules in P_i are simulated by P_1' on nonterminals of the form $[A, i]$. The components P_2', P_3' change only the second terms of such symbols, for all the nonterminals in the sentential form (P_2' goes from odd numbers to even numbers, P_3' goes from even numbers to odd numbers). Thus, during their simulation by P_1', the rules in P_i are never mixed with rules in P_j, for $i \ne j$. Hence Γ and Γ' generate the same language.

Three components suffice, but note that the first one, P_1', is as complex (in the number of productions) as the whole initial system Γ. This suggests that in comparing systems one should take into account not only the number of components but also the *size* of components (expressed, e.g., by the number of productions).

Let us denote by $CD_{n,m}(f)$ the family of languages generated by λ-free CD grammar systems working in the f mode, with at most n components, each of which contains at most m productions. We add the letter D in front of CD when only *deterministic* systems are used (a CD grammar system is deterministic when for each component P_i, if $A \to x_1, A \to x_2$ are in P_i, then $x_1 = x_2$).

Theorem 3.2.

(i) $CD_{\infty,\infty}(f) = CD_{\infty,1}(f) = CF$, for all $f \in \{= 1, \ge 1, *\} \cup \{\le k \mid k \ge 1\}$.

(ii) $CD_{\infty,\infty}(= k) = CD_{\infty,k}(= k)$, $CD_{\infty,\infty}(\ge k) = CD_{\infty,2k-1}(\ge k)$, for $k \ge 2$.

(iii) $CF \subset CD_{\infty,1}^\lambda(t) \subset CD_{\infty,2}^\lambda(t) \subseteq CD_{\infty,3}^\lambda(t) \subseteq CD_{\infty,4}^\lambda(t) \subseteq CD_{\infty,5}^\lambda(t)$
$= CD_{\infty,\infty}^\lambda(t) = ET0L$.
$CF_{fin} \subset (CD_{\infty,1}^\lambda(t) = DCD_{\infty,1}^\lambda(t)) \subseteq DCD_{\infty,2}^\lambda(t) \subseteq DCD_{\infty,3}^\lambda(t) \subseteq$
$DCD_{\infty,4}^\lambda(t) = DCD_{\infty,\infty}^\lambda(t) = EDT0L.$

(iv) $CD_{n,m}(f) \subset CD_{n+1,m}(f)$, $CD_{n,m}(f) \subset CD_{n,m+1}(f)$, for $f \in \{*, t\}$.

In the above, CF_{fin} denotes the family of finite index context-free languages. Especially interesting relations here are: $CD^\lambda_{\infty,\infty}(t) \subseteq CD^\lambda_{\infty,5}(t)$, $DCD^\lambda_{\infty,\infty}(t) \subseteq DCD^\lambda_{\infty,4}(t)$ (see [17]) and those in (iv).

For $f = *$ it is proved in [45] that

$$L_{(n+1)m} = \{a^{2^i} \mid 1 \le i \le (n+1)m\} \in CD_{n+1,m}(*) - CD_{n,m}(*), \ n, m \ge 1,$$

whereas for $f = t$ one uses

$$L_{(n+1)m} = \{a_i \mid 1 \le i \le (n+1)m\} \in CD_{n+1,m}(t) - CD_{n,m}(t), \ n, m \ge 1.$$

Open problems: Which of the inclusions from Theorems 3.1 and 3.2 that are not given as proper inclusions ($\subset$) are proper? Are the bounds 5 and 4 from Theorem 3.2 (iii) optimal? These bounds were obtained using λ-rules. What about the case of λ-free systems? Can the results from Theorem 3.2 (ii) be improved? Are the strict inclusions from Theorem 3.2 (iv) true also for the derivation modes $= k, \ge k$? Are the hierarchies $CD_n(f)$, for $f \in \{= k, \ge k \mid k \ge 2\}$, infinite? (Are systems with $n+1$ components always more powerful than systems with n components?)

Many other problems are still open in this area (see [17]). Most of them are related to the same, technical, missing tool: counterexamples, necessary conditions (mainly for families $CD_n(f), f \in \{= k, \ge k \mid k \ge 2\}$). The examples Γ_3 and Γ_4 in Section 3.2 indicate that the question is not trivial.

3.4 Hybrid systems

The systems considered in the previous sections were *homogeneous*, all their components were supposed to work in the same mode. It is perhaps more realistic to consider systems consisting of components working in different modes. This leads to the notion of a *hybrid CD grammar system*, which is a construct

$$\Gamma = (N, T, S, (P_1, f_1), \ldots, (P_n, f_n)), n \ge 1,$$

where $(N, T, S, P_1, \ldots, P_n)$ is a usual CD grammar system and $f_i \in D, 1 \le i \le n$; f_i is the derivation mode in the i-th component of Γ. The language generated by Γ is

$$L(\Gamma) = \{w \in T^* \mid S \Longrightarrow^{f_{i_1}}_{P_{i_1}} w_1 \Longrightarrow^{f_{i_2}}_{P_{i_2}} \ldots \Longrightarrow^{f_{i_m}}_{P_{i_m}} w_m = w,$$

$$m \ge 1, 1 \le i_j \le n, 1 \le j \le m\}.$$

We denote by $HCD_n, n \ge 1$, the family of languages generated by λ-free (context-free) hybrid CD grammar systems with at most n components; we put $n = \infty$ when no restriction is imposed on the number of components. All the results in this section hold true also for systems which are allowed to use λ-rules, but we do not explicitly state the results in this form.

Of course, $CD_n(f) \subseteq HCD_n$ for $n \ge 1, f \in D$. Moreover, [112]:

Theorem 3.3. *If a hybrid CD grammar system Γ has the following two properties, then $L(\Gamma) \in CF$:*

(1) There is no component in Γ working in a mode $f \in \{= k, \geq k \mid k \geq 2\}$,
(2) There are at most two components of Γ working in the t mode.

Conversely, there are hybrid systems which do not satisfy one of conditions (1), (2) which generate non-context-free languages.

The second assertion is proved by systems like Γ_1, Γ_2 in Section 3.2: one of P_1, P_2 in Γ_1 can be allowed to work in any mode in D, if the other one works in the $= 2$ or in the ≥ 2 mode, then the generated language will be $\{a^n b^n c^n \mid n \geq 1\}$.

Surprisingly enough, the number of components does not induce an infinite hierarchy. More exactly, combining the results in [95] and [112], we have:

Theorem 3.4.
(i) $CF = HCD_1 \subset HCD_2 \subseteq HCD_3 \subseteq HCD_4 = HCD_\infty \subseteq MAT_{ac}$.
(ii) $ET0L \subset HCD_4$, $CD_\infty(=) \subset HCD_3$,
$CD_\infty(=) \subseteq HCD_\infty(\text{fin-}t) \subset (HCD_4 \cap MAT)$,

where $HCD_\infty(\text{fin-}t)$ is the family of languages generated by hybrid CD grammar systems of arbitrary degree but such that the components working in the t mode are used a bounded number of times only.

The equality $CF = HCD_1$ is immediate, the strict inclusion $CF \subset HCD_2$ follows from $CF \subset CD_2(= k) \subseteq HCD_2, k \geq 2$. The non-trivial relation here is $HCD_\infty \subseteq HCD_4$. This is proved in [95] as follows.

Take a hybrid CD grammar system of degree $n, n \geq 4$, $\Gamma = (N, T, S, (P_1, f_1), \ldots, (P_n, f_n))$. We shall construct a system

$$\Gamma = (N', T, S', (P'_1, t), (P'_2, t), (P'_3, t), (P'_4, = k))$$

such that $L(\Gamma) = L(\Gamma')$. This can be done using the following facts.

1. Given a hybrid CD grammar system Γ_1, an equivalent system Γ_2 can be constructed containing at most one component working in one of the modes $*, = 1, \geq 1, \leq k$, for $k \geq 1$. This component works in the $*$-mode (simply put together the rules of components working as above).
2. Given a hybrid CD grammar system Γ_1, an equivalent system Γ_2 can be constructed containing at most three components working in the t-mode (and preserving the other components of Γ_1). (The technique used is the same as the one used in proving $CD_n(t) \subseteq CD_3(t), n \geq 3$.)
3. Given a hybrid CD grammar system Γ_1, an equivalent system Γ_2 can be constructed such that, if it contains two components i, j working in the modes $= k_i, = k_j$, then $k_i = k_j$ (all the components working in the $= k$-mode have the same value for k; the new k is a multiple of the initial k_i's).

4. Given a hybrid CD grammar system Γ_1, an equivalent system Γ_2 can be constructed, containing only components working in the t-mode and in the $= k$-mode, for a given k. (The $\geq k$ components can be simulated by components working in the $= k'$ mode for $k \leq k' \leq 2k - 1$.)

5. Given a hybrid CD grammar system Γ_1, an equivalent system Γ_2 can be constructed containing at most four components (namely, three of them working in the t mode and one working in the $= k$ mode). (Use the above steps and repeat some of them whenever necessary; e.g., after 4 we apply 3 again.)

Technical details needed above can be found in [95] and in [17]. As a by-product of these proofs, we get Theorem 3.1 (v): $CD_\infty(\geq) \subseteq CD_\infty(=)$.

Here is a hybrid system (of degree 4) generating a non-ET0L language (see [112]).

$$\Gamma = (\{S, A, B, C, X, Y\}, \{a, b, c\}, S, (P_1, t), (P_2, = 2), (P_3, t), (P_4, t)),$$
$$P_1 = \{S \to ABS, S \to ABX, C \to B, Y \to X\},$$
$$P_2 = \{X \to Y, A \to a\},$$
$$P_3 = \{X \to X, B \to bC\},$$
$$P_4 = \{X \to c, B \to c\}.$$

Let us examine the possible derivations in Γ. We have to start by using the component P_1. This leads to a string $(AB)^m X$, $m \geq 1$. From now on the rules $S \to ABS, S \to ABX$ will never be used. The component P_3 cannot be used for strings containing the symbol X, because the derivation cannot be correctly terminated in the t-mode. Using P_2 means either to replace two occurrences of A by a, or to replace X by Y and one occurrence of A by a. In the latter case, using P_1 we can replace Y by X, hence the use of P_2 can be iterated. Therefore, any number of occurrences of A can be replaced by a. In order to use P_3, we have to replace X by Y; by P_3 each occurrence of B will introduce a symbol b (and the nonterminal C). In order to use P_3 again we have to use first P_1, thus turning each C back to B and each Y back to X, and then P_2 again, in order to replace the trap-symbol (for P_3) X by Y. Consequently, between every two uses of P_3, hence between every two increases of the number of occurrences of b, we have to introduce at least one occurrence of a. This can be repeated at most m times, where m is the number of occurrences of A we have started with. This implies that we can produce each sentential form of the type $(ab^n B)^m X$, with $1 \leq n \leq m$. Using P_4 we get $(ab^n c)^m c$, $1 \leq n \leq m$. Consequently,

$$L(\Gamma) = \{(ab^n c)^m c \mid 1 \leq n \leq m\}.$$

This is not an ET0L language: $ET0L$ is a full AFL; erasing the symbol c by a morphism we get the language $\{(ab^n)^m \mid 1 \leq n \leq m\}$, which is known not to be an ET0L language (see Corollary V.2.2 on p. 248 in [129]).

A number of problems are still **open** in this area. To start with we have the relations not specified in Theorem 3.4. Now, let $HCD_n(f_1, f_2, \ldots, f_n)$ be the family of languages generated by hybrid CD grammar systems $\Gamma = (N, T, S, (P_1, f_1), \ldots, (P_n, f_n))$. By Theorem 3.4 (i), $HCD_n(f_1, \ldots, f_n) \subseteq HCD_4(=, t, t, t)$. What can one say about families $HCD_n(f_1, \ldots, f_n)$ for $n = 2, 3, 4$? How many of them are distinct? (Theorem 3.3 characterizes the families which equal CF; what about the other ones?) Is the inclusion $ET0L \subseteq HCD_3$ proper?

3.5 Increasing the power by teams

In the blackboard model, as well as in CD grammar systems considered so far, at each moment only one component is enabled. Removing this restriction we obtain the notion of a *team CD grammar system*, as introduced in [71]. We use here the presentation in [121].

A CD grammar system *with (prescribed) teams (of variable size)* is a construct

$$\Gamma = (N, T, S, P_1, \ldots, P_n, R_1, \ldots, R_m), \ n, m \geq 1,$$

where $(N, T, S, P_1, \ldots, P_n)$ is a usual CD grammar system, and $R_1, \ldots, R_m$ are subsets of $\{P_1, \ldots, P_n\}$ (called *teams*). A team $R_i = \{P_{j_1}, \ldots, P_{j_s}\}$ is used in derivations as follows:

$$x \Longrightarrow_{R_i} y \text{ iff } x = x_1 A_1 x_2 A_2 \ldots x_s A_s x_{s+1}, y = x_1 y_1 x_2 y_2 \ldots x_s y_s x_{s+1},$$
$$x_l \in (N \cup T)^*, 1 \leq l \leq s+1, A_r \to y_r \in P_{j_r}, 1 \leq r \leq s.$$

(Note that because a team is a *set*, no order of components is assumed in a derivation step as above.)

Having defined the one step derivation, we can define derivations in R_i of k steps, at most k steps, at least k steps, and of any number of steps, denoted by $\Longrightarrow_{R_i}^{=k}, \Longrightarrow_{R_i}^{\leq k}, \Longrightarrow_{R_i}^{\geq k}, \Longrightarrow_{R_i}^{*}$, respectively. For maximal derivations in a team R_i we can consider three variants:

$$x \Longrightarrow_{R_i}^{t_0} y \text{ iff } x \Longrightarrow_{R_i}^{\geq 1} y \text{ and there is no } z \text{ such that } y \Longrightarrow_{R_i} z,$$

$$x \Longrightarrow_{R_i}^{t_1} y \text{ iff } x \Longrightarrow_{R_i}^{\geq 1} y \text{ and for no component } P_{j_r} \in R_i \text{ and for no } z$$
$$\text{there is a derivation } y \Longrightarrow_{P_{j_r}} z,$$

$$x \Longrightarrow_{R_i}^{t_2} y \text{ iff } x \Longrightarrow_{R_i}^{\geq 1} y \text{ and there is a component } P_{j_r} \in R_i$$
$$\text{for which no derivation } y \Longrightarrow_{P_{j_r}} z \text{ is possible.}$$

In the t_0 mode the team as a whole cannot perform any further steps, in the t_1 mode no component of the team can apply any of its rules, whereas in the case of t_2 at least one component cannot rewrite any symbol of the current string. The mode t_0 is considered in [56], t_1 in [71], and t_2 in [121]. Note that the three t modes are different from each other: a derivation in the team $\{\{A \to a\}, \{A \to b\}\}$ leading to, say, abA, is correctly terminated in the t_0

mode, but not in the others; with the same string, $\{\{A \to a\}, \{B \to b\}\}$ can finish the work in the t_2 mode, but not in the t_1 mode.

If each subset of $\{P_1, \ldots, P_n\}$ can be a team, then we say that Γ has *free teams*; when all teams have the same number of components, then we say that Γ has *teams of constant size*. For the case of teams of constant size we consider a finite set $W \subseteq (N \cup T)^*$ instead of an axiom $S \in N$, in order not to produce artificial counterexamples when using λ-free rules (strings of length less that s, where s is the size of teams, cannot be generated). However, we require that W contains only one nonterminal string.

We denote by $PT_s CD(f)$ the family of languages generated by λ-free CD grammar systems with prescribed teams of constant size s in the derivation mode $f \in \{*, t_0, t_1, t_2\} \cup \{\leq k, = k, \geq k \mid k \geq 1\}$. When dealing with free teams, the letter P is omitted. When the size of teams is not constant, we replace s with $*$. When λ-rules are allowed we add as usual the superscript λ.

Summarizing the results from [56], [71] and [121], we obtain the following result.

Theorem 3.5.
(i) $PT_s CD(f) = PT_* CD(f) = MAT$, for all $s \geq 2$ and $f \in \{*\} \cup \{\leq k, = k, \geq k \mid k \geq 1\}$.
(ii) $T_s CD(f) = PT_s CD(f) = PT_* CD(f) = T_* CD(f) = MAT_{ac}$, for all $s \geq 2$ and $f \in \{t_0, t_1, t_2\}$.
(iii) All the results above hold true also for the case of using λ-rules (MAT and MAT_{ac} are then replaced by MAT^λ and MAT_{ac}^λ).

These equalities suggest a number of interesting corollaries: teams of size two suffice for obtaining the maximal generative capacity, enhancing the cooperation by using teams increases considerably the power of CD grammar systems (remember that $ET0L \subset MAT_{ac}$ and that $CD_\infty(f) = CF$ for $f \in \{= 1, \geq 1, *\} \cup \{\leq k \mid k \geq 1\}$). All the three variants of maximal derivation are equivalent, when λ-rules are allowed we get new characterizations of recursively enumerable languages ($MAT_{ac}^\lambda = RE$, [41]).

In order to illustrate the work of teams and in particular to show how the power of CD grammar systems is increased by increasing the cooperation among components, we recall the following example from [71].

Consider the system

$$\Gamma = (\{A, B, A', B'\}, \{a, b, c, d, e\}, \{AB\}, P_1, P_2, P_3, P_4, P_5, P_6),$$
$$P_1 = \{A \to A'A'\},$$
$$P_2 = \{B \to aBc, B \to bBd, B \to aB'c, B \to bB'd\},$$
$$P_3 = \{A' \to AA\},$$
$$P_4 = \{B' \to aB'c, B' \to bB'd, B' \to aBc, B' \to bBd\},$$
$$P_5 = \{A \to e, A' \to e, B \to B, B' \to B'\},$$

$$P_6 = \{B \rightarrow B, B \rightarrow ac, B \rightarrow bd, B' \rightarrow B', B' \rightarrow ac,$$
$$B' \rightarrow bd, A \rightarrow A, A' \rightarrow A'\}.$$

The language generated by Γ, with free teams of size 2, working in the t_1 mode, is

$$L_{t_1}(\Gamma, 2) = \{e^{2^n} w f(w) \mid n \geq 0, w \in \{a, b\}^+, |w| = 2^n\},$$

where $f(w) = h(mi(w))$, with mi denoting the mirror image, and h the morphism defined by $h(a) = c, h(b) = d$.

Indeed, when P_5 is used, neither B nor B' can be present in the current sentential form when this step is completed, when P_6 is used, neither A nor A' can be present in the current sentential form when this step is completed. Therefore P_5, P_6 can participate in a team only together, and such a team is used at the last step of a derivation. Now P_1 cannot be in a team containing P_3, and P_2 cannot be in a team containing P_4. Starting from a string containing only symbols A and B (initially we have AB), besides the team $\{P_5, P_6\}$, only the team $\{P_1, P_2\}$ is applicable, and its use leads to a string containing only symbols A' and B'. Now only $\{P_3, P_4\}$ can be applied, yielding a string containing the nonterminals A, B. In such a cycle the number of occurrences of A is doubled and each time that either the rule $A \rightarrow A'A'$ or the rule $A' \rightarrow AA$ is used, one rule from either P_2 or P_3 is used. Therefore the number of symbols a, b and c, d, respectively, equals the number of occurrences of A, A' minus one. Finally, every A, A' is replaced by e, while rules $B \rightarrow B, B' \rightarrow B'$ are used in P_6, except at the last step when one of the rules $B \rightarrow ac, B \rightarrow bd, B' \rightarrow ac, B' \rightarrow bd$ is used. In this way, the number of occurrences of a, b (hence also of c, d) equals the number of occurrences of e. Hence the language $L_{t_1}(\Gamma, 2)$ is generated.

Let us now use a morphism to erase the symbol e. If $L_{t_1}(\Gamma, 2) \in ET0L = CD(t)$, then the obtained language, $\{wf(w) \mid w \in \{a, b\}^+, |w| = 2^n, n \geq 0\}$, would be in $ET0L$. But according to Theorem V.2.10 in [129], $\{w \in \{a, b\}^+ \mid |w| = 2^n, n \geq 0\} \in ET0L$ (the mapping f is a bijection from $\{w \in \{a, b\}^+ \mid |w| = 2^n, n \geq 0\}$ to $\{w \in \{c, d\}^+ \mid |w| = 2^n, n \geq 0\}$), which contradicts Corollary IV.3.4 in [129]. Thus, $L_{t_1}(\Gamma, 2) \notin ET0L$.

3.6 Descriptional complexity

The main motivation behind grammar systems was the modeling, in grammatical fashion, of distributed information processing, e.g., as it takes place within blackboard systems. From the language theoretic point of view grammar systems offer an elegant language generating mechanism. As illustrated in this chapter many times already cooperation leads often to quite remarkable *increase* of generative power. Another attractive feature of grammar systems is a possible *decrease* of the complexity of language specification. In order to discuss this aspect of grammar systems we will consider now three measures of descriptional complexity which were well investigated within the framework of context-free grammars (see [60]).

Let $\Gamma = (N, T, S, P_1, \ldots, P_n)$ be a CD grammar system and consider the following "natural" measures of descriptional complexity of Γ:

$$Var(\Gamma) = card(N),$$

$$Prod(\Gamma) = \sum_{i=1}^{n} card(P_i),$$

$$Symb(\Gamma) = \sum_{i=1}^{n} \sum_{A \to x \in P_i} (|x| + 2).$$

For a measure $M \in \{Var, Prod, Symb\}$ and a class Y of grammar systems we define

$$M_Y(L) = \min\{M(\Gamma) \mid L(\Gamma) = L \text{ and } \Gamma \in Y\},$$

for L being a language in the family generated by elements of Y.

Obviously, for each context-free language L, for each $M \in \{Var, Prod, Symb\}$, and for each class Y of CD grammar systems (containing context-free grammars), we have

$$M_Y(L) \le M_{CF}(L).$$

Consider now a more general setting. Let X, Y be two classes of language generating mechanisms and let $\mathcal{L}$ be the family of languages which can be generated both by an element of X and by an element of Y. Let M be a descriptional complexity measure, and assume that $M_Y(L) \le M_X(L)$ for all $L \in \mathcal{L}$. Then, the change in descriptional complexity (with respect to M) when going from Y to X can be captured in one of the following manners.

$$Y = X \ (M) \quad \text{iff} \quad M_Y(L) = M_X(L) \text{ for all } L \in \mathcal{L},$$

$$Y <_1 X \ (M) \quad \text{iff} \quad \text{there is a language } L \in \mathcal{L} \text{ such that}$$
$$M_Y(L) < M_X(L),$$

$$Y <_2 X \ (M) \quad \text{iff} \quad \text{for all } n \ge 1, \text{ there is an } L_n \in \mathcal{L} \text{ such that}$$
$$M_X(L_n) - M_Y(L_n) > n,$$

$$Y <_3 X \ (M) \quad \text{iff} \quad \text{there are } L_n \in \mathcal{L} \text{ and } n \ge 1, \text{such that}$$
$$lim_{n \to \infty} \frac{M_Y(L_n)}{M_X(L_n)} = 0,$$

$$Y <_4 X \ (M) \quad \text{iff} \quad \text{there are } p \ge 1 \text{ and languages } L_n \in \mathcal{L}, n \ge 1,$$
$$\text{such that} M_X(L_n) > n \text{ and } M_Y(L_n) \le p.$$

Clearly, $<_i$ implies $<_{i-1}$ for $i \in \{2, 3, 4\}$.

Summarizing the results from [43] and [17], we obtain the following table: at the intersection of the row $M, M \in \{Var, Prod, Symb\}$, with the column $f, f \in \{*, t, \le k, = k, \ge k\}$, we find the relation ρ if $CD_\infty(f) \ \rho \ CF \ (M)$ holds.

	$*$	t	$\leq k$	$= k$	$\geq k$
Var	$=$	$<_4$	$=$	$<_4$	$<_4$
$Prod$	$=$	$<_3$	$=$	$<_4$	$<_4$
$Symb$	$=$	$<_3$	$=$	$<_3$	$<_3$

The following context-free language proves the entries of this table for $M = Prod$, and $f \in \{= k, \geq k\}, k \geq 2$:

$$L_{n,k} = \{a^i b a^j d c^{(k-1)n} \mid 0 \leq i + j \leq n\}, \ k \geq 2, n \geq 1.$$

We have

$$Prod_{CF}(L_{n,k}) \geq \log_2(n + 1)$$

(this can be proved as in [7]). On the other hand

$$Prod_{CD.CF(f)}(L_{n,k}) \leq 16, \text{ for } f \in \{= k, \geq k \mid k \geq 2\},$$

because $L_{n,k} = L_f(\Gamma)$, for

$$\Gamma = (\{S, A, A', B, B', C\}, \{a, b, c, d\}, S, P_1, P_2, P_3, P_4, P_5),$$

with

$$
\begin{aligned}
P_1 &= \{S \to S, S \to ABC^{(k-1)n}\}, \\
P_2 &= \{A \to aA', C \to c, A \to ab, A \to b\}, \\
P_3 &= \{B \to aB', C \to c, B \to ad, B \to d\}, \\
P_4 &= \{A' \to A, A \to A, B' \to B, B \to B\}, \\
P_5 &= \{C \to C, C \to c\}.
\end{aligned}
$$

The reader can easily check that $L_{n,k}$ is generated by Γ in either $= k$ or $\geq k$ mode.

There are in [17] also results concerning the *index* of CD grammar systems. As expected, the systems of finite index generate strict subfamilies of the families generated without restrictions about the index (CF_{fin} when using derivation modes $*, = 1, \geq 1, \leq k$, for $k \geq 1$, and MAT_{fin} in derivation modes $t, = k, \geq k$, for $k \geq 2$).

The proofs of the equality $CD_3(t) = ET0L$ and of relations with families like CF and MAT are effective. In certain cases, also results on the complexity of some decision problems about CD grammar systems are available. For instance, the membership problem for $CD_3(t)$ is NP-complete. The NP-completeness of a special problem concerning CD grammar systems with $=1$ mode of derivation has been proved in [84].

The complexity of the membership and emptiness problems for $CD_\infty (= k)$ and $CD_\infty (\geq k), k \geq 2$, is still **open** (these problems are decidable for MAT, hence also for the considered families).

3.7 Other classes of CD grammar systems

We have already mentioned grammar systems with a control graph, which specifies the order of the enabling of components. If arbitrary graphs can be used, then the generative power of CD grammar systems is increased, but for some sorts of graphs (e.g., a ring) it might be the case that the generative power decreases [30], [17]. One can also consider grammar systems with start conditions formulated as random context conditions: a component may start working when (the common sentential form is available and) certain symbols are present while other symbols are not present in the sentential form. In fact, both start and stop conditions of this type can be considered. Moreover, one can check the presence/absence of symbols as above, or of certain strings associated with the components, or one can check whether or not the whole sentential form is a member of a given regular language. In general, when strings can be checked as above, characterizations of context-sensitive languages are obtained if λ-rules are not allowed (hence characterizations of recursively enumerable languages are obtained if λ-rules are allowed). Details can be found in [19], [17] and [29].

A similarly powerful variant is that of CD grammar systems with the communication aided by a generalized sequential machine (gsm). Assume that the agents "speak" different languages, hence they need the help of a "translator": the sentential form has to be translated into the component language before the component starts rewriting. Technically, a gsm is added to the system and the gsm works in between each two derivation steps corresponding to components enabling. Because in this way we implicitly have an iterated application of the gsm, and iterated gsm's are known to characterize CS (or RE when erasings are allowed), the CD grammar systems with "intermediate" gsm's also characterize CS (or RE), [94].

A surprising characterization of the family RE is obtained in [46], using CD grammar systems with right-linear components and *two multiplicative registers*. The idea of using registers is related to the idea of regulating the work of usual grammars by using *valences*, [102]: associate elements of a given group ("valences") to the rules of a system and allow each component to stop the work only when the total of the valences used in the derivation is equal to the identity of the group. When using integers, we stop at 0 (we say that we have additive valences/registers), when using positive rational numbers, we stop at 1 (and we have multiplicative valences/registers), etc. Systems with one register, working as above, and also with two registers, were considered in [42]. In the latter case two numbers are associated with each rule, hence two registers. At the end of a component's working step, the first register must be empty (equal to the group identity) while the second one is not restricted. However a new component starts working with the registers having the contents interchanged (hence with a non-empty contents of the first register and the empty contents of the second one). The whole derivation starts with both registers empty.

The interplay of registers proves to be very powerful, at least for the multiplicative case: the obtained family of languages (even when using only right-linear rules) is closed under intersection and morphisms, and it contains the Dyck language, [46]. Hence, according to the characterization of RE languages as a morphic image of the intersection of two context-free languages, this leads to a characterization of RE languages in terms of CD grammar systems with registers.

We close our discussion here by mentioning one further variant, that of a *colony*, [73]. This is a special case of a CD grammar system, with the components generating finite languages. We write the system in the form $\Gamma = (N, T, w, F_1, \ldots, F_n)$, where $w \in (N \cup T)^*$, $N = \{S_1, \ldots, S_n\}$, S_i is a nonterminal associated with F_i, which in turn is a finite subset of $(N \cup T - \{S_i\})^*$, $1 \leq i \leq n$. Rewriting S_i means to replace it with an element of F_i. Motivations for considering such devices can be found in [17], [73], results can be found in [17], [75], [110]. Most of the problems still open for CD grammar systems (including hierarchies on the number of components) are solved for colonies.

Quite a number of other variants of and problems about CD grammar systems appear in the literature: systems with a sort of appearance checking feature ([44]), systems with separated terminal and nonterminal alphabets for each component ($\Gamma = (T, S, (N_1, T_1, P_1), \ldots, (N_n, T_n, P_n))$, without imposing $N_i \cap T_j = \emptyset$, for $i \neq j$, [17], [22]), hierarchical systems [98], [16], systems with Lindenmayer components [15], [141], systems with "similar" components in the sense of grammar forms theory [18], [97], deterministic systems [45], associated Szilard languages [40], systems with "fair" activation of components [39], systems of push-down automata [36], [37], and many others. The reader is referred to the mentioned papers for details and more pointers to the literature.

4. PC grammar systems

4.1 Definitions

Definition 4.1. *A PC grammar system of degree $n, n \geq 1$, is an $(n+3)$-tuple*

$$\Gamma = (N, K, T, (S_1, P_1), \ldots, (S_n, P_n)),$$

where N is a nonterminal alphabet, T is a terminal alphabet, $K = \{Q_1, Q_2, \ldots, Q_n\}$ (the sets N, T, K are mutually disjoint), P_i is a finite set of rewriting rules over $N \cup K \cup T$, and $S_i \in N$, for all $1 \leq i \leq n$.

Let $V_\Gamma = N \cup K \cup T$.

The sets $P_i, 1 \leq i \leq n$, are called the *components* of the system, and the elements $Q_1, \ldots, Q_n$ of K are called *query symbols*; the index i of Q_i points to the i-th component P_i of Γ.

If we want to give explicitly grammars as components of Γ, then we can write $\Gamma = (N, K, T, G_1, \ldots, G_n)$, with $G_i = (N \cup K, T, S_i, P_i), 1 \leq i \leq n$.

As in the case of CD grammar systems, we can consider also "independent" grammars as components, $\Gamma = (K, G_1, \ldots, G_n)$, with $G_i = (N_i \cup K, T_i, S_i, P_i), 1 \leq i \leq n$, without any assumptions about the relationships between N_i and $T_j, i \neq j$ (hence allowing that the terminals of one component are rewritten in another component). Such variants were investigated in [91], but we will not consider them here.

Definition 4.2. *Given a PC grammar system* $\Gamma = (N, K, T, (S_1, P_1), \ldots, (S_n, P_n))$ *as above, for two n-tuples* $(x_1, x_2, \ldots, x_n)$, $(y_1, y_2, \ldots, y_n)$, *with* $x_i, y_i \in V_\Gamma^*$, $1 \leq i \leq n$, *where* $x_1 \notin T^*$, *we write* $(x_1, \ldots, x_n) \Longrightarrow (y_1, \ldots, y_n)$ *if one of the following two cases holds.*

(i) *For each* $i, 1 \leq i \leq n$, $|x_i|_K = 0, 1 \leq i \leq n$, *and for each* $i, 1 \leq i \leq n$, *we have either* $x_i \Longrightarrow y_i$ *by a rule in* P_i, *or* $x_i = y_i \in T^*$.

(ii) *There is* $i, 1 \leq i \leq n$, *such that* $|x_i|_K > 0$. *Let, for each such* i, $x_i = z_1 Q_{i_1} z_2 Q_{i_2} \ldots z_t Q_{i_t} z_{t+1}$, $t \geq 1$, *for* $z_j \in (N \cup T)^*$, $1 \leq j \leq t+1$. *If* $|x_{i_j}|_K = 0$, *for all* $j, 1 \leq j \leq t$, *then* $y_i = z_1 x_{i_1} z_2 x_{i_2} \ldots z_t x_{i_t} z_{t+1}$ *and* $y_{i_j} = S_{i_j}, 1 \leq j \leq t$. *If for some* $j, 1 \leq j \leq t$, $|x_{i_j}|_K \neq 0$, *then* $y_i = x_i$. *For all* $i, 1 \leq i \leq n$, *such that* y_i *is not specified above, we have* $y_i = x_i$.

An n-tuple $(x_1, \ldots, x_n)$ with $x_i \in V_\Gamma^*$ for all $i, 1 \leq i \leq n$, is called a *configuration* (of Γ).

Thus, a configuration $(x_1, x_2, \ldots, x_n)$ directly yields a configuration $(y_1, y_2, \ldots, y_n)$ if either

(i) no query symbol appears in $x_1, \ldots, x_n$, and then we have a componentwise derivation, $x_i \Longrightarrow y_i$, in each component P_i, $1 \leq i \leq n$ (one rule is used in each component P_i), except for the case when x_i is terminal, $x_i \in T^*$; then $y_i = x_i$, or

(ii) query symbols occur in some x_i. Then a *communication step* is performed: each occurrence of Q_j in x_i is replaced by x_j, providing x_j does not contain query symbols. More precisely, a component x_i (containing query symbols) is modified only when all occurrences in it of query symbols refer to strings without occurrences of query symbols. In a communication step, the communicated string x_j replaces the query symbol Q_j (we say that Q_j is *satisfied* in this way). After that, the grammar G_j resumes rewriting beginning again from its axiom. The communication has priority over the effective rewriting: no rewriting is possible as long as at least one query symbol is present. If some query symbols are not satisfied at a given communication step, then they may be satisfied at the next step (providing they ask then for strings without query symbols).

Note that rules $x_1 Q_i x_2 \to x$ are never used, hence we shall asume that such rules do not appear in the PC grammar systems we work with. Also note that $(x_1, \ldots, x_n) \Longrightarrow (y_1, \ldots, y_n)$ is not defined when $x_1 \in T^*$.

We have denoted in the same way, by $\Longrightarrow$, both the componentwise derivation steps and the communication steps. As usual, by $\Longrightarrow^*$ we shall denote the reflexive and transitive closure of this relation, which corresponds to sequences of possibly interleaved derivations and communication steps.

A PC grammar system deadlocks in two cases: (1) when no query symbol is present, a component x_i of the current configuration $(x_1, \ldots, x_n)$ is not a terminal string and no rule of P_i can be applied to it (this can happen both after a rewriting and after a communication), and (2) when a *circular query* appears: P_{i_1} introduces Q_{i_2}, P_{i_2} introduces Q_{i_3}, and so on until $P_{i_{k-1}}$ introduces Q_{i_k}, and P_{i_k} introduces Q_{i_1}; no derivation is possible (the communication has priority), but no communication (in this cycle) is possible (only strings without occurrences of query symbols are communicated).

Definition 4.3. *The language generated by a PC grammar system* Γ *as above is*

$$L(\Gamma) = \{x \in T^* \mid (S_1, S_2, \ldots, S_n) \Longrightarrow^* (x, \alpha_2, \ldots, \alpha_n),$$
$$\alpha_i \in V_\Gamma^*, 2 \le i \le n\}.$$

Hence, we start from the n-tuple of axioms, $(S_1, \ldots, S_n)$, and proceed by repeated rewriting and communication steps, until the component P_1 produces a terminal string. Notice that in $L(\Gamma)$ we retain the strings generated in this way on the first component, independently of the form of the strings generated by $P_2, \ldots, P_n$ (in particular, they may contain also query symbols). Moreover, the system stops when the first component produces a terminal string.

The component P_1 is called the *master* of the system.

In the above we have considered PC grammar systems in which no restriction is imposed on the use of query symbols, more exactly, each component P_i is allowed to introduce any symbol Q_j. (Obviously, when P_i introduces the symbol Q_i, the derivation is blocked by the circularity of this query.) A classification can be obtained by considering the shape of the *communication graph*, that is the graph defined by the rules which introduce query symbols (hence such a graph is "static"). We consider here only two basic classes of PC grammar systems.

Definition 4.4. *Let* $\Gamma = (N, K, T, (S_1, P_1), \ldots, (S_n, P_n))$ *be a PC grammar system. If only* P_1 *is allowed to introduce query symbols (formally,* $P_i \subseteq (N \cup T)^* \times (N \cup T)^*$ *for* $2 \le i \le n$), *then we say that* Γ *is a centralized PC grammar system; otherwise* Γ *is non-centralized.*

A PC grammar system is said to be returning *(to axiom) if, after communicating, each component which has sent its string to another component returns to axiom. A PC grammar system is* non-returning *if point (ii) of Definition 4.2 is modified by removing the condition "and $y_{i_j} = S_{i_j}$". Thus, after communicating, the component P_{i_j} does not return to its axiom S_{i_j}, but rather it continues to process the current string.*

In terms of the classroom model, the master grammar is the team leader, the other components are the team "processors". In CD grammar systems, all components are "equal" within the system. In a PC grammar system we already have a hierarchy with two levels – the leader and the group members. This difference becomes quite drastic in centralized systems: only the leader can ask for communication.

A PC grammar system is said to be regular, linear, context-free, context-sensitive, λ-free, etc. when the rules in its components are of the corresponding types, where we call a system *regular* if its rules are right-linear, that is of the forms $A \rightarrow xB, A \rightarrow x$, with A, B nonterminals and x an arbitrary terminal string.

Notations. Because the returning and the non-returning modes of derivation can be used for the same system, we denote by $L_r(\Gamma)$ the language generated by Γ in the returning mode and by $L_{nr}(\Gamma)$ the language generated in the non-returning mode. If necessary, we also write $\Longrightarrow_r$ for denoting a returning derivation step and $\Longrightarrow_{nr}$ for a non-returning one. As for language families, we denote by PC_nX the family of languages generated by non-centralized PC grammar systems of degree at most n, with components of type X, in the returning mode; when only centralized PC grammar systems are used, we add the letter C, thus obtaining the families CPC_nX. When the non-returning mode of derivation is considered, we add the symbol N in the front of PC, CPC, thus obtaining the families $NPC_nX, NCPC_nX$; X can be $REG, LIN, CF, CS, CS^\lambda$, where REG indicates that right-linear rules are used and CS^λ indicates that rules of arbitrary type are used. The subscript n is replaced by ∞ when systems of arbitrary degree are considered.

Note that the regular, linear and context-free PC grammar systems considered here are λ-free.

4.2 Examples

Let us consider the following PC grammar systems:

$$\Gamma_1 = (\{S_1, S_1', S_2, S_3\}, K, \{a, b\}, (S_1, P_1), (S_2, P_2), (S_3, P_3)),$$
$$P_1 = \{S_1 \to abc, S_1 \to a^2b^2c^2, S_1 \to aS_1', S_1 \to a^3Q_2,$$
$$S_1' \to aS_1', S_1' \to a^3Q_2, S_2 \to b^2Q_3, S_3 \to c\},$$
$$P_2 = \{S_2 \to bS_2\},$$
$$P_3 = \{S_3 \to cS_3\},$$
$$\Gamma_2 = (\{S_1, S_2\}, K, \{a, b\}, (S_1, P_1), (S_2, P_2)),$$
$$P_1 = \{S_1 \to S_1, S_1 \to Q_2Q_2\},$$
$$P_2 = \{S_2 \to aS_2, S_2 \to bS_2, S_2 \to a, S_2 \to b\},$$
$$\Gamma_3 = (\{S_1, S_2\}, K, \{a, b\}, (S_1, P_1), (S_2, P_2)),$$
$$P_1 = \{S_1 \to S_1, S_1 \to Q_2Q_2\},$$
$$P_2 = \{S_2 \to aS_2, S_2 \to S_2b, S_2 \to ab\}.$$

Then

$$L_r(\Gamma_1) = L_{nr}(\Gamma_1) = \{a^nb^nc^n \mid n \geq 1\},$$
$$L_r(\Gamma_2) = L_{nr}(\Gamma_2) = \{xx \mid x \in \{a, b\}^+\},$$
$$L_r(\Gamma_3) = L_{nr}(\Gamma_3) = \{a^nb^ma^nb^m \mid n, m \geq 1\}.$$

Let us briefly examine the work of Γ_1. We start with (S_1, S_2, S_3). Using the third rule in P_1, then the fifth one, and the unique rules in P_2, P_3 for $n \geq 0$ steps, we get

$$(S_1, S_2, S_3) \Longrightarrow_r (aS_1', bS_2, cS_3) \Longrightarrow_r^* (a^{n+1}S_1', b^{n+1}S_2, c^{n+1}S_3).$$

Eventually, the sixth rule of P_1 will be used:

$$(a^{n+1}S_1', b^{n+1}S_2, c^{n+1}S_3) \Longrightarrow_r (a^{n+4}Q_2, b^{n+2}S_2, c^{n+2}S_3).$$

Because the query symbol Q_2 is present, we have to perform a communication step: $b^{n+2}S_2$ is sent to the first component, replacing Q_2

$$(a^{n+4}Q_2, b^{n+2}S_2, c^{n+2}S_3) \Longrightarrow_r (a^{n+4}b^{n+2}S_2, S_2, c^{n+2}S_3).$$

In a deterministic way, we have to perform now the following steps

$$(a^{n+4}b^{n+2}S_2, S_2, c^{n+2}S_3) \Longrightarrow_r (a^{n+4}b^{n+4}Q_3, bS_2, c^{n+3}S_3)$$
$$\Longrightarrow_r (a^{n+4}b^{n+4}c^{n+3}S_3, bS_2, S_3) \Longrightarrow_r (a^{n+4}b^{n+4}c^{n+4}, b^2S_2, cS_3).$$

Therefore, all strings $a^nb^nc^n, n \geq 4$, can be produced in this way. By a derivation

$$(S_1, S_2, S_3) \Longrightarrow_r (a^3 Q_2, b S_2, c S_2) \Longrightarrow_r (a^3 b S_2, S_2, c S_3) \Longrightarrow_r$$
$$\Longrightarrow_r (a^3 b^3 Q_3, b S_2, c^2 S_3) \Longrightarrow_r (a^3 b^3 c^2 S_3, b S_2, S_3) \Longrightarrow (a^3 b^3 c^3, b^2 S_2, c S_3),$$

we can obtain the string $a^3 b^3 c^3$, whereas the strings $abc, a^2 b^2 c^2$ are produced directly by the master component P_1.

Because there is only one query of P_1 to P_2 and only one to P_3, for Γ_1 (and the same is true for Γ_2 and Γ_3) the returning and the non-returning modes coincide.

The systems Γ_2 and Γ_3 work as follows: P_1 does nothing for a number of steps while P_2 generates a string, then P_1 introduces $Q_2 Q_2$. Hence the string produced by P_2 is sent to P_1 and duplicated (it must be terminal, otherwise the system is blocked).

Note that all the three systems above are centralized.

Hence, quite simple PC grammar systems can generate some classic examples of non-context-free languages.

However, the last language represents a particular case of a "crossed agreement", where the general case is encoded by the language

$$L = \{a^n b^m c^n d^m \mid n, m \geq 1\}.$$

We could not give a *centralized* (returning or non-returning) PC grammar system with context-free components that generates L.[1] On the other hand, we can generate L with *non-centralized* systems, both in the returning and in the non-returning case.

Consider, for instance, the following returning systems:

$$\begin{aligned}
\Gamma_4 \ = \ & (N, K, \{a, b, c, d\}, (S_1, P_1), (S_2, P_2), (S_3, P_3)), \\
& N = \{S_1, S_2, S_3, S_1', A, B, B', D, D'\}, \\
& P_1 = \{S_1 \to S_1, S_1 \to Q_3, B' \to B, D' \to D, S_1 \to S_1', \\
& \qquad S_1' \to Q_2, B' \to b, D' \to d\}, \\
& P_2 = \{S_2 \to S_2, S_2 \to Q_1, B \to bB', D \to dD'\}, \\
& P_3 = \{S_3 \to AD', A \to aAc, A \to aB'c\},
\end{aligned}$$

$$\begin{aligned}
\Gamma_5 \ = \ & (N, K, \{a, b, c, d\}, (S_1, P_1), \ldots, (S_{10}, P_{10})), \\
& N = \{S_i \mid 1 \leq i \leq 10\} \cup \{A, D\} \cup \{A_i, D_i \mid 1 \leq i \leq 4\}, \\
& P_1 = \{S_1 \to S_1, S_1 \to Q_2, S_1 \to Q_6, S_1 \to Q_{10}\}, \\
& P_2 = \{S_2 \to S_2, S_2 \to aAcD, S_2 \to a^2 Ac^2 D, S_2 \to a^3 Ac^3 D, \\
& \qquad S_2 \to a^4 Ac^4 D, A \to aAc, A \to aQ_3 c, D \to Q_4, \\
& \qquad S_3 \to b^3, S_4 \to d\}, \\
& P_3 = \{S_3 \to bS_3\},
\end{aligned}$$

---

[1] Such PC grammar systems were recently found by A. Chiţu, which has proved that $L \in CPC_4 CF \cap NCPC_4 CF$.

$$P_4 = \{S_4 \rightarrow dS_4\},$$
$$P_5 = \{S_5 \rightarrow Q_2, S_2 \rightarrow Q_2, D \rightarrow Q_2, S_4 \rightarrow A\},$$
$$P_6 = \{S_6 \rightarrow S_6, S_6 \rightarrow AbDd, S_6 \rightarrow Ab^2Dd^2, S_6 \rightarrow Ab^3Dd^3,$$
$$S_2 \rightarrow Ab^4Dd^4, D \rightarrow bDd, D \rightarrow bQ_7d, A \rightarrow Q_8,$$
$$S_6 \rightarrow c^3, S_8 \rightarrow a\},$$
$$P_7 = \{S_7 \rightarrow cS_7\},$$
$$P_8 = \{S_8 \rightarrow aS_8\},$$
$$P_9 = \{S_9 \rightarrow Q_6, S_6 \rightarrow Q_6, A \rightarrow Q_6, S_7 \rightarrow A\},$$
$$P_{10} = \{S_{10} \rightarrow a^iD_i, D_i \rightarrow bD_id, D_i \rightarrow bc^id,$$
$$S_{10} \rightarrow A_id^i, A_i \rightarrow aA_ic, A_i \rightarrow ab^ic \mid i = 1, 2, 3, 4\}.$$

Then $L_r(\Gamma_4) = L_{nr}(\Gamma_5) = L$; details can be found in [116].

Let us consider three further examples that turn out to be very useful:

$$\Gamma_6 = (\{S_1, S_2\}, K, \{a\}, (S_1, P_1), (S_2, P_2)),$$
$$P_1 = \{S_1 \rightarrow aQ_2, S_2 \rightarrow aQ_2, S_2 \rightarrow a\},$$
$$P_2 = \{S_2 \rightarrow aS_2\},$$
$$\Gamma_7 = (\{S_1, S_2, A, B\}, K, \{a, b\}, (S_1, P_1), (S_2, P_2)),$$
$$P_1 = \{S_1 \rightarrow aS_1, S_1 \rightarrow S_1a, S_1 \rightarrow ba, S_1 \rightarrow ab\},$$
$$P_2 = \{S_2 \rightarrow A^{p-1}, A \rightarrow B\}, \ p \geq 2,$$
$$\Gamma_8 = (\{S_1, S_2, S_3, A, B\}, K, \{a\}, (P_1, S_1), (P_2, S_2), (P_3, S_3)),$$
$$P_1 = \{S_1 \rightarrow aA, S_1 \rightarrow aQ_2, B \rightarrow aA, B \rightarrow a\},$$
$$P_2 = \{S_2 \rightarrow aQ_1, A \rightarrow aQ_3\},$$
$$P_3 = \{S_3 \rightarrow aQ_1, A \rightarrow aB\}.$$

Then

$$L_r(\Gamma_6) = \{a^{2m+1} \mid m \geq 1\},$$
$$L_{nr}(\Gamma_6) = \{a^{\frac{(m+1)(m+2)}{2}} \mid m \geq 1\},$$
$$L_r(\Gamma_7) = L_{nr}(\Gamma_7) = \{a^iba^j \mid 1 \leq i + j \leq p\},$$
$$L_r(\Gamma_8) = \{a^{7 \cdot 2^m - 6} \mid m \geq 1\}.$$

The system Γ_6 shows the important differences between the returning and the non-returning modes of derivation, Γ_8 illustrates again the intricate work of non-centralized systems: a one-letter non-regular language can be produced by a PC grammar system with three regular (in the restricted sense) components.

An interesting feature is illustrated by Γ_7. The component P_2 does not directly contribute to a generated string, however it limits the length of the derivation: because P_2 can work at most p steps, the system itself can work only at most p steps.

4.3 On the generative capacity

We recall, without proofs, a number of results from [17].

Theorem 4.1.
(i) $Y_n CS^\lambda = RE$, for all n and $Y \in \{PC, CPC, NPC, NCPC\}$; $CPC_\infty CS = NCPC_\infty CS = CS$..
(ii) $Y_n REG - LIN \neq \emptyset$, $Y_n LIN - CF \neq \emptyset$, for all $n \geq 2$, and $Y_n REG - CF \neq \emptyset$, for all $n \geq 3$ and for all $Y \in \{PC, CPC, NPC, NCPC\}$.
(iii) $Y_n REG - CF \neq \emptyset$, for all $n \geq 2$ and for all $Y \in \{NPC, NCPC\}$.
(iv) $LIN - (CPC_\infty REG \cup NCPC_\infty REG) \neq \emptyset$.
(v) $CPC_2 REG \subset CF$, $PC_2 REG \subseteq CF$.
(vi) $CPC_\infty REG$ contains only semilinear languages.
(vii) $CPC_n CF$, for all $n \geq 2$, contains non-semilinear languages.
(viii) $PC_n REG$, for all $n \geq 3$, and $NPC_n REG$, $NCPC_n REG$, for all $n \geq 2$, contain one-letter non-regular languages.
(ix) If $L \subseteq V^*, L \in CPC_n REG$, then there is a constant q such that each $z \in L, |z| > q$, can be written in the form $z = x_1 y_1 x_2 y_2 \ldots x_m y_m x_{m+1}$, for $1 \leq m \leq n, y_i \neq \lambda, 1 \leq i \leq m$, and for all $k \geq 1$, $x_1 y_1^k x_2 y_2^k \ldots x_m y_m^k x_{m+1} \in L$.
      Consequence: $CPC_n REG \subset CPC_{n+1} REG$, for all $n \geq 1$.
(x) $PC_n REG \subset PC_{n+1} REG$, for all $n \geq 1$.
(xi) $CPC_n REG \subset CPC_n LIN \subset CPC_n CF$, for all $n \geq 1$.

The first equality in (i) is obvious, the others are proved in [17] (direct, non-constructive, simulation of a centralized PC grammar system by a type-0 grammar with bounded workspace); (ii), (iii), (vii), (viii) are proved by examples (some of them are mentioned in the previous section), (iv) uses the linear language $\{a^n b^m cb^m a^n \mid m, n \geq 1\}$, the first relation in (v) appears already in [124], but the second one is obtained in [140] using a long construction; for (vi) it is shown that $L_r(\Gamma)$ is (the gsm image of) a matrix language of finite index, the pumping property in (ix) is proved in [72], where also a direct, combinatorial, proof of (x) is given; (xi) is obtained by combining the previous results.

Here are some other important results.

Theorem 4.2. (i) $NPC_\infty CF \subseteq PC_\infty CF$,   (ii) $MAT \subset PC_\infty CF$,   (iii) $CPC_\infty REG \subset MAT_{fin}$.

The first relation indicates the fact that the non-returning feature can be simulated by the returning feature, in the powerful framework of non-centralized systems. The proof appears in [47], and it implies, as special cases, the inclusions $NPC_\infty X \subseteq PC_\infty X$, for $X \in \{REG, LIN\}$. The same relation for LIN was also proved in [139]; both [47] and [139] improve an earlier result of [88], where it is proved that $NCPC_\infty X \subseteq PC_\infty X$, for $X \in \{REG, LIN, CF\}$.

The second inclusion, proved in [85], confirms the power of non-centralization (and, according to Theorem 3.1 (ii), it implies that $CD_\infty(f) \subset PC_\infty CF$, $f \in \{= k, \geq k \mid k \geq 2\}$). The third inclusion is proved in [90].

The underlying idea of the proof in [88] of the inclusion $NCPC_\infty CF \subseteq PC_\infty CF$ is as follows (we describe it because when starting from centralized systems the construction is simpler than that used [47] and [139] for the non-centralized case). Starting from a system with components $P_1, \ldots, P_n$, centralized and working in the non-returning mode, one constructs a system which contains three components, $P_{i,1}, P_{i,2}, P_{i,3}$ associated with each P_i. $P_{i,3}$ is used only at the beginning of derivations just to synchronize, while "partners" $P_{i,1}, P_{i,2}$ simulate the work of P_i. When one of them uses a rule from P_i, the other gets ready to receive the current string of the partner in the case when P_0 (a new component associated with P_1 which is the master of the new system) asks for the string generated by P_i. Then the string is sent to P_0 and also at the same time to the partner, thus saving a copy of the string. In this way, although working in the returning mode, the communicated string will be processed farther. Some additional components are used for synchronizing the process and as "garbage collectors".

The strictness of (ii) is obtained by using the result from [61], that each one-letter matrix language is regular (from Theorem 4.1 (viii) we know that this is not the case for PC_3REG). The strictness of (iii) follows from Theorem 4.1 (iv) ($LIN \subset MAT_{fin}$).

Point (iii) of Theorem 4.2 implies point (vi) of Theorem 4.1, because matrix languages of finite index are known to be semiliniar.

Theorem 4.3. $LIN \subset PC_\infty REG$.

Since systems with right-linear components generate the strings "from left to right", one would think that linear languages of the form

$$L = \{wc \ mi(w) \mid w \in \{a, b\}^*\}$$

cannot be generated by such systems. This is true for centralized systems (see the proof for point (iv) of Theorem 4.1, in [17], using the linear language $\{a^n b^m cb^m a^n \mid n, m \geq 1\}$) and this is probably also true for non-centralized non-returning systems. However, the returning centralized framework provides tools for simulating linear grammars by systems with right-linear components.

Instead of presenting the proof of Theorem 4.3 from [48], we apply the construction from this proof to show that the language $L = \{wc \ h(mi(w))c \mid w \in \{a, b\}^*\}$, where h is the morphism defined by $h(a) = b, h(b) = a$, is in $PC_\infty REG$.

Consider the following linear grammar for L: $G = (\{S, A\}, \{a, b, c\}, S, P)$, with

$$P = \{p_0 : S \rightarrow Ac, \ p_1 : A \rightarrow aAb, \ p_2 : A \rightarrow bAa, \ p_3 : A \rightarrow c\}.$$

Let Γ be the following PC grammar system:

$$\Gamma = (N, \{a, b, c\}, K, (S_0, P_0), (S_1, P_1), (S_1', P_1'), (S_2, P_2), (S_2', P_2')),$$
$$N = \{S_0, S_{0,1}, S_{0,2}, S_1, S_{1,1}, S_1', S_{1,1}', S_1'', S_2, S_{2,1}, S_2', S_{2,1}', S_2'',$$
$$A, A', A'', \overline{A}\},$$
$$P_0 = \{S_0 \to c\overline{A}, \; S_0 \to S_{0,1}, \; S_{0,1} \to S_{0,2}, \; S_{0,2} \to Q_1',$$
$$S_{0,2} \to Q_2', \; A'' \to \overline{A}, \; A'' \to c\},$$
$$P_1 = \{S_1 \to S_{1,1}, \; S_{1,1} \to S_1, \; S_1 \to Q_0, \; \overline{A} \to bA\},$$
$$P_1' = \{S_1' \to S_{1,1}', \; S_{1,1}' \to S_1', \; S_1' \to S_1'', \; S_1'' \to aQ_1,$$
$$A \to A', \; A' \to A''\},$$
$$P_2 = \{S_2 \to S_{2,1}, \; S_{2,1} \to S_2, \; S_2 \to Q_0, \; \overline{A} \to aA\},$$
$$P_2' = \{S_2' \to S_{2,1}', \; S_{2,1}' \to S_2', \; S_2' \to S_2'', \; S_2'' \to bQ_2,$$
$$A \to A', \; A' \to A''\}.$$

The components P_1, P_1' are associated to the rule $p_1 : A \to aAb$, P_2, P_2' are associated to the rule $p_2 : A \to bAa$, and the rule $p_3 : A \to c$ is simulated by P_0, which also introduces the rightmost occurrence of c in the strings of L.

For the following derivation in G:

$$S \Longrightarrow_{p_0} Ac \Longrightarrow_{p_1} aAbc \Longrightarrow_{p_2} abAabc \Longrightarrow_{p_2} abbAaabc \Longrightarrow_{p_3} abbcaabc,$$

we obtain the following derivation in Γ (which simulates the rules in the reversed order):

$$(S_0, S_1, S_1', S_2, S_2') \Longrightarrow (c\overline{A}, S_{1,1}, S_{1,1}', Q_0, S_2'')$$
$$\Longrightarrow (S_0, S_{1,1}, S_{1,1}', c\overline{A}, S_2'') \Longrightarrow (S_{0,1}, S_1, S_1', caA, bQ_2)$$
$$\Longrightarrow (S_{0,1}, S_1, S_1', S_2, bcaA) \Longrightarrow (S_{0,2}, S_{1,1}, S_{1,1}', S_{2,1}, bcaA')$$
$$\Longrightarrow (Q_2', S_1, S_1', S_2, bcaA'') \Longrightarrow (bcaA'', S_1, S_1', S_2, S_2')$$
$$\Longrightarrow (bca\overline{A}, S_{1,1}, S_{1,1}', Q_0, S_2'') \Longrightarrow (S_0, S_{1,1}, S_{1,1}', bca\overline{A}, S_2'')$$
$$\Longrightarrow (S_{0,1}, S_1, S_1', bcaaA, bQ_2) \Longrightarrow (S_{0,1}, S_1, S_1', S_2, bbcaaA)$$
$$\Longrightarrow (S_{0,2}, S_{1,1}, S_{1,1}', S_{2,1}, bbcaaA') \Longrightarrow (Q_2', S_1, S_1', S_2, bbcaaA'')$$
$$\Longrightarrow (bbcaaA'', S_1, S_1', S_2, S_2') \Longrightarrow (bbcaa\overline{A}, Q_0, S_1'', S_{2,1}, S_{2,1}')$$
$$\Longrightarrow (S_0, bbcaa\overline{A}, S_1'', S_2, S_{2,1}, S_{2,1}') \Longrightarrow (S_{0,1}, bbcaabA, aQ_1, S_2, S_2')$$
$$\Longrightarrow (S_{0,1}, S_1, abbcaabA, S_2, S_2') \Longrightarrow (S_{0,2}, S_{1,1}, abbcaabA', S_{2,1}, S_{2,1}')$$
$$\Longrightarrow (Q_1', S_1, abbcaabA'', S_2, S_2') \Longrightarrow (abbcaabA'', S_1, S_1', S_2, S_2')$$
$$\Longrightarrow (abbcaabc, S_{1,1}, S_{1,1}', S_{2,1}, S_{2,1}').$$

It is instructive to note the role of the "waiting rules" $S_i \to S_{i,1}, S_{i,1} \to S_i$, $S_i' \to S_{i,1}', S_{i,1}' \to S_i'$, $i = 1, 2$, and the way how P_i, P_i' cooperate in order to simulate the rule p_i. The non-centralization is very essential.

The results from Theorems 4.1 – 4.3 concerning the relationships between families $Y_\infty X$ with $Y \in \{PC, CPC, NPC, NCPC\}$ and $X \in \{REG, CF\}$, and between these families and REG, LIN, CF, MAT, CS, RE, are given in Figure 1 (an arrow represents an inclusion, not necessarily proper). Note the symmetry of the diagram, and the fact that the four families REG, LIN, CF, MAT are included in the four families $CPC_\infty REG$, $PC_\infty REG$, $CPC_\infty CF$, $PC_\infty CF$, respectively. In the right hand part of the figure, two arrows are missing, indicating two important **open problems**: which of the inclusions $LIN \subseteq NPC_\infty REG$ and $MAT \subseteq NPC_\infty CF$ hold? We *conjecture* that the first inclusion does not hold, and in particular, that the linear language $\{wc\ mi(w) \mid w \in \{a, b\}^*\}$ is not in $NPC_\infty REG$.

Many other problems about PC grammar systems are still **open**. They are similar to those concerning CD grammar systems. The main one concerns the hierarchies on the number of components. An intriguing open problem concerns the relationship of CS to $PC_\infty CF$. and $NPC_\infty CF$.[2]

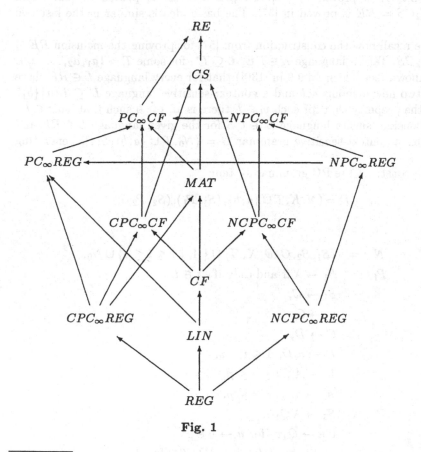

Fig. 1

---

[2] Recently, Şt. Bruda has proved that $NPC_\infty CF \subseteq CS$.

4.4 The context-sensitive case

We discuss the context-sensitive case separately, because it is completely settled (in a quite interesting way).

Theorem 4.4.
(i) $CS = Y_nCS = Y_\infty CS$, for all $n \geq 1$ and $Y \in \{CPC, NCPC\}$.
(ii) $CS = PC_1CS = PC_2CS \subset PC_3CS = PC_\infty CS = RE$.
(iii) $CS = NPC_1CS \subset NPC_2CS = NPC_\infty CS = RE$.

Non-centralized systems with three components working in the returning mode or with two components only, working in the non-returning mode, are powerful enough for generating any recursively enumerable language. In the centralized case, one does not exceed the power of context-sensitive grammars. This is again an illustration of the power of non-centralization.

Point (i) is proved in [17], $PC_3CS = RE$ is proved in [127], and $NPC_2CS = RE$ is proved in [57]. The basic idea is similar in the last two cases.

We recall now the construction from [57] for proving the inclusion $RE \subseteq NPC_2CS$. Take a language $L \in RE$, $L \subseteq T^*$, for some $T = \{a_1, a_2, \ldots, a_p\}$. It is known (see Theorem 9.9 in [130]) that for every language $L \in RE$ there exist two new symbols a, b and a context-sensitive language $L' \subseteq L\{a\}\{b\}^*$ with the property that for each $w \in L$ there is an $i \geq 0$ such that $wab^i \in L'$. Now consider such a language $L' \in CS$ for the given language $L \in RE$ and consider a context-sensitive grammar $G = (N_0, T \cup \{a, b\}, S, P)$ generating L'.

We construct the PC grammar system

$$\Gamma = (N, K, T \cup \{a, b\}, (S_1, P_1), (S_2, P_2)),$$

where

$$
\begin{aligned}
N &= \{S_1, S_2, C, D, X, Z\} \cup \{A_i \mid 1 \leq i \leq p\} \cup N_0, \\
P_1 &: \quad S_1 \rightarrow \lambda \text{ if and only if } \lambda \in L, \\
&\quad\quad S_1 \rightarrow C, \\
&\quad\quad C \rightarrow C, \\
&\quad\quad C \rightarrow D, \\
&\quad\quad D \rightarrow a_iD, \ 1 \leq i \leq p, \\
&\quad\quad D \rightarrow A_i, \ 1 \leq i \leq p, \\
&\quad\quad A_i \rightarrow a_i, \ 1 \leq i \leq p, \\
P_2 &: \quad S_2 \rightarrow XQ_1S, \\
&\quad\quad Cu \rightarrow Q_1v, \text{ for } u \rightarrow v \in P, \\
&\quad\quad C\alpha \rightarrow \alpha Q_1, \text{ for } \alpha \in N \cup T \cup \{a, b\},
\end{aligned}
$$

$$\alpha C \rightarrow Q_1\alpha, \text{ for } \alpha \in N \cup T \cup \{a,b\},$$
$$XD \rightarrow XQ_1Z,$$
$$a_iDZa_i \rightarrow a_iDa_iQ_1Z, \ 1 \le i \le p,$$
$$A_iZa_ia \rightarrow A_ia_iZa, \ 1 \le i \le p.$$

The reader can verify that $L_{nr}(\Gamma) = L$. In the first phase of a derivation, P_1 uses the rule $C \rightarrow C$ only and in the meantime P_2 generates a string wab^i with $w \in L$. After that, P_1 introduces D and starts producing a string $z \in T^*$. Now, P_2 checks symbol by symbol whether or not $z = w$ and only in the affirmative case the work of the system stops with a terminal string on the first component.

Observe that the first component of the PC grammar system in the previous construction is right-linear and that this component never asks for the strings generated by the second component.

In the above proof, the rule $S_1 \rightarrow \lambda$ in the first component is used only to generate λ whenever $\lambda \in L$. However if we allow rules $S_i \rightarrow \lambda$ for $i \ge 2$, then we can generate all recursively enumerable languages as follows. Consider an arbitrary Chomsky grammar $G = (N, T, S, P)$ and construct the PC grammar system Γ with two components, consisting of the following sets of rules:

$$P_1 \ : \ u \rightarrow v, \text{ for all } u \rightarrow v \in P, \text{ with } |u| \le |v|,$$
$$u \rightarrow vQ_2^m, \text{ for all } u \rightarrow v \in P \text{ with } |u| - |v| = m > 0,$$
$$P_2 \ : \ S_2 \rightarrow \lambda.$$

Obviously, $L_r(\Gamma) = L_{nr}(\Gamma) = L(G)$, and Γ is a centralized system.

Consequently, denoting by X_nCS', $n \ge 1$, the families of languages generated by systems with context-sensitive components where the rules $S_i \rightarrow \lambda$, $1 \le i \le n$, are allowed and $X \in \{PC, CPC, NPC, NCPC\}$, we obtain

Theorem 4.5. $CS = X_1CS' \subset X_2CS' = X_nCS' = RE$, *for all* $n \ge 2$ *and each* $X \in \{PC, CPC, NPC, NCPC\}$.

In this case, the hierarchy induced by the number of components reduces to two levels.

4.5 Non-synchronized PC grammar systems

An essential feature of PC grammar systems is the synchronization of the rewriting steps: there is a universal clock which marks the time (ticks) in the same way for all components and in each time unit (if no communication takes place) each component must use one of its rules, except for components whose strings are already terminal. What if such a synchronization is not required (which ammounts to adding rules $A \rightarrow A$ for all $A \in N$ to each component)? We get then unsynchronized systems.

We denote by $L_{u,r}(\Gamma)$ and $L_{u,nr}(\Gamma)$ the language generated by a PC grammar system Γ in the unsynchronized returning and unsynchronized non-returning modes, respectively. We also denote by UY_nX the family of languages generated by unsynchronized PC grammar systems, corresponding to families Y_nX discussed in previous sections. Here are some results concerning these families (for proofs, see [104], [108], [17]).

Theorem 4.6.
(i) $UCPC_\infty X = X$, for $X \in \{REG, LIN\}$.
(ii) $UPC_2 REG - REG \neq \emptyset$, $UPC_\infty REG \subseteq CF$,
      $UPC_2 LIN - CF \neq \emptyset$, $UCPC_2 CF - CF \neq \emptyset$.
(iii) $UNCPC_2 REG$ contains non-semilinear languages,
      $UNCPC_2 CF$ contains one-letter non-regular languages.
(iv) $(CPC_\infty REG \cap NCPC_\infty REG) - UCPC_\infty CF \neq \emptyset$.

Points (i) and (iv) show that desynchronizing decreases considerably the generative capacity, while points (ii) and (iii) show that unsynchronized PC grammar systems are still very powerful. For instance, for the systems

$$\Gamma_1 = (\{S_1, S_2, A, B\}, K, \{a, b\}, (S_1, P_1), (S_2, P_2)),$$
$$P_1 = \{S_1 \rightarrow bQ_2, B \rightarrow bQ_2, B \rightarrow b\},$$
$$P_2 = \{S_2 \rightarrow aS_2, S_2 \rightarrow aA, A \rightarrow aB\},$$
$$\Gamma_2 = (\{S_1, S_2, A, B\}, K, \{a, b\}, (S_1, P_1), (S_2, P_2)),$$
$$P_1 = \{S_1 \rightarrow aQ_2, S_1 \rightarrow aA, B \rightarrow bA\},$$
$$P_2 = \{S_2 \rightarrow aQ_1, A \rightarrow bB, A \rightarrow b\},$$

we have

$$L_{u,nr}(\Gamma_1) = \{(ba^n)^m b \mid m \geq 1, n \geq 2\} \text{ (non-semilinear)},$$
$$L_{u,r}(\Gamma_2) = \{a^{2m} a^{2s+1} b^{2t+1} \mid m, s, t \geq 1, s \geq t\} \text{ (non-regular)}.$$

4.6 Descriptional and communication complexity

The usefulness of PC grammar systems from the point of view of the descriptional complexity of context-free languages has been investigated (as in the case of CD grammar systems) for the three basic measures, Var, $Prod$, $Symb$. The definitions of $Prod(\Gamma)$ and $Symb(\Gamma)$, for a PC grammar system Γ, are obvious, however in the case of Var we have to consider also the used query symbols. If $\Gamma = (N, K, T, (S_1, P_1), \ldots, (S_n, P_n))$, then

$$Var(\Gamma) = card(N) + card(\{Q_i \mid \text{there is } A \rightarrow uQ_iv \text{ in some } P_j, 1 \leq j \leq n\}).$$

The basic results are as expected: a significant decrease of complexity, in the sense of relations $<_i$, $i = 1, 2, 3, 4$, considered in Section 3.6. The table given below summarizes the results from [106] and [17] (at the intersection of the row marked with $M \in \{Var, Prod, Symb\}$ with the column marked with $X \in \{PC, CPC, NPC, NCPC\}$ we have the relation ρ if $X_\infty CF \rho CF (M)$).

	PC	CPC	NPC	$NCPC$
Var	<4	<4	<4	<4
$Prod$	<4	<4	<4	<4
$Symb$	<1	<1	<2	<2

For instance, the system

$$\Gamma_n = (\{S_1, S_2\}, K, \{a, b\}, (S_1, P_1), (S_2, P_2)),$$
$$P_1 = \{S_1 \to S_1\} \cup \{S_1 \to Q_2^k b^k Q_2 \mid 1 \le k \le n\},$$
$$P_2 = \{S_2 \to aS_2, \ S_2 \to a\},$$

generates both in the returning and the non-returning mode the language

$$L_n = \bigcup_{k=1}^{n} \{a^{k_i} b^k a^i \mid i \ge 1\},$$

for which we have $Var_{CF}(L_n) = n + 1$. However, $Var(\Gamma_n) = 3$.

For PC grammar systems, a specific complexity measure, intrinsically related to the mode of working of these systems, is *the number of communications.*

Consider a PC grammar system $\Gamma = (N, K, T, (S_1, P_1), \ldots, (S_n, P_n))$ and let D be a derivation in Γ, in mode $f \in \{r, nr\}$,

$$D : (S_1, \ldots, S_n) \Longrightarrow_f (w_{1,1}, \ldots, w_{1,n}) \Longrightarrow_f \cdots \Longrightarrow_f (w_{k,1}, \ldots, w_{k,n}).$$

Let

$$Com(w_{i,1}, \ldots, w_{i,n}) = \sum_{j=1}^{n} |w_{i,j}|_K, 1 \le i \le k, \text{ and}$$

$$Com(D) = \sum_{i=1}^{k} Com(w_{i,1}, \ldots, w_{i,n}).$$

For $x \in L_f(\Gamma)$, define

$$Com(x, \Gamma) = \min\{Com(D) \mid D : (S_1, \ldots, S_n) \Longrightarrow_f^* (x, \alpha_2, \ldots, \alpha_n)\}.$$

Then

$$Com(\Gamma) = \sup\{Com(x, \Gamma) \mid x \in L_f(\Gamma)\},$$

and, for a language L and a class $X_\infty CF$ of PC grammar systems, $X \in \{PC, CPC, NPC, NCPC\}$,

$$Com_X(L) = \min\{Com(\Gamma) \mid L = L_f(\Gamma), \Gamma \in X_\infty CF\}.$$

In what follows we shall consider only centralized returning systems, hence we do not specify the class $X_\infty CF$ and write simply Com for $Com_{CPC_\infty CF}$.

Here are some results from [106] and [17].

Theorem 4.7.

(i)   For each $n \geq 1$, there is $L_n \in CPC_\infty CF$ such that $Com(L_n) = n$. (In terms of [60], Com is a connected measure.)

(ii) $Com(\Gamma)$ and $Com(L_r(\Gamma))$ cannot be algorithmically computed for an arbitrary context-free (centralized and returning) PC grammar system Γ.

(iii) It is not decidable whether or not $Com(\Gamma) = Com(L_r(\Gamma))$ for an arbitrary context-free (centralized and returning) PC grammar system Γ.

(iv) The measure Com is incompatible with each of Var, Prod, Symb (that is, there are languages L such that $Com(\Gamma)$ and $M(\Gamma), M \in \{Var, Prod, Symb\}$, cannot be simultaneously minimized, for $L = L_r(\Gamma)$).

Point (i) is proved using the language

$$L_n = \{b(a^i b a^i)^{2n+1} b \mid i \geq 1\},$$

for which $Com(L_n) = n$. Point (ii) is proved considering languages of the form

$$L' = L(G)\{c, d, e\}^+ \cup \{a, b\}^+ L,$$

where G is an arbitrary context-free grammar with the terminal alphabet $\{a, b\}$ and $L = \{c^n d^m c e^m \mid m \geq n \geq 1\} \in CPC_\infty CF - CF$. Then $L' \in REG$ if and only if $L(G) = \{a, b\}^+$, which is undecidable. When $L' \notin REG$, then $L' \notin CF$, hence $Com(L') > 1$. Thus $Com(L')$ is not computable (and the same holds for a given system Γ generating L' and having $Com(\Gamma) = 0$ if and only if $L_r(\Gamma) = \{a, b\}^+ \{c, d, e\}^+$).

Point (iii) can be obtained in the same way, whereas point (iv) is proved using the language

$$L = \{a^n b^n c b^n c b^n c a^n \mid n \geq 1\}.$$

The reader can check that $Com(L) = 2$, but for all systems Γ with $Com(\Gamma) = 2, L_r(\Gamma) = L$, we have $M(\Gamma) > M_{CPC_\infty CF}(L), M \in \{Var, Prod, Symb\}$.

Therefore, Com behaves very much like the index of context-free grammars and languages (the index is also not algorithmically computable [60]). As in the case of the index, one may allow only a bounded number of communications. There are two possibilities: to consider systems Γ such that $Com(\Gamma) \leq k$, for a given k, or to consider arbitrary Γ and to select from $L_f(\Gamma), f \in \{r, nr\}$, only those strings that can be generated by a bounded number of communications. In both cases, the obtained families of languages have many nice (context-free-like) properties. It turns out that bounding the number of communications is a very strong restriction.

For instance, let us denote by $X_n^k Y$ and $X_\infty^k Y$, for $X \in \{PC, CPC, NPC, NCPC\}$, $Y \in \{REG, LIN, CF\}, n \geq 1, k \geq 1$, the family of languages L in $X_n Y$ and $X_\infty Y$, respectively, such that $Com_{X_\infty Y}(L) \leq k$. Obviously,

$$X_n^k Y \subseteq X_m^l Y,$$

for all X, Y as above, $k \leq l, n \leq m$.

The following relations were proved in [70], [137], and [101], completing point (i) of Theorem 4.7.

Theorem 4.8.
(i) $CPC_n^{k-1}REG \subset CPC_n^k REG$, $n \geq 2, k \geq n$.
(ii) $CPC_\infty^k REG \subset CPC_\infty^{k+1} REG$, $k \geq 1$.

The following pumping result (similar to point (ix) in Theorem 4.1) is used when proving these strict inclusions.

Theorem 4.9. *For every $L \in CPC_\infty^k REG$ there is a constant p such that every $w \in L$ with $|w| \geq p$ can be written in the form*

$$w = w_1 \alpha_1 w_2 \alpha_2 \ldots w_{k+1} \alpha_{k+1} w_{k+2},$$

where
(i)   *there is j, $1 \leq j \leq k+1$, with $|\alpha_j| > 0$,*
(ii) *$|w_{k+1}| < p$,*
(iii) *for all $i \geq 1$, $w_1 \alpha_1^i w_2 \alpha_2^i \ldots w_{k+1} \alpha_{k+1}^i w_{k+2} \in L$.*

Here is another result illustrating the power of restricting the number of communications.

Theorem 4.10. $CPC_n^1 Y = CPC_2^1 Y$, $Y \in \{REG, LIN\}$.

A proof (and related results) can be found in [137] and [125].

This theorem and the inclusion $CPC_2 REG \subset CF$ (point (v) of Theorem 4.1) yield an interesting corollary: any system Γ, with any number of regular components, generates a context-free language when each string of $L_r(\Gamma)$ can be produced by derivations using at most one communication.

All the results above deal with centralized (and returning) systems, hence with systems having a very particular communication graph: a star. A number of other particular communication graphs were investigated in [64], [65], and [100], and infinite hierarchies corresponding to the number of communication steps were in general obtained for the corresponding families of languages. Among the considered communication graphs are: trees, one-way rings, two-way rings, one-way arrays, two-way arrays, and directed acyclic graphs. The reader is referred to the papers cited above for details.

4.7 PC grammar systems with communication by command

In the PC grammar systems considered in the previous sections, a request for communication is initiated by the receiving component by introducing query symbols. We say that this is a communication *on request*. Easy to be formalized and mathematically fruitful, this is not the only possibility. From the practical point of view, at least equally important is the communication *by command*, initiated by the component of a network which sends messages.

PC grammar systems with communication by command modeling the WAVE paradigm, [50], [51], [132], used by a series of parallel machines (some of them commercially available, such as the Connection Machine, [142]) were introduced in [25].

The basic idea is to consider a system consisting of several grammars, like in a usual PC grammar system, working separately on their sentential forms, where each of the grammars has its own regular (or other type) language playing the role of an "entry filter". In certain situations, the rewriting is interrupted and each component sends its current sentential form to other components, namely to those for which the entry filter contains the sentential form. As in the usual PC grammar systems, the set of terminal strings generated in this way by a *master* grammar is the language generated by the system.

Several issues have to be settled here before a formal model based on such a communication can be formulated:

1. When a rewriting process has to be interrupted, and the sentential forms to be checked whether or not they can be communicated? A possibility is to do this after every single derivation step (after using a rule in each component, synchronously), or after a maximal derivation step in each component (all components work as long as they can).
2. What should be communicated: the whole current sentential form or only a part of it? (In the latter case: which part of it?)
3. What is the next sentential form of a component which have communicated its sentential form without receiving other strings? (Hence again: returning or non-returning?)
4. What if multiple messages are communicated to the same component? A number of solutions are possible here, some of them inspired by the way of solving such conflicts in practical cases, [62], [142], where a function is computed over the messages (e.g., a binary addition). Most natural seem to be two variants: to select one of the received messages (nondeterministically, or according to a given priority ordering of the components), or to concatenate all received messages, in the order of the components.

We discuss now a variant, investigated in [25], [68]. These are systems working with *maximal* derivations as rewriting steps, communicating *without splitting* the strings, *replacing* the string of the target component by a *concatenation* of the received messages, in the order of the system components, and *returning* to axioms after communicating; the generated language will be the language of the first component (which is the *master*).

Formally, a PC grammar system *with communication by command* (a CCPC grammar system, for short) is a construct of the form

$$\Gamma = (N, T, (S_1, P_1, R_1), \ldots, (S_n, P_n, R_n)), \ n \geq 1,$$

where N is the *nonterminal* alphabet, T is the *terminal* alphabet (N and T are disjoint), and $(S_i, P_i, R_i), 1 \leq i \leq n$, are the *components* of the system, where $S_i \in N$ is the *axiom*, P_i is the set of *production rules* over $N \cup T$, and R_i, with $R_i \subseteq (N \cup T)^*$, is the *selector language* of the component i.

We consider here only λ-free context-free systems (hence each $P_i \subseteq N \times (N \cup T)^+$).

A *rewriting step* in Γ is defined by

$$(x_1, \ldots, x_n) \Longrightarrow (y_1, \ldots, y_n) \text{ iff for each } i, 1 \le i \le n, x_i \Longrightarrow^* y_i \text{ in } P_i$$
$$\text{and there is no } z_i \in (N \cup T)^* \text{ such that } y_i \Longrightarrow z_i \text{ in } P_i.$$

Aa *communication step*, denoted by $(x_1, \ldots, x_n) \vdash (y_1, \ldots, y_n)$, is defined as follows.

Let, for $1 \le i, j \le n$,

$$\delta(x_i, j) = \begin{cases} \lambda, & \text{if } x_i \notin R_j \text{ or } i = j, \\ x_i, & \text{if } x_i \in R_j \text{ and } i \ne j. \end{cases}$$

Let, for $1 \le j \le n$,

$$\Delta(j) = \delta(x_1, j)\delta(x_2, j) \ldots \delta(x_n, j),$$

(this is the "total message" to be received by the j-th component), and let, for $1 \le i \le n$,

$$\delta(i) = \delta(x_i, 1)\delta(x_i, 2) \ldots \delta(x_i, n),$$

(this is the "total message" sent by the i-th component; if $\delta(i) = x_i^k$, then the i-th component has sent a message to k components).

Then, for $1 \le i \le n$, we define

$$y_i = \begin{cases} \Delta(i), & \text{if } \Delta(i) \ne \lambda, \\ x_i, & \text{if } \Delta(i) = \lambda \text{ and } \delta(i) = \lambda, \\ S_i, & \text{if } \Delta(i) = \lambda \text{ and } \delta(i) \ne \lambda. \end{cases}$$

Thus, y_i is either the concatenation of the messages received by the i-th component, if it receives at least one message, or it is the previous string, when the i-th component is not involved in communications, or it is equal to S_i, if the i-th component sends messages but it does not receive messages. Observe that a component cannot send messages to itself.

The generated language is defined as follows:

$$L(\Gamma) = \{w \in T^* \mid (S_1, \ldots, S_n) \Longrightarrow (x_1^{(1)}, \ldots, x_n^{(1)}) \vdash (y_1^{(1)}, \ldots, y_n^{(1)})$$
$$\Longrightarrow (x_1^{(2)}, \ldots, x_n^{(2)}) \vdash (y_1^{(2)}, \ldots, y_n^{(2)}) \Longrightarrow \ldots$$
$$\Longrightarrow (x_1^{(s)}, \ldots, x_n^{(s)}), \text{ for } s \ge 1 \text{ and } w = x_1^{(s)}\}.$$

In what follows we denote by $CCPC_n X$ the family of languages $L(\Gamma)$, generated by CCPC grammar systems with at most n components of the type X, $n \ge 1$. Here we shall consider $X \in \{REG, CF\}$, where REG indicates the use of right-linear rules, whereas CF indicates the use of λ-free context-free rules. When the number of components is not specified, we write $CCPC_\infty X$.

Example 4.1. Let

$$\Gamma_1 = (N, \{a, b, c\}, (S_1, P_1, R_1), (S_2, P_2, R_2), (S_3, P_3, R_3)),$$
$$N = \{S_1, S_2, S_2', S_3, S_3', X\},$$
$$P_1 = \{S_1 \to aS_1, S_1 \to bS_1, S_1 \to X\}, \ R_1 = \{a, b\}^* c,$$
$$P_2 = \{S_2 \to S_2', X \to c\}, \ R_2 = \{a, b\}^* X,$$
$$P_3 = \{S_3 \to S_3', X \to c\}, \ R_3 = \{a, b\}^* X.$$

We start from (S_1, S_2, S_3). A maximal componentwise derivation is of the form

$$(S_1, S_2, S_3) \Longrightarrow (xX, S_2', S_3'),$$

for some $x \in \{a, b\}^*$. The string xX will be communicated to both the second and the third component, hence we have

$$(xX, S_2', S_3') \vdash (S_1, xX, xX) \Longrightarrow (yX, xc, xc) \vdash$$
$$\vdash (xcxc, yX, yX) \Longrightarrow (xcxc, yc, yc),$$

for some $y \in \{a, b\}^*$. The string $xcxc$ is terminal, hence we obtain

$$L(\Gamma_1) = \{xcxc \mid x \in \{a, b\}^*\}.$$

Therefore, the very simple CCPC system Γ_1, with only three right-linear components, is able to generate the non-context-free (replication) language above. Observe that each derivation in Γ_1 contains exactly two communication steps (and three rewriting steps, the last one being considered only for the sake of consistency with the definition of $L(\Gamma)$, where the last step is supposed to be a rewriting one).

The following theorem summarizes the results from [25], [69] and [68]:

Theorem 4.11.
(i) $REG = CCPC_1REG = CCPC_2REG \subset CCPC_3REG = CCPC_4REG$
$= \ldots = CCPC_\infty REG = CS.$
(ii) $CF = CCPC_1CF \subset CCPC_2CF = CCPC_3CF = \ldots = CCPC_\infty CF$
$= CS.$

Observe the fact that the hierarchies on the number of components collapse in both cases. Systems with two context-free components and systems with three regular components can simulate systems with arbitrarily many components; characterizations of the family CS are obtained (characterizations of RE are obtained when λ-rules are allowed). The surprising relations here are

$$CS \subseteq CCPC_3REG, \text{ and } CS \subseteq CCPC_2CF.$$

A proof of the first inclusion can be found in [69]. We give now the main part of the proof from [25] for the second inclusion.

Let L be a context-sensitive language generated by a grammar $G = (N, T, S, P)$ in (weak) Kuroda normal form, that is, having productions of the form $AB \to CD$, $A \to BC$, $A \to B$, and $A \to a$, where A, B, C, D are nonterminals and a is a terminal symbol. Without loss of generality, we may assume that $A \neq B$ in rules of the form $AB \to CD$. If $r : AA \to CD \in P$, then we replace it by $A \to (A, r), (A, r)A \to CD$. We construct a CCPC grammar system $\Gamma = (\overline{N}, T, (S_1, P_1, R_1), (S, P_2, R_2))$ with context-free components which simulates the derivations in G and generates L.

Let S_1, S_1' be symbols not in $V = N \cup T$, and let $V' = N' \cup T$, where N' denotes the set of primed version of symbols in N.

Now we define Γ as follows:

$$\overline{N} = N \cup N' \cup \{S_1, S_1'\}$$
$$\cup \{A^{(r)} \mid A \in N, r : A \to x \in P, \text{ or}$$
$$r : AB \to CD \in P, \text{ or } r : BA \to CD \in P\},$$

$$P_1 = \{S_1 \to S_1'\}$$
$$\cup \{A' \to A \mid A \in N\}$$
$$\cup \{A^{(p)} \to a \mid p : A \to a \in P\}$$
$$\cup \{A^{(r)} \to BC \mid r : A \to BC \in P\}$$
$$\cup \{A^{(s)} \to B \mid s : A \to B \in P\}$$
$$\cup \{A^{(q)} \to C, B^{(q)} \to D \mid q : AB \to CD \in P\},$$

$$R_1 = \{\alpha A^{(p)} \beta \mid \alpha, \beta \in V'^*, p : A \to a \in P\}$$
$$\cup \{\alpha A^{(r)} \beta \mid \alpha, \beta \in V'^*, r : A \to BC \in P\}$$
$$\cup \{\alpha A^{(s)} \beta \mid \alpha, \beta \in V'^*, s : A \to B \in P\}$$
$$\cup \{\alpha A^{(q)} B^{(q)} \beta \mid \alpha, \beta \in V'^*, q : AB \to CD \in P\},$$

$$P_2 = \{A \to A' \mid A \in N\}$$
$$\cup \{A \to A^{(p)} \mid p : A \to a \in P\}$$
$$\cup \{A \to A^{(r)} \mid r : A \to BC \in P\}$$
$$\cup \{A \to A^{(s)} \mid s : A \to B \in P\}$$
$$\cup \{A \to A^{(q)}, B \to B^{(q)} \mid q : AB \to CD \in P\},$$

$$R_2 = V^*.$$

The second component indicates the application of some production of G by rewriting each $A \in N$ occurring in its sentential form, x_2, either to the corresponding primed version or to a version superscripted by a label of a corresponding production. The so obtained string, y_2, is communicated if and only if (1) either it has exactly one occurrence of either $A^{(p)}$ or $A^{(r)}$ or $A^{(s)}$, for some productions $p : A \to a$, $r : A \to BC$, $s : A \to B$ in P and the other nonterminal letters are primed versions of elements of N, or (2) y_2 has exactly one substring of the form $A^{(q)}B^{(q)}$ for some production $q : AB \rightarrow CD$ and the remaining nonterminal symbols are again from N'.

The first component, P_1, simulates the application of the corresponding production and rewrites the symbols from N' to their corresponding elements from N.

The reader may verify that $L(G) = L(\Gamma)$, hence indeed $CS \subseteq CCPC_2CF$.

4.8 Further variants and results

In a PC grammar system, a derivation ends when the master grammar produces a terminal string. Another possibility is to collect in the language generated by the system all the terminal words produced by the components of the system. Two variants can be considered: the derivation stops when (at least) one component produces a terminal string, or the derivation can be continued. Results about PC grammar systems with such *competitive* derivations (the first component which reaches a terminal string contributes to the system language and the derivation terminates) and context-free or L-components can be found in [138].

PC grammar systems with L-components were considered in [107], where it is proved that PC systems with 0L or DT0L components generate languages not in $ET0L$. However, centralized PC systems with ED0L components, working in the returning mode, generate only EDT0L languages (the tables can compensate the cooperation in the PC grammar systems style).

A promising idea is to consider PC grammar systems with communication on request but using *query words* for initiating the communication. Here each component has associated a finite/regular language of query words; when a component i produces a string having as a substring a query word associated with a component j, then the sentential form of the component j is communicated to the component i and it replaces the occurrence of the query word in the sentential form of the component i. Because the query words can be produced in a large number of steps, possibly involving also communications, the work of such systems seems to be quite intricate. And (like in the case of PC grammar systems with communication by command), because by query words we can check context conditions, easy characterizations of recursively enumerable languages can be obtained in this way; see [93].

Further types of synchronization are discussed in [108] and [17], adding to the usual synchronization (one rule per each time unit) restrictions concerning the used rules/nonterminals.

We have not discussed yet results about the closure and decidability properties of PC grammar systems and language families. Thus, e.g., it is known [17] that the family $PC_\infty CF$ is closed under union, concatenation, Kleene +, and substitution by λ-free regular languages.

The closure under intersection by regular languages as well as the closure properties of other families of languages generated by PC grammar systems are still *open*. For instance, it is conjectured in [17] that $CPC_\infty CF$ is not closed under concatenation and Kleene +.

Only a few decidability results are given in [17] and some other appear in [136]. Among them, the *circularity problem* is worth mentioning, since it is specific to PC grammar systems. By definition, when a cycle of queries appears in a derivation, the system gets stuck. Of course, the problem concerns only non-centralized systems. Is it decidable whether or not circular queries can appear in the work of an arbitrary system? The problem is still *open* for returning PC grammar systems with context-free components. It is decidable for returning or non-returning systems with linear components as well as for non-returning systems with context-free components (see [136]) and it is not decidable for systems with context-sensitive components.

In [49] one proves that the pattern languages in the sense of [3] are contained in all families $Y_\infty X, Y \in \{PC, CPC, NPC, NCPC\}, X \in \{REG, CF\}$, except for $CPC_\infty REG$ and $NCPC_\infty REG$. Theorem 3.6 in [3] proves that the uniform membership problem is NP-complete for pattern languages. It follows then that the uniform membership problem is at least NP-complete for all families mentioned above. On the other hand, if a PC grammar system Γ is given, then the problem whether or not an arbitrary string is in $L_f(\Gamma)$, for $f \in \{r, nr\}$, can be solved in polynomial time for Γ linear [9]. (That is, the non-uniform membership problem can be solved in polynomial time for linear PC grammar systems.)

Very few results are known concerning the difference between PC grammar systems with right-linear rules and regular rules in the restricted sense. For centralized returning systems, the right-linear rules are strictly more powerful [48] but for other cases the problem is still unsolved.

Related to this topic is the question whether or not λ-rules can be removed without decreasing the generative power. In general, not much is known about normal forms for PC grammar systems. Also very little is known about *automata characterizing* families of languages generated by various classes of PC grammar systems.

Several variants of PC grammar systems were proposed also in [92]: systems with infinitely many components (hence using query words; a restriction on the complexity of each component – on *Symb*, for instance – is necessary, otherwise any language can be generated in this way), *hybrid* systems, with the components working in different modes (returning or non-returning, maybe also using different numbers of rules in each time unit, not exactly one each as in usual systems), and systems with *partial activity* (take a collection of sets of components and in each time unit let exactly one such set to be active; a situation intermediate between synchronized and unsynchronized systems is obtained), etc.

Other variants of PC grammar systems can be found in [120] and [138].

5. Related models

5.1 Eco-grammar systems

An eco-grammar (EG) system is a recently introduced grammatical model aiming at the modeling of the interplay between *environment* and *agents* in complex systems such as ecosystems. An EG system can be viewed as a generalization of both CD and PC grammar systems. The underlying idea in defining the notion of an EG system were various descriptions of life as considered in the area of Artificial Life: "Life is a pattern in spacetime", with "interdependence of parts" and a "functional interaction with the environment" [77], [52]. A general frame where such statements obviously apply is that of an ecosystem, where the basic relation is the relation between *environment* and *agents* (e.g., animals). Under the fundamental assumption that the state of the components of such a system can be described by *strings* over given alphabets, we are led to a representation as shown in Figure 2. A detailed discussion concerning the motivation, as well as various formal properties of EG systems can be found, e.g., in [24] (where the EG systems were introduced), in [23] and in [14].

Let us discuss now the conceptual construction of an EG system as presented in Figure 2. We distinguish the environment, described by a string w_E over some alphabet V_E and developing according to a set P_E of 0L rules,

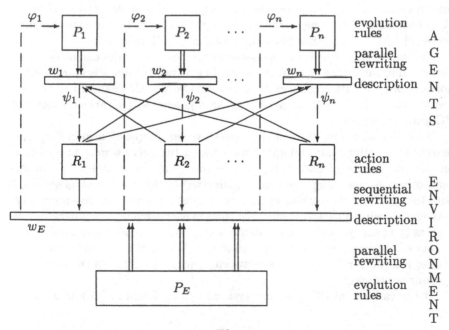

Fig. 2

and n agents $A_1, \ldots, A_n$. Each agent A_i is described by a string w_i, over an alphabet V_i and it evolves according to a set P_i of 0L rules. At a given moment, only a subset $\varphi_i(w_E)$ of P_i is *active*, depending on the current state of the environment. Moreover, each agent A_i has an associated set R_i of rewriting rules by which A_i acts, locally, on the environment or on another agent. These rules are used in a sequential manner (using $x \to y$ means to replace exactly one occurrence of x by y). At a given moment, a subset $\psi_i(w_i)$ of R_i is active, depending on the current state of A_i. A rule $x \to y$, with x, y consisting of symbols from V_E, will be used for acting on the environment; a rule with x, y over an alphabet V_j will be used for acting on the j-th agent. The whole "life" of the system is supposed to be governed by a universal clock, dividing the time in unit intervals: in every time unit, the agents act on the environment or on other agents (using exactly one action rule), then the evolution rules rewrite, in parallel, all the remained symbols in the strings describing the environment and the agents. Thus, the action has priority over evolution.

Note the essential difference between environment and agents and the generality of the model.

Thus, formally, an EG system of degree $n, n \geq 1$, is a construct

$$\Sigma = (E, A_1, A_2, \ldots, A_n),$$

such that

$$E = (V_E, P_E) \text{ is the environment}$$
$$A_i = (V_i, P_i, R_i, \varphi_i, \psi_i), \ 1 \leq i \leq n, \text{ are the agents}$$

where

V_E, V_i alphabets $(V_E \cap V_i = \emptyset), 1 \leq i \leq n$,

P_E a finite and complete set of 0L rules over V_E,

P_i a finite and complete set of 0L rules over $V_i, 1 \leq i \leq n$,

R_i a finite set of rewriting rules either over V_E or over $\bigcup\limits_{j \neq i} V_j$,

$1 \leq i \leq n$,

$\varphi_i : V_E^* \longrightarrow 2^{P_i}, \ 1 \leq i \leq n$,

$\psi_i : V_i^* \longrightarrow 2^{R_i}, \ 1 \leq i \leq n$.

An n-tuple $\sigma = (w_E, w_1, \ldots, w_n)$, with $w_E \in V_E^*$ and $w_i \in V_i^*, 1 \leq i \leq n$, is called a *state* of the system.

Starting from an initial state σ_0, Σ will *develop* a sequence of states, $\sigma_0 \Longrightarrow_\Sigma \sigma_1 \Longrightarrow_\Sigma \sigma_2 \Longrightarrow_\Sigma \ldots$, where $\Longrightarrow_\Sigma$ denotes the transition between states. For simplicity, we define the state transition only for the case when R_i contain only rules over V_E (no interaction between agents; in the papers quoted above only such systems were investigated). Thus we write $(w_E, w_1, \ldots, w_n) \Longrightarrow_\Sigma (w'_E, w'_1, \ldots, w'_n)$ if the following conditions hold:

1. $w_i \Longrightarrow w'_i, 1 \leq i \leq n$, is an L rewriting using the rules from $\varphi_i(w_E)$,
2. $w_E = z_1 x_1 z_2 x_2 \ldots z_n x_n z_{n+1}$,
   $w'_E = z'_1 y_1 z'_2 y_2 \ldots z'_n y_n z'_{n+1}$, where
   $z_r \Longrightarrow z'_r, 1 \leq r \leq n+1$, is an L rewriting using the rules in P_E,
   $x_j \to y_j \in \psi_{i_j}(w_{i_j}), 1 \leq j \leq n, \{i_1, \ldots, i_n\} = \{1, 2, \ldots, n\}$

Thus the next state of an agent A_i is defined by its evolution rules, as selected by the mappings φ_i depending on the environment. The state of the environment is modified first by the action rules from the sets $\psi_i(w_i)$ of all agents – each of them uses exactly one rule – then by the evolution rules in the set P_E, used in the L mode on the remaining symbols.

We denote by $Seq(\Sigma, \sigma_0)$ the set of all sequences $\sigma_0 \Longrightarrow_\Sigma \sigma_1 \Longrightarrow_\Sigma \ldots$ as above.

A number of "natural" problems (related to Artificial Life) can be considered in this framework. Here are some of them. Under what conditions is $Seq(\Sigma, \sigma_0)$ infinite? How does one reach a blocking state, for which no further step is possible? Given a state σ, can we find a predecessor state σ' such that $\sigma' \Longrightarrow \sigma$? Are there "garden of Eden" states, for which no precedessors can be found? What do the infinite sequences of states look like? Must they be periodical or not? Must they repeat arbitrarily long sequences of states (with specific properties)?

The blocking of a system can appear either when the sets $\varphi_i(w_E)$ do not contain enough rules for rewriting the strings w_i (w_i contains symbols for which there are no rules in $\varphi_i(w_E)$), or when the environment does not contain enough symbols for applying the action rules of the agents. The first case corresponds to an influence of the environment (through the selection mapping φ_i) which is harmful for the agent, the second one can be interpreted as *overpopulation*.

We can consider much more specific life-like aspects. An agent which either cannot evolve (as above) or has the corresponding string empty, can be considered *dead*. A possible variant is to associate with each A_i a "starvation constant", s_i, and to allow A_i not to evolve, or not to act on the environment or on other agents for at most s_i time units in a row. A dead agent is removed from the system. On the other hand, we can introduce the *birth*, by considering a special symbol in alphabets V_i, say ⊔ (already used with the same purpose in the so-called "L systems with fragmentation", see, e.g., [128]). When ⊔ appears, it splits the string in parts, each part describing the state of a newly born agent. Thus, e.g., $w_i = w'_i \sqcup w''_i$ leads to two new agents, described by w'_i and w''_i, respectively. All the new agents introduced in this way will have the same evolution rules P_i, the same action rules R_i, and the same selection mappings φ_i, ψ_i (this corresponds to the inherited characteristics in real life).

Agents not changing their state for a "long" interval can be considered *adult*. If they remain constant forever, we can call them *living fossils*. If the stagnation period is well determined and it depends on the environment state, we can model in this way *hibernating* animals. Agents acting only on another agents, not on environment, can be considered *carnivorous*; when an agent A_i acts systematically on a specific another agent A_j then we say that A_i is *parasitic*.

Depending on a statistics of symbols appearing in the environment description, we may speak about *pollution, changes of seasons*, and so on. Many other possibilities to capture real life-like features in an EG system are discussed in [24] and [14]. A promising idea is to consider hierarchical EG systems, with the agents of one level being the environment for the agents of the level above them. Trophic chains can be modelled in this way.

From the above it should be clear that the model is powerful indeed: it can capture at its *artificial* level many aspects of *real* life. We will discuss now some formal language theoretic problems and results related to EG systems; many of these problems and results are relevant for the environment-agents interplay.

With an EG system Σ and an initial state σ, we can associate the language of all strings w_E describing the environment in all states reachable by Σ. This language *generated* by Σ is denoted by $L(\Sigma, \sigma_0)$ and by $LEG(n)$ we denote the family of all languages generated by systems with at most n agents.

We refer to [24], [34], and [111] for results concerning families $LEG(n)$ – we present here only one such result.

Like in L systems, also in EG grammar systems we can consider extended symbols. Then, besides $L(\Sigma, \sigma_0)$, we associate with Σ and σ_0 the language $L_E(\Sigma, \sigma_0) = L(\Sigma, \sigma_0) \cap T^*$, for a given alphabet T of terminal symbols. We denote by $ELEG(n)$ the corresponding family of languages and we use $ELEG_{reg}(n)$ to denote the family of languages generated by EG systems with extended symbols, without interaction between agents, and with *regularly defined* mappings φ_i, ψ_i. We say that φ_i is regularly defined if $\varphi_i^{-1}(P) = \{w \mid \varphi_i(w) = P\}$ is a regular language for each $P \subseteq P_i$; similarly for $\psi_i, i = 1, 2, \ldots$.

The following result is from [24].

Theorem 5.1. $ELEG_{reg}(1) = RE$.

This result is quite significant: EG systems with regularly defined selection mappings (but with extended symbols) and with only one agent are as powerful as Turing machines. This is due mainly to the "self-referenced" structure of the system and to the generality of mappings φ_i, ψ_i, which, although regularly defined, can check whether or not the string w_E has a specific form. In this way one can influence the rewriting of w_E by the form of w_E itself, and simulate non-context-free rewriting by context-free action rules controlled in an appropriate way by φ_i, ψ_i.

By the Turing–Church thesis, $ELEG_{reg}(n) \subseteq RE$ for all n. Altogether we get $ELEG_{reg}(1) = ELEG_{reg}(n)$, for all $n \geq 1$. Hence systems with one agent only are as powerful as systems with an arbitrary number of agents.

In EG systems considered above the agents are acting only locally on the environment. This is not the case when the agents are powerful polluting sources, volcanos, damaged nuclear power plants, and so on, when we may suppose that the agents influence the very evolution of their environment. This was the basic idea of considering in [26], [27] the so-called *conditional tabled* EG systems. The structure of such a system is considered in Figure 3. Again the environment and the agents are described by strings and evolve according to given sets of L rules. However, we no longer consider action rules (the agents act already on the environment, influencing its evolution). As before, the active rules of agents depend, via the mappings φ_i, on the state of the environment, but furthermore, also the active rules of the environment are selected according to the current state of all agents, via a mapping δ. These mappings are defined in a *conditional* way: one gives certain *permitting* strings and certain *forbidding* strings for each mapping and for each set of rules (a table, in the terminology of L systems). A table is used only when the associated permitting string appears in the current checked string and the forbidding string does not appear in the current string. The presence/absence of a string x in another string w can be defined in various modes: x is a *block* of w, x is a *scattered* substring of w, or x is a *permuted* scattered substring of w. We denote by $CTEG_{(i,j;\alpha)}(n)$ the family of environmental languages generated by such conditional tabled EG systems with at most n agents, with permitting contexts of length at most i and with forbidding contexts of length at most j, where the conditions are checked in the mode $\alpha \in \{b, s, p\}$ (b = block, s = scattered, p = permutted scattered). Whenever i, j or n is not specified we replace it by ∞. When $i \leq 1$ and $j \leq 1$, α is not specified, because the three modes b, s, p coincide.

Various results about conditional EG systems can be found in [26], [27], and [86]. We recall now three of them.

Theorem 5.2.
(i) $CTEG_{(\infty,0;\alpha)}(n) = CTEG_{(\infty,0;\alpha)}(1)$, for $\alpha \in \{s, p\}$. *(Again one agent systems are as powerful as arbitrary systems.)*
(ii) The families $CTEG_{(1,0)}(1)$ and $CTEG_{(0,2;\alpha)}(1)$, for $\alpha \in \{s, p\}$, contain languages which are not ET0L languages, but $CTEG_{(0,1)}(\infty) \subset ET0L$.
(iii) Every recursively enumerable language is the morphic image of a language in the family $CTEG_{(1,2;b)}(1)$.

Further variants of EG systems were considered in [96] (with components based on *patterns*, in the sense of [3], rather than L systems), in [54] (where multi-dimensional arrays rather than strings are considered), in [76] (*unary* systems, where the environment is described by strings over a one-letter alphabet).

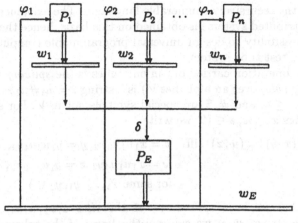

Fig. 3

It is, of course, premature to evaluate the *relevance* of eco-grammar systems from the Artificial Life point of view. In the above we have tried to demonstrate that they are *adequate* in the sense that they are able to capture many real life-like features. Hopefully they will turn out to be useful in understanding real systems involving environment-agents interactions. Providing that a real system can be satisfactorily described in terms of an EG system, we can then ask various questions such as discussed above. One can also consider computer simulations of EG systems in order to discover properties which cannot be found directly, by analytical methods. The difficulties arising in such simulations are discussed in [79].

Much more specific applications are discussed in [31] and [87]. For instance, [31] considers an EG system model of the famous MIT robot Herbert, whereas [87] uses EG systems as models of various problem solving architectures in AI which involve a *global database* (environment), manipulated by certain *operations* (rewriting rules), under the control of a general *strategy* (selection mappings), having as goal to generate, in a *problem space* (the possible states of the system), the *solution* of a problem (a terminal string). A specific case of such a problem, modelled by EG systems, is discussed in detail in [87]: the well-known "eight puzzle" (slide 8 square tiles in a 3 × 3 frame in order to obtain a prescribed arrangement).

5.2 Test tube systems

Following [20], we present here a symbol processing mechanism having the architecture of a PC grammar system with communication by command, but with the components being test tubes working as splicing schemes in the sense of [63], [122]. The communication is performed by redistributing the contents of tubes, in a similar way to the *separate* and *merge* operations from [2] and [78]. Such devices prove to be computationally complete, they characterize the recursively enumerable languages. The existence of universal test tube distributed systems is obtained on this basis, hence the proof of the theoretical possibility to design universal programmable computers with the structure of a test tube system.

The basic operation carried out in our tubes is the *splicing*.

A *splicing rule* (over an alphabet V) is a string $r = u_1\#u_2\$u_3\#u_4$, where $u_i \in V^*, 1 \leq i \leq 4$, and $\#, \$$ are special symbols not in V. For such a rule r and the strings $x, y, w, z \in V^*$ we write

$$(x, y) \vdash_r (w, z) \quad \text{iff} \quad x = x_1u_1u_2x_2, y = y_1u_3u_4y_2,$$
$$w = x_1u_1u_4y_2, z = y_1u_3u_2x_2,$$
$$\text{for some } x_1, x_2, y_1, y_2 \in V^*.$$

We say that we have *spliced* x, y at the *sites* u_1u_2, u_3u_4, respectively, obtaining the strings w, z; we call x, y the *terms* of the splicing. When understood from the context, we omit the index r from $\vdash_r$.

A *splicing scheme* (or an H scheme) is a pair $\sigma = (V, R)$, where V is an alphabet and R is a set of splicing rules (over V). For a language $L \subseteq V^*$, we define

$$\sigma(L) = \{w \in V^* \mid (x, y) \vdash_r (w, z) \text{ or } (x, y) \vdash_r (z, w), \text{ for } x, y \in L, r \in R\}.$$

Then, we define

$$\sigma^*(L) = \bigcup_{i \geq 0} \sigma^i(L),$$

where

$$\sigma^0(L) = L,$$
$$\sigma^{i+1}(L) = \sigma^i(L) \cup \sigma(\sigma^i(L)), i \geq 0.$$

Thus, $\sigma^*(L)$ is the smallest language containing L and closed under the splicing operation.

An *extended H system* is a quadruple

$$\gamma = (V, T, A, R),$$

where V is an alphabet, $T \subseteq V$ (the terminal alphabet), $A \subseteq V^*$ (the set of axioms), and $R \subseteq V^*\#V^*\$V^*\#V^*$. The pair $\sigma = (V, R)$ is called the *underlying H scheme* of γ. The language generated by γ is defined by

$$L(\gamma) = \sigma^*(A) \cap T^*.$$

An H system $\gamma = (V, T, A, R)$ is said to be *of type* F_1, F_2, for two families of languages F_1, F_2, if $A \in F_1, R \in F_2$.

We use $EH(F_1, F_2)$ to denote the family of languages generated by extended H systems of type (F_1, F_2).

An H system $\gamma = (V, T, A, R)$ with $V = T$ is said to be *non-extended*; the family of languages generated by non-extended H systems of type (F_1, F_2) is denoted by $H(F_1, F_2)$. Obviously, $H(F_1, F_2) \subseteq EH(F_1, F_2)$.

The splicing operation is introduced in [63] for finite sets of rules; the case of arbitrarily large sets of splicing rules is considered in [117]; the extended H systems were introduced in [122].

In [28], [126] it is proved that

$$H(FIN, FIN) \subseteq REG.$$

(The inclusion is, in fact, proper.) Using this relation, it is proved in [122] that

$$EH(FIN, FIN) = REG.$$

Moreover, it is proved in [118] that the extended H systems with finite sets of axioms and regular sets of splicing rules are computationally complete, that is,

$$EH(FIN, REG) = RE.$$

Therefore, such systems are as powerful as the Turing machines. However, from a practical point of view, it is not realistic to deal with infinite – even regular – sets of rules. Several ways to handle this problem were proposed: finite sets of rules controlled by random context-like conditions [55], working with multisets [119], and organizing the work in a distributed way, similar to a PC grammar system with communication by command [20] – we will discuss this model now.

A *test tube* (TT) *system* (of degree $n, n \geq 1$) is a construct

$$\Gamma = (V, (A_1, R_1, V_1), \ldots, (A_n, R_n, V_n)),$$

where V is an alphabet, $A_i \subseteq V^*$, $R_i \subseteq V^* \# V^* \$ V^* \# V^*$, and $V_i \subseteq V$, for each $1 \leq i \leq n$.

Each triple (A_i, R_i, V_i) is called a *component* of the system, or a *tube*; A_i is the set of axioms of the tube i, R_i is the set of splicing rules of the tube i and V_i is the *selector* of the tube i.

Let

$$B = V^* - \bigcup_{i=1}^{n} V_i^*.$$

The pair $\sigma_i = (V, R_i)$ is the underlying H scheme associated with the i-th component of the system.

An n-tuple $(L_1, \ldots, L_n), L_i \subseteq V^*, 1 \leq i \leq n$, is called a *configuration* of the system; L_i is also called the *contents* of the i-th tube.

For two configurations $(L_1, \ldots, L_n), (L_1', \ldots, L_n')$, we define

$$(L_1, \ldots, L_n) \implies (L_1', \ldots, L_n') \text{ iff, for each } i, 1 \leq i \leq n,$$

$$L_i' = \bigcup_{j=1}^{n} (\sigma_j^*(L_j) \cap V_i^*) \cup (\sigma_i^*(L_i) \cap B).$$

This means that the contents of each tube is spliced according to the associated set of rules (we pass from L_i to $\sigma_i^*(L_i), 1 \leq i \leq n$), and the result is redistributed among the n tubes according to the selectors $V_1, \ldots, V_n$. The part which cannot be redistributed (does not belong to some $V_i^*, 1 \leq i \leq n$) remains in the tube. Because we have imposed no restrictions on the alphabets V_i (e.g., we did not require that they are pairwise disjoint), when a string in $\sigma_j^*(L_j)$ belongs to several languages V_i^*, then copies of this string will be distributed to all tubes i with this property.

The *language generated* by Γ is

$$L(\Gamma) = \{w \in V^* \mid w \in L_1^{(t)} \text{ for some}$$
$$(A_1, \ldots, A_n) \Longrightarrow^* (L_1^{(t)}, \ldots, L_n^{(t)}), t \geq 0\},$$

where $\Longrightarrow^*$ is the reflexive and transitive closure of the relation $\Longrightarrow$.

Given two families of languages, F_1, F_2, we denote by $TT_n(F_1, F_2)$ the family of languages $L(\Gamma)$, for $\Gamma = (V, (A_1, R_1, V_1), \ldots, (A_m, R_m, V_m))$, with $m \leq n$, $A_i \in F_1, R_i \in F_2$, for each $i, 1 \leq i \leq m$. We also say that Γ is of type (F_1, F_2). When n is not specified, we replace it by ∞.

We illustrate the definitions above by the following **example**.

Consider the TT system

$$\Gamma = (\{a, b, c, d, e\}, (\{cabc, ebd, dae\}, \{b\#c\$e\#bd, da\#e\$c\#a\}, \{a, b, c\}),$$
$$(\{ec, ce\}, \{b\#d\$e\#c, c\#e\$d\#a\}, \{a, b, d\}).$$

Then

$$L(\Gamma) \cap ca^+b^+c = \{ca^n b^n c \mid n \geq 1\} \qquad (*)$$

Indeed, the only possible splicings in the first component are

$$(\alpha a^i b^j | c, e | bd) \vdash_1 (\alpha a^i b^{j+1} d, ec), \quad \alpha \in \{c, d\},$$
$$(da | e, c | a^i b^j \alpha) \vdash_2 (da^{i+1} b^j \alpha, ce), \quad \alpha \in \{c, d\}.$$

One further occurrence of a and one further occurrence of b can be added in this way; only when both one a and one b are added, hence the obtained string is $da^{i+1}b^{j+1}d$, we can move this string to the second tube. Here the possible splicings are

$$(\alpha a^i b^j | d, e | c) \vdash_1 (\alpha a^i b^j c, ed), \quad \alpha \in \{c, d\},$$
$$(c | e, d | a^i b^j \alpha) \vdash_2 (ca^i b^j \alpha, de), \quad \alpha \in \{c, d\}.$$

Only the string $ca^i b^j c$ can be moved to the first tube, hence the process can be iterated. No splicing in tube 1 can involve a string of the form $da^i b^j d$.

Consequently, we have the equality $(*)$, and so $L(\Gamma)$ is not a regular language.

The following results are from [20].

Theorem 5.3.

(i) $TT_2(FIN, FIN) - REG \neq \emptyset$, $TT_3(FIN, FIN) - CF \neq \emptyset$, $TT_6(FIN, FIN) - CS \neq \emptyset$.

(ii) $TT_1(FIN, FIN) \subset REG \subset TT_2(FIN, FIN) \subseteq TT_3(FIN, FIN) \subseteq \ldots$
$\subseteq TT_\infty(FIN, FIN) = TT_\infty(F_1, F_2) = RE$, for all F_1, F_2 such that $FIN \subseteq F_i \subseteq RE$, $i = 1, 2$.

(iii) $TT_6(FIN, FIN)$ contains non-recursive languages.

The relation $RE \subseteq TT_\infty(FIN, FIN)$ is the core result of point (ii) above, and the construction used in its proof is also important for obtaining universal test tube systems.[3]

Take a type-0 grammar $G = (N, T, S, P)$ with $T = \{a_1, \ldots, a_n\}$ and $N = \{Z_1, \ldots, Z_m\}, Z_1 = S$. Consider the new symbols A, B and replace in each rule of P each occurrence of Z_i by $BA^iB, 1 \leq i \leq m$. We obtain a set P' of rules such that $S \Longrightarrow^* w, w \in T^*$, according to the grammar G if and only if $BAB \Longrightarrow^* w$ using the rules in P'. For uniformity, let $A = a_{n+1}$ and $B = a_{n+2}$. Let Γ be the following TT system:

$$\Gamma = (V, (A_1, R_1, V_1), (A_2, R_2, V_2), (A_{3,1}, R_{3,1}, V_{3,1}), \ldots,$$
$$(A_{3,n+2}, R_{3,n+2}, V_{3,n+2}), (A_4, R_4, V_4), (A_5, R_5, V_5),$$
$$(A_6, R_6, V_6), (A_7, R_7, V_7)),$$

with

$$V = T \cup \{A, B, X, X', Y, Z, Z'\} \cup \{Y_i \mid 1 \leq i \leq n + 2\},$$

and

$$
\begin{aligned}
A_1 &= \emptyset, \\
R_1 &= \emptyset, \\
V_1 &= T, \\
A_2 &= \{XBA^{m+1}BBABY, Z'Z\} \cup \\
     &\quad \{ZvY \mid u \to v \in P'\} \cup \\
     &\quad \{ZY_i \mid 1 \leq i \leq n + 2\}, \\
R_2 &= \{\#uY\$Z\#vY \mid u \to v \in P'\} \cup \\
     &\quad \{\#a_iY\$Z\#Y_i \mid 1 \leq i \leq n + 2\} \cup \\
     &\quad \{Z'\#Z\$XBA^{m+1}B\#\}, \\
V_2 &= T \cup \{A, B, X, Y\}, \\
A_{3,i} &= \{X'a_iZ\}, \\
R_{3,i} &= \{X'a_i\#\$X\# \mid 1 \leq i \leq n + 2\}, \\
V_{3,i} &= T \cup \{A, B, X, Y_i\}, \text{ for } 1 \leq i \leq n + 2, \\
A_4 &= \{ZY\}, \\
R_4 &= \{\#Y_i\$Z\#Y \mid 1 \leq i \leq n + 2\}, \\
V_4 &= T \cup \{A, B, X'\} \cup \{Y_i \mid 1 \leq i \leq n + 2\}, \\
A_5 &= \{XZ\}, \\
R_5 &= \{X\#Z\$X'\#\}, \\
V_5 &= T \cup \{A, B, X', Y\}, \\
\end{aligned}
$$

---

[3] Recently, Cl. Zandron has proved that $RE = TT_{10}(FIN, FIN)$.

$$A_6 = \{ZZ\},$$
$$R_6 = \{\#Y\$ZZ\#\},$$
$$V_6 = T \cup \{Y, Z'\},$$
$$A_7 = \{ZZ\},$$
$$R_7 = \{\#ZZ\$Z'\#\},$$
$$V_7 = T \cup \{Z'\}.$$

It is proved in [20] that $L(\Gamma) = L(G)$. Here is the basic intuition behind this proof. The first tube selects the terminal strings, the second one simulates rules in P' on the right-hand end of the current string, and also starts "rewinding" the string: the rightmost symbol of the string, a_i, is removed, the marker Y_i memorizes that, then tube (3.i) introduces a_i at the left-hand end of the string. Iterating this procedure, the string of G is circularly permuted, which allows the simulation of any rule in P' at the right-hand end of a string produced in Γ. The actual beginning of the string is marked with $BA^{m+1}B$, and only when this substring is in the leftmost position, P_1 can start finishing the derivation; the last two tubes remove the auxiliary symbols Y, Z'. If the string is terminal, it can be communicated to the first tube and accepted.

On the basis of the above construction we can infer the existence of universal TT systems, where the notion of a universal TT system is understood in the same way as for Turing machines (and other equivalent models of computation). One specifies all but one components of a universal system and then given an arbitrary system M one encodes it on the remaining component in such a way that the resulting totally specified system simulates M.

The construction of the TT system Γ in the above proof depends on the elements of the starting grammar G. If the grammar G is a universal type-0 grammar, then Γ will be a universal TT system.

A universal type-0 grammar is a construct $G_U = (N_U, T, -, P_U)$, where N_U is the nonterminal alphabet, T is the terminal alphabet, and P_U is the set of rewriting rules. Then for a fixed encoding w and an arbitrary grammar $G = (N, T, S', P)$ the grammar $G'_U = (N_U, T, w(G), P_U)$ is equivalent with G, meaning that $L(G) = L(G'_U)$.

A universal type-0 grammar can be obtained from a universal Turing machine, [134], using the standard transition from Turing machines to Chomsky grammars. A direct construction of a universal type-0 grammar can be found in [10].

This suggests the following definition. A universal TT system for a given alphabet T is a construct $\Gamma_U = (V_U, (A_{1,U}, R_{1,U}, V_{1,U}), \ldots, (A_{n,U}, R_{n,U}, V_{n,U}))$ with $V_{1,U} = T$, with the components as in a TT system, all of them being fixed, and with the following property: there is a specified $i, 1 \leq i \leq n$, such that if we take an arbitrary TT system Γ, then there is a set $A_\Gamma \subseteq V^*$ such that for the system

$$\Gamma'_U = (V_U, (A_{1,U}, R_{1,U}, V_{1,U}), \ldots, (A_{i,U} \cup A_\Gamma, R_{i,U}, V_{i,U}), \ldots,$$
$$(A_{n,U}, R_{n,U}, V_{n,U}))$$

we have $L(\Gamma'_U) = L(\Gamma)$.

Otherwise stated: by encoding Γ as new axioms to be added to the ith component of Γ_U, one obtains a system equivalent with Γ.

The following result is proved in [20].

Theorem 5.4. *For every alphabet T, there are universal TT systems of degree $card(T) + 8$ and of type (FIN, FIN).*

Proof. Begin with the above given construction of the system Γ from a universal type-0 grammar. The alphabet V of Γ is fixed,

$$V = T \cup \{A, B, X, X', Y, Z, Z'\} \cup \{Y_i \mid 1 \le i \le card(T) + 2\}.$$

Similarly, all other components of Γ are fixed. Denote by Γ_U the obtained system. Because G_U contains no axiom, the axiom $XBA^{m+1}BBABY$ of the component A_2 of Γ_U will be omitted, and this is the place where we will add the new axioms which encode a given TT system to be simulated.

More precisely, given an arbitrary TT system Γ_0, by Theorem 5.3 (ii), there is a type-0 grammar $G_0 = (N, T, S, P)$ such that $L(\Gamma_0) = L(G_0)$. Take $w(G)$ which is the code of G_0 constructed as in [10]. Add to A_2 the set $A_{\Gamma_0} = \{XBA^{m+1}Bw'(G_0)Y\}$, where $w'(G_0)$ is obtained from $w(G_0)$ by replacing each nonterminal Z_i by its "code" BA^iB. We obtain then the system Γ'_U such that $L(\Gamma'_U) = L(G_0)$. Indeed, for $G'_U = (N_U, T, w(G_0), P_U)$ we have $L(G'_U) = L(G_0)$. From the construction above we have $L(G'_U) = L(\Gamma'_U)$. As G_0 is equivalent with the arbitrarily given TT system Γ, we have $L(\Gamma'_U) = L(\Gamma)$. Hence Γ_U is indeed universal. □

Observe that the "program" of the particular TT system Γ introduced in the universal TT system (which behaves like a computer) consists of only one string which is added as an axiom of the second component of the universal system.

References

[1]   S. Abraham, Compound and serial grammars, *Inform. Control*, 20 (1972), 432–438.

[2]   L. M. Adleman, On constructing a molecular computer, Manuscript in circulation, January 1995.

[3]   D. Angluin, Finding patterns common to a set of strings, *Journal of Computer and System Sciences*, 21 (1980), 46–62.

[4]   A. Atanasiu, V. Mitrana, The modular grammars, *Intern. J. Computer Math.*, 30 (1989), 101–122.

[5]   H. Bordihn, H. Fernau, Accepting grammars with regulation, *Intern. J. Computer Math.*, 53 (1994), 1–18.

[6]   H. Bordihn, H. Fernau, M. Hölzer, Accepting grammar systems, *Computers and AI*, 1996.

[7]   W. Bucher, K. Culik II, H. A. Maurer, D. Wotschke, Concise description of finite languages, *Theor. Computer Sci.*, 14 (1981), 227–246.

[8]   L. Cai, The computational complexity of PCGS with regular components, *Proc. of Developments in Language Theory Conf.*, Magdeburg, 1995, 209–219.

[9]   L. Cai, The computational complexity of PCGS with linear components, *Computers and AI*, 15, 2-3 (1996), 199–210..

[10]   C. Calude, Gh. Păun, Global syntax and semantics for recursively enumerable languages, *Fundamenta Informaticae*, 4, 2 (1981), 254–254.

[11]   E. Csuhaj-Varju, Some remarks on cooperating grammar systems, in *Proc. Conf. Aut. Lang. Progr. Syst.* (I. Peak, F. Gecseg, eds.), Budapest, 1986, 75–86.

[12]   E. Csuhaj-Varju, Cooperating grammar systems. Power and parameters, *LNCS* 812, Springer-Verlag, Berlin, 1994, 67–84.

[13]   E. Csuhaj-Varju, Grammar systems: a multi-agent framework for natural language generation, in [109], 63–78.

[14]   E. Csuhaj-Varju, Eco-grammar systems: recent results and perspectives, in [111], 79–103.

[15]   E. Csuhaj-Varju, J. Dassow, On cooperating distributed grammar systems, *J. Inform. Process. Cybern.*, *EIK*, 26 (1990), 49–63.

[16]   E. Csuhaj-Varju, J. Dassow, J. Kelemen, Gh. Păun, Stratified grammar systems, *Computers and AI*, 13 (1994), 409–422.

[17]   E. Csuhaj-Varju, J. Dassow, J. Kelemen, Gh. Păun, *Grammar Systems. A Grammatical Approach to Distribution and Cooperation*, Gordon and Breach, London, 1994.

[18]   E. Csuhaj-Varju, J. Dassow, V. Mitrana, Gh. Păun, Cooperation in grammar systems: similarity, universality, timing, *Cybernetica*, 4 (1993), 271–282.

[19]   E. Csuhaj-Varju, J. Dassow, Gh. Păun, Dynamically controlled cooperating distributed grammar systems, *Information Sci.*, 69 (1993), 1–25.

[20]   E. Csuhaj-Varju, L. Kari, Gh. Păun, Test tube distributed systems based on splicing, *Computers and AI*, 15, 2-3 (1996), 211–232.

[21]   E. Csuhaj-Varju, J. Kelemen, Cooperating grammar systems: a syntactical framework for the blackboard model of problem solving, in *Proc. Conf. Artificial Intelligence and Information-Control Systems of Robots, '89* (I. Plander, ed.), North-Holland, Amsterdam 1989, 121–127.

[22]   E. Csuhaj-Varju, J. Kelemen, On the power of cooperation: a regular representation of r.e. languages, *Theor. Computer Sci.*, 81 (1991), 305–310.

[23]   E. Csuhaj-Varju, J. Kelemen, A. Kelemenova, Gh. Păun, Eco-grammar systems: A preview, in *Cybernetics and Systems 94* (R. Trappl, ed.), World Sci. Publ., Singapore, 1994, 941–949.

[24]   E. Csuhaj-Varju, J. Kelemen, A. Kelemenova, Gh. Păun, Eco-grammar systems: A grammatical framework for life-like interactions, *Artificial Life*, 3,1 (1996), 1–28.

[25]   E. Csuhaj-Varju, J. Kelemen, Gh. Păun, Grammar systems with WAVE-like communication, *Computers and AI*, 1996.

[26]   E. Csuhaj-Varju, Gh. Păun, A. Salomaa, Conditional tabled eco-grammar systems, in [111], 227–239.

[27]   E. Csuhaj-Varju, Gh. Păun, A. Salomaa, Conditional tabled eco-grammar systems versus (E)T0L systems, *JUCS*, 1, 5 (1995), 248–264.

[28]   K. Culik II, T. Harju, Splicing semigroups of dominoes and DNA, *Discrete Appl. Math.*, 31 (1991), 261–277.

[29]   J. Dassow, Cooperating distributed grammar systems with hypothesis languages, *J. Exper. Th. AI*, 3 (1991), 11–16.

[30]   J. Dassow, A remark on cooperating distributed grammar systems controlled by graphs, *Wiss. Z. der TU Magdeburg*, 35 (1991), 4–6.

[31] J. Dassow, An example of an eco-grammar system: a can collecting robot, in [111], 240–244.

[32] J. Dassow, J. Kelemen, Cooperating distributed grammar systems: a link between formal languages and artificial intelligence, *Bulletin of EATCS*, 45 (1991), 131–145.

[33] J. Dassow, J. Kelemen, Gh. Păun, On parallelism in colonies, *Cybernetics and Systems*, 24 (1993), 37–49.

[34] J. Dassow, V. Mihalache, Eco-grammar systems, matrix grammars and E0L systems, in [111], 210–226.

[35] J. Dassow, V. Mitrana, Splicing grammar systems, *Computers and AI*, 15, 2-3 (1996), 109–122.

[36] J. Dassow, V. Mitrana, Cooperating distributed pushdown automata systems, submitted, 1995.

[37] J. Dassow, V. Mitrana, Deterministic pushdown automata systems, submitted, 1995.

[38] J. Dassow, V. Mitrana, "Call by name" and "call by value" in cooperating distributed grammar systems, *Ann. Univ. Buc., Matem.-Inform. Series*, 45, 1 (1996), 29–40.

[39] J. Dassow, V. Mitrana, Fairness in grammar systems, submitted, 1995.

[40] J. Dassow, V. Mitrana, Gh. Păun, Szilard languages associated to cooperating distributed grammar systems, *Stud. Cercet. Mat.*, 45, 5 (1993), 403–413.

[41] J. Dassow, Gh. Păun, *Regulated Rewriting in Formal Language Theory*, Springer-Verlag, Berlin, Heidelberg, 1989.

[42] J. Dassow, Gh. Păun, Cooperating distributed grammar systems with registers, *Found. Control Engineering*, 15 (1990), 19–38.

[43] J. Dassow, Gh. Păun, On the succinctness of descriptions of context-free languages by cooperating distributed grammar systems, *Computers and AI*, 10 (1991), 513–527.

[44] J. Dassow, Gh. Păun, On some variants of cooperating distributed grammar systems, *Stud. Cercet. Mat.*, 42, 2 (1990), 153–165.

[45] J. Dassow, Gh. Păun, St. Skalla, On the size of components of cooperating grammar systems, *LNCS* 812, Springer-Verlag, Berlin, 1994, 325–343.

[46] S. Dumitrescu, *CD and PC grammar systems*, PHD Thesis, Univ. of Bucharest, Faculty of Mathematics, 1996.

[47] S. Dumitrescu, Non-returning parallel communicating grammar systems can be simulated by returning systems, *Theor. Computer Sci.*, to appear.

[48] S. Dumitrescu, Gh. Păun, On the power of parallel communicating grammar systems with right-linear components, submitted, 1995.

[49] S. Dumitrescu, Gh. Păun, A. Salomaa, Pattern languages versus parallel communicating grammar systems, *Intern. J. Found, Computer Sci.*, to appear.

[50] L. Errico, *WAVE: An Overview of the Model and the Language*, CSRG, Dept. Electronic and Electr. Eng., Univ. of Surrey, UK, 1993.

[51] L. Errico, C. Jesshope, Towards a new architecture for symbolic processing, in *Proc. Conf. Artificial Intelligence and Information-Control Systems of Robots '94* (I. Plander, ed.), World Sci. Publ., Singapore, 1994, 31–40.

[52] J. D. Farmer, A. d'A Belin, *Artificial Life: the coming evolution*, in *Proc. in Cellebration of Muray Gell-Man's 60th Birthday*, Cambridge Univ. Press., Cambridge, 1990.

[53] H. Fernau, M. Holzer, H. Bordihn, Accepting multi-agent systems: the case of CD grammar systems, *Computers and AI*, 15, 2-3 (1996), 123–140.

[54] R. Freund, Multi-level eco-array grammars, in [111], 175–201.

[55] R. Freund, L. Kari, Gh. Păun, DNA computing based on splicing: the existence of universal computers, submitted, 1995.

[56] R. Freund, Gh. Păun, A variant of team cooperation in grammar systems, *JUCS*, 1, 3 (1995), 105–130.

[57] R. Freund, Gh. Păun, C. M. Procopiuc, O. Procopiuc, Parallel communicating grammar systems with context-sensitive components, in [111], 166–174.

[58] G. Georgescu, The generative power of small grammar systems, in [111], 152–165.

[59] G. Georgescu, A sufficient condition for the regularity of PCGS languages, *J. Automata, Languages, Combinatorics*, to appear.

[60] J. Gruska, Descriptional complexity of context-free languages, *Proc. MFCS 73*, High Tatras, 1973, 71–84.

[61] D. Hauschild, M. Jantzen, Petri nets algorithms in the theory of matrix grammars, *Acta Informatica*, 31 (1994), 719–728.

[62] W. D. Hillis, *The Connection Machine*. The MIT Press, Cambridge, Mass., 1985.

[63] T. Head, Formal language theory and DNA: an analysis of the generative capacity of specific recombinant behaviors, *Bull. Math. Biology*, 49 (1987), 737–759.

[64] J. Hromkovič, Descriptional and computational complexity measures for distributive generation of languages, in *Proc. of Developments in Language Theory Conf.*, Magdeburg, 1995.

[65] J. Hromkovič, J. Kari, L. Kari, Some hierarchies for the communicating complexity measures of cooperating grammar systems, *Theor. Computer Sci.*, 127 (1994), 123–147.

[66] J. Hromkovič, J. Kari, L. Kari, D. Pardubska, Two lower bounds on distributive generation of languages, *Proc. MFCS 94*, LNCS 841, Springer-Verlag, Berlin, 1994, 423–432.

[67] J. Hromkovič, D. Wierzchula, On nondeterministic linear time, real time and parallel communicating grammar systems, in [109], 184–190.

[68] L. Ilie, Collapsing hierarchies in parallel communicating grammar systems with communication by command, *Computers and AI*, 15, 2-3 (1996), 173–184.

[69] L. Ilie, A. Salomaa, 2-testability and relabeling produce everything, submitted, 1995.

[70] C. M. Ionescu, O. Procopiuc, Bounded communication in parallel communicating grammar systems, *J. Inform. Process. Cybern.*, EIK, 30 (1994), 97–110.

[71] L. Kari, A. Mateescu, Gh. Păun, A. Salomaa, Teams in cooperating grammar systems, *J. Exper. Th. AI*, 7 (1995), 347–359.

[72] J. Kari, L. Sântean, The impact of the number of cooperating grammars on the generative power, *Theor. Computer Sci.*, 98 (1992), 621–633.

[73] J. Kelemen, A. Kelemenova, A subsumption architecture for generative symbol systems, in *Cybernetics and Systems Research 92* (R. Trappl, ed.), World Sci. Publ., Singapore, 1992, 1529–1536.

[74] J. Kelemen, R. Mlichova, Bibliography of grammar systems, *Bulletin of EATCS*, 49 (1992), 210–218.

[75] A. Kelemenova, E. Csuhaj-Varju, Languages of colonies, *Theor. Computer Sci.*, 134 (1994), 119–130.

[76] A. Kelemenova, J. Kelemen, From colonies to eco-grammar systems. An overview, *LNCS 812*, Springer-Verlag, Berlin, 1994, 213–231.

[77] C. Langton, Artificial life, in *Artificial Life* (C. Langton, ed.), Santa Fe Institute Studies in the Sciences of Complexity, Addison Wesley Publ., 1989.

[78] R. J. Lipton, Speeding up computations via molecular biology, Manuscript in circulation, December 1994.

[79] M. Maliţa, Gh. Ştefan, The eco-chip: a physical support for artificial life systems, in [111], 260–275.

[80] A. Mateescu, A survey of teams in cooperating distributed grammar systems, in [111], 137–151.

[81] A. Mateescu, Teams in cooperating distributed grammar systems; an overview, Proc. Decentralized Intelligent and Multi-Agent Systems, DIMAS '95, Cracow, 1995, 309–323.

[82] R. Meersman, G. Rozenberg, Cooperating grammar systems, Proc. MFCS '78, LNCS 64, Springer-Verlag, Berlin, 1978, 364–374.

[83] R. Meersman, G. Rozenberg, D. Vermeir, Cooperating grammar systems, Techn. Report, 78 - 12, Univ. Antwerp, Dept. of Math., 1978.

[84] M. Middendorf, Supersequences, runs and CD grammar systems, in Developments in Theoretical Computer Science (J. Dassow, A. Kelemenova, eds.), Gordon and Breach, London, 1994, 101–113.

[85] V. Mihalache, Matrix grammars versus parallel communicating grammar systems, in [109], 293–318.

[86] V. Mihalache, Extended conditional tabled eco-grammar systems, J. Inform. Processing Cybern., 30, 4 (1994), 213–229.

[87] V. Mihalache, General Artificial Intelligence systems as eco-grammar systems, in [111], 245–259.

[88] V. Mihalache, On parallel communicating grammar systems with context-free components, in [113], 258–270.

[89] V. Mihalache, Szilard languages associated to parallel communicating grammar systems, Proc. of Developments in Language Theory Conf., Magdeburg, 1995, 247–256.

[90] V. Mihalache, On the generative capacity of parallel communicating grammar systems with regular components, Computers and AI, 15, 2-3 (1996), 155–172.

[91] V. Mihalache, Terminal versus nonterminal symbols in parallel communicating grammar systems, Rev. Roum. Math. Pures Appl., 1996.

[92] V. Mihalache, Variants of parallel communicating grammar systems, Mathematical Linguistics Conf., Tarragona, 1996.

[93] V. Mihalache, Parallel communicating grammar systems with query words, Ann. Univ. Buc., Matem.-Inform. Series, 45, 1 (1996), 81–92.

[94] Grammar-gsm pairs, Ann. Univ. Buc., Ser. Matem.-Inform., 39-40 (1991-1992), 64–71.

[95] V. Mitrana, Hybrid cooperating distributed grammar systems, Computers and AI, 2 (1993), 83–88.

[96] V. Mitrana, Eco-pattern systems, in [111], 202–209.

[97] V. Mitrana, Similarity in grammar systems, Fundamenta Informaticae, 24 (1995), 251–257.

[98] V. Mitrana, Gh. Păun, G. Rozenberg, Structuring grammar systems by priorities and hierarchies, Acta Cybernetica, 11 (1994), 189–204.

[99] P. H. Nii, Blackboard systems, in The Handbook of AI, vol. 4 (A. Barr, P. R. Cohen, E. A. Feigenbaum, eds.), Addison-Wesley, Reading, Mass., 1989.

[100] D. Pardubska, On the power of communication structure for distributive generation of languages, in Developments in Language Theory (G. Rozenberg, A. Salomaa, eds.), World Sci. Publ., Singapore, 1994, 90–101.

[101] D. Pardubska, Communication complexity hierarchies for distributive generation of languages, Fundamenta Informaticae, to appear.

[102] Gh. Păun, A new type of generative device: valence grammars, Rev. Roum. Math. Pures Appl., 25, 6 (1980), 911–924.

[103] Gh. Păun, Parallel communicating grammar systems: the context-free case, Found. Control Engineering, 14, 1 (1989), 39–50.

[104] Gh. Păun, On the power of synchronization in parallel communicating grammar systems, *Stud. Cerc. Matem.*, 41, 3 (1989), 191–197.

[105] Gh. Păun, Non-centralized parallel communicating grammar systems, *Bulletin of EATCS*, 40 (1990), 257–264.

[106] Gh. Păun, On the syntactic complexity of parallel communicating grammar systems, *Kybernetika*, 28 (1992), 155–166.

[107] Gh. Păun, Parallel communicating systems of L systems, in *Lindenmayer systems. Impacts on Theoretical Computer Science, Computer Graphics, and Developmental Biology* (G. Rozenberg, A. Salomaa, eds.), Springer-Verlag, Berlin, 1992, 405–418.

[108] Gh. Păun, On the synchronization in parallel communicating grammar systems, *Acta Informatica*, 30 (1993), 351–367.

[109] Gh. Păun (ed.), *Mathematical Aspects of Natural and Formal Languages*, World Sci. Publ., Singapore, 1994.

[110] Gh. Păun, On the generative capacity of colonies, *Kybernetika*, 31 (1995), 83–97.

[111] Gh. Păun (ed.), *Artificial Life. Grammatical Models*, The Black Sea Univ. Press, Bucharest, 1995.

[112] Gh. Păun, On the generative capacity of hybrid CD grammar systems, *J. Inform. Process. Cybern.*, *EIK*, 30, 4 (1994), 231–244.

[113] Gh. Păun (ed.), *Mathematical Linguistics and Related Topics*, The Publ. House of the Romanian Academy, Bucharest, 1995.

[114] Gh. Păun, Grammar systems: A grammatical approach to distribution and cooperation, *Proc. ICALP '95 Conf.*, *LNCS 944* (1995), 429–443.

[115] Gh. Păun, Generating languages in a distributed way: grammar systems, *XIth Congress on Natural and Formal Languages*, Tortosa, 1995, 45–67.

[116] Gh. Păun, Parallel comunicating grammar systems. A survey, *XIth Congress on Natural and Formal Languages*, Tortosa, 1995, 257–283.

[117] Gh. Păun, On the splicing operation, *Discrete Appl. Math.*, 70 (1996), 57–79.

[118] Gh. Păun, Regular extended H systems are computationally universal, *J. Automata, Languages, Combinatorics*, 1,1 (1996), 27–36.

[119] Gh. Păun On the power of the splicing operation, *Intern. J. Computer Math.*, 59 (1995), 27–35.

[120] Gh. Păun, L. Polkowski, A. Skowron, Parallel communicating grammar systems with negociation, *Fundamenta Informaticae*, 1996.

[121] Gh. Păun, G. Rozenberg, Prescribed teams of grammars, *Acta Informatica*, 31 (1994), 525–537.

[122] Gh. Păun, G. Rozenberg, A. Salomaa, Computing by splicing, *Theor. Computer Sci.*, 1996, 168, 2 (1996), 321–336.

[123] Gh. Păun, A. Salomaa, S. Vicolov, On the generative capacity of parallel communicating grammar systems, *Intern. J. Computer Math.*, 45 (1992), 49–59.

[124] Gh. Păun, L. Sântean, Parallel communicating grammar systems: the regular case, *Ann. Univ. Buc., Ser. Matem.-Inform.*, 38 (1989), 55–63.

[125] Gh. Păun, L. Sântean, Further remarks about parallel communicating grammar systems, *Intern. J. Computer Math.*, 34 (1990), 187–203.

[126] D. Pixton, Regularity of splicing languages, *Discrete Appl. Math.*, 1995, 69 (1996), 101–124.

[127] O. Procopiuc, C. M. Ionescu, F. L. Ţiplea, Parallel communicating grammar systems: the context-sensitive case, *Intern. J. Computer Math*, 49 (1993), 145–156.

[128] G. Rozenberg, K. Ruohonen, A. Salomaa, Developmental systems with fragmentation, *Intern. J. Computer Math.*, 5 (1976), 177–191.

[129] G. Rozenberg, A. Salomaa, *The Mathematical Theory of L Systems*, Academic Press, New York, 1980.

[130] A. Salomaa, *Formal Languages*, Academic Press, New York, 1973.

[131] L. Sântean, Parallel communicating systems, *Bulletin of EATCS*, 42 (1990), 160–171.

[132] P. S. Sapaty, *The WAVE Paradigm*, Internal Report 17/92, Dept. Informatics, University of Karlsruhe, 1992.

[133] S. Skalla, On the number of active nonterminals of cooperating distributed grammar systems, in *Artificial Intelligence and Information-Control Systems of Robots 94* (I. Plander, ed.), World Sci. Publ., Singapore, 1994, 367–374.

[134] A. M. Turing, On computable numbers, with an application to the Entscheidungsproblem, *Proc. London Math. Soc.*, Ser. 2, 42 (1936), 230–265; a correction, 43 (1936), 544–546.

[135] F. L. Ţiplea, C. Ene, A coverability structure for parallel communicating grammar systems, *J. Inform. Process. Cybern.*, EIK, 29 (1993), 303–315.

[136] F. L. Ţiplea, C. Ene, C. M. Ionescu, O. Procopiuc, Some decision problems for parallel communicating grammar systems, *Theor. Computer Sci.*, 134 (1994), 365–385.

[137] F. L. Ţiplea, O. Procopiuc, C. M. Procopiuc, C. Ene, On the power and complexity of parallel communicating grammar systems, in [111], 53–78.

[138] G. Vaszil, Parallel communicating grammar systems without a master, *Computers and AI*, 15, 2-3 (1996), 185–198.

[139] G. Vaszil, The simulation of a non-returning parallel communicating grammar system with a returning system in case of linear component grammars, submitted, 1995.

[140] S. Vicolov, Non-centralized parallel grammar systems, *Stud. Cercet. Matem.*, 44 (1992), 455–462.

[141] D. Wätjen, On cooperating distributed limited 0L systems, *J. Inform. Process. Cybern. EIK*, 29 (1993), 129–142.

[142] *Connection Machine Model CM-2 Technical Summary*, Thinking Machines Corporation, Cambridge, Mass., 1987.

Contextual Grammars and Natural Languages

Solomon Marcus

The year 1957: two complementary strategies

The systematic investigation of natural languages by means of algebraic, combinatorial and set-theoretic models begun in the 1950s concomitantly in Europe and the U.S.A. An important year in this respect seems to be 1957, when Chomsky published his pioneering book [6] concerning the new generative approach to syntactic structures and some Russian mathematicians proposed a conceptual framework for the study of general morphological categories [10], of the category of grammatical case (A. N. Kolmogorov; see [74]), and of the category of part of speech [75], giving the start in the development of analytical mathematical models of languages.

Each of these two trends involved specific mathematical tools. While generative grammars are inspired from combinatorics, formal systems and the theory of free monoids, analytical models are based on some ideas coming from set theory, binary relations, free monoids, and closure operators.

Besides their differences concerning the mathematical aspect, these two trends are contrasting in two more respects:

a) They have different linguistic traditions. Chomsky invokes the Indian Panini grammar and what he calls "Cartesian Linguistics" (related to René Descartes), while analytical models, initially motivated by the work done in the field of automatic translation of languages, are directly formalizing ideas coming from structural linguistics (Prague Circle, American descriptive linguistics, etc.).

b) Generative grammars and analytical models have complementary strategies. The former are looking for generative devices approximating the process of obtaining well formed strings. The latter assume as given a certain level of well-formedness and analyze the corresponding structure. Chomskian grammars use some means external to language (conceived as a set of string on a terminal alphabet), such as the auxiliary alphabet, while analytical models adopt an internal, intrinsic view point.

Generative grammars are a rupture from the linguistic tradition of the first half of 20th century, while analytical models are just the development, the continuation of this tradition. It was natural to expect an effort to bridge this gap. This effort came from both parts and, as we shall see, contextual grammars are a component of this process.

The origin of contextual grammars

Contextual grammars (shortly CGs) have their origin in the attempt to transform in generative devices some procedures developed within the framework of analytical models. The idea to connect in this way the analytical study with the generative approach to natural languages was one of the main problems investigated in mathematical linguistics in the period from 1957 to 1970. In this order of ideas, we can give the examples of Lambek's syntactic calculus [26], of categorial grammars following the tradition of the Polish school of logic, of dependency grammars, having their origin in Tesnière's book [72], and of configurational grammars studied by Gladkij [14]; see also the corresponding chapters in [33].

CGs try to exploit two ideas that were at the very beginning of the tradition of descriptive distributional linguistics in U.S.A., in both the 1940s and the 1950: the idea of a string on a given finite non-empty alphabet A and the idea of a context on A, conceived as an ordered pair (u, v) of strings over A. The term *context* is, in many cases, not effectively used, some equivalent terms such as *environment* or *neighbourhood* being preferred. The main aim of descriptive linguistics was to describe natural languages by syntactic means, by avoiding as much as possible the involvement of semantic aspects. A typical example in this respect is Harris's book [15], where the author starts with the idea to consider as the only semantic factor accepted in his analyses the distinction between well-formed strings (wfs) and non-well-formed strings (nwfs). This distinction is assumed as an axiom, as a primitive convention, having an intuitive-empirical base, resulting from the experience of any native speaker of a language. The main problem for Harris was to observe and describe the interplay of strings and their substrings, on the one hand, and the corresponding contexts, on the other hand, and to establish what contextual and concatenation operations, applied to wfs, lead to wfs; in other words, what transformations of a string keep invariant its well-formedness.

Let us recall that such a belief in the great capacities of syntactic and contextual operations to cover the quasi-totality of the needs of a linguistic analysis was an implicit presupposition in the approach adopted by the first researchers in automatic translation of languages; later it became clear that the semantic and pragmatic factors are much more important than was initially assumed.

Motivation of simple contextual grammars
and of contextual grammars with choice

There is general agreement about the fuzzy status of the set of wfs in a natural language (it seems that there is no general agreement concerning the fuzzy status of wfs in a programming language). Many necessary conditions for a string to be well-formed (wf) are known, but no such sufficient condition (and, a fortiori, no necessary and sufficient condition) in this respect is known.

There is also a general agreement about the (countable) infinity of the set of wfs in a natural language. So, any effective linguistic analysis approximates the total set of wfs by a precise subset of it; this subset is usually called a *level of grammaticality* (lg).

Given a certain lg, i.e., a precise set L of wfs on A, we say that a string s on A is accepted in L by a context (u, v) over A if the string usv is in L. We assume that it is more convenient to define first a lg (the set L), and then to distinguish various types of strings and of contexts, according to the test of acceptance in L. But, if we want to reach a more interesting lg, in the sense of infinity of L, a natural question arises: could we get an infinite set L starting from a lower lg, expressed by a finite set of wfs? This question immediately provokes another one: could we obtain the transformation of a lower lg into a higher lg by exclusively contextual operations (i.e., placing some strings in some contexts) applied a finite number of times? Looking for the answer to these questions, we reach, successively, the concept of a simple contextual grammar (scg) and the concept of a contextual grammar with choice (cgc). Let us follow this itinerary.

There are some very standard strings, such as "This house is beautiful" or "All men are mortal", that are accepted as wfs without any doubt, by any native speaker of English. So, we may assume that a finite set of such strings can represent our starting point; we could call these strings *primitive wfs*. Let us observe that the finite set L_0 of primitive wfs can be choosed in various ways. This set is under two conflicting trends: on the one hand, the desire to have L_0 as small as possible, in order to simplify the process of placing strings from L_0 in various possible contexts; on the other hand, the desire to have L_0 as large as possible, in order to can include in it a richer variety of types of wfs.

Once the set L_0 is fixed, the next question concerns the choice of a suitable finite set C_0 of contexts to be applied to strings in L_0 and to strings obtained from L_0 in this way. In other words, the operation of applying contexts from C_0 starts with strings in L_0, but can be iteratively applied a finite number of times. This situation will lead to a set L of strings which is the smallest possible to satisfy the following requirements: any string in L_0 is in L; for any string x in L and for any context (u, v) in C_0 the string uxv is in L.

As a matter of fact, the finite sets L_0 and C_0 should be selected not consecutively, but concomitantly, because strings and contexts appear together and they depend on each other. The double constraint acting on the set L_0 is equally acting on C_0 and to some extent each of the sets C_0 and L_0 can be improved at the expense of the other but at the same time an increase in complexity of L_0 requires a corresponding increase in complexity of C_0.

Experience with natural languages shows that any string selects a class of contexts and any context selects a class of strings. For instance, a qualificative adjective a selects the set of all noun groups that could contain a, while a context of the form $(\omega, noun)$, where ω is the empty string, selects all qualificative adjectives accepted in English by this context. Needless to say that this selection is made in respect to a lg; in the example above, the considered lg was formed by all English noun groups. A detailed contextual analysis of English, with respect to different lg, was done by Du Feu and Marcus [12]. A similar analysis of French is due to Vasilescu [76], [77], [78]. Hungarian and Romanian, as well as some programming languages are considered under this aspect in [41], [42]. The main idea guiding this process is the fact that, despite the fuzzy nature of the property of well-formedness, the approximation of the fuzzy set of all wfs by means of a precise subset of wfs can be always improved.

As a matter of fact, there is no evidence that the status of wfs is reflected in the best way by the concept of a fuzzy set considered by Zadeh [80]. It seems that the concept of a rough set introduced by Pawlak [57] is a better candidate in this respect; this conjecture remains to be checked.

The duality between strings and contexts and the Sestier closure

The concept of a cgc takes into account the capacity of any string to select a class of preferential contexts. However, this capacity is only a part of a more comprehensive phenomenon, the duality between strings and contexts.

Given a language L over A, we can associate to any set S_0 of strings over A the class $C(S_0)$ of contexts accepting in L any string in S_0. In a similar way, we can associate to any set C_0 of contexts over A the set $S(C_0)$ of all strings accepted in L by any context in C_0. It is easy to see that if S is contained in the set T of strings, then $C(S)$ contains the set $C(T)$, while if C is contained in the set D of contexts, then $S(C)$ contains the set $S(D)$ (so, both operators we have just defined are antimonotonous). Moreover, S_0 is contained in $S(C(S_0))$, while C_0 is contained in $C(S(C_0))$. This situation shows that the interplay between strings and contexts is ruled by a Galois correspondence or a Galois connection. We recall (see, for instance, [62]) that given two ordered sets X and Y, a mapping f from X into Y and a mapping g from Y into X, the pair (f, g) defines a *Galois connection* between X and Y when the following requirements are satisfied: i) if u and v are in X and u precedes v, then $f(v)$ precedes $f(u)$; ii) if y and z are in Y and y precedes z, then $g(z)$ precedes $g(y)$; iii) x in X always precedes $g(f(x))$, while y in Y always precedes $f(g(y))$. Now let us take as X the family of various sets of strings over A and as Y the family of various sets of contexts over A. Let us associate to any set S of strings over A the set $f(S)$ of contexts accepting in L any string in S; let us associate to any set C of contexts over A the set $g(C)$ of strings accepted in L by any context in C. Take as order relation in both X and Y the relation of set-theoretic inclusion. In view of the above considerations, it follows that the pair (f, g) of mappings we have just defined is a Galois connection between X and Y, i.e., between the family of sets of strings and the family of sets of contexts over A. This Galois connection is relative to a given language L over A. Since the set of languages over A is not only infinite but uncountable, it follows that the set of Galois connections we can define between sets of strings and sets of contexts over A is uncountable (here too we may remark that the family of sets of strings and the family of sets of contexts over A are both uncountable).

The above duality between strings and contexts was for the first time exploited by Sestier [68]. For every language L over A he defines the contextual operator c associating to each set X of strings over A the set $c(L, X)$ of all contexts accepting in L all strings in X. Similarly, Sestier defines an "inverse" operator s associating to each set Y of contexts over A the set $s(L, Y)$ of strings accepted in L by any context in Y. Then, Sestier considers the composition $\varphi(X, L) = s(L, c(L, X))$, called today the *Sestier closure* of X in respect to L. Indeed, it is a closure operator, as, for instance, defined by Ward [79]: Given an ordered set Z, a closure operator γ on Z is a mapping γ of Z into Z satisfying the conditions: i) any x in Z precedes (in the sense of order) the element $\gamma(x)$; ii) if x and y are in Z and x precedes y, then $\gamma(x)$ precedes $\gamma(y)$ (property of monotony); iii) for any x in Z, $\gamma(\gamma(x)) = \gamma(x)$ (the iteration of γ becomes stationary from the first step of its application). Moreover, if Z has an infimum 0, it is required that $\gamma(0) = 0$.

In our case, Z is the set of all possible languages over A and the order relation is the inclusion relation. The quality of the Sestier operator φ to be a closure operator can be checked directly, but it also follows from a theorem due to Ore [52]: Given two ordered sets X and Y and a Galois connection (f, g) between X and Y, the composition, in any order, of f and g, defines a closure operator in X (if we begin with f) or in Y (if we begin with g).

The nice fact with the Sestier operator is that, despite the property of the mapping s to be, in some sense, the inverse of the mapping c, these two mappings don't neutralize each other (as happens with addition and substraction, or with multiplication and division) and, generally speaking, $\varphi(X)$ is different from X. When $\varphi(X) = X, X$ is said to be a closed set (we can replace $\varphi(X, L)$ by $\varphi(X)$ when no ambiguity exists about the considered language L), i.e., a closed language.

Steps in modelling morphological categories

Sestier was guided by algebraic motivations associated with an implicit feeling of what could be linguistically relevant. However, as was shown later, in several steps [33], [36], [37], [38], Sestier closures are the most comprehensive model of morphological categories, as they were investigated within the framework of analytical models of language. In order to reach this result, we have to consider the work done independently of Sestier, in direct relation with problems appearing in natural languages.

The first model of morphological categories was proposed by Dobrushin [10]. His starting point is (although not explicitly) the concept of a distributional class, considered in American descriptive linguistics. For each a in A, Dobrushin defines the distributional class (dc) of a in respect to L as the set $D_L(a)$ containing all elements b in A such that a and b are accepted in L exactly by the same contexts. Then he considers what he calls the domination relation: given a and c in A, a dominates c in respect to the language L if any context accepting a in L accepts c in L too. For instance, if A is the French vocabulary and L is the set of noun groups in French, then the qualificative adjective *grand* dominates the qualificative adjectives *petit* and *mince*. However, while *petit* dominates *grand* in respect to L (so, *petit* belongs to the distributional class of *grand*), *mince* does not dominate *grand*, because, for instance, in the well-formed noun group *une femme mince* the replacement of *mince* by *grand* leads to a non-well-formed string.

In a next step, Dobrushin defines the concept of an initial distributional class (idc), as a dc such that any element in A dominating it belongs to it. For instance, keeping the previous interpretation of L, $D(grand)$ is an idc, but $D(mince)$ is not, because *grand* dominates *mince* but does not belong to $D(mince)$. This example, like many others considered in [41], [42], shows that the domination relation between words permits to compare them from the view point of their morphological homonymy (mh): the homonymy among flexional forms of *grand* is smaller than the homonymy among flexional forms of *mince* (the latter has the same form for masculine and feminine, both at singular and at plural, while the former does not). More generally: two words in the same dc have the same type of mh; if a dominates b but the converse is not true, then the mh of b is larger than that of a; if neither a dominates b nor b dominates a, then a and b are not comparable.

The last step in Dobrushin's approach now appears very natural: define an elementary morphological category (emc) as the union between an idc $D(a)$ and the set of all elements in A that are dominated by $D(a)$. For instance, in the case of $a = grand$, this will be the set of all flexional forms of qualificative adjectives that have values of masculine singular.

However, many other types of morphological categories that are not "elementary" need a corresponding generalization. The systematic investigation was done, in this respect, by Marcus [34] by replacing elements in A with strings on A and idc by arbitrary sets of strings. In this way, the model obtained transgressed the borders of morphology and became a general procedure to capture phenomena of contextual homonymy [41], [42].

But even remaining at the morphological level of flexional forms, it was shown by Kunze [25] that Dobrushin's domination relation, so efficient for Slavic languages, for Latin, French, and Romanian, is less efficient for German, where, for instance, a set such as $\{ich, du\}$ (personal pronouns of first and second person) requires a more general domination relation. If for Marcus [28] the grammatical category (gc) $G(X)$ generated by a set X of strings is the union of X with the set of all strings dominated by any string in X, for Kunze [25] the category $K(X)$ generated by X, let us call it the Kunze category, is the union of X with all sets Y of strings such that every context accepting in L all strings in X also accepts in L all strings in Y.

The gc, the Kunze category, and the Sestier closure are in strong relations. The following theorems are valid [38]:

Given the alphabet A and the language L on A, the Kunze category $K(X)$ in respect to L generated by a set X of strings over A is identical to the Sestier closure $\varphi(X)$ in respect to L. The gc $G(X)$ in respect to L is always contained in $K(X)$, so the Kunze category is broader than the gc (but, when X is a dc, we have $G(X) = K(X)$). For any set X of strings there exists a set Y of strings such that $\varphi(X) = G(Y)$; in other words, any Sestier closure is a gc and, in view of the first theorem, any Kunze category is a gc. The converse is not true.

These theorems show that gc and Sestier closure cover a large variety of morphological phenomena and of more general grammatical phenomena related to the contextual behavior of strings and to their contextual ambiguity. Let us observe that neither in traditional nor in structural or generative linguistics does there exist a clear conceptual status of what a general morphological category or a general grammatical category could be.

The contextual approach in a generative perspective

In view of the important linguistic significance of Sestier closures, and of gc, as it follows from the above considerations, it is worth investigating their generative counterpart.

Take first the languages over A that are gc in respect to a given language L; as we have seen, among these languages we will find all Sestier closures too. For any language X over A, we have $G(G(X)) = G(X)$ [34]. It follows that a necessary and sufficient condition for a language X over A to be a gc with respect to L (i.e., there exists a language U such that $X = G(U)$) is $X = G(X)$. This means that the contextual homonymy of strings in X is in some sense maximal; no string outside X can exist whose contextual homonymy is equal to or larger than the contextual homonymy of any string in X. If, following Dobrushin, we would add, for the gc generated by a language X, the condition to be an initial gc (i.e., to require from X to be initial: any string dominating all strings in X belongs to X), then the condition $X = G(X)$ would mean that X is comparable, from the view point of contextual homonymy, to no other language. If we imagine a graph whose nodes are languages over A and an arrow goes from X to Y when each string in X dominates each string in Y (with respect to a fixed given language L), then any gc X would be a terminal node, while any initial gc would be an isolated node.

A legitimate question arises: What is the place in the Chomsky hierarchy (and in its further refinements) of those languages that are gc or initial gc? The exact formulation of this question is: Given a language L over A, are there gc and initial gc in respect to L in any class of the enlarged Chomsky hierarchy (including, for instance, not only regular, context-free, and context-sensitive, but also their known refinements)? The answer could depend, in turn, on the place of L in the enlarged Chomsky hierarchy.

Let us call two languages over A *grammatically conjugate* if each of them is a gc in respect to the other; let us call them *initially grammatically conjugate* if they are both grammatically conjugate and initially conjugate (i.e., each of them is initial in respect to the other). We can extend these concepts, in an obvious way, to arbitrary sets of languages: a set of languages over A is grammatically conjugate (initially grammatically conjugate) if any pair of languages in the set is grammatically conjugate (initially grammatically conjugate).

Since any Sestier closure is a closure operator, we have, in respect to a fixed language L over A, $\varphi(\varphi(X)) = \varphi(X)$, for any language X over A. It easily follows that X is a Sestier closure (i.e., there exists U such that $X = \varphi(U)$) if and only if $X = \varphi(X)$. We called such languages *closed languages*. Let us remember that Sestier closures are more comprehensive linguistic categories than grammatical categories, because, for any X, $G(X)$ is contained in $\varphi(X)$, while sometimes (see the situation in German), $\varphi(X)$ is not contained in $G(X)$. So, it would be linguistically relevant to find out how closed languages are distributed among different classes of the enlarged Chomsky hierarchy. Here too, the answer could depend on the place of the given language L in this hierarchy. Moreover, since any closed language is a gc, the distribution of the closed languages, their generative typology, are correlated with the generative typology of gc.

We consider two languages as *closed conjugate* if each of them is a closed language with respect to the other. A *set of languages* is said to be *closed conjugate* if every language in the set is closed conjugate to any other language in the set. We could also introduce the concept of an *initially closed language* (that would mean: it is both closed and initial; but here, *initial* is no longer in the Dobrushin sense, it is in Kunze's sense: X is Kunze initial if any language Y that Kunze-dominates X – in respect to the given language L – is contained in X). We could consider *initially closed conjugate languages* as sets of languages where each pair of languages is such that every language in the pair is initially closed in respect to the other language in the pair.

A special case was left aside: the possibility, for a language, to be conjugate (in the various respects considered above) to itself. Let us observe the special role here of the empty context. We could, in some situations, exclude it from our considerations.

Interesting questions appear concerning conjugate sets of languages. Here are some examples (the word *conjugate* has to be interpreted, consecutively, in each of the four meanings introduced above: grammatically, initially grammatically, closed, initially closed conjugate sets of languages). For what values of n can we find conjugate sets of n languages in every class of the enlarged Chomsky hierarchy? For what values of n can we find conjugate sets of n languages situated in at least two different classes of the enlarged Chomsky hierarchy? The same question when *two* is replaced by *three, four,* or *five.* The simplest questions of this type would be whether there exist conjugate pairs of languages belonging to different generative types and whether there exist conjugate triples of such languages.

Contextual grammars can generate both strings and contexts

In view of the essentially contextual procedures used in the definition of gc, Kunze domination and Sestier closures, we should expect a relatively simple link between them, on the one hand, and contextual grammars (in their various variants proposed by Marcus [39], and other authors; see [55], [56]), on the other hand. This means: the identification of those types of contextual grammars that generate mc, gc and closed languages, initially cg and initially closed languages (with respect to a given language, discussed in respect to its contextual type or its Chomskian type); the characterization of those pairs or sets of contextual grammars that generate various types of conjugate languages and of conjugate sets of languages (particularly, to what extent two conjugate languages can be generated by contextual grammars of the same type and to what extent can they be generated by contextual grammars of different types; similar questions for conjugate sets of languages).

A more specific aspect of contextual grammars is their natural way to lead to two different types of generative processes: one generating strings, the other generating contexts. Given an alphabet A, let us consider a pair of mappings (f, g), where f associates to each set S of strings over A a set $f(S)$ of contexts over A and g associates to each set C of contexts over A a set $g(C)$ of strings over A. Under what conditions on f and g does there exist a language L over A, in respect to which the pair (f, g) defines a Galois connection between the family of sets of strings and the family of sets of contexts over A? Is the existence of L always unique?

Let us call (f, g) a string-context Galois connection (in respect to L). A contextual grammar associated to (f, g) should generate, on the one hand, a set S of strings, on the other hand, a set C of contexts, such that $f(S) = C$. Such a grammar is *f-compatible with the Galois connection* (f, g). If $g(C) = S$, then the grammar is *g-compatible* with (f, g). If we have both $f(S) = C$ and $g(C) = S$, then the grammar is *fg-compatible* with (f, g). Given a string-context Galois connection (f, g), what restrictions do operate in the family $\mathcal{F}$ of contextual grammars that are f-compatible, g-compatible, fg-compatible with (f, g)? What is the possible typology of $\mathcal{F}$? Could it be empty? Could it cover all asociated pairs $(S, f(S)), (g(C), C), (g(C), f(S))$?

We gave enough motivations to answer now the basic question: how can a contextual grammar generate both strings and contexts? Let us first consider an scg $G = (A, L_0, C_0)$, where L_0 is a finite language over A, while C_0 is a finite set of contexts over A. Let us recall that the language generated by G is the minimal set L of strings over A such that L_0 is contained in L and for any string x in L and any context (u, v) in C_0 the string uxv is in L_0. Taking profit from the similarity of situation, in G, of L_0 and C_0, we can change their roles in the generative process and, replacing strings by contexts and contexts by strings, we get: the set of contexts generated by G is the minimal set C of contexts such that C_0 is contained in C and, for any context (u, v) in C and for any string x in L_0, at least one of the contexts $(ux, v), (u, xv), (xu, v), (u, vx)$ is in C.

We can be more specific and restrictive, by defining *simple left internal contextual grammars* (shortly, slicg), when the variant (ux, v) is considered, *simple right internal contextual grammars* (sricg), when we choose the variant (u, xv), *simple left external contextual grammars* (slecg) corresponding to the variant (xu, v) and *simple right external contextual grammars* (srecg) associated with the variant (u, vx). Then, we can define *simple internal cg* (sicg) as grammars that have at least one of the contexts $(ux, v), (u, xv)$ in C; *strong simple internal cg* (ssicg), if they are both slicg and sricg; *simple external cg* (secg) when at least one of the contexts $(xu, v), (u, vx)$ is in C; *strong simple external cg* (ssecg) if they are both slecg and srecg, *simple left cg* (slcg) if at least one of the contexts $(ux, v), (xu, v)$ is in C; *simple right cg* (srcg) if at least one of the contexts $(u, xv), (u, vx)$ is in C; *strong simple left cg* (sslcg) if they are both slicg and slecg; *strong simple right cg* (ssrcg) if they are both sricg and srecg; *strong simple cg* (sscg) if they are both ssicg and ssecg or, equivalently, if they are both sslcg and ssecg. In all these variants, the structure of the grammar remains the same; only the generative process is changing, with possible different generated sets of contexts.

Now the problem is: Given an scg G of one of the above types; denoting by S the set of strings generated by G and by C the set of contexts generated by G, does there exists a language L such that G is f-compatible, g-compatible, or fg-compatible with the Galois connection (f, g) associated to L (i.e., to have $f(S) = C, g(C) = S$ or both of these equalities)? If such a language L exists, is it unique? If it is not unique, what can we say about the family of all languages L with this property? Obviously, the answer may depend on the type of scg we are considering, i.e., on the type of generative process leading to the contexts generated by the grammar.

Each context (u, v) has two components; the *left*, u, and the *right*, v. Any left or right component is said to be a *unilateral component*. We can define the left (right, unilateral) Chomskian generative type of a set of contexts as the Chomskian type of the set of all left (right, unilateral) components of these contexts. In this way, we may associate to every cg G two Chomskian types: one related to the set of strings generated by G, the other related to the set of contexts generated by G. While the former set is unique, the latter depends on the type of scg we are using. It would be interesting to determine to what extent these two Chomskian types can be different. We can also consider the Chomskian type of the set of left contexts and the Chomskian type of the set of right contexts and investigate to what extent could they be different.

If, instead of referring to the Chomskian type, we refer to the typology of contextual grammars, as established, for instance, by [56], then we are again confronted with problems similar to the above ones.

Interplay of strings, contexts and contextual grammars with choice

Let us now consider the contextual grammars with choice (cgc). In contrast with scg, whose definition is symmetric in respect to strings and contexts, the definition of cgc no longer has this symmetry property. So, together with the type $G = (A, L_0, C_0, f)$, where (A, L_0, C_0) is a scg, while f is a mapping associating to each string s in L_0 a subset $f(s)$ of C_0, we will introduce a dual type of cgc, as a system $H = (A, L_0, C_0, g)$, where the first three objects are the same as above, while g is a mapping associating to each context c in C_0 a subset $g(c)$ of L_0. The set $C(H)$ of contexts generated by H is, by definition, the smallest set C' of contexts such that C_0 is contained in C' and if $c = (u, v)$ belongs to C' and the string x belongs to $g(c)$, then at least one of the contexts $(ux, v), (u, xv), (xu, v), (u, vx)$ belongs to C'. As in the case of scg, we can consider different other variants, by imposing one or some of these four variants of contexts. This time, however, we will partially change the possible typology and this change (leading to a larger variety of possibilities) can be considered also for scg. So, we have left internal cgc (licgc), left external cgc (lecgc), right internal cgc (ricgc), right external cgc (recgc), bilateral internal cgc (bicgc) as being both licgc and ricgc, bilateral external cgc (becgc) as being both lecgc and recgc, left cgc (lcgc) as being both licgc and lecgc, right cgc (rcgc) as being both ricgc and recgc; weak internal cgc (wicgc) when all generated contexts are internal, some of them at left, others at right; weak external cgc (wecgc) obtained from wicgc by changing "internal" with "external"; weak left cgc (wlcgc) when all contexts are generated at left, certain of them internal, others external; weak right cgc (wrcgc), obtained from wlcgc by changing "left" with "right".

In a further step, we may raise, with respect to $L(G)$ and $C(H)$, questions similar to those formulated for scg. Under what conditions does there exist a language M such that $L(G)$ and $C(H)$ are corresponding to each other within the framework of the string-contexts Galois connection in respect to M? What about the unicity of M? Since the Galois connection is given by a pair (h, p) of functions, $L(G)$ and $C(H)$ may correspond to each other in three different ways and we get three kinds of compatibility between the pair (G, H) of cgc and the Galois connection (h, p): h-compatibility, p-compatibility and hp-compatibility. We may ask more specific questions: how could we express in terms of the selection functions f from G and g from H the condition of existence of a language M such that the pair (G, H) of cgc is h-compatible, p-compatible or hp-compatible with respect to the string-contexts Galois connection associated to M? If M is not unique, we have to look for the simplest possible M, for instance, M regular will be preferred to

M non-regular, but context-free. We may obtain a hierarchy of pairs (G, H) in respect to the complexity of the language M determining a string-contexts Galois connection g-compatible, p-compatible or hp-compatible to (G, H). Some suggestions coming from English or French can be obtained when examining distributional classes and contextual classes in these languages, at various levels of grammaticality (see [41]). There should be a correspondence between $f(s)$ and $g(c)$, when s is in $g(c)$ and c is in $f(s)$.

Let us now refer to the cgc in their more general form ([55], [56]): $G = (A, B, (S_1, C_1), \ldots, (S_n, C_n))$, where A is the alphabet, B is a finite language over A, $S_1, \ldots, S_n$ are languages over A and $C_1, \ldots, C_n$ are finite sets of contexts over A. A natural interpretation of $S_1, \ldots, S_n$ is that of a sequence of different levels of grammaticality (lg) and, in this respect, the most interesting case, in the study of natural languages, is that in which $S_1, \ldots, S_n$ form an ascending chain: each S_i is contained in the next lg S_{n+1} (i between 1 and $n - 1$). This is what happens in [12] and [41], [42]. As a preliminary step, we could sometimes work with pairwise disjoint sets $S_1, \ldots, S_n$, but in a further step we have to take successively their union under the form $S_1, S_1 \cup S_2, \ldots, S_1 \cup S_2 \cup S_3, \ldots$ and finally we reach an ascending chain of lg.

The contextual counterpart of the above generalized form of cgc (gcgc) can be defined as a system $H = (A, C_0, (C_1, S_1), \ldots, (C_n, S_n))$, where A is the alphabet; C_0 is a finite set of contexts over A; $C_1, \ldots, C_n$ are sets of contexts over A, while $S_1, \ldots, S_n$ are finite sets of strings over A. The set C of contexts generated by H is obtained as the minimal set C of contexts satisfying the following requirements: C_0 is contained in C; given a context $c = (u, v)$ in C and a string x in some selector S_i, we can internally extend (u, v) by means of x, if $u = u'u_i, v = v_iv'$, where (u_i, v_i) is a context in C_i; the extended context belonging to C will be either (ux, v) or (u, xv). Similarly we can define the extended context in other ways, by using, as in the previous cases, the combinatorics of distinctions left-right, internal-external, weak-strong.

The conditions concerning the maximal local use of selectors (Mlus), maximal global use of selectors (Mgus) in a gcgc as well as the corresponding minimal conditions (mlus, mgus) were introduced in [46] and can be expressed as follows. Mlus: for each i from 1 to n, no string-selector from S_i can be substring of another string-selector from S_i; Mgus: for each i, no string-selector from S_i can be substring of another string-selector from $S_1 \cup S_2 \cup \ldots \cup S_n$; mlus: for each i, no string-selector from S_i can be super-string of another string-selector from S_i; mgus: for each i, no string-selector from S_i can be super-string of another string-selector from $S_1 \cup S_2 \cup \ldots \cup S_n$. These conditions express natural requirements of economy and simplicity. As a matter of fact, we would like to have both Mgus and mgus and this is what really happens in the contextual analysis of English by Du Feu, Marcus [12], where three different levels of grammaticality were considered (so, $i = 1, 2, 3$). Since the usual way in a linguistic analysis is to take $S_1, \ldots, S_n$ such that they form an ascending chain, Mgus is equivalent, in this case, to Mlus and mgus is equivalent to mlus, so the distinction local-global is no longer relevant (better to say that it is neutralized).

Going deeper in the interplay strings-contexts

All phenomena related to gcgc have their correspondent with respect to their contextual counterpart discussed in the preceding section. The topics we pointed out about cgc can be extended to gcgc and, *mutatis mutandis*, to their contextual counterpart. This transfer is a part of a broader principle, the duality between strings and contexts. Given a language L over A and two contexts c_1 and c_2 over A, c_1 L-dominates c_2 if any string accepted by c_1 in L is also accepted by c_2 in L. The grammatical category $G(C)$ generated by the set C of contexts over A is, by definition, the union of C with the set of all contexts L-dominated by any context in C. On the other hand, denoting by $S(C)$ the set of all strings accepted in L by any context in C, we define the Sestier closure $\varphi(C)$ as the set of all contexts accepting in L any string in $S(C)$. So, all problems related to gc and Sestier closure have their equivalent when strings are replaced by contexts.

However, we can proceed in a more radical way and replace the language L, which is a set of strings, by a set K of contexts over A. The domination relation between contexts will have, in this case, a large variety. For instance, if $c_1 = (u_1, v_1)$ and $c_2 = (u_2, v_2)$, we can define: the left internal domination of c_2 by c_1, when for any string x over A, such that $(u_1 x, v_1)$ is in K, $(u_2 x, v_2)$ is in K; the left external domination of c_2 by c_1, when for any string x, such that $(x u_1, v_1)$ is in K, $(x u_2, v_2)$ is in K; the right internal domination and the right external domination are obtained correspondingly. We can also define, in an obvious way, the bilateral internal domination, the bilateral external domination, the left domination (when both internal and external left domination take place), the right domination, and some other types of domination. For each of them we obtain a corresponding concept of grammatical category. On the other hand, given a set C of contexts over A, we can define a large variety of Sestier closures of C with respect to K: left internal, left external, right internal, right external, etc. Indeed, taking into account the identity between Sestier closure and Kunze categories, the Sestier closure of C will be the union of C with all sets of contexts that are Kunze dominated by C in respect to K. However, the Kunze domination between C and another set D of contexts can be conceived in different ways, following the combinatorics of distinctions left-right, internal-external, strong-weak. For instance, the left internal Kunze domination between C and D is given by the following formulation: Given a string x such that (ux, v) is in K for any (u, v) in C, we will have $(u'v, v')$ in K for any (u', v') in D.

Obviously, this multiplicity of possibilities leads to a very rich typology of contextual situations and of Galois connections. But this richness is supported by the reality of natural languages. Since a similar rich typology belongs to contextual grammars too, it follows that for each type separately we can raise the problems already discussed for sgc and cgc.

There is also the possibility to consider, following [46], the scattered use of selectors and to enlarge the notion of a context to a finite ordered set of strings.

Besides the interpretation of $S_1, \ldots, S_n$, respectively of $C_1, \ldots, C_n$ as levels of grammaticality defined, in the first case, by means of sets of strings and, in the second case, by means of sets of contexts, there are also other possible interpretations, going beyond natural languages. It is enough, in this respect, to adopt a new interpretation of the elements in the alphabet A. For instance, if these elements are symptoms of various possible illnesses, then strings over A may define various possible syndromes (Celan-Marcus [5], Popescu [59]). A clinical examination involves a degree of ambiguity in establishing the diagnostic; pathognomonic syndromes (those devoid of ambiguity) are only some times available. Usually, a syndrome leads to a state of hesitation among n different possible illnesses and the selectors $S_1, \ldots, S_n$ are modelling just these different illnesses, each of them represented by the set of corresponding syndromes.

Finally, let us observe that the request expressed in [43] to have selectors as simple as possible (finite or, at most, regular) is fully satisfied in both interpretations of selectors considered above: levels of grammaticality and sets of syndromes of some definite illnesses. Practically, we work, in natural languages, with finite levels of grammaticality and then, by grammatical inference, we proceed to an approximation with infinite sets; in this respect, see [63] for a more careful analysis of the infinity of natural languages.

A higher level of abstraction: parts of speech

Higher levels of abstraction and new linguistic aspects of contextual grammars can be obtained by exploiting the idea presented in the last part of the preceding section: to reinterpret in a significant way the elements of the alphabet A. In most parts of the discussion above, the elements of A were interpreted as morphemes or flexional forms. Let us now consider the interpretations of elements in A as flexional paradigms. This means that each element in A will be the set of all flexional forms of a lexical unit. For instance, a noun-form will be in the same set with all its flexional forms reflecting the variability in respect to various morphological categories (case, number, etc.). Obviously, this operation is less relevant in English, in view of its very poor morphology. French also, due to its rather analytic declension, has poor flexional classes, but its verb paradigms are richer. German is more interesting in this respect. Most relevant are the languages with a very synthetic morphology, such as Latin and various Slavic languages. This situation explains why the first mathematical model having as departure point the partition of A in flexional classes was proposed by a Russian scholar, Kulagina [24], and further developed by various authors from Slavic countries (Uspenskij [75],

Revzin [60], [61], Shreider [69], [70], Trybulec [73], Semeniuk-Polkowska [67]; a further mathematical development of this model was proposed by Marcus [28] who applied it to the typology of languages [30], [32] and to the comparative analysis of grammatical gender in Romance and in Slavic languages [40].

Obviously, by considering that flexional forms lead to a partition of the vocabulary we adopt a simplification of the linguistic reality, because some flexional forms are common to different paradigms; this fact is very frequent in English, but less frequent in languages with synthetic morphology, for which the considered model was mainly conceived.

The important discovery made by Kulagina [24] is the fact that, when elements of A are interpreted as flexional paradigms, the corresponding distributional classes model the parts of speech (noun, verb, qualificative adjective, etc.). Starting from a partition P of A, Kulagina defines the derivative P' of P in respect to L, as a new partition of A, whose terms are unions of L-equivalent terms of P; in other words, the terms of P' are the distributional classes in respect to L, when elements of the alphabet are no longer the elements of A, but the terms of P. A string $P_1 P_2 \ldots P_n$ on this new alphabet is L-well-formed (L is a given language on A) if there exists a string $a_1 a_2 \ldots a_n$ on A that is in L and such that a_1 is in P_1, a_2 in $P_2, \ldots, a_n$ in P_n. If we try to iterate this operation, we fail, because an important theorem by Kulagina asserts that $P'' = P'$.

For a mathematical refinement of the derivative of a partition see Solomon [71]. By means of Kulagina's model, contextual grammars of various types and their relations to closure operations and Galois connections may receive new linguistic interpretations, that cover not only the contextual behavior of linguistic strings, but also important paradigmatic aspects.

Other aspects of Galois connections and of closure operators are considered by Pană [53], Coardoş [7], Schwartz [65]. A general concept of contextual domination, including all known particular notions in this respect was proposed by Schwarz [66]. A general concept of language system was proposed by Nebeský [49]. Other contextual aspects were considered by Dobrushin [11], Ito [20], [21], Dincă [9], Calude [2]. Other aspects of domination relations were analyzed by Mayoh [47], Nebeský [48], Cazimir [3], [4]. A complete characterization of classes of Chomsky in terms of analytical models was given by Novotný [50], who also gave a general framework for the general study of languages [51].

Generative power of contextual grammars

The only contributions concerning the generative relevance of contextual grammars in respect to natural languages are [43], [44], [45]; it is shown that internal contextual grammars with maximal use of selectors are the most appropriate (in respect to other classes of contextual grammars) to model the generative capacity of natural languages (and of some artificial languages too), because they are able to describe all the usual restrictions appearing in such languages and leading to their non-context-freeness: reduplication, long distance dependencies, and agreements.

The main candidates for modelling the generative aspects of natural languages were so far considered the indexed grammars [1], head grammars [58], and the tree adjunct grammars ([23]; see also [22]), as well as some of their generalizations. Comparing the arguments brought in [43], [44], [45] with those brought in favor of the above types of grammars by Gazdar and Pullum [13] and Partee, Meulen, and Wall [54], it follows that contextual grammars are not transgressed by the other ones, concerning their relevance for the generative capacities of natural languages. Let us also take into account the fact that, besides their generative capacities, contextual grammars are, in their initial motivations and by their strong relations with the morphological, distributional, and contextual aspects in a genuine interaction with natural languages. (See also Manaster Ramer [27].)

Further suggestions: restricted contextual grammars, grammar systems and splicing contextual schemes

Let us define the concept of a restricted simple contextual grammar (rscg) as a system $G = (A, \mathcal{L}, \mathcal{C})$, where $\mathcal{L}$ is a finite set of finite languages $L_1, L_2, \ldots, L_n$ on A, while $\mathcal{C}$ is a finite set of finite sets $C_1, C_2, \ldots, C_p$ of contexts on A. This grammar generates both strings and contexts. In order to generate strings, we define the language $L(G)$ generated by G as the smallest language L over A such that: each L_i $(1 \leq i \leq n)$ is contained in L; if $x \in L$, then there exists i between 1 and p and j between 1 and n, such that the string $uxyv$ is in L for any (u, v) in C_i and any y in L_j. In order to generate contexts, we define the set $C(G)$ of contexts generated by G as the smallest set C of contexts on A such that: each C_i $(1 \leq i \leq p)$ is contained in C; if (u, v) is in C, then there exists i between 1 and n for which the context (ux, v) is in C for any x in L_i.

We have defined above a rscg working in the left internal mode, so we can call this grammar a left internal rscg (lirscg). We can define similarly the other variants related to the distinctions left-right, internal-external, strong-weak, etc. One can also extend the restricted type to contextual grammars with choice and their various generalizations.

The linguistic motivation of the restricted type of contextual grammars is related to the fact that the elementary units of contextual analysis are the distributional classes of strings (i.e., strings having the same contextual behavior) and the equivalence classes of contexts (i.e., contexts accepting exactly the same strings); these two types of situations are reflected by the sets $\mathcal{L}$ and $\mathcal{C}$.

Now, how could one bridge these ideas with that of a grammar system (see [8])? If the basic idea of a grammar system is that of a chain of grammars where outputs of one of them become inputs for another, then this idea is potentially contained in that of a Sestier closure. Indeed, the operator c in $\varphi(X) = s(c(X))$ has as inputs various languages over A and as outputs sets of contexts on A, while the next operator s has as inputs just the outputs of the first operator c. In this way, c and s are mediating the inputs and the outputs of the Sestier closure operator φ. The corresponding contextual grammar system should simulate this situation. We have to take into consideration these alternatives of the Sestier closure too: define the operators $e(C)$ = the set of strings accepted by at least one context in C and $f(S)$ = the set of contexts accepting at least one string in S and then consider the composed operators $s(f(S)), e(f(S))$ and $e(c(S))$.

Another possible perspective for contextual grammars is related to the concept of a *splicing scheme*, introduced by Head, [16] (see also [18]). A splicing scheme is a triple $S = (A, T, P)$, where A is a finite set, T is a finite subset of $A^* \times A^* \times A^*$ and P is a binary relation on T which satisfies the following condition: If p, x, q, u, y, v are in A^*, (p, x, q) and (u, y, v) are in T, and $(p, x, q)P(u, y, v)$, then $x = y$. The set A is called the *alphabet*, T the *triples* and P the *pairing* relation of the scheme. Informally expressed, the splicing scheme (A, T, P) acts on each pair of strings of the form $hpxqk$ and $wuxvz$, where each of the letters is a string in A^* and $(p, x, q)P(u, x, v)$ to produce the pair $hpxvz$ and $wuxqk$.

One can give to the above definitions from [16] a contextual version, permitting to take advantage of the aparatus related to algebraic contextual analysis (Galois connections, Sestier closures, various types of contextual dominations, etc). Let A be a non-empty alphabet, C a finite set of contexts on A, L a finite language obtained by placing some strings in A^* in some contexts in C and P a binary relation in L such that if $pxqPuyv$ (with (p, q) and (u, v) in C), then $x = y$. Let us call the system $G = (A, C, L, P)$ a *contextual splicing scheme*. The language $L(G)$ generated by G is the set of all pairs of strings of the form $kpxvz$ and $wuxqk$, where (p, q) and (u, v) are in C, and pxq and uxv are strings in L that are in the relation P, while $h, k, w,$ and z are strings on A. It is easy to see that contextual splicing schemes are equivalent to splicing schemes, in their initial form, proposed by Head, [16], [17]. It remains to check how can we exploit this contextual version of a splicing scheme. For instance, the binary relation P could be introduced in C instead to be defined in L: two contexts (p, q) and (u, v) in C are in relation P if for any string x accepted by (p, q) in L and for any string y accepted in L by (u, v) we have $x = y$.

Let us add, as a final remark, that the monograph in two volumes by Helden [19] develops, along more than thousand pages, a careful analysis of various natural languages, by using just the contextual tools discussed above.

References

[1]   A. V. Aho, Indexed grammars – an extension of context-free grammars, *Journal of the ACM*, 15 (1968), 647–671.

[2]   C. Calude, On metrizability of a free monoid, *Discrete Mathematics*, 15 (1976), 307–310.

[3]   B. Cazimir, Analyse contextuelle par rapport à deux langages, *Revue Roum. Math. Pures Appl.*, 20, 2 (1975), 201–211.

[4]   B. Cazimir, Opérateurs contextuels et catégories morphologiques par rapport à deux langages, *Bull. Math. Soc. Sci. Math. Roumanie*, 17(65), 3 (1975), 247–269.

[5]   E. Celan, S. Marcus, Le diagnostic comme langage, *Cahiers Ling. Théorique et Appl.*, 10, 2 (1973), 163–173.

[6]   N. A. Chomsky, *Syntactic Structures*, Mouton, The Hague, 1957.

[7]   V. Coardoş, Formal languages and Galois connections, *Revue Roum. Ling. – Cahiers Ling. Théor. Appl.*, 13, 2 (1976), 665–668.

[8]   E. Csuhaj-Varju, J. Dassow, J. Kelemen, Gh. Păun, *Grammar Systems. A Grammatical Approach to Distribution and Cooperation*, Gordon and Breach, London, 1994.

[9]   A. Dincă, The metric properties of the semigroups and the languages, *Lecture Notes in Computer Science*, 45, Springer-Verlag, Berlin, 1976, 260–264.

[10]   R. L. Dobrushin, The elementary grammatical category (in Russian), *Byulleten Objedinenija po Problemam Mashinogo Perevoda*, 5 (1957), 19–21.

[11]   R. L. Dobrushin, Mathematical methods in linguistics. Applications (in Russian), *Matematicheskie Prosveshchenie*, 6 (1961), 52–59.

[12]   V. Du Feu, S. Marcus, English grammatical homonymy and grammatical categories from the view point of algebraic linguistics, *Revue Roum. de Ling.*, 18, 3 (1973), 215–241.

[13]   G. Gazdar, G. K. Pullum, Computationally relevant properties of natural languages and their grammars, *New Generation Computing*, 3 (1985), 273–306.

[14]   A. V. Gladkij, Configuration characteristics of languages (in Russian), *Problemy Kibernetiki*, 10 (1963), 251–260.

[15]   Z. S. Harris, *Methods in Structural Linguistics*, Chicago Univ. Press, 1951.

[16]   T. Head, Formal language theory and DNA: an analysis of the generative capacity of specific recombinant behaviors, *Bull. Math. Biology*, 49 (1987), 737–759.

[17]   T. Head, Splicing schemes and DNA, in *Lindenmayer Systems: Impacts on Theoretical Computer Science and Developmental Biology* (G. Rozenberg, A. Salomaa, eds.), Springer-Verlag, Berlin, 1992, 371–383.

[18]   T. Head, Gh. Păun, D. Pixton, Language theory and molecular genetics: generative mechanisms suggested by DNA recombination, in this *Handbook*.

[19]   W. A. van Helden, *Case and Gender-Concept Formation Between Morphology and Syntax* (2 volumes), Rodopi, Amsterdam-Atlanta, Georgia, 1993.

[20]   M. Ito, Quelques remarques sur l'espace contextuel, *Mathematical Linguistics*, 57 (1971), 18–28.

[21]   M. Ito, Sur l'extension et la restriction d'un langage associé à l'espace contextuel, *Proc. Japan Academy*, 48 (1972), 94–97.

[22]   A. K. Joshi, Tree adjoining grammars: How much context-sensitivity is required to provide reasonable structural descriptions? in *Natural Language Processing: Psycholinguistic, Computational and Theoretic Perspectives* (D. R. Dowty et al., eds.), Cambridge Univ. Press, New York, 1985, 206–250.

[23] A. K. Joshi, L. Levy, M. Takahashi, Tree adjunct grammars, *Journal of the Computer and System Sciences*, 10, 1 (1975), 136–163.

[24] O. S. Kulagina, On a way to define the grammatical concepts by means of set-theory (in Russian), *Problemy Kibernetiki*, 1 (1958), 203–214.

[25] J. Kunze, Versuch eines objektivierten Grammatikmodells II, *Zeitschrift für Phonetik, Sprach. und. Komm.*, 20, 5/6 (1967), 415–448.

[26] J. Lambek, On the calculus of syntactic types, in *Structure of Language and its Mathematical Aspects* (R. Jakobson, ed.), Proc. of the 12th Symp. Appl. Math., Providence, R. I., 1961, 166–178.

[27] A. Manaster Ramer, Uses and misuses of mathematics in linguistics, *Proc. of the Xth Congress of Natural and Formal Languages*, Sevilla, 1994.

[28] S. Marcus, On a logical model of the part of speech (in Romanian), *Studii şi Cercetări Matematice*, 13, 1 (1962), 137–162.

[29] S. Marcus, Sur un modèle logique de la catégorie grammaticale élémentaire, I, *Revue Roum. Math. Pures Appl.*, 7, 1 (1962), 91–107.

[30] S. Marcus, *Mathematical Linguistics* (in Romanian), Ed. Didactică şi Pedagogică, Bucureşti, 1963.

[31] S. Marcus, Modèles mathématiques pour la catégorie grammaticale du cas, *Revue Roum. Math. Pures Appl.*, 8, 4 (1963), 585–610.

[32] S. Marcus, Typologie des langues et modèles logiques, *Acta Math. Acad. Sci. Hungaricae*, 14, 3-4 (1963), 269–281.

[33] S. Marcus, *Algebraic Linguistics; Analytical Models*, Academic Press, New York, 1967.

[34] S. Marcus, *Introduction mathématique à la linguistique structurale*, Dunod, Paris, 1967.

[35] S. Marcus, Catégories de Dobrushin, fermetures de Sestier, voisinages de Sakai, *Glossa*, 1, 1 (1967), 59–67.

[36] S. Marcus, Catégories morphologiques, *IIème Conférence Internationale sur le Traitement Automatique des Langues*, Grenoble, 1967, communication 39, 1–8.

[37] S. Marcus, Opérateurs contextuels et catégories morphologiques, *Bull. Math. Soc. Sci. Math. Roumanie*, 10, 3 (1968), 65–72.

[38] S. Marcus, Sur la domination au sens de Kunze dans la linguistique algébrique, *Revue Roum. Math. Pures Appl.*, 14, 3 (1969), 378–396.

[39] S. Marcus, Contextual grammars, *Revue Roum. Math. Pures Appl.*, 14, 10 (1969), 1473–1482.

[40] S. Marcus, Les modèles mathématiques et l'opposition romane-slave dans la typologie du genre grammatical, *Actes du XIIème Congress Intern. de Linguistique et de Philologie Romane*, vol. 1, Ed. Academiei, Bucureşti, 1970, 247–252.

[41] S. Marcus (ed.), *Contextual Ambiguities in Natural and in Artificial Languages*, I, Ghent (Belgium), *Communication and Cognition*, 1981.

[42] S. Marcus (ed.), *Contextual Ambiguities in Natural and in Artificial Languages*, II, Ghent (Belgium), *Communication and Cognition*, 1983.

[43] S. Marcus, C. Martin-Vide, Gh. Păun, Contextual grammars and natural languages, *Intern. Conf. Math. of Natural Languages*, Philadelphia, Penn., 1995.

[44] S. Marcus, C. Martin-Vide, Gh. Păun, On the internal contextual grammars with maximal use of selectors, *Automata and Formal Languages Conf.*, Salgotarjan, 1996.

[45] S. Marcus, C. Martin-Vide, Gh. Păun, Contextual grammars versus natural languages, submitted, 1996.

[46] C. Martin-Vide, A. Mateescu, J. Miquel-Verges, Gh. Păun, Internal contextual grammars: minimal, maximal and scattered use of selectors, *BISFAI 95 Conf.* (M. Kappel, E. Shamir, eds.), Jerusalem, 1995, 132–142.

[47] B. Mayoh, Grammatical categories, *Revue Roum. Math. Pures Appl*, 12, 6 (1967), 843–848.

[48] L. Nebeský, Conditional replacement of words, *Prague Bull. Math. Linguistics*, 3 (1965), 3–12.

[49] L. Nebeský, Language systems, *Revue Roum. Math. Pures Appl.*, 13, 3 (1968), 371–375.

[50] M. Novotný, Complete characterizations of classes of Chomsky by means of configurations, *Acta Facultatis Rerum Naturalium Univ. Comenianae. Mathematica*, special issue, 1971, 63–71.

[51] M. Novotný, On some relations defined by languages, *Prague Studies in Math. Linguistics*, 4 (1972), 157–170.

[52] O. Ore, Galois connections, *Trans. Amer. Math. Soc.*, 55 (1944), 493–513.

[53] D. Pană, Closure operators and the relation of domination in algebraic linguistics (in Romanian), *Studii și Cercetări Matematice*, 27, 4 (1975), 461–470.

[54] B. H. Partee, A. T. Meulen, R. A. Wall, *Mathematical Methods in Linguistics*, Kluwer, Dordrecht, 1990.

[55] Gh. Păun, *Contextual Grammars* (in Romanian), Ed. Academiei, București, 1982.

[56] Gh. Păun, Marcus contextual grammars. After 25 years, *Bulletin EATCS*, 52 (1994), 263–273.

[57] Z. Pawlak, Rough sets, *Intern. J. Computer and Information Sciences*, 11 (1982), 341–356.

[58] C. J. Pollard, *Generalized Structure Grammars, Head Grammars, and Natural Languages*, PhD Thesis, Stanford University, California, 1984.

[59] M. Popescu, Sur un modèle linguistico-mathématique du changement du diagnostic, *Bull. Math. Soc. Sci. Math. Roumanie*, 21, 1-2 (1977), 113–122.

[60] I. I. Revzin, *Models of Language* (in Russian), Izdatelstvo Akademii Nauk, Moskow, 1962.

[61] I. I. Revzin, *The Modeling Method and the Typology of Slavic Languages* (in Russian), Nauka, Moskow, 1967.

[62] J. Riguet, Relations binaires, fermetures, correspondences de Galois, *Bull. Soc. Math. France*, 76 (1948), 114–155.

[63] W. J. Savitch, Why it might pay to assume that languages are infinite, *Annals of Math. and Artificial Intelligence*, 8 (1993), 17–25.

[64] L. Schwartz, On Trybulec domination in algebraic linguistics, *Revue Roum. Math. Pures Appl.*, 17, 3 (1972), 425–433.

[65] L. Schwartz, Binary relations, Galois connections and topological closure operators (in Romanian), *Studii și Cercetări Matematice*, 34, 1 (1982), 73–83.

[66] D. Schwarz, A general notion of domination relation in algebraic linguistics, *Revue Roum. Ling. – Cahiers Ling. Théor. Appl.*, 13, 1 (1976), 305–313.

[67] M. Semeniuk-Polkowska, Syntactic pseudo-configurations based on Trybulec domination, *Revue Roum. Math. Pures Appl.*, 27, 6 (1982), 709–717.

[68] A. Sestier, Contributions à une théorie ensembliste des classifications linguistiques, *Actes du Ier Congrès de l'AFCAL*, Grenoble, 1960, 293–305.

[69] Ju. A. Shreider, The algebra of binary relations (in Russian). Annex to S. Marcus, *Set-Theoretic Models of Languages* (in Russian), Nauka, Moskow, 1970.

[70] Ju. A. Shreider, On the O. S. Kulagina concept of derived relation, *Computational Linguistics*, 9 (1973), 241–253.

[71] S. Y. Solomon, The n-derivative of a partition, *Recueil Linguistique de Bratislava*, 4 (1973), 237–243.

[72] L. Tesnière, *Elélements de syntaxe structurale*, Klincksieck, Paris, 1959.

[73] A. Trybulec, The decomposition of the vocabulary in languages with given paradigms (in Russian), *Nauchno Technicheskaja Informacija*, 12 (1967), 40–44.

[74] V. A. Uspenskij, On the definition of case by A. N. Kolmogorov, *Byulleten Obyedinenija po Problemam Mashinogo Perevoda*, 5 (1957), 11–18.

[75] V. A. Uspenskij, On the definition of parts of speech in the set-theoretic system of language (in Russian), *ibidem*, 22–26.

[76] L. Vasilescu, Analyse distributionelle algébrique des substantifs français, *Cahiers Ling. Théor. Appl.*, 10, 2 (1973), 251–259.

[77] L. Vasilescu, Analyse distributionelle algébrique des formes verbales du français, *Cahiers Ling. Théor. Appl.*, 11, 1 (1974), 137–169.

[78] L. Vasilescu, Analyse distributionelle algébrique des formes adjectivales du français, *Revue Roum. Ling.*, 20, 3 (1975), 233–248.

[79] M. Ward, The closure operators of a lattice, *Annals of Math.*, 43, 2 (1942), 191–196.

[80] L. Zadeh, Fuzzy sets, *Information and Control*, 8 (1965), 338–353.

Contextual Grammars and Formal Languages

Andrzej Ehrenfeucht, Gheorghe Păun, and Grzegorz Rozenberg

1. Introduction

Contextual grammars were introduced by S. Marcus in 1969 [29], in an attempt to build a bridge between analytical and generative models of natural languages. In particular, contextual grammars were "translating" the central notion of context from the analytical models into the framework of generative grammars. The chapter by S. Marcus in this handbook [31] gives a lucid account of the motivation behind contextual grammars from the natural point of view.

Since their introduction by Marcus, contextual grammars have been intensely investigated by formal language theorists, see, e.g., [60], [68], and their references, and by today it is clear that contextual grammars offer quite novel insight into a number of issues central to formal language theory.

First of all, they provide an important tool in the study of *recursion* in formal language theory. In a context-free grammar the fact that a nonterminal A may derive (possibly in a number of steps) a string xAy is interpreted that if a string z is in the syntactic class A, then also the string xzy is in this class. In contextual grammars this recursion is given *explicitly* by associating the context (x, y) with the string z. In a sense, the research on contextual grammars is a study of recursion in its pure form.

Also, contextual grammars offer a very convenient framework for the systematic study of the basic language theoretic operations, such as insertion and deletion (see, e.g., [26]), and the operation of *splicing* (see, e.g., [16], [17]) which provide a nice bridge between formal language theory and DNA computing.

Finally it should be stressed that contextual grammars form a major contribution to our understanding of *pure grammars*, i.e., grammars which generate languages without the use of nonterminal (auxiliary) symbols.

In this chapter we consider contextual grammars from the formal language theory point of view.

2. Contextual grammars with unrestricted choice

2.1 Preliminaries

We assume that the reader is familiar with basic notions of formal language theory. Throughout this chapter we will use the following, rather standard, set-theoretic and language-theoretic notation.

2^X = the power set of X,
$card(X)$ = the cardinality of the set X,
$\emptyset$ = the empty set,
V^* = the free monoid generated by V,
λ = the empty string,
$V^+ = V^* - \{\lambda\}$,
$|x|$ = the length of the string x,
$|x|_a$ = the number of occurrences of the symbol $a \in V$ in the string $x \in V^*$,
$|x|_U$ = the number of occurrences of symbols in $U \subseteq V$ in the string $x \in V^*$,
$length(L) = \{|x| \mid x \in L\}$, the length set of L,
REG, LIN, CF, CS, RE = the families of regular, linear, context-free, context-sensitive, and of recursively enumerable languages, respectively,
FIN = the family of finite languages,
LIN_1 = the family of minimal linear languages (generated by linear grammars with only one nonterminal),
Ψ_V = the Parikh mapping associated to the alphabet V,
$L_1 \backslash L_2 = \{w \mid$ there is $x \in L_1$ such that $xw \in L_2\}$, the left quotient of L_2 with respect to L_1,
L_2/L_1 = the right quotient of L_2 with respect to L_1,
$\partial_x^l(L), \partial_x^r(L)$ = the left and the right derivatives of $L \subseteq V^*$ with respect to $x \in V^*$,
$Pref(L)$ = the set of prefixes of strings in L,
$Sub(L)$ = the set of substrings of strings in L,
$Suf(L)$ = the set of suffixes of strings in L,
$PPref(L)$ = the set of proper prefixes of strings in L,
$PSub(L)$ = the set of proper substrings of strings in L,
$PSuf(L)$ = the set of proper suffixes of strings in L,
$alph(L)$ = the set of symbols appearing in the words of L,
$mi(x)$ = the mirror image of x,
$D_{a,b}$ = the Dyck language over the alphabet $\{a, b\}$.

2.2 Definitions

Definition 2.1. *A total contextual grammar is a system*

$$G = (V, B, C, \varphi),$$

where V is an alphabet, B is a finite language over V, C is a finite subset of $V^*\$V^*$, where $\$ is a special symbol not in V, and $\varphi : V^* \times V^* \times V^* \longrightarrow 2^C$.

The strings in B are called *axioms*, the elements $u\$v$ in C are called *contexts*, and φ is the *choice mapping*. Here we do not impose restrictions on φ – we even do not assume that φ is computable.

Definition 2.2. *For a total contextual grammar* $G = (V, B, C, \varphi)$ *we define the relation* $\Longrightarrow_G$ *on* V^* *as follows:*

$$x \Longrightarrow_G y \quad \text{iff} \quad x = x_1 x_2 x_3, \ y = x_1 u x_2 v x_3, \ \text{for some}$$
$$x_1, x_2, x_3 \in V^*, u\$v \in C, \ \text{such that } u\$v \in \varphi(x_1, x_2, x_3).$$

The subscript G *may be omitted from* $\Longrightarrow_G$ *whenever* G *is understood. Denoting by* $\Longrightarrow^*$ *the reflexive and transitive closure of* $\Longrightarrow$, *the language generated by* G *is*

$$L(G) = \{x \in V^* \mid w \Longrightarrow^* x, \ \text{for some } w \in B\}.$$

Note that, by definition, $B \subseteq L(G)$.

We say that x *directly derives* y *in* G *or that* x *derives* y *in* G *whenever* $x \Longrightarrow y$ *or* $x \Longrightarrow^* y$ *holds, respectively.*

In a total contextual grammar, a context is adjoined depending on the whole current string. Two special cases of total contextual grammars are very natural and have been extensively investigated.

Definition 2.3.
(i) *A total contextual grammar* $G = (V, B, C, \varphi)$ *is called external if* $\varphi(x_1, x_2, x_3) = \emptyset$ *for all* $x_1, x_2, x_3 \in V^*$ *such that* $x_1 x_3 \neq \lambda$.
(ii) *A total contextual grammar* $G = (V, B, C, \varphi)$ *is called internal if* $\varphi(x_1, x_2, x_3) = \varphi(x_1', x_2, x_3')$ *for all* $x_1, x_1', x_2, x_3, x_3' \in V^*$.

Note that in both cases, the adjoining of a context $u\$v \in \varphi(x_1, x_2, x_3)$ to the string $x_1 x_2 x_3$ depends on x_2 only. Therefore, in these cases we can simplify the choice mapping φ by having $\varphi : V^* \longrightarrow 2^C$. The so modified total contextual grammars are called *contextual grammars*.

Since in this way there is no difference between external and internal grammars, we shall distinguish between derivation relations, defining:

$$x \Longrightarrow_{ex} y \quad \text{iff} \quad y = uxv \text{ for } u\$v \in \varphi(x);$$
$$x \Longrightarrow_{in} y \quad \text{iff} \quad x = x_1 x_2 x_3, \ y = x_1 u x_2 v x_3, \ \text{for some}$$
$$x_1, x_2, x_3 \in V^*, u\$v \in \varphi(x_2).$$

We call $\Longrightarrow_{ex}$ an external (direct) derivation and $\Longrightarrow_{in}$ an internal (direct) derivation. Accordingly we associate with a contextual grammar G two languages:

$$L_\alpha(G) = \{x \in V^* \mid w \Longrightarrow_\alpha^* x \text{ for some } w \in B\},$$

for $\alpha \in \{ex, in\}$.

For a contextual grammar $G = (V, B, C, \varphi)$ the relation $u\$v \in \varphi(x)$ can be interpreted as a rewriting rule $x \to uxv$. Therefore, a contextual grammar with internal derivation can be considered to be a pure grammar with arbitrarily many (length-increasing) productions of the form $x \to uxv$, where there exists a bound k such that, for each x and each production $x \to uxv$, $|uv| \leq k$ holds. (Pure grammars are grammars that do not use nonterminals; see, e.g. [37].)

This suggests the following *compact presentation* of a contextual grammar: a system $G = (V, B, P)$, where V is an alphabet, B is a finite language over V, and P is a finite set of pairs (D, C), with $D \subseteq V^*$, $D \neq \emptyset$, and C a finite subset of $V^*\$V^*$, for $\$$ a special symbol not in V. The pairs (D, C) are called *productions*, D is the *selector* of such a production and C is its set of *contexts*. The interpretation is equivalent to $C \subseteq \varphi(x)$ for all $x \in D$: the contexts in C can be adjoined to strings in D. When presenting a grammar in the form $G = (V, B, P)$ we often use the mapping φ defined by $\varphi(x) = \{u\$v \mid u\$v \in C$ for some production $(D, C) \in P$ such that $x \in D\}$. For $x \in V^* - \{y \in D \mid (D, C) \in P\}$ we put $\varphi(x) = \emptyset$.

In general, when considering a grammar in the *functional presentation*, $G = (V, B, C, \varphi)$, we will specify the mapping φ only for strings x such that $\varphi(x) \neq \emptyset$; it is implicitly understood that $\varphi(x) = \emptyset$ for all unspecified x.

In examples and proofs we shall use the most convenient presentation of contextual grammars for a given purpose; even in the case of the compact presentation we shall use the choice mapping φ defined as above, without explicitly defining it.

A further simplification of (total) contextual grammars is to ignore the selection (then the total contextual grammars coincide with the internal contextual grammars):

Definition 2.4. *A contextual grammar $G = (V, B, C, \varphi)$ is said to be without choice if $\varphi(x) = C$ for all $x \in V^*$.*

In such a case, the mapping φ can be ignored, and we write the grammar in the form $G = (V, B, C)$. Again we define two derivation relations:

$$x \Longrightarrow_{ex} y \quad \text{iff} \quad y = uxv \text{ for some } u\$v \in C, \text{ and}$$
$$x \Longrightarrow_{in} y \quad \text{iff} \quad x = x_1 x_2 x_3, y = x_1 u x_2 v x_3 \text{ for some } u\$v \in C.$$

Five basic families of languages are obtained in this way.

TC = the family of languages generated by total contextual grammars,

ECC = the family of languages externally generated by contextual grammars,

ICC = the family of languages internally generated by contextual grammars,

EC = the family of languages externally generated by contextual grammars without choice,

IC = the family of languages internally generated by contextual grammars without choice.

In the general case of contextual grammars it is natural to require that φ is a computable mapping. The corresponding families are denoted by TC_c, ECC_c, ICC_c, respectively.

2.3 Examples

The following examples illustrate the above definitions; also, they will be used in the sequel of this chapter.

Example 2.1. Every finite language, B, belongs to every family of contextual languages: for $G = (V, B, \emptyset)$ we have $L_{ex}(G) = L_{in}(G) = B$.

Example 2.2. For each alphabet V, the language V^* belongs to every family of contextual languages: for $G = (V, \{\lambda\}, \$V)$ we have $L_{ex}(G) = L_{in}(G) = V^*$.

If we take $G' = (V, V, \$V)$, then $L_{ex}(G) = L_{in}(G) = V^+$.

In general, for $G_k = (V, V^k, \$V), k \geq 0$, we obtain $L_{ex}(G_k) = L_{in}(G_k) = V^k V^*$.

Example 2.3. Consider the grammar $G = (\{a, b\}, \{\lambda\}, \{a\$b\})$. Obviously, $L_{ex}(G) = \{a^n b^n \mid n \geq 1\}$, whereas $L_{in}(G) = D_{a,b}$. This second equality can be proved by induction on the length of the strings.

Note that $L_{ex}(G)$ is not regular and $L_{in}(G)$ is not linear.

Example 2.4. For $G = (\{a, b, c\}, \{\lambda\}, \{\alpha\beta\$\gamma, \ \alpha\$\beta\gamma \mid \{\alpha, \beta, \gamma\} = \{a, b, c\}\})$, we obtain

$$L_{in}(G) = \{x \in \{a, b, c\}^* \mid |x|_a = |x|_b = |x|_c\}.$$

The inclusion $\subseteq$ is obvious: each context increases the number of a, b, c occurrences by one. The inverse inclusion can be proved by induction on the length of strings (each $x \in \{a, b, c\}^*$ with $|x|_a = |x|_b = |x|_c$ contains a substring $\alpha\beta$, for $\alpha, \beta \in \{a, b, c\}$, $\alpha \neq \beta$, and also an occurrence of γ, before or after $\alpha\beta$, such that $\{\alpha, \beta, \gamma\} = \{a, b, c\}$; removing these symbols α, β, γ – they correspond to one context of G – we obtain a shorter string, x', with $|x'|_a = |x'|_b = |x'|_c$).

Note that the above language is not context-free.

Example 2.5. The grammar $G = (\{a\}, \{a^3\}, \{\$a, \$a^2\}, \varphi)$, with φ defined by

$$\varphi(a^i) = \begin{cases} \$a, & \text{if } i+1 \neq 2^k, k \geq 0, \\ \$a^2, & \text{if } i+1 = 2^k, k \geq 0, \end{cases}$$

for all $i \geq 1$, generates

$$L_{ex}(G) = \{a^n \mid n \geq 1, n \neq 2^k, k \geq 0\},$$
$$L_{in}(G) = a^+ - \{a, a^2\}$$

(in the internal mode, each string $a^i, i \geq 3$, can be prolonged by one more occurrence of a, using $\$a \in \varphi(a^2)$, for some subword a^2 of a^i).

Example 2.6. Consider an arbitrary language $L \subseteq V^*$ and a symbol $c \notin V$. Then for the grammar

$$G_L = (V \cup \{c\}, \{\lambda\}, \{(V^* - L, \$V), (L, \$V \cup \{\$c\})\})$$

we have $L_{ex}(G_L) = V^* \cup L\{c\}$.

Note that if L is not recursive, then the associated mapping φ is not computable. Moreover, if $L \notin F$, for a family F of languages closed under intersection with regular sets and right derivative, then $L_{ex}(G) \notin F$, too.

Example 2.7. Let $G = (\{a, b\}, \{a\}, \{\$a, \$b\}, \varphi)$, with

$$\varphi(a^i) = \begin{cases} \{\$a, \$b\}, & \text{if } i = 2^p 3^q 5^r 7^s, p^s + q^s = r^s, s \geq 3, \\ \{\$a\}, & \text{otherwise}, \end{cases}$$

for all $i \geq 1$. Then

$$L_{ex}(G) = a^+ \cup \{a^n b \mid n = 2^p 3^q 5^r 7^s, (p, q, r, s) \text{ corresponds}$$
$$\text{to a counterexample to Fermat last theorem}\}.$$

Example 2.8. Consider the grammar

$$G = (\{a, b, c\}, \{babccab\}, \{\$a, a\$a, b\$b, c\$c\}, \varphi),$$
$$\varphi(ab^n cca) = \{a\$a\}, \ n \geq 1,$$
$$\varphi(bcca^n b) = \{b\$b\}, \ n \geq 1,$$
$$\varphi(ba^n b^m c) = \{\$a\}, \ n, m \geq 1,$$
$$\varphi(a^n b^n caca^m b^m) = \{c\$c\}, \ n, m \geq 1.$$

Then we have

$$L_{in}(G) = \{ba^n b^m ca^p ca^n b^m \mid n, m \geq 1, p \geq 0\} \cup$$
$$\cup \{bc^i a^n b^n caca^n b^n c^i \mid n, i \geq 1\}.$$

We conclude this section with a more intricate example.

Example 2.9. For

$$G = (\{a, b, c, d\}, \{a\}, \{(\{a\}, \{\$b, \$c\}), (\{bc\}, \$d), (\{dc\}, c\$d)\}),$$

we have

$$length(L_{in}(G) \cap ab(cd)^+) = \{2^n \mid n \geq 2\}. \tag{*}$$

Because $\varphi(x) \neq \emptyset$ only for finitely many strings x, the grammar G is equivalent with the pure grammar $G' = (\{a, b, c, d\}, a, P)$ where P consists of the following productions:

$$\pi_1 : a \to ab, \ \pi_2 : a \to ac, \ \pi_3 : bc \to bcd, \ \pi_4 : dc \to cdcd.$$

Let us examine the words in the set $L = L(G') \cap ab(cd)^+$. For producing a string $z = ab(cd)^m, m \geq 1$, we have to use one time the production π_1 (there is only one occurrence of b in z), and the equal number of times the productions π_2 and π_3. But π_2 cannot be used after using π_1 (b must remain adjacent to a) and π_3 cannot be used before using π_1. Therefore we start by using n times π_2, then π_1, then π_3 (n times). After using π_1 we get the string $abc^n, n \geq 1$. Production π_3 can be applied only for adjoining d to bc, and bc is never changed; the leftmost occurrence of c remains the same during the whole derivation. Since we may change the order of using π_3 and π_4 without modifying the generated string, we may assume that after producing abc^n we proceed to $abcd^n c^{n-1}$ using π_3. From now on only production π_4 can be used (otherwise we obtain a string not in L).

One can prove by induction on n that if a string obtained from $abcd^n c^{n-1}$ is in $ab(cd)^+$, then it is equal to $ab(cd)^{2^n - 1}$. This concludes the proof of $(*)$.

2.4 Necessary conditions and counterexamples

In fact, some of the conditions considered here are both necessary and sufficient for a language to be of certain type in the above classification of contextual languages.

We say that a language $L \subseteq V^*$ has the *external bounded step* (EBS) property if there is a constant p such that for each $x \in L, |x| > p$, there is $y \in L$ such that $x = uyv$ and $0 < |uv| \leq p$.

A language $L \subseteq V^*$ has the *internal bounded step* (IBS) property if there is a constant p such that for each $x \in L, |x| > p$, there is $y \in L$ such that $x = x_1 u x_2 v x_3, y = x_1 x_2 x_3$, and $0 < |uv| \leq p$.

A language $L \subseteq V^*$ has the *bounded length increase* (BLI) property if there is a constant p such that for each $x \in L$ there is $y \in L$ with $-p \leq |x| - |y| \leq p$.

Clearly, if a language has either the EBS or the IBS property, then it has also the BLI property.

Convention 2.1. The empty context, $u\$v, uv = \lambda$, is useless in contextual grammars of the types considered up to now, hence it is ignored. In what follows we always assume that all used contexts are different from $.

Lemma 2.1. *A language is in the family ECC if and only if it has the EBS property.*

Proof. For a contextual grammar $G = (V, B, C, \varphi)$, let

$$p_1 = \max\{|x| \mid x \in B\}, \quad p_2 = \max\{|uv| \mid u\$v \in C\}.$$

Then $L_{ex}(G)$ has the EBS property for $p = \max\{p_1, p_2\}$. (If $z \in L_{ex}(G), |z| > p$, then $z \notin B$, hence $z = uz'v$ for some $u\$v \in \varphi(z'), z' \in L_{ex}(G)$.)

Conversely, consider a language $L \subseteq V^*$ having the EBS property for a constant p. Construct the grammar $G = (V, B, C, \varphi)$ with

$$B = \{x \in L \mid |x| \leq p\},$$
$$C = \{u\$v \mid u, v \in V^*, 0 < |uv| \leq p\},$$
$$\varphi(x) = \{u\$v \in C \mid uxv \in L\}, x \in V^*.$$

The equality $L = L_{ex}(G)$ easily follows from the definition of B, C, φ.   □

The following lemmas are easy to prove (using the finiteness of the sets of axioms and of contexts).

Lemma 2.2. *A language is in the family TC if and only if it has the IBS property.*

Corollary 2.1.
(i) All contextual languages have the BLI property.
(ii) Each language in the family ICC has the IBS property.

Corollary 2.2.

$$L_1 = \{a^{2^n} \mid n \geq 1\} \notin TC \qquad \text{(BLI property)},$$
$$L_2 = ab^+a \notin ECC \qquad\qquad \text{(EBS property)},$$
$$L_3 = \{a^n ba^n ba^n \mid n \geq 1\} \notin TC \quad \text{(IBS property)}.$$

Note that both L_2 and L_3 have the BLI property, hence this property is not sufficient for a language to be in TC.

We say that a language $L \subseteq V^*$, $\text{card}(V) \geq 2$, has the *mix* property if there is a constant p such that if there are $a, b \in V$ and $x, y \in L$ with $|x|_a > p, |y|_b > p$, then for each $n \geq 1$ there is $z_n \in L$ with $|z_n|_a \geq n, |z_n|_b \geq n$.

Lemma 2.3. *Each language in $EC \cup IC$ has the mix property.*

Corollary 2.3. $L = a^+ \cup b^+ \notin EC \cup IC$.

Note that this language has the EBS and the IBS properties. As we have seen in Example 6, for each language $L \subseteq V^*$, and $c \notin V$, the language $L' = V^* \cup L\{c\}$ is in ECC. This implies that languages in ECC (and hence also in TC) do not have pumping properties (consider $L = \{a^{2^n}c \mid n \geq 1\}$; here no string $a^{2^n}c$ can be pumped, whichever type of usual pumping is considered). However, we have the following result.

Lemma 2.4. *If $L \subseteq V^*, L \in ICC$, then there are two constants p, q such that every $z \in L, |z| > p$, can be written in the form $z = uvwxy$ with $u, v, w, x, y \in V^*, 0 < |vx| \leq q$, and $uv^i wx^i y \in L$ for all $i \geq 0$.*

Corollary 2.4. $L = \{a^n b^n c^n \mid n \geq 1\} \notin ICC$.

Of course, the pumping property from Lemma 2.4 implies the IBS property (hence the BLI property, as well), but the converse does not hold: the language $L = a^+ \cup \{a^n ba^m ba^p \mid n + m + p = 2^k, k \geq 1\}$ has the IBS property, but no pumping is possible in a string $a^n ba^m ba^p$ from L.

Lemma 2.5. *If $L \subseteq V^*, L \in EC$, then there are two constants p, q such that every $z \in L, |z| > p$, can be written in the form $z = uyv$ with $u, y, v \in V^*, 0 < |uv| \leq q$, and $u^i yv^i \in L$ for all $i \geq 0$.*

The following property of external contextual languages is related to the EBS property. For a language $L \subseteq V^*$ we denote

$$Min_0(L) = \{x \in L \mid PSub(x) \cap L = \emptyset\},$$
$$Min_i(L) = Min_{i-1}(L) \cup Min_0(L - Min_{i-1}(L)), \ i \geq 1.$$

Lemma 2.6. *If $L \subseteq V^*, L \in ECC$, then all sets $Min_i(L), i \geq 0$, are finite.*

Proof. Let $G = (V, B, C, \varphi)$ be a contextual grammar such that $L = L_{ex}(G)$. Consider the sets

$$K_i(G) = \{u_j \ldots u_1 wv_1 \ldots v_j \mid 0 \leq j \leq i, w \in B, u_1\$v_1 \in \varphi(w),$$
$$u_k\$v_k \in \varphi(u_{k-1} \ldots u_1 wv_1 \ldots v_{k-1}), 2 \leq k \leq j\},$$

for $i \geq 0$ (when $i = 0$ we have $K_0(G) = B$).

All these sets are finite (B and C are finite). By induction on i, we can prove that $Min_i(L) \subseteq K_i(G), i \geq 0$, which implies that also $Min_i(L)$ is finite.   □

Corollary 2.5. $L = a^*ba^*ba^* \notin ECC$. *(We have $Min_0(L) = ba^*b$.)*

Every language over a one-letter alphabet has the property from Lemma 2.6. Since not all one-letter languages are in ECC, this condition is not sufficient for a language to be in ECC.

We present now a necessary condition of a different type.

A language $L \subseteq V^*$ is said to be *k-slender*, for some integer k [75], if $card(L \cap V^n) \leq k$ for all $n \geq 0$. A language is *slender* if it is k-slender for some k. A 1-slender language is called *thin*.

Lemma 2.7. *If $L \in IC$ is an infinite language such that $alph(L) \geq 2$, then L is not slender.*

Proof. Let $G = (V, B, C)$ be a grammar such that $L_{in}(G) = L, card(V) \geq 2$.

(1) If $C \subseteq a^*\$a^*$ for some $a \in V$, then we take $u\$v \in C$, with $|uv| = a^i, i \geq 1$, and $w \in B, alph(w) \neq \{a\}$. All strings

$$z_{n,m} = (uv)^m w(uv)^{n-m}, \ 0 \leq m \leq n, n \geq 1,$$

are in $L, |z_{n,m}| = |w| + n|uv|$ and $z_{n,m} \neq z_{n,p}$, for $0 \leq m, p \leq n, m \neq p$. Therefore, L is not slender.

(2) If there is $u\$v \in a^*\$a^*$ and $u'\$v' \in b^*\$b^*$, for $a \neq b$, then we consider the strings in L

$$z_{n,m} = (uv)^{n-m|u'v'|}(u'v')^{m|uv|}w, \ 0 \leq m \leq \left[\frac{n}{|u'v'|}\right], n \geq 1,$$

for some $w \in B$. Again $|z_{n,m}| = |w| + n|uv|$, $z_{n,m} \neq z_{n,p}$ for all $m \neq p$, hence L is not slender.

(3) If there is $u\$v \in C$ with $u = a^i, v = b^j, i, j \geq 1$, and $a \neq b$, then we consider the strings in L

$$z_{n,m} = (uv)^m u^{n-m} v^{n-m} w, \ 0 \leq m \leq n, n \geq 1,$$

for some $w \in B$. We have $|z_{n,m}| = |w| + n|uv|$ and $z_{n,m} \neq z_{n,p}, m \neq p$ ($z_{n,m} = (a^i b^j)^m a^{i(n-m)} b^{j(n-m)} w$), hence L is not slender.

(4) If none of the previous cases holds, then there is $u\$v \in C$ with one of u, v containing occurrences of two symbols $a, b \in V, a \neq b$. Assume that $v = v_1 a b^i, i \geq 1$. When u has this property the argument is the same. Consider the strings in L

$$z_{n,m} = w(uv)^m u^{n-m} (v_1 a)^{n-m} (b^i)^{n-m}, \ 0 \leq m \leq n, n \geq 1,$$

for $w \in B$. Because $|z_{n,m}| = |w| + n|uv|$ and $z_{n,m} \neq z_{n,p}$ for $m \neq p$, it follows that L is not slender.

This concludes the analysis of all cases.   □

Corollary 2.6. $L_1 = a \cup b^+ \notin IC$, $L_2 = \{a^n b^n \mid n \geq 1\} \notin IC$.

Note that all other families of languages contain slender (even thin) languages: for instance, L_2 is both in EC and in ICC (for $G = (\{a, b\}, \{ab\}, \{((ab), \{a\$b\})\})$ we have $L_2 = L_{in}(G)$).

We close this section with an example of a language L not in ICC, where the above given necessary conditions do not suffice to prove that $L \notin ICC$.

Lemma 2.8. *The language* $L = \{x \ mi(x) \mid x \in \{a, b\}^*\}$ *is not in the family* ICC.

Proof. Assume to the contrary that $L = L_{in}(G)$ for a contextual grammar $G = (\{a, b\}, B, P)$. Clearly, if $u\$v \in \varphi(x)$ for some $x \in Sub(L)$, then $|uv|_a = |uv|_b$. Take such a *useful* context, $u\$v \in \varphi(x), x \in Sub(L)$.

Assume that $|v|_a > 0$; the case $|v|_b > 0$ is symmetric. Consider the string $w = xb^k b^k \ mi(x)$, for $k > 2|uv|$. Since $w \in L = L_{in}(G)$ and $w \Longrightarrow_{in}^* w' = uuxvvb^k b^k \ mi(x)$, we have $w' \in L_{in}(G)$. However, $w' \notin L$: if $w' = y \ mi(y)$, then $y = uuxvvb^{k-|uv|}$, $mi(y) = b^{|uv|} b^k \ mi(x)$, and so $uuxvvb^{k-|uv|} = xb^{k+|uv|}$, that is $v \in b^+$ (we have $k + |uv| \geq k - |uv| + 2|v|$); a contradiction.

If $v = \lambda$, then $|u|_a > 0, |u|_b > 0$. We take the string $w = mi(x)b^k b^k x$, for $k > 2|u|$ and we obtain $w \Longrightarrow_{in}^* w' = mi(x)b^k b^k uux \in L_{in}(G)$. However, $w' \notin L$: as above, we get $u \in b^+$; a contradiction. Thus L is not in ICC.   □

2.5 Generative capacity

We shall compare the eight families of contextual languages with each other and with families in Chomsky hierarchy.

Lemma 2.9. $EC = LIN_1$.

Proof. Obvious mutual simulations. $\qquad\square$

Lemma 2.10. $REG \subseteq ICC_c$.

Proof. Let L be a regular language and let $A = (V, Q, \delta, q_0, F)$ be the minimal deterministic finite automaton recognizing L.

For each $w \in V^*$, define the mapping $\rho_w : Q \longrightarrow Q$ by

$$\rho_w(q) = q' \text{ iff } \delta(q, w) = q', \quad q \in Q.$$

Obviously, if $x_1, x_2 \in V^*$ are such that $\rho_{x_1} = \rho_{x_2}$, then for every $u, v \in V^*$, ux_1v is in L if and only if ux_2v is in L.

The set of mappings from Q to Q is finite, hence the set of mappings ρ_w as above is finite. Let n_0 be their number. We construct the contextual grammar $G = (V, B, C, \varphi)$ with

$$B = \{w \in L \mid |w| \leq n_0 - 1\},$$
$$C = \{u\$v \mid 1 \leq |uv| \leq n_0\},$$
$$\varphi(w) = \begin{cases} \{u\$v \mid |uwv| \leq n_0 \text{ and } \rho_w = \rho_{uwv}\}, & \text{for } |w| \leq n_0 - 1, \\ \emptyset, & \text{otherwise.} \end{cases}$$

From the definition of mappings ρ_w and the definitions of B, C, φ, it follows immediately that $L_{in}(G) \subseteq L$.

Assume that the inverse inclusion is not true and let $x \in L - L_{in}(G)$ be a string of minimal length with this property. Thus $x \notin B$, hence $|x| \geq n_0$. Let $x = zz'$ with $|z| = n_0$ and $z' \in V^*$. If $z = a_1 a_2 \ldots a_{n_0}$, then it has $n_0 + 1$ prefixes, namely $\lambda, a_1, a_1 a_2, \ldots, a_1 \ldots a_{n_0}$. There are only n_0 different mappings ρ_w. Therefore there are two prefixes u_1, u_2 of z such that $u_1 \neq u_2$ and $\rho_{u_1} = \rho_{u_2}$. Without loss of generality we may assume that $|u_1| < |u_2|$. By substituting u_2 by u_1 we obtain a string x' which is also in L. As $|x'| < |x|$ and x was of minimal length in $L - L_{in}(G)$, we obtain $x' \in L_{in}(G)$. However, $|u_2| - |u_1| \leq |u_2| \leq n_0$, hence if $u_2 = u_1 u_3$, then $\$u_3 \in C$ and $\$u_3 \in \varphi(u_1)$. This implies that $x' \Longrightarrow_{in} x$, that is $x \in L_{in}(G)$, a contradiction. In conclusion, $L \subseteq L_{in}(G)$. $\qquad\square$

Note that in the previous proof only contexts of the form $\$v$ are necessary, and that the mapping φ is computable and has nonempty values only for finitely many strings.

Lemma 2.11. $CF \subseteq TC_c$.

Proof. As a consequence of pumping lemmas, every context-free language has the IBS property; according to Lemma 2.2 it is then in TC. Because the membership question is solvable for context-free languages, the mapping φ defined as in the proof of Lemma 2.1 is computable. Thus every context-free language is in TC_c.   □

Theorem 2.1. *The relations in the diagram in Figure 1 hold, where each arrow indicates a strict inclusion and every two non-linked families are incomparable.*

Proof. The inclusions between families of contextual languages follow from definitions, the inclusions between families of contextual languages and families in Chomsky hierarchy are proved in Lemmas 2.9–2.11, or are straightforward ($IC \subseteq CS, TC_c \subseteq RE$).

The strictness of these inclusions as well as most of the incomparabilities follow from examples in Section 2.3 and counterexamples in Section 2.4, or using the necessary conditions in Section 2.4.

For example, $a^*ba^*ba^*$ is in IC (it is generated by $G = (\{a,b\}, \{bb\}, \{\$a\}))$, but not in ECC (Lemma 2.6), whereas $\{x\ mi(x) \mid x \in \{a,b\}^*\}$ is in EC (it is generated by $G = (\{a,b\}, \{\lambda\}, \{a\$a, b\$b\}))$, but not in ICC (Lemma 2.8).

If in Example 2.6 we start from a recursive non-context-sensitive language L, then $L_{ex}(G_L)$ will be recursive non-context-sensitive. If L is not

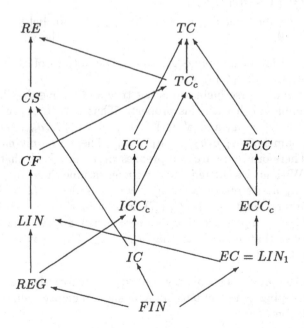

Fig. 1

recursively enumerable, then $L_{ex}(G_L)$ is not recursively enumerable. It follows that $ECC_c - CS \neq \emptyset$ and $ECC - RE \neq \emptyset$. The inverse relations are obvious: there are regular languages not in ECC. The corresponding assertions for families of internally generated languages follow in a similar way: for every $L \subseteq V^*$ and $c, d \notin V$, construct the grammar $G_L = (V \cup \{c, d\}, \{cc\}, \{(\{cc\}, \{\$a \mid a \in V\}), (\{c\}L, \{d\$d\})\})$. We obtain $L_{in}(G_L) = \{cc\}V^* \cup \{cd^n cx_1 dx_2 d \ldots x_n d \mid n \geq 1, x_1 \ldots x_i \in L, 1 \leq i \leq n\}$. If L does not belong to a family of languages F closed under intersection with regular sets, right and left derivative (the families in Chomsky hierarchy, including the family of recursive languages, have these closure properties), then $L_{in}(G_L) \notin F$: $L = \partial_{cdc}^l (\partial_d^r (L_{in}(G_L) \cap \{cdc\}V^*\{d\}))$.

On the other hand, the language $L = a^+ \cup \{a^n b^n \mid n \geq 1\}$ is not in ICC: if $L = L_{in}(G)$ for some $G = (\{a, b\}, B, P)$, then we must have a context $u\$v, uv \in a^+$, such that $u\$v \in \varphi(a^i), i \geq 0$ (such a context is necessary in order to generate strings in a^+ of arbitrarily large length); using such a context we can obtain $a^{i+1}b^{i+1} \Longrightarrow_{in} ua^i vab^{i+1}$, which is not in L, a contradiction.

The reader can easily prove all other relationships in the theorem.   □

Consider now the case of one-letter languages.

We have seen that ECC contains non-regular languages over $V = \{a\}$ (Example 2.5). Because every regular one-letter language has the EBS property, it follows that each such language is in the family ECC. On the other hand, the restrictions of EC and IC to the one-letter alphabet coincide, and there are one-letter regular languages not in $EC \cup IC$: take $L = \{a^2\} \cup \{a^{3n} \mid n \geq 1\}$. If $L = L_{ex}(G) (= L_{in}(G))$ for some $G = (\{a\}, B, C)$, then each context $u\$v$ used for producing strings a^{3n} (there is such a context) must have $uv = a^{3i}, i \geq 1$; adding such a context to a^2 we obtain the string $a^{3i+2}, i \geq 1$, which does not belong to the considered language.

Moreover (the proof is left to the reader), we have

Theorem 2.2. *A one-letter language is regular if and only if it is in the family ICC.*

2.6 Closure properties

The closure properties of families X_c are the same as those of families $X, X \in \{ICC, ECC, TC\}$, therefore we shall discuss here only the families IC, EC, ICC, ECC, TC. Except for TC, these families have poor closure properties (this is somewhat expected, in view of the fact that the contextual grammars do not use nonterminal symbols in derivations).

Theorem 2.3. *The closure properties in the table in Figure 2 hold, where Y indicates the closure and N indicates the non-closure of the corresponding family under the corresponding operation.*

Operation	IC	EC	ICC	ECC	TC
Union	N	N	N	Y	Y
Intersection	N	N	N	N	N
Complement	N	N	N	N	N
Concatenation	N	N	N	N	Y
Kleene closure	N	N	N	N	Y
Morphisms (λ-free)	Y	Y	N	Y	Y
Finite substitution	Y	Y	N	Y	Y
Substitution	N	N	N	N	Y
Intersection with regular sets	N	N	N	N	N
Inverse morphisms	N	N	N	N	N
Shuffle	N	N	N	N	N
Mirror image	Y	Y	Y	Y	Y

Fig. 2

Proof. The negative results can be proved using examples and counterexamples from the previous sections (sometimes in a modified form). We give some details only for the case of inverse morphisms.

Consider the language $L = \{c\}\{bc, ba\}^* \in EC$ and the morphism $h : \{a, b, c\}^* \longrightarrow \{a, b, c\}^*$ defined by $h(a) = abc, h(b) = ab, h(c) = cb$. We have $h^{-1}(L) = \{c\}\{b, c\}^*\{a\}$. (The inclusion $\supseteq$ is obvious, because $\{cb\}\{ab, cb\}^*\{abc\} \subseteq L$. Conversely, if $y \in h^{-1}(L)$, that is $h(y) \in L$, then y cannot contain substrings aa, ab, ac, as otherwise $h(y)$ will contain substrings $abcabc, abcab, abccb$, hence substrings ca, cc, which is impossible. Therefore $|y|_a = 1$. Clearly, y must begin with c, hence $y = cxa, x \in \{b, c\}^*$.) As $Min_0(h^{-1}(L))$ is infinite, $h^{-1}(L) \notin ECC$.

For the grammar $G = (\{a, b, c, d\}, \{bacada\}, \{((\{b^i ac^i \mid i \geq 1\}, \{b\$cad\})\})$ we obtain

$$L_{in}(G) = \{b^m ac^{m_1}(ad)^{n_1} c^{m_2}(ad)^{n_2} \ldots c^{m_r}(ad)^{n_r} a \mid$$
$$r \geq 1, m \geq 1, 1 \leq m_r \leq n_r, n_i, m_i \geq 1, 1 \leq i \leq r - 1,$$
$$\text{and } m = \sum_{i=1}^{r} n_i = \sum_{i=1}^{r} m_i\}.$$

Take also the morphism $h : \{a, b, c, d\}^* \longrightarrow \{a, b, c, d\}^*$ defined by $h(a) = a, h(b) = b, h(c) = c, h(d) = da$. The strings $h(x)$ do not contain the subword dc, hence $h^{-1}(y)$, for $y \in L_{in}(G)$, is defined only for $y = b^m ac^m(ad)^m a$ (hence with $r = 1$ in the above specification of $L_{in}(G)$). Therefore $h^{-1}(L_{in}(G)) = \{b^m ac^m ad^m \mid m \geq 1\}$ and so by Lemma 2.2 it is not in TC.

We consider now the family IC. Consider the grammar $G = (\{a, b, c, d\},$ $\{dd\}, \{\$abc\})$, and the morphism $h : \{a, b, c, d\}^* \longrightarrow \{a, b, c, d\}^*$ defined by $h(a) = abc, h(b) = da, h(c) = bca, h(d) = bcd$. Let us analyse the form of the strings x such that $h(x) \in L_{in}(G)$. The strings in $L_{in}(G)$ start with a or with d. If $h(x)$ starts with a, then it starts with abc. Assume that we have $h(x) = (abc)^i z$ for some $i \geq 0$, and this is the maximal prefix of $h(x)$ of this form. The string z cannot start with b or c (these symbols must have a corresponding occurrence of a to their left). Substrings aa, bb, cc, ac, ba are not possible in $h(x)$. If z starts with a, then again we must have abc, contradicting the maximality of i above. Consequently, z must start with d and continue with a (the image of b). Let us assume that we have $j \geq 0$ strings bca, hence $h(x) = (abc)^i da(bca)^j w$. Take the greatest j with this property. The string w can begin with b only and after b we can have only c; after c we cannot have a again (this contradicts the choice of j) or b (we do not have a corresponding occurrence of a to the left). Consequently, $h(x) = (abc)^i da(bca)^j bcdu$. The only possibility is to have $u = (ab)^k, k \geq 0$. Consequently, $h^{-1}(L_{in}(G)) = \{a^i bc^j da^k \mid i, j, k \geq 0\}$. This language is not in IC: a context introducing the symbol c can be used for introducing the symbol c before the occurrence of b, hence producing a parasitic string.

As far as the positive closure properties from the table in Figure 2 are concerned, the closure of ECC under union and of all families under mirror image are obvious. The other positive results follow from the next two lemmas. □

Lemma 2.12. *The family TC is closed under substitution.*

Proof. Take $L \subseteq V^*, L \in TC$, and let $s : V^* \longrightarrow 2^{U^*}$ be a substitution such that $s(a) \in TC$ for each $a \in V$. Let $V = \{a_1, \ldots, a_n\}$. According to Lemma 2.2, all languages $L, s(a_i), 1 \leq i \leq n$, have the IBS property. Let $p_0, p_1, \ldots, p_n$ be the associated constants, respectively, and define

$$p = p_0 \cdot \max\{p_i \mid 1 \leq i \leq n\}.$$

The language $s(L)$ has the IBS property, with respect to the constant p. Indeed, take a string $x \in s(L), |x| > p$. Consider $y \in L$ such that $x \in s(y)$. If there is a symbol a_i appearing in y which is substituted in x by a string z such that $|z| > p_i$, then $z = z_1 u z_2 v z_3$ for $1 \leq |uv| \leq p_i$ and $z' = z_1 z_2 z_3 \in s(a_i)$. It follows that z can be replaced in x by z' and this is exactly the IBS property: the obtained string is in $s(L)$ and $|uv| \leq p$. If all symbols a_i of y are replaced in x by strings z_i with $|z_i| \leq p_i$, then $|y| > p_0$. It follows that $y = y_1 u y_2 v y_3$ for $1 \leq |uv| \leq p_0$, $y' = y_1 y_2 y_3 \in L$. Let y_1', u', y_2', v', y_3' be the substrings of x corresponding to y_1, u, y_2, v, y_3 in y. It follows that $y_1' y_2' y_3' \in s(y') \subseteq s(L)$ and $|u'v'| \leq |uv| \cdot \max\{p_i \mid 1 \leq i \leq n\} \leq p_0 \cdot \max\{p_i \mid 1 \leq i \leq n\} \leq p$. The IBS property holds again.

Therefore, according to Lemma 2.2, the language $s(L)$ is in TC. □

Lemma 2.13. *The families IC, EC, ECC are closed under finite substitution (hence also under morphisms).*

Proof. Consider a contextual grammar $G = (V, B, C, \varphi)$ and a finite substitution $s : V^* \longrightarrow 2^{U^*}$. Extend it with $s(\$) = \{\$\}$. We construct the grammar $G' = (U, s(B), s(C), \varphi')$, where

$$\varphi'(x) = s(\varphi(x)), \text{ for } x \in s(V^*).$$

We leave to the reader the proof of $L_{ex}(G') = s(L_{ex}(G))$.

The proof is the same for the case of contextual grammars without choice, both with external and internal derivation (just ignore the mappings φ, φ' above); in the internal derivation case the construction of G' is the same.   □

Thus, ICC is an anti-AFL, but the other families are neither AFL's (no one is closed under intersection with regular sets), nor anti-AFL's (they are closed under arbitrary morphisms).

In view of these poor closure properties it is natural to look for operations that are "more related" to contextual grammars, in the hope to obtain further positive closure properties. Consider, e.g., the following operation.

The *right prolongation* of a language $L_1 \subseteq V^*$ with respect to a language $L_2 \subseteq V^*$, denoted $rp(L_1, L_2)$, is the smallest language $L \subseteq V^*$ having the following properties:

(i) $L_1 \subseteq L$,
(ii) if $x \in L, x = x_1 x_2$, with $x_2 \in V^+$, and $x_2 a \in L_2$, for $a \in V$,
then $xa \in L$.

Theorem 2.4. *The families IC, EC, ICC are not closed under the operation rp, but ECC and TC are.*

Proof. The closure of ECC and TC under rp follows from Lemmas 2.1 and 2.2: if L_1 has either the EBS or the IBS property, irrespective of the type of L_2, the language $rp(L_1, L_2)$ has also this property.

Because $rp(\{a, b\}, \{aa, bb\}) = a^+ \cup b^+$, it follows that IC and EC are not closed under this operation.

Consider now the grammar $G = (\{a, b\}, \{babab\}, \{(\{aba\}, \{a\$a\})\})$. We have $rp(L_{in}(G), \{aa, ba\}) = \{ba^n ba^n ba^m \mid n \geq 1, m \geq 0\}$. This language is not in the family ICC: in order to generate strings $ba^n ba^n b$ with arbitrarily large n we need $a^i \$ a^i \in \varphi(a^j ba^k)$ for $i \geq 1, j, k \geq 0$; then $ba^{j+1} ba^{j+1} ba^k \Longrightarrow_{in} ba^{j+1} ba^{j+i+1} ba^{k+i}$, which is not in our language, a contradiction.   □

A natural problem appearing in this framework is to investigate the smallest family containing a given family of contextual languages and closed under certain operations (e.g., the smallest AFL containing ICC). We shall investigate such problems in Sections 3.2 and 3.3, for restricted types of internal and external contextual languages.

2.7 Decidability properties

We shall consider now the basic decision problems in language theory: emptiness, finiteness, equivalence, inclusion, and membership, with the usual definitions. The results established in a series of lemmas will be summarized in Theorem 2.5 at the end of this section.

Lemma 2.14. *The emptiness is decidable for all classes of contextual grammars.*

Proof. If $G = (V, B, C, \varphi)$ is a total contextual grammar, then $L(G) \neq \emptyset$ if and only if $B \neq \emptyset$. $\qquad\square$

Lemma 2.15. *The finiteness is decidable for grammars corresponding to families IC, ICC_c, EC.*

Proof. If $G = (V, B, C)$ is a grammar without choice, then $L_{in}(G)$ and $L_{ex}(G)$ are infinite if and only if $B \neq \emptyset, C \neq \emptyset$ (we have assumed that the empty context, \$, is not present).

If $G = (V, B, C, \varphi)$, with computable $\varphi : V^* \longrightarrow 2^C$, then $L_{in}(G)$ is infinite if and only if there is $u\$v \in C, u\$v \in \varphi(x)$ for some $x \in Sub(B)$. As $Sub(B)$ is finite, the existence of such a context is decidable. $\qquad\square$

Lemma 2.16. *The membership is decidable for grammars corresponding to families $IC, ICC_c, EC, ECC_c, TC_c$.*

Proof. It suffices to prove the assertion for total contextual grammars. Take $G = (V, B, C, \varphi), \varphi : V^* \times V^* \times V^* \longrightarrow 2^C$ computable, and construct the sets

$$H_0(G) = B,$$
$$H_i(G) = H_{i-1}(G) \cup \{x \in V^* \mid w \Longrightarrow x \text{ for some } w \in H_{i-1}(G)\}, i \geq 1.$$

We have $x \in L(G)$ if and only if $x \in H_{|x|}(G)$. Indeed (assuming that $\$ \notin C$), each derivation for x, if any, must have at most $|x|$ steps.

The sets $H_i(G)$ can be algorithmically constructed and they are finite, hence we can decide whether or not $x \in H_{|x|}(G)$. $\qquad\square$

We move now to negative decidability properties.

Lemma 2.17. *The membership is undecidable for grammars with arbitrary choice mapping.*

Proof. The families ICC, ECC, TC contain languages which are not recursively enumerable (Theorem 2.1). $\qquad\square$

As often done, in order to obtain undecidability results, we reduce the considered problems to the *Post Correspondence Problem* (PCP). More exactly, for two n-tuples of non-empty strings over $V = \{a, b\}$, $x = (x_1, \ldots, x_n), y = (y_1, \ldots, y_n)$, we consider the languages

$$L(z) = \{ba^{i_k} \ldots ba^{i_1} c z_{i_1} \ldots z_{i_k} \mid k \geq 1, 1 \leq i_j \leq n, 1 \leq j \leq k\},$$

for $z \in \{x, y\}$. We have $L(x) \cap L(y) \neq \emptyset$ if and only if $PCP(x, y)$ has a solution, which is undecidable. Moreover, $L(x) \cap L(y)$ is infinite if and only if it is non-empty.

The languages $L(x), L(y)$ are in $EC \cap ICC$: for

$$G(z) = (\{a, b, c\}, \{ba^i c z_i \mid 1 \leq i \leq n\}, \{ba^i \$ z_i \mid 1 \leq i \leq n\}),$$

$$G'(z) = (\{a, b, c\}, \{ba^i c z_i \mid 1 \leq i \leq n\}, \{(\{c\}, \{ba^i \$ z_i \mid 1 \leq i \leq n\})\}),$$

we have $L_{ex}(G(z)) = L_{in}(G'(z)) = L(z), z \in \{x, y\}$. Consequently, we have the following result.

Lemma 2.18. *It is undecidable whether or not the intersection of two languages in a family of contextual languages, F, different from IC, is empty, finite, or infinite.*

Of course, "the intersection of two languages in F" means "the intersection of two languages generated by two arbitrary grammars corresponding to the family F"; we shall use this kind of formulation also below.

Lemma 2.19. *The finiteness is undecidable for families $ICC, ECC_c, ECC, TC_c, TC$.*

Proof. Take two n-tuples of non-empty words over $\{a, b\}$ and consider the grammar $G(y)$ as above, such that $L_{ex}(G(y)) = L(y)$. Construct the sets $K_i(G(y))$ as in the proof of Lemma 2.6, $K_i(G(y)) = \{w \in L_{ex}(G(y)) \mid$ there is a derivation of w in the grammar $G(y)$ using at most i contexts$\}, i \geq 1$.

These sets are finite and they can be effectively constructed. Moreover, the language $L(y)$ is linear, hence it has a solvable membership problem. Clearly, we have

$$L_{ex}(G(y)) = \bigcup_{i=0}^{\infty} K_i(G(y)).$$

Construct the contextual grammar

$$G = (\{a, b, c\}, \{ba^i c y_i \mid 1 \leq i \leq n\}, \{ba^i \$ y_i \mid 1 \leq i \leq n\}, \varphi),$$

$$\varphi(w) = \{ba^i \$ y_i \mid 1 \leq i \leq n\}, \text{ for } w \in K_r(G(y)) \text{ such that}$$

$$K_r(G(y)) \cap L(x) = \emptyset, r \geq 0.$$

If $K_r(G(y)) \cap L(x) \neq \emptyset$, then $K_p(G(y)) \cap L(x) \neq \emptyset$ for all $p \geq r$ (we have $K_r(G(y)) \subseteq K_p(G(y))$ for $r \leq p$). Therefore, if there is r such that $K_r(G(y)) \cap L(x) \neq \emptyset$, then $L_{ex}(G)$ is finite. Conversely, $L_{ex}(G) = L(y)$ when

$(\cup_{i=0}^{\infty} K_r(G(y))) \cap L(x) = \emptyset$. Consequently, $L_{ex}(G)$ is infinite if and only if $L(y) \cap L(x) = \emptyset$, which is undecidable.

Note that the mapping φ considered above is computable.

For grammars $G = (V, B, C, \varphi)$ with non-computable φ we cannot decide whether or not $L_{in}(G)$ is infinite without deciding whether or not $\varphi(x) \neq \emptyset$ for at least one string $x \in Sub(B)$: take $G_{x,y} = (\{a\}, \{\lambda\}, \{\$a\}, \varphi)$, with $\varphi(\lambda) = \{\$a\}$ iff $PCP(x,y)$ has a solution, and $\varphi(z) = \emptyset$ for all z otherwise.

$\square$

Lemma 2.20. *The equivalence is undecidable for families* $ICC_c, ICC, ECC_c, ECC, TC_c, TC$.

Proof. For two n-tuples x, y of non-empty words over $\{a, b\}$, we construct the grammar

$$G = (\{a, b, c\}, \{ba^i cy_i \mid 1 \leq i \leq n\}, \{ba^i \$y_i \mid 1 \leq i \leq n\}, \varphi),$$

$$\varphi(z) = \{ba^i \$y_i \mid ba^i zy_i \notin L(x)\}, z \in L(y).$$

Clearly, $L_{ex}(G) = L(y)$ if and only if $L(y) \cap L(x) = \emptyset$, which is undecidable. The mapping φ is computable.

Consider now the grammars

$$G'(y) = (\{a, b, c, d\}, \{dcd\}, \{(\{c\}, \{ba^i \$y_i \mid 1 \leq i \leq n\})\}),$$

$$G''(y) = (\{a, b, c, d\}, \{dcd\}, \{(\{c\}, \{ba^i \$y_i \mid 1 \leq i \leq n\}),$$

$$(\{d\}L(x)\{d\}, \{d\$d\})\}).$$

We obtain $L_{in}(G'(y)) = \{d\}L(y)\{d\}$ whereas $L_{in}(G''(y)) = L_{in}(G'(y))$ if and only if the context $d\$d$ can never be used, hence if and only if $L(x) \cap L(y) = \emptyset$, which is undecidable. The mappings φ', φ'' are computable. $\square$

Theorem 2.5. *The decidability results in the table in Figure 3 hold, where D indicates a decidable problem, U indicates an undecidable problem, and the question mark points to an open problem.*

Many other decidability questions can be formulated about contextual grammars. For instance, it is undecidable whether or not, for a grammar $G = (V, B, C, \varphi)$, we have $L_{ex}(G) \in EC$: consider the grammar

$$G = (\{a, b, c\}, \{c\}, \quad \{((L(x) - L(y)) \cup \{c\}, \{ba^i \$x_i \mid 1 \leq i \leq n\}),$$

$$(L(x) \cap L(y), \{ba^i \$x_i \mid 1 \leq i \leq n\} \cup \{c\$c\})\}).$$

We have $L_{ex}(G) = L(x) \in EC$ if and only if the context $c\$c$ is never used, hence if and only if $L(x) \cap L(y) = \emptyset$, which is undecidable. Indeed, if the context $c\$c$ is used, the derivation cannot continue, the obtained string is of the form czc for $z \in L(x) \cap L(y)$. The string z can be arbitrarily large (if $L(x) \cap L(y) \neq \emptyset$, then $L(x) \cap L(y)$ is infinite), hence we need a context for introducing the symbol c. In a grammar without choice, this context can be used an arbitrarily number of times, hence strings not in $L_{ex}(G)$ are produced. Consequently, $L_{ex}(G) \notin EC$, but this not decidable.

Problem	IC	ICC_c	ICC	EC	ECC_c	ECC	TC_c	TC
Emptiness	D	D	D	D	D	D	D	D
Finiteness	D	D	U	D	U	U	U	U
Equivalence	?	U	U	?	U	U	U	U
Inclusion	?	U	U	?	U	U	U	U
Membership	D	D	U	D	D	U	D	U

Fig. 3

3. Contextual grammars with restricted choice

As defined in Section 2.2, the contextual grammars have an infinitistic definition: the choice mapping φ in the functional presentation is defined for all strings in V^* (correspondingly, the selectors in the compact presentation can be infinite sets). However, the notion of a "grammar" assumes some sort of finite mechanism describing the syntax of a language. Restrictions on the general definition of a contextual grammar (on the mapping φ and on selectors) are thus necessary. The most natural idea is to impose certain regularity on the definition of choice. For instance, we can ask to have selectors in a given family F.

Particularly interesting are the cases when the selectors are finite or regular (the membership problem for them can be solved in real time). In this section we shall consider grammars with selectors of any given type in Chomsky hierarchy.

3.1 Definitions and basic results

Definition 3.1. *Let F be a family of languages. A contextual grammar with F choice is a grammar $G = (V, B, \{(D_1, C_1), \ldots, (D_n, C_n)\})$, with $D_i \in F$ for all $1 \le i \le n$.*

We denote by $ICC(F), ECC(F)$ the families of languages generated in the internal and in the external mode, respectively, by contextual grammars with F choice. We consider here F one of families FIN, REG, CF, CS, RE, with emphasis on $F \in \{FIN, REG\}$.

Many of the examples used in the previous sections are grammars with F choice with $F \in \{FIN, REG\}$. On the other hand, $ICC(F) \subseteq ICC$ and $ECC(F) \subseteq ECC$, for all F, hence all necessary conditions for families ICC, ECC hold true also for $ICC(F), ECC(F)$, respectively. Some of these conditions have sharper forms for families $ICC(F), ECC(F)$ with particular F. Thus, e.g., we have the following result.

Lemma 3.1. *If $L \subseteq V^*, L \in ICC(FIN)$, then there are three constants p, q, r such that every $z \in L, |z| > p$, can be written in the form $z = uvwxy$ with $u, v, w, x, y \in V^*, 0 < |vx| \leq q, |w| \leq r$, and $uv^i w x^i y \in L$ for all $i \geq 0$.*

Proof. The same as the proof of Lemma 2.4, taking $r = \max\{|w| \mid w \in D, (D, C) \in P\}$, for a grammar $G = (V, B, P)$ with finite choice. $\square$

The following relations are easy to be proved (in the case of RE we can use Turing-Church thesis).

Lemma 3.2. $ICC(F) \subseteq F, ECC(F) \subseteq F$, for $F \in \{CS, RE\}$.

Lemma 3.3. $IC \subseteq ICC(REG), EC \subseteq ECC(REG)$.

As it is expected, we have the following.

Lemma 3.4. $ICC(FIN) \subset ICC(REG) \subset ICC(CF) \subset ICC(CS) \subset ICC(RE)$.

Proof. The inclusions are obvious, we have to prove only their strictness.
(1) For the grammar

$$G_1 = (\{a, b\}, \{abab\}, \{(ab^+a, \{a\$a\}), (ba^+b, \{b\$b\})\}),$$

we have $L_{in}(G_1) = \{a^n b^m a^n b^m \mid n, m \geq 1\}$. This language does not have the property from Lemma 3.1. Hence $L_{in}(G_1) \notin ICC(FIN)$, and $ICC(REG) - ICC(FIN) \neq \emptyset$.

(2) Consider the grammar G_2 from Example 2.8, Section 2.3. It is easy to see that it has CF-choice, hence $L_{in}(G_2) \in ICC(CF)$. Assume that $L_{in}(G_2) = L_{in}(G)$ for some $G = (\{a, b, c\}, B, \{(D_1, C_1), \ldots, (D_n, C_n)\})$, with regular sets $D_i, 1 \leq i \leq n$.
If $w = bc^i a^n b^n caca^n b^n c^i \in B$ and $w \Longrightarrow_{in} w'$, then we must have $w' = bc^j a^{n'} b^{n'} caca^{n'} b^{n'} c^j$ with $n' = n$ (if $n' > n$, then there exists $a^{n'-n} b^{n'-n} \$ a^{n'-n} b^{n'-n} \in \varphi(b^n caca^n)$ and we can obtain also the derivation $bc^i a^{n+1} bb^n caca^n ab^{n+1} c^i \Longrightarrow_{in} ba^i a^{n+1} ba^{n'-n} b^{n'-n} b^n caca^n a^{n'-n} b^{n'-n} a b^{n+1} c^i$, which is not in $L_{in}(G_2)$, a contradiction). Therefore, infinitely many strings in the form as w, w' above are obtained starting from strings in B of the form $ba^n ba^m caca^n ba^m, n, m \geq 1$. Consider a derivation step when the first new occurrences of c are added to such a string, $ba^n b^m caca^n b^m \Longrightarrow_{in} bc^i a^{n'} b^{m'} caca^{n'} b^{m'} c^i, i \geq 1$. Clearly, the first string $b^{m'}$ must be equal to b^m and the last string $a^{n'}$ must be equal to a^n, hence $n' = n, m' = m$. As $n' = m'$, we also have $n = m$. There is a context $c^i \$ c^i$ such that $c^i \$ c^i \in \varphi(a^n b^m caca^n b^m)$ for infinitely many $n, m \geq 1$. If $c^i \$ c^i$ is associated with some D_j containing infinitely many strings $a^n b^m caca^n b^m$, then $D_j \cap \{a^n b^m caca^n b^m \mid n, m \geq 1\}$ is an infinite subset of $\{a^n b^n caca^n b^n \mid n \geq 1\}$. This is not possible when D_j is regular, contradiction. Consequently, $L_{in}(G_2) \notin ICC(REG)$ and $ICC(CF) - ICC(REG) \neq \emptyset$.

(3) For the grammar

$$G_3 = (\{a, b, c\}, \{cbac\}, \{(\{cb\}, \{\$a\}), (\{b\}, \{a\$\}), (\{caba^{2^n}c \mid n \geq 1\}, \{c\$c\})\}),$$

we obtain

$$
\begin{aligned}
L_{in}(G_3) \;=\; & \{cba^n c \mid n \geq 1\} \cup \\
& \cup \; \{ca^n ba^m c \mid n, m \geq 1\} \cup \\
& \cup \; \{c^i a^n ba^{2^m} c^i \mid n, m \geq 1, i \geq 2\}.
\end{aligned}
$$

Assume that $L_{in}(G_3) = L_{in}(G)$ for some $G = (\{a, b, c\}, B, \{(D_1, C_1), \ldots, (D_n, C_n)\})$, with context-free sets $D_i, 1 \leq i \leq n$. Consider the strings of the form $w = c^2 aba^{2^m} c^2, m \geq 1$. As the number of contexts is finite, for large enough m we cannot have $z \Longrightarrow_{in} w$ for some $z = c^2 aba^{2^p} c^2$, hence w must be produced from a string $z' = caba^q c$. Therefore there exists $c\$a^i c \in \varphi(caba^{2^m - i}), i \geq 0$, or $c\$c \in \varphi(caba^{2^m} c)$. In the first case we can also have $caba^{2^m - i} a^p c \Longrightarrow_{in} ccaba^{2^m} a^p cc$, for all $p \geq 1$, leading to strings not in the language. In the second case, because arbitrarily many strings $caba^{2^m} c$ must belong to D_j, for some context-free D_j, we must have in D_j also strings $caba^p c$ with $p \neq 2^k, k \geq 1$ (intersect D_j with $caba^+ c$, then erase with a left derivative the prefix cab and with a right derivative the suffix c; we must obtain a regular sublanguage of a^+). Then $caba^p c \Longrightarrow_{in} ccaba^p cc$, again a parasitic string. In conclusion, $L_{in}(G_3) \notin ICC(CF)$ and $ICC(CS) - ICC(CF) \neq \emptyset$.

(4) For a language $L \subseteq V^+, L \in RE - CS$, and two symbols c, d not in V, construct the grammar

$$G_L = (V \cup \{c, d\}, \{ccac \mid a \in V\}, \{(\{cc\}, \$V), (\{c\}L\{c\}, \{d\$d\})\}).$$

We obtain

$$L_{in}(G_L) = \{cc\}V^+\{c\} \cup \{cd^i cxcd^i \mid i \geq 1, x \in L\}.$$

Because $L = \partial^l_{cdc}(\partial^r_{cd}(L_{in}(G_L) \cap \{cdc\}V^+\{cd\}))$, we have $L_{in}(G_L) \notin CS$, hence $L_{in}(G_L) \notin ICC(CS)$ (Lemma 3.2). Consequently, we have $ICC(RE) - ICC(CS) \neq \emptyset$.   □

The reader can prove in a similar way the counterpart of the previous lemma for external grammars.

Lemma 3.5. $ECC(FIN) \subset ECC(REG) \subset ECC(CF) \subset ECC(CS) \subset ECC(RE)$.

The proof of Lemma 2.10 shows that $REG \subseteq ICC(FIN)$. Moreover,

Lemma 3.6. $ECC(REG) \subseteq LIN$.

Proof. Consider a grammar $G = (V, B, \{(D_1, C_1), \ldots, (D_n, C_n)\})$ with regular selectors $D_i, 1 \leq i \leq n$. Take finite automata (not necessarily deterministic)

$A_i = (V, Q_i, \delta_i, q_{0,i}, F_i)$ such that $D_i = L(A_i), 1 \leq i \leq n$. Assume, without loss of generality, that $Q_i \cap Q_j = \emptyset$ for all $i \neq j$, and construct the linear grammar $G = (N, V, S, P)$ with

$$N = \{S\} \cup \{M \mid M \subseteq \bigcup_{i=1}^{n} (Q_i \times Q_i)\},$$

(the sets M are considered nonterminals) and P containing the following rules.

1. $S \rightarrow uMv$,
   for $u\$v \in C_i, M = \{(q_{0,i}, q_f)\}$, for some $q_f \in F_i, 1 \leq i \leq n$.

(We add the context $u\$v$ only to strings in D_i and we memorize this fact in the pair $(q_{0,i}, q_f)$, which must be completed to a non-deterministically guessed correct recognition of a string in D_i.)

2. $M \rightarrow uM'v$, where
   - $u\$v \in C_i$ for some $i, 1 \leq i \leq n$, and $M \in N$,
   - if $(q_1, q_2) \in M, q_1, q_2 \in Q_j$ for some $j, 1 \leq j \leq n$, then M' contains a pair $(q_1', q_2'), q_1', q_2' \in Q_j$, such that $q_1' \in \delta_j(q_1, u), q_2 \in \delta_j(q_2', v)$,
   - besides the pairs (q_1', q_2') defined above, we add to M' a pair $(q_{0,i}, q_f)$, $q_f \in F_i$, if it is not already in M'.

(We simulate in this way the adjoining of the context $u\$v$, by guessing at the same time paths in automata $A_1, \ldots, A_n$ which can bring forward the pairs of states already present in the current nonterminal M. A new pair is added – if it is not already present – for checking the correct adjoining of the last considered context.)

3. $M \rightarrow x$, where
   - $x \in B, M \in N$,
   - for all $(q, q') \in M$ such that $q, q' \in Q_i$ for some $i, 1 \leq i \leq n$, we have $q' \in \delta_i(q, x)$.

(The pairs of states in M are matched on a string in B, hence all pairs of states $(q_{0,i}, q_f)$ introduced when adjoining contexts are linked by correct paths in the corresponding automata.)

4. $S \rightarrow x, x \in B$.

From the previous explanations it is easy to see that $L_{ex}(G) = L(G')$, hence $L_{ex}(G) \in LIN$. □

The following theorem summarizes the above lemmas.

Theorem 3.1. *The relations from the diagram in Figure 4 hold, where an arrow indicates a strict inclusion and the non-linked families are incomparable.*

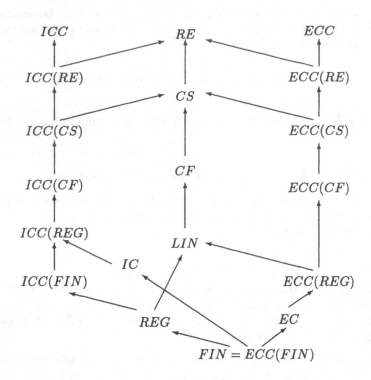

Fig. 4

Proof. For incomparabilities, use the fact that $REG - ECC \neq \emptyset$ and that $L = \{x\ mi(x) \mid x \in \{a,b\}^*\} \in EC - ICC$, as proved in Sections 2.4, 2.5. $\square$

The equality $FIN = ECC(FIN)$ implies that the contextual grammars with FIN-choice working in the external mode are not interesting. This shows in fact that the finite choice is not appropriately defined for such grammars. A more natural variant is to let the context which is adjoined to depend on a finite prefix and a finite suffix of the current string.

Definition 3.2. *A contextual grammar with (external) bounded choice is a construct*

$$G = (V, B, \{(D_1, C_1), \ldots, (D_n, C_n)\}), n \geq 1,$$

where V is an alphabet, B is a finite subset of V^*, D_i are finite subsets of $V^* \times V^*$, and C_i are finite subsets of $V^*\$V^*, 1 \leq i \leq n$.

For such a grammar we define the relation $\Longrightarrow_{ex}$ on V^* by

$$x \Longrightarrow_{ex} y \quad \text{iff} \quad x = x_1 x_2 x_3, y = uxv, \text{ where}$$

$$x_1, x_2, x_3 \in V^*, (x_1, x_3) \in D_i, u\$v \in C_i, \text{ for some } 1 \leq i \leq n.$$

Therefore, the context $u\$v \in C_i$ is adjoined only to strings having a pair of a prefix and a suffix in D_i. The generated language, $L_{ex}(G)$, is defined in the usual way. We denote

$$k = \max\{|x| \mid (x,y) \in D_i \text{ or } (y,x) \in D_i, 1 \leq i \leq n\}.$$

This is called the (selectivity) *depth* of G.

We denote by $ECC(k), k \geq 0$, the family of languages externally generated by grammars with bounded choice as above of depth at most k. Let $ECC(\infty) = \bigcup_{k \geq 0} ECC(k)$.

Theorem 3.2. $EC = ECC(0) \subset ECC(1) \subset \ldots \subset ECC(\infty) \subset ECC(REG).$

Proof. Only the strictness of the inclusions needs a proof.

Because $a^+ \cup b^+$ can be generated by $G = (\{a,b\}, \{a,b\}, \{(\{(\lambda,a)\}, \{\$a\}), (\{(\lambda,b)\}, \{\$b\})\})$, we have $ECC(1) - ECC(0) \neq \emptyset$.

For $k \geq 2$ we obtain that $ECC(k) - ECC(k-1) \neq \emptyset$ using the language

$$L_k = a^k b(aa^{k-1})^+ \cup a^k b(ba^{k-1})^+,$$

whereas the language

$$L = \{a^n ba^n, ba^n ba^n b, a^n bba^n, bba^n bba^n bb \mid n \geq 1\}$$

is in $ECC(REG) - ECC(\infty)$. □

The most interesting families $ICC(F), ECC(F)$ are the lowest ones in Figure 4, that is $ICC(FIN)$ and $ECC(REG)$. We will consider each of them separately now.

3.2 Internal contextual grammars with finite choice

We know already the place of the family $ICC(FIN)$ in Chomsky hierachy – see Figure 5, which presents in another way the information from Figure 4.

We also recall that, see Section 2.6, ICC is an anti-AFL, whereas IC is closed under morphisms, finite substitutions and mirror image. For $ICC(FIN)$ we have

Theorem 3.3. $ICC(FIN)$ *is an anti-AFL.*

Proof. Again, the only involved case is that of inverse morphisms.

Consider the contextual grammar $G = (\{a,b,c,d\}, \{ccbcdc\}, \{(\{ccbcdc\}, \{ab\$ab\}), (\{cc\}, \{\$ba\}), (\{c\}, \{d\$a\})\})$, and the morphism $h : \{a,b\}^* \longrightarrow \{a,b,c,d\}^*$ defined by $h(a) = ab, h(b) = cdc$. Every string in $L_{in}(G)$ contains exactly four occurrences of the symbol c, hence all strings x such that $h(x) \in L_{in}(G)$ must contain exactly two occurrences of b. Therefore, $h(x)$ is of the form $x_1 cdcx_2 cdcx_3$, with $x_1, x_2, x_3 \in \{a,b\}^*$. This means that the context

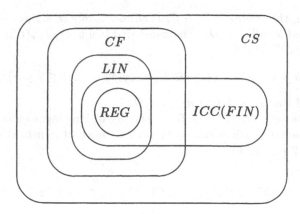

Fig. 5

$d\$a$ is used only once, hence $h(x) = (ab)^n cdc(ab)^m cdc(ab)^n, n \geq 0, m \geq 1$. The derivation of such a string proceeds as follows:

$$ccbcdc \implies_{in}^* (ab)^n ccbcdc(ab)^n \implies_{in}^* (ab)^n cc(ba)^{m-1} bcdc(ab)^n$$
$$\implies_{in}^* (ab)^n cdca(ba)^{m-1} bcdc(ab)^n = (ab)^n cdc(ab)^m cdc(ab)^n.$$

Consequently, $h^{-1}(L_{in}(G)) = \{a^n ba^m ba^n \mid n \geq 0, m \geq 1\}$, a language which is not in the family $ICC(FIN)$ (not in ICC, either). □

In view of these (negative) closure properties, it is interesting to consider the completion of $ICC(FIN)$ with respect to certain AFL operations. The result is spectacular: the smallest full trio (family closed under morphisms, inverse morphisms, and intersection with regular sets) containing $ICC(FIN)$ is RE. In fact, a still stronger result is true (not involving the closure under intersection with regular sets).

Theorem 3.4. *Every recursively enumerable language, L, can be written in the form $L = h_1(h_2^{-1}(L'))$, where $L' \in ICC(FIN)$, h_1 is a weak coding and h_2 is a morphism.*

Proof. Take $L \subseteq T^*, L \in RE$, and a type-0 grammar $G = (N, T, S, P)$ for L. Consider the new symbols, $[,], \vdash$, and construct the contextual grammar $G' = (V, B, P')$, with

$$V = N \cup T \cup \{[,], \vdash\},$$
$$B = \{S\},$$

and the set P' consisting of the following productions:

1. $(\{u\}, \{[\$]v\})$, for each $u \to v \in P$,
2. $(\{\alpha[u]\}, \{\vdash \$\alpha\})$, for $\alpha \in N \cup T, u \to v \in P$,
3. $(\{\alpha \vdash \beta\}, \{\vdash \$\alpha\})$, for $\alpha, \beta \in N \cup T$.

Consider a new symbol, b_w, associated with each string w in the set

$$R = \{[u] \mid u \rightarrow v \in P\} \cup \{\vdash \alpha \mid \alpha \in N \cup T\}.$$

Denote by Z the set of all such symbols b_w and define the weak coding $h_1 : (Z \cup T)^* \longrightarrow T^*$ by

$$h_1(b_w) = \lambda, w \in R, \quad h_1(a) = a, a \in T,$$

and the morphism $h_2 : (Z \cup T)^* \longrightarrow (N \cup T \cup \{[,],\vdash\})^*$ by

$$h_2(b_w) = w, w \in R, \quad h_2(a) = a, a \in T.$$

We obtain then $L = h_1(h_2^{-1}(L_{in}(G')))$.

The intuition behind the construction of G' (and of h_1, h_2) is as follows. The symbols between [and], and the symbol directly to the right of $\vdash$ are considered *dead*, all other symbols in a string of $L_{in}(G)$, except for [,], $\vdash$, are *alive*. (The symbols [,], $\vdash$ are also called *killers*.) The alive symbols in a string $z \in L_{in}(G')$ correspond to a sentential form produced by G. We simulate a derivation in G on the alive symbols of z, using productions in group 1 of P'. In this way, new alive and new dead symbols are obtained. Their position can be changed using productions in groups 2, 3 above (the alive symbols can be moved to the right, crossing blocks $w \in R$, of dead symbols). The inverse morphism h_2^{-1} can be applied only when a string in $(R \cup T)^*$ is obtained. This implies that all nonterminals are dead, the corresponding derivation in G is a terminal one. The coding h_1 removes all symbols b_w, hence we obtain a string in T^*.

Using these explanations, the reader should be able to prove the equality $L = h_1(h_2^{-1}(L_{in}(G')))$.   □

The previous construction can be modified in order to get

Corollary 3.1. *Every language* $L \in RE$ *can be written in the form* $L = L_1 \backslash L_2$, *for* $L_1 \in REG, L_2 \in ICC(FIN)$.

Proof. Starting from a type-0 grammar $G = (N, T, S, P)$, we construct the grammar G' as above, then we proceed as follows:

– add the new symbol # to the alphabet of G',
– put $B = \{S\#\}$ instead of $B = \{S\}$,
– add the productions of the form

$$4. (\{\alpha\#\}, \{\vdash, \alpha\}), \alpha \in T.$$

Denote by G'' the so obtained contextual grammar. We get

$$L = (R^*\#) \backslash L_{in}(G''),$$

where R is the regular language in the previous proof.   □

An interesting consequence of these representations is the following result.

Theorem 3.5. *The family $ICC(FIN)$ is incomparable with each family F such that $LIN \subseteq F \subset RE$ and F is closed*

1) under left quotient with regular sets, or
2) under weak codings and inverse morphisms.

Proof. We know that $LIN - ICC \neq \emptyset$, hence $F - ICC(FIN) \neq \emptyset$ for all F as above. Conversely, if $ICC(FIN) \subseteq F$, because F has the required closure properties, it follows from Theorem 3.4 and from its Corollary that $RE \subseteq F$, which contradicts the strict inclusion $F \subset RE$. Hence F is incomparable with $ICC(FIN)$.   □

For instance, a family F fulfilling the conditions in Theorem 3.5 is MAT^λ, the family of languages generated by matrix grammars with arbitrary context-free rules (but without appearance checking). This family is a full semi-AFL and it is strictly included in RE. According to the previous theorem, $ICC(FIN) - MAT^\lambda \neq \emptyset$. As $ICC(FIN) \subseteq CS$, we obtain in this way that $CS - MAT^\lambda \neq \emptyset$, a result that has been proved recently. Another important family F as above is $ET0L$, hence $ICC(FIN) - ET0L \neq \emptyset$.

3.3 External contextual grammars with regular choice

The relationship between $ECC(REG)$ and other relevant families of languages from Figure 4 is given in Figure 6.

From the proof of Theorem 2.3, it follows that $ECC(REG)$ is not closed under concatenation, Kleene closure, intersection with regular sets, and inverse morphisms. Moreover, we have the following result.

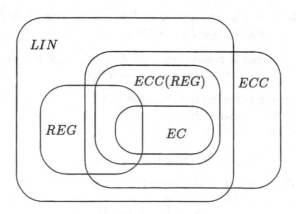

Fig. 6

Lemma 3.7. *The family $ECC(REG)$ is not closed under morphisms.*

Proof. For the grammar

$$G = (\{a,b,c,d\}, \{b\}, \{(d^*bd^*, \{d\$, \$d\}), (\{b\}, \{a\$\}),$$
$$(a^+ba^*, \{a\$, a\$a\}), (a^+ba^+, \{\$c\})\}),$$

we have

$$L_{ex}(G) = d^*bd^* \cup \{a^nba^m, a^nba^mc \mid n > m \geq 1\}.$$

Consider the morphism $h : \{a,b,c,d\}^* \longrightarrow \{a,b,c\}^*$ defined by $h(a) = a, h(b) = b, h(c) = c, h(d) = a$. We will prove that the language $h(L_{ex}(G)) = a^*ba^* \cup \{a^nba^mc \mid n > m \geq 1\}$ is not in $ECC(REG)$. Assume to the contrary that $h(L_{ex}(G))$ is in $ECC(REG)$. Consider $G' = (\{a,b,c\}, B, \{(D_1, C_1), \ldots, (D_r, C_r)\})$ such that $L_{ex}(G') = h(L_{ex}(G))$. Every string in $L_{ex}(G')$ contains the symbol b, hence all strings in B contain a symbol b. From strings of the form a^iba^mc we can produce only strings of the form a^nba^mc. Consequently, for all sufficiently large m, the strings a^nba^mc are produced by derivations starting from axioms of the form $a^iba^j, i, j \geq 0$. Consider the step when the symbol c is introduced, $a^{n_1}ba^{m_1} \Longrightarrow_{ex} a^{n_2}a^{n_1}ba^{m_1}a^{m_2}c$. This means that there is a production (D_j, C_j) in G' with $a^{n_1}ba^{m_1} \in D_j$, $a^{n_2}\$a^{m_2}c \in C_j$. As $m = m_1 + m_2$ can be arbitrarily large and m_2 is bounded, it follows that m_1 can be arbitrarily large. Using a pumping lemma for the regular language D_j, we find a constant $s \geq 1$ such that $a^{n_1}ba^{m_1+ts} \in D_j$ for all $t \geq 0$. All strings $a^{n_1}ba^{m_1+ts}$ are in $h(L_{ex}(G))$, hence we can perform the derivation $a^{n_1}ba^{m_1+ts} \Longrightarrow_{ex} a^{n_1+n_2}ba^{m_1+m_2+ts}c$. If t is sufficiently large so that $n_1 + n_2 < m_1 + m_2 + ts$, then such a string is not in $h(L_{ex}(G))$, a contradiction. Hence $h(L_{ex}(G)) \notin ECC(REG)$. $\square$

The family ECC is closed under union. This result is essentially based on the fact that we have a finite set of axioms (rather than one axiom only). For instance, the language $\{a, b\}$ cannot be generated by a contextual grammar with one axiom only. This suggests to consider the smallest family of languages generated by grammars with regular choice, generated by grammars with one axiom. Denote this family by $ECC_1(REG)$. By the above, this family is an anti-AFL (the proofs of the corresponding properties from Theorem 2.3 start from languages in $ECC_1(REG)$).

We shall investigate here the completion of this family by finite unions, intersection with regular sets, and morphisms. More precisely, we consider languages of the form

$$h((L_1 \cup L_2 \cup \ldots \cup L_n) \cap R),$$

where h is a morphism, R is a regular language, and $L_1, \ldots, L_n$ are languages in $ECC_1(REG)$. We consider the following cases (the notation for each family is given in parentheses):

(A1) $n = 1$ and $L_1 \in EC_1(REG)$ $(Sa = \text{single axiom})$
(A2) $n = 1$ and $L_1 \in ECC(REG)$ $(Fa = \text{finite set of axioms})$
(A3) n arbitrary, $L_i \in EC_1(REG)$, $(Fu = \text{finite union})$
$$1 \leq i \leq n$$

(B1) $R = V^*$ $(Pl = \text{plain squeezing})$
(B2) R arbitrary $(Re = \text{regular squeezing})$

(C1) h is the identity $(Id = \text{identity})$
(C2) h is a coding $(Co = \text{coding})$
(C3) h is a weak coding $(Wc = \text{weak coding})$
(C4) h is non-erasing $(Ne = \text{non-erasing morphism})$
(C5) h is arbitrary $(Mo = \text{arbitrary morphism})$

Consequently, we obtain $3 \times 2 \times 5 = 30$ different cases, leading to 30 families of languages. In our notation we identify these families by the 3-tuples of abbreviations of the involved cases A, B, C; hence by $\mathcal{L}(X_1, X_2, X_3)$, where (X_1, X_2, X_3) is an element of

$$\{Sa, Fa, Fu\} \times \{Pl, Re\} \times \{Id, Co, Wc, Ne, Mo\}.$$

For instance, $ECC_1(REG) = \mathcal{L}(Sa, Pl, Id)$, and $ECC(REG) = \mathcal{L}(Fa, Pl, Id)$.

All 30 families considered here are included in LIN, because $ECC_1(REG) \subseteq ECC(REG) \subseteq LIN$ and LIN is closed under union, arbitrary morphisms and intersection with regular sets.

However, many of these families equal the family of linear languages. In the sequel we investigate the interrelationships between these families, also their relationship to LIN.

It is easy to see that $REG \subseteq \mathcal{L}(Sa, Re, Id)$, $FIN \subseteq \mathcal{L}(Fa, Pl, Id)$, and that all families above contain non-regular languages. Moreover, we know that

$$L_1 = ab^+a \notin \mathcal{L}(Fa, Pl, Id),$$
$$L_2 = \{a, b\} \notin \mathcal{L}(Sa, Pl, Id),$$
$$L_3 = a^*ba^* \cup \{a^n ba^m c \mid n > m \geq 1\} \in \mathcal{L}(Sa, Pl, Co) - \mathcal{L}(Fa, Pl, Id).$$

These facts as well as the following variants or stronger forms of them will be useful in forming the diagram of the relationships between all the families.

$$L_1 \in \mathcal{L}(Sa, Re, Id) - \mathcal{L}(Fu, Pl, Mo),$$
$$L_2 \in \mathcal{L}(Fa, Pl, Id) - \mathcal{L}(Sa, Pl, Mo),$$
$$L_3 \in \mathcal{L}(Fu, Pl, Id).$$

When considering a family $\mathcal{L}(X_1, X_2, X_3)$ it often happens that a feature X_i is strong enough so that one can replace another feature X_j by a weaker one, X'_j, and still obtain a language family that is not smaller than the original one.

Lemma 3.8. *For all $X \in \{Co, Wc, Ne, Mo\}$ we have $\mathcal{L}(Fu, Pl, X) \subseteq \mathcal{L}(Fa, Pl, X)$.*

Proof. For a language $L \in \mathcal{L}(Fu, Pl, X)$, consider the grammars $G_i = (V_i, \{w_i\}, P_i)$, $1 \le i \le n$, and a morphism $h : (\bigcup_{i=1}^{n} V_i)^* \longrightarrow U^*$ such that $L = h(\bigcup_{i=1}^{n} L_{ex}(G_i))$.

For each $i, 1 \le i \le n$, let $[V_i, i] = \{[a, i] \mid a \in V_i\}$, and define the codings $g_i : (V_i \cup \{\$\}) \longrightarrow ([V_i, i] \cup \{\$\})$ by $g_i(a) = [a, i]$, for all $a \in V_i$, $g_i(\$) = \$$.

Consider the grammar $G = (V, B, P)$ with

$$V = \bigcup_{i=1}^{n} [V_i, i],$$
$$B = \{g_1(w_1), \dots, g_n(w_n)\} \text{ and }$$
$$P = \{(g_i(D), g_i(C)) \mid (D, C) \in P_i, 1 \le i \le n\}.$$

We have $L_{ex}(G) = \bigcup_{i=1}^{n} g_i(L_{ex}(G_i))$. This is a consequence of the fact that g_i are one-to-one codings which ensures that each derivation in G starting from $g_i(w_i)$ will use only productions $(g_i(D), g_i(C))$ for $(D, C) \in P_i$.

Consider now $h' : (\bigcup_{i=1}^{n}[V_i, i])^* \longrightarrow U^*$ defined by

$$h'([a, i]) = h(a), \text{ for all } a \in V_i, \text{ and } 1 \le i \le n.$$

We clearly have $h'([a, i]) = h(g_i^{-1}([a, i]))$ for all $a \in V_i$ and all $1 \le i \le n$.

It is easily seen that $h'(L_{ex}(G)) = L$. Since the mappings h, h' are of the same type (g_i^{-1} is a coding), it follows that $L \in \mathcal{L}(Fa, Pl, X)$. □

Lemma 3.9. $\mathcal{L}(Fa, Re, X) \subseteq \mathcal{L}(Sa, Re, X)$ *for all* $X \in \{Id, Co, Wc, Ne, Mo\}$.

Proof. Let $G = (V, B, P)$ be a grammar with $B = \{w_1, \dots, w_k\}$, let R be a regular language, and let $h : V^* \longrightarrow U^*$ be a morphism. We construct the grammar $G' = (V, \{\lambda\}, P')$ with

$$P' = P \cup \{(\{\lambda\}, \{\$w_i\}) \mid 1 \le i \le k\}.$$

Obviously, $L_{ex}(G') = L_{ex}(G) \cup \{\lambda\}$.

If $\lambda \in L_{ex}(G)$ (this means in fact that some strings w_i are empty), then we let R unchanged. Clearly, $h(L_{ex}(G) \cap R) = h(L_{ex}(G') \cap R)$.

If $\lambda \notin L_{ex}(G)$, then we take $R' = R - \{\lambda\}$ and we have then $h(L_{ex}(G) \cap R) = h(L_{ex}(G') \cap R')$ (the empty string λ is eliminated by the intersection).

Consequently, $h(L_{ex}(G) \cap R) \in \mathcal{L}(Sa, Re, X)$, where X depends on the form of h. □

Lemma 3.10. $\mathcal{L}(X, Pl, Mo) \subseteq \mathcal{L}(X, Pl, Wc)$ *for* $X \in \{Sa, Fa\}$.

Proof. Take a grammar $G = (V, \{w_1, \ldots, w_k\}, P)$ and a morphism h : $V^* \longrightarrow U^*$. Consider new symbols [and], as well as a copy a' for each $a \in V$. Let $V' = \{a' \mid a \in V\}$ and define the morphism g : $(V \cup \{\$\})^* \longrightarrow (U \cup V' \cup \{[,], \$\})^*$ by

$$g(a) = [a'h(a)] \text{ for all } a \in V, \quad g(\$) = \$.$$

Construct the grammar $G_h = (V_h, B_h, P_h)$ with

$$V_h = U \cup V' \cup \{[,]\},$$
$$B_h = \{g(w_1), \ldots, g(w_k)\}, \text{ and}$$
$$P_h = \{(g(D), g(C)) \mid (D, C) \in P\}.$$

Consider now the weak coding h' : $V_h \longrightarrow U \cup \{\lambda\}$, defined by

$$h'(a) = a, \text{ for all } a \in U, \text{ and}$$
$$h'(a') = h'([) = h'(]) = \lambda, \text{ for all } a \in V.$$

Then it is easily seen that $h(L_{ex}(G)) = h'(L_{ex}(G_h))$ (the effect of h is "hidden" in G_h by the coding g, and h' is used for removing the auxiliary symbols). $\qquad\square$

Lemma 3.11. $\mathcal{L}(X, Y, Ne) \subseteq \mathcal{L}(X, Y, Co)$ *for all* $X \in \{Sa, Fa\}, Y \in \{Pl, Re\}$.

Proof. Consider a grammar $G = (V, \{w_1, \ldots, w_k\}, P)$ with $k \geq 1$ ($k = 1$ corresponds to the case $X = Sa$), a non-erasing morphism h : $V^* \longrightarrow U^*$, and a regular language $R \subseteq V^*$. Consider again the copy symbols a' for all $a \in V$, the alphabet

$$V_h = U \cup V' \times U \cup U \times V' \cup V' \times U \times V',$$

and the morphism g : $(V \cup \{\$\})^* \longrightarrow (V_h \cup \{\$\})^*$ such that $g(\$) = \$$ and for each $a \in V$

$$g(a) = \begin{cases} (a', b, a'), & \text{if } h(a) = b \in U, \\ (a', b_1)b_2 \ldots b_{n-1}(b_n, a'), & \text{if } h(a) = b_1 \ldots b_n \text{ with } n \geq 2 \text{ and } b_i \in U \\ & \text{for all } 1 \leq i \leq n. \end{cases}$$

Construct now the grammar $G_h = (V_h, B_h, P_h)$, with

$$B_h = \{g(w_1), \ldots, g(w_k)\}, \text{ and}$$
$$P_h = \{(g(D), g(C)) \mid (D, C) \in P\}.$$

Let $R_h = g(R)$ and consider the coding h' : $V_h \longrightarrow U$ defined by

$$h'(a) = a, \text{ for all } a \in U,$$
$$h'((a', b)) = h'((b, a')) = h'((a', b, a')) = b, \text{ for all } a \in V \text{ and } b \in U.$$

The equality

$$h(L_{ex}(G) \cap R) = h'(L_{ex}(G_h) \cap R_h),$$

can be proved similarly to Lemma 3.10 (again the effect of the morphism h is introduced by g in G_h, the block of letters corresponding to a symbol $a \in V$ is marked by primed occurrences of a in symbols $(a', b), (b, a'), (a', b, a')$; these pairs and triples of symbols are mapped to symbols in U by the coding h').

When $R = V^*$, we cover the case $Y = Pl$.   □

In what follows we will demonstrate that twelve of the families of languages considered above are equal to LIN. Clearly, it suffices to prove that LIN is included in those families, because it is known that all the families that we consider are included in LIN. The notion of *regular separation* is central in the proofs. It is defined as follows.

For a context-free grammar $G = (N, V, S, P)$ and each $A \in N$, we denote by $contr_G(A)$ the language $L(G_A)$, where $G_A = (N, V, A, P)$ (hence $L(G_A)$ is the set of strings generated by G starting from the nonterminal A).

Definition 3.3. *A context-free grammar $G = (N, V, S, P)$ is said to be regularly separable if for each nonterminal $X \in N$ there is a regular language $R_X \subseteq V^*$ such that:*
  1. *$R_X \cap R_Y = \emptyset$ for all $X, Y \in N$ such that $X \neq Y$, and*
  2. *$contr_G(X) \subseteq R_X$ for each $X \in N$.*

Definition 3.4. *A linear language L is separable if there exist a finite language F and a linear grammar $G = (N, V, S, P)$ which is regularly separable and*
  1. *$L = F \cup L(G)$,*
  2. *$F \cap R_X = \emptyset$ for each $X \in N$.*

Definition 3.5. *A linear grammar $G = (N, V, S, P)$ is said to have the 3nt-property if it contains only right-linear and left-linear rules, it contains no chain rule and no λ-rule, and $N = N_l \cup N_r \cup N_f$, where N_l, N_r, N_f are pairwise disjoint sets, such that*

$$N_l = \{B \mid B \rightarrow uX \in P, u \in V^+, X \in N\},$$
$$N_r = \{C \mid C \rightarrow Xv \in P, v \in V^+, X \in N\}, \text{ and}$$
$$N_f = \{D \mid D \rightarrow w \in P, w \in V^+\}.$$

The nonterminals in N_l are called left-pushers, those in N_r are called right-pushers and the nonterminals in N_f are called final.

Lemma 3.12. *For every linear language L there is a finite language F and a linear grammar G with the 3nt-property, such that $L = F \cup L(G)$.*

Proof. Let $L \subseteq V^*$ be a linear language, with $V = \{a_1, \ldots, a_n\}$. We have

$$L = \{x \in L \mid |x| \leq 1\} \cup \bigcup_{i=1}^{n} \{a_i\}(\partial_{a_i}^l(L) - \{\lambda\}).$$

Each language $L_i = \partial_{a_i}^l(L) - \{\lambda\}$ is linear (and λ-free). Let $G_i = (N_i, V, S_i, P_i)$ be a linear grammar for $L_i, 1 \leq i \leq n$. Without loss of generality we may assume that G_i contains no chain rule and no λ-rule. Moreover, replacing each rule $p : X \to uYv, u, v \in V^+$, by $X \to uY_p, Y_p \to Yv$, where Y_p is a new nonterminal associated with this rule p, we obtain an equivalent grammar. Therefore we may assume that all nonterminal rules in P_i are either left-linear or right-linear. Assume that all sets $N_i, 1 \leq i \leq n$, are pairwise disjoint and construct the grammar $G_0 = (N, V, S, P)$ with

$$N = \{S\} \cup \bigcup_{i=1}^{n} N_i, \quad S \text{ a new symbol,}$$

$$P = \{S \to a_i S_i \mid 1 \leq i \leq n\} \cup \bigcup_{i=1}^{n} P_i.$$

It is easy to see that $L = F \cup L(G_0)$, for $F = \{x \in L \mid |x| \leq 1\}$, and that S appears only in "left-pushing" rules.

Replace now each rule of the form $X \to uY$ in P by

$$X^l \to uY^l, \; X^l \to uY^r, \; X^l \to uY^f,$$

each rule of the form $X \to Yv$ by

$$X^r \to Y^l v, \; X^r \to Y^r v, \; X^r \to Y^f v,$$

and each rule $X \to w$ by

$$X^f \to w.$$

Denote by P' the set of rules obtained in this way, and let $G = (\{X^l, X^r, X^f \mid X \in N\}, T, S^l, P')$. It is easy to see that $L(G_0) = L(G)$ (the derivations in the two grammars are identical modulo the superscripts l, r, f). Moreover, G has the 3nt-property: consider the sets (obviously disjoint): $N_\alpha = \{X^\alpha \mid X \in N\}$, $\alpha \in \{l, r, f\}$. $\qquad \square$

Lemma 3.13. *Every linear language is a coding of a separable linear language.*

Proof. Let $L \subseteq V^*$ be a linear language. According to the previous lemma, there exist a finite set $F \subseteq V \cup \{\lambda\}$ and a grammar $G_0 = (N, V, S, P)$ with the 3nt-property, such that $L = F \cup L(G_0)$ and for every $x \in L(G_0)$, we have $|x| \geq 2$. Let N_l, N_r, N_f be the sets of nonterminals of the three categories considered above.

We construct the grammar $G = (N', V', S_0, P')$ with

$$N' = \{B_i \mid B \in N_l \cup N_r, 0 \le i \le 2\} \cup N_f,$$
$$V' = V \times N,$$

and P' obtained as follows. For each $X \in N'$, define the morphism $h_X :$ $V^* \to V'^*$ by $h_X(a) = (a, X), a \in V$. Then

- for each left-pusher production $B \to uX$ in P and for each $i, 0 \le i \le 2$, we introduce in P' the rules

$$B_i \to h_{B_i}(u)X_i, \text{ if } X \in N_l,$$
$$B_i \to h_{B_i}(u)X_{i+1(mod\ 3)}, \text{ if } X \in N_r,$$
$$B_i \to h_{B_i}(u)X, \text{ if } X \in N_f;$$

- for each right-pusher production $C \to Xv$ in P and for each $i, 0 \le i \le 2$, we introduce in P' the rules

$$C_i \to X_{i+1(mod\ 3)}h_{C_i}(v), \text{ if } X \in N_l,$$
$$C_i \to X_i h_{C_i}(v), \text{ if } X \in N_r,$$
$$C_i \to X h_{C_i}(v), \text{ if } X \in N_f;$$

- for each final production $D \to w$ in P we introduce

$$D \to h_D(w).$$

Consider also the coding $h' : V \cup V' \longrightarrow V$ defined by $h'((a, X)) = a$ for all $a \in V$ and $X \in N$, and $h'(a) = a$ for $a \in V$.

From the construction of G it follows that $h'(L(G)) = L(G_0)$ (the derivations in the two grammars coincide, modulo the subscripts 0, 1, 2 of nonterminals and the pairing of terminals with nonterminals; when the pairing is removed by the coding h', we get the same string). Since $h'(F) = F$, we have $L = h'(F \cup L(G))$. It remains to prove that G is a separable grammar and that $F \cap R_X = \emptyset$ for all $X \in N'$. The latter property is obvious, because $F \subseteq V \cup \{\lambda\}$ and $R_X \subseteq V'^+$ for all $X \in N'$.

In order to prove that G is separable, let us examine the form of strings in $contr_G(X)$, for each symbol $X \in N'$. Each string in $L(G)$ is of the form $(a_1, Z_1)(a_2, Z_2)\ldots(a_m, Z_m)$. If such a string is generated starting from X, then $Z_1 \in \{X_0, X_1, X_2\}$ if X is a left-pusher, and $Z_m \in \{X_0, X_1, X_2\}$ if X is a right-pusher. Consequently, the pair (Z_1, Z_m) will identify, together with the type of X, the strings produced by X in G. The table in Figure 7 contains all possible classes of pairs (Z_1, Z_2). We denote Z_1 by F_2 (from "the 2nd component of the first pair") and Z_m by L_2 (from "the 2nd component of the last pair").

Row	Type of X	F_2	L_2
1	left-pusher	$X_i, 0 \leq i \leq 2$	any $D \in N_f$
2	left-pusher	$X_i, 0 \leq i \leq 2$	any $B_j, B \in N_l, 0 \leq j \leq 2$
3	left-pusher	$X_i, 0 \leq i \leq 2$	any $C_{i+1(mod\ 3)}, C \in N_r$
4	right-pusher	any $D \in N_f$	$X_i, 0 \leq i \leq 2$
5	right-pusher	any $C_j, C \in N_r, 0 \leq j \leq 2$	$X_i, 0 \leq i \leq 2$
6	right-pusher	any $B_{i+1(mod\ 3)}, B \in N_l$	$X_i, 0 \leq i \leq 2$
7	final	D	C

Fig. 7

The sets R_X, R_{X_i} can now be defined as follows:

– if $X \in N_l$, and $0 \leq i \leq 2$, then

$$R_{X_i} = \{(a, X_i) \mid a \in V\}V'^*\{(a, L_2) \mid a \in V, L_2 \text{ as on rows } 1, 2, 3$$
$$\text{of the table}\};$$

– if $X \in N_r$, and $0 \leq i \leq 2$, then

$$R_{X_i} = \{(a, F_2) \mid a \in V, F_2 \text{ as on rows } 4, 5, 6 \text{ of the table}\}V'^*\{(a, X_i) \mid$$
$$a \in V\};$$

– if $X \in N_f$, then

$$R_X = \{(a, X) \mid a \in V\}V'^*\{(a, X) \mid a \in V\} \cup \{(a, X) \mid a \in V\}.$$

Claim 1. *The sets R_{X_i} and R_X are pairwise disjoint.*

Proof. The grammar G_0 has the 3nt-property and this holds also for G, hence a nonterminal is either a left-pusher, or a right-pusher, or final. The pairs (F_2, L_2) in the previous classification (hence in the definition of sets R_{X_i}, R_X) are different in different rows of the table. This is clear for all pairs of rows, due to the 3nt-property, except for the case of rows 3 and 6, when F_2 is a left-pusher and L_2 is a right-pusher in both cases. However, in row 3 we have $F_2 = X_i$ and $L_2 = Y_{i+1(mod\ 3)}$ for all $0 \leq i \leq 2$, hence L_2 has the subscript greater than i (modulo 3), whereas in row 6 we have $F_2 = Y_{i+1(mod\ 3)}, L_2 = X_i, 0 \leq i \leq 2$, that is F_2 has a greater subscript (modulo 3). Consequently, also this case leads to disjoint sets of strings.

Claim 2. *For each $X \in N', contr_G(X) \subseteq R_X$.*

Proof. Let us analyse the possible forms of derivations in G, starting from a symbol $X \in N'$.

Case 1. X is a left-pusher, $X = B_i$ for $0 \leq i \leq 2$ and $B \in N_l$.

If a derivation does not use right-pushing rules, then it is of the form

$$B_i \implies h_{B_i}(u_1)B_i^{(1)} \implies h_{B_i}(u_1)h_{B_i^{(1)}}(u_2)B_i^{(2)} \implies \ldots$$
$$\implies h_{B_i}(u_1)\ldots h_{B_i^{(n)}}(u_{n+1})D$$
$$\implies h_{B_i}(u_1)\ldots h_{B_i^{(n)}}(u_{n+1})h_D(w) = (a_1, B_i)z(a_s, D).$$

Consequently, $F_2 = B_i$ and $L_2 = D$, which corresponds to row 1 in the table from Figure 7.

If a derivation uses a right-pushing rule, then consider the first step when a right-pushing nonterminal is introduced:

$$B_i \implies h_{B_i}(u_1)B_i^{(1)} \implies \ldots \implies h_{B_i}(u_1)\ldots h_{B_i^{(n)}}(u_{n+1})C_{i+1(mod\ 3)} \implies$$
$$\implies h_{B_i}(u_1)\ldots h_{B_i^{(n)}}(u_{n+1})C_{i+1(mod\ 3)}^{(1)}h_{C_{i+1(mod\ 3)}}(v_1) \implies \ldots$$
$$\ldots \implies h_{B_i}(u_1)\ldots h_D(w)\ldots h_{C_{i+1(mod\ 3)}}(v_1) = (a_1, B_i)z(a_s, C_{i+1(mod\ 3)}).$$

We have $F_2 = B_i$ and $L_2 = C_{i+1(mod\ 3)}$, which corresponds to row 3 in the table from Figure 7.

Case 2. X is a right-pusher, $X = C_i$ for $0 \le i \le 2$ and $C \in N_r$.

When a derivation does not use left-pushing rules, then we obtain a string corresponding to row 4 of the table. When also left-pushers appear, we obtain a string corresponding to row 6 of the table (the argument is similar to that used in Case 1).

Case 3. X is a final nonterminal, $X = D$ for some $D \in N_f$.

We have only one-step derivations and they lead to strings corresponding to row 7 of the table.

Consequently, G is separable, and this completes the proof. $\square$

Note that no terminal string generated by the above grammar G corresponds to rows 2 or 5 in the classification table from Figure 7 (only the nonterminal strings can have both F_2 and L_2 to be left-pushers or right-pushers).

Lemma 3.14. *If L is a separable linear language, then $L \in \mathcal{L}(Fa, Re, Id)$.*

Proof. Let F be a finite set and $G_0 = (N, V, S, P_0)$ be a regularly separable linear grammar such that $L = F \cup L(G_0)$, with $F, R_X, X \in N$, as in Definitions 7 and 8.

Construct the contextual grammar $G = (V, B, P)$, with

$$B = \{w \in V^* \mid X \to w \in P\} \cup F,$$

and P containing all productions

$(R_Y, \{u\$v\})$ for $X \to uYv \in P$, with $u, v \in V^*$, such that $uv \neq \lambda$,

(for each production, either u or v is empty, but we prefer this compact notation for all types of rules).

Consider also the regular language $R = R_S \cup F$. The reader can verify that we have $L = L_{ex}(G) \cap R$. $\square$

Combining all the relationships we have proved already, we get the following result.

Theorem 3.6. *The relationships in the diagram from Figure 8 hold. Here a double arrow from a family F_1 to a family F_2 indicates that F_1 is strictly included in F_2 and a single arrow indicates that $F_1 \subseteq F_2$ and we do not know whether the inclusion is proper or not. Two families not related by a path in this diagram are not necessarily incomparable, with the exception of the pairs involving one of families FIN, REG: these two families are incomparable with all families to which they are not linked in the diagram. The families that we have writen down in one node of the diagram are equal.*

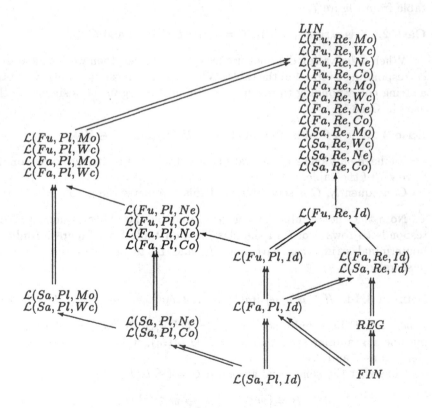

Fig. 8

4. Variants of contextual grammars

Many variants of the grammars considered in the previous sections were already investigated in the literature. We present here some of them, giving only some specific results.

4.1 Deterministic grammars

A total contextual grammar $G = (V, B, C, \varphi)$ is said to be *deterministic* if $card(\varphi(x_1, x_2, x_3)) \leq 1$ for all $x_1, x_2, x_3 \in V^*$. (In the compact writing, a grammar $G = (V, B, \{(D_1, C_1), \ldots, (D_n, C_n)\})$ is deterministic when the associated mapping φ has the property above.)

The families of languages generated by deterministic grammars are denoted by $DICC, DECC, DTC$, etc. (with the subscript c when computable choice mappings are used).

As expected, the determinism decreases strictly the generative power of grammars of all types. For instance, the following result is easy to prove.

Theorem 4.1. *Every language in the family DECC is slender.*

The languages in $DICC$ are not necessarily slender: a^+b^+ can be generated by $G = (\{a, b\}, \{ab\}, \{(\{a\}, \{\$a\}), (\{b\}, \{\$b\})\})$. However, the following holds.

Theorem 4.2. $ICC(CF) - DTC \neq \emptyset$.

Proof. Consider the grammar

$$G = (\{a, b, c\}, \{abab\}, \{(\{b^n a^n \mid n \geq 1\}, \{ab\$ab, c\$c\})\}).$$

The language $L_{in}(G)$ is not in the family DTC. Assume to the contrary that $L_{in}(G) = L(G')$, for some $G' = (\{a, b, c\}, B, C, \varphi)$ with $card(\varphi'(x_1, x_2, x_3)) \leq 1$ for all $x_1, x_2, x_3 \in \{a, b, c\}^*$.

Consider the strings $a^n b^n a^n b^n$ and $a^n c b^n a^n c b^n, n \geq 1$. All such strings are in $L_{in}(G) = L(G')$. If some $z \in L(G')$ generates such a string in a direct derivation, $z \Longrightarrow a^n b^n a^n b^n$ or $z \Longrightarrow a^n c b^n a^n c b^n$, then the only possibility is to have $z = a^m b^m a^m b^m$ and the used contexts are

$$(a^{n-m} b^{n-m}, a^{n-m} b^{n-m}) \in \varphi'(a^m, b^m a^m, b^m), \qquad (*)$$

in the first case, and

$$(a^{n-m} c b^{n-m}, a^{n-m} c b^{n-m}) \in \varphi'(a^m, b^m a^m, b^m) \qquad (*)$$

in the second case.

Moreover, because G' is deterministic, for each string $z = a^m b^m a^m b^m$ there is at most one derivation $z \Longrightarrow a^n b^n a^n b^n, z \Longrightarrow a^n c b^n a^n c b^n$ as above.

(Further derivations $z \Longrightarrow w$ are possible, but for decompositions $z = x_1 x_2 x_3$ different from those in the derivations $(*)$.)

Let

$$k = \max\{n \mid a^n b^n a^n b^n \in B, \text{ or } a^n cb^n a^n ab^n \in B\}.$$

We have $k \geq 1$, because at least $abab$ is in B (this is the shortest string in $L(G')$).

Consider now the sets

$$M_1 = \{a^n b^n a^n b^n \mid 1 \leq n \leq 2k+1\},$$
$$M_2 = \{a^n cb^n a^n cb^n \mid 1 \leq n \leq 2k+1\},$$

and

$$B' = B \cap (M_1 \cup M_2).$$

Clearly, $M_1 \cup M_2 \subseteq L(G')$ and from the above discussion it follows that for every $y \in (M_1 \cup M_2) - B'$ there is $x \in M_1$ such that $x \Longrightarrow y$. However,

$$card(M_1 \cup M_2) = 4k + 2, \quad card(B') \leq 2k,$$

and so we must have at least $2k + 2$ derivations of the form $x \Longrightarrow y$, $x \in M_1, y \in (M_1 \cup M_2) - B'$. Because $card(M_1) = 2k + 1$, there is $x \in M_1$ for which two derivations $x \Longrightarrow y_1, x \Longrightarrow y_2$, for $y_1 \neq y_2, y_1, y_2 \in (M_1 \cup M_2) - B'$, are possible, both of the form $(*)$. This contradicts the determinism of G'. Hence $L_{in}(G) \notin DTC$. $\square$

Corollary 4.1. *Both inclusions* $DICC \subset ICC, DTC \subset TC$ *are proper.*

Note that the grammars from Examples 2.5 and 2.8 in Section 2.3 are deterministic, hence $DECC_c$ and $DICC_c$ contain non-context-free languages.

Somewhat surprising, some languages which "look nondeterministic" at the first sight can be generated by deterministic grammars. Here is an example.

Consider

$$G = (\{a, b, c\}, \{xc \ mi(x) \mid x \in \{a, b\}^*, |x| \leq 2\}, \{a\$a, b\$b, ba\$ab\}, \varphi),$$
$$\varphi(c) = \{ba\$ab\},$$
$$\varphi(xc \ mi(x)) = \begin{cases} \{a\$a\}, & \text{if } x \in \{a, b\}^*, |x| = 2k+1, k \geq 0, \\ \{b\$b\}, & \text{if } x \in \{a, b\}^*, |x| = 2k, k \geq 1. \end{cases}$$

Then $L_{in}(G) = \{xc \ mi(x) \mid x \in \{a, b\}^*\}$. To see this consider a string $z = wc \ mi(w)$, choose the leftmost occurrence of b on an odd position, counting from the central c. Write $z = b^j(ab)^i bxc \ mi(x)b(ba)^i b^j, j \in \{0, 1\}, i \geq 0, |x| = 2k, k \geq 1$; then $z' = b^j(ab)^i xc \ mi(x)(ba)^i b^j$ can be generated from $xc \ mi(x)$, and z can be generated from z'. Hence the problem is reduced to generating the shorter string $xc \ mi(x)$, and so, inductively, we either reach an axiom string, or a string $(ba)^i c(ab)^i, i \geq 1$, which can be produced using the context $ba\$ab$.

4.2 One-sided contexts

In many examples considered in the preceding sections we have used grammars containing contexts with one member being empty.

For any family X of contextual languages, we denote by $1X$ the corresponding family of languages generated by grammars with contexts of the form $\$v$ and by $11X$ the family of languages generated by grammars using both right contexts $\$v$ and left contexts $u\$$ (but not $u\$v$ with $u \neq \lambda, v \neq \lambda$).

Variants of most of the previous results can be obtained for one-sided grammars. For instance, each language $L \in 11TC$ has the following 1IBS property: there is a constant p such that for each $x \in L, |x| > p$, there is $y \in L$ such that $x = x_1ux_2, y = x_1x_2$, and $0 < |u| \leq p$.

We collect in the next theorem, without proofs, some results about the generative power of one-sided grammars.

Theorem 4.3.
1. $EC - 11TC \neq \emptyset$, $IC - 11TC \neq \emptyset$,
2. $1EC \subset 11EC \subseteq (REG \cap EC)$,
3. $1IC = 11IC \subset (CF \cap IC)$, $1IC - LIN \neq \emptyset$,
4. $1ECC(REG) \subset REG$, $11ECC(REG) \subset LIN$,
5. $1ECC - CF \neq \emptyset$, $11ECC(REG) - REG \neq \emptyset$,
6. $REG \subset 1ICC \subset 11ICC$,
7. $11TC = 1TC$.

Consider now internal contextual grammars with finite choice, as in Section 3.2, with one-sided contexts only. The following counterparts of Theorems 3.4 and 3.5 can be obtained.

Theorem 4.4. *Every recursively enumerable language $L \subseteq V^*$ can be written in the form $L = (R \backslash L') \cap V^*$, for $L' \in 1ICC(CF)$ and $R \in REG$.*

Proof. According to [76], a *grammatical transformation* is a triple $\tau = (N, T, P)$, where N, T are disjoint alphabets and P is a finite set of rewriting rules over $N \cup T$. For a language $L \subseteq N^*$, we define

$$\tau(L) = \{x \in T^* \mid y \Longrightarrow_P^* x \text{ for some } y \in L\}.$$

Theorem 3 in [76] says that each language $L \in CS, L \subseteq V^*$, can be written in the form $L = \tau(L_0)$, for $L_0 \in REG, L_0 \subseteq N^*$, and $\tau = (N, V, P)$ with P containing productions of the form

$$A \to B, \ AB \to AC, \ A \to a, \text{ for } A, B, C \in N, a \in V.$$

Let us now take a language $L \in RE, L \subseteq V^*$. There are two new symbols, $b, c \notin V$, and a language $L' \subseteq b^*cL, L' \in CS$, such that for every $w \in L$ there is a string b^icw in $L', i \geq 0$ (Theorem 9.9 in [77]). Let $V' = V \cup \{b, c\}$. For this language L', consider $L_0 \in REG$ and τ as above: $\tau = (N_0, V', P), L_0 \subseteq$

$N_0^*, N_0 \cap V' = \emptyset$, and $L' = \tau(L_0)$. Take a grammar $G_0 = (N_1, N_0, X_0, P_0)$ with $N_1 \cap N_0 = \emptyset, N_1 \cap V' = \emptyset$, generating L_0. Without loss of generality we may assume that $N_1 = N_1' \cup \{X_f\}$, and P_0 contains rules of the forms

a) $X_1 \to X_2 A$, for $X_1 \in N_1', X_2 \in N_1, A \in N_0$,
b) $X_f \to \lambda$.

(Such a grammar always exists for a regular language: take a left-regular grammar for L_0 and replace each terminal rule $X \to A$ by $X \to X_f A$, then add the rule $X_f \to \lambda$.)

We construct the contextual grammar $G = (W, \{X_0\}, P')$ with

$$W = N_1 \cup N_0 \cup V' \cup \{], \#\},$$

and the set P' containing the following productions:

(1) $(\{X_1\}, \{\$]X_2 A\})$, for $X_1 \to X_2 A \in P_0, X_1 \in N_1', X_2 \in N_1, A \in N_0$,
(2) $(\{X_f\}, \{\$]\})$,
(3) $(\{A\}, \{\$]\alpha\})$, for $A \to \alpha \in P, A \in N_0, \alpha \in N_0 \cup V'$,
(4) $(\{AB\}, \{\$]]AC\})$, for $AB \to AC \in P, A, B, \in N_0$,
(5) $(\{\alpha x]^{|x|} \mid x \in (N_0 \cup V')^+\}, \{\$]\alpha\})$, for $\alpha \in N_0 \cup V'$,
(6) $(\{X_f]\}\{x]^{|x|} \mid x \in (N_0 \cup V')^+\}^+ b^* c, \{\$\#\})$.

Consider also the regular language

$$R = \{X] \mid X \in N_1'\}^+ \{X_f]\}(N_0 \cup V' \cup \{]\})^* \{\#\}.$$

Then we have $L = (R \backslash L_{in}(G)) \cap V^*$.

The intuition behind the previous construction is as follows. The symbol $]$ is a *killer*. Each occurrence of $]$ kills a symbol α in $N_0 \cup N_1 \cup V'$ according to the following rules:

1. if $x = x_1\alpha]x_2$, for $\alpha \in N_0 \cup N_1 \cup V'$, then the specified occurrence of α is killed by the specified occurrence of $]$;
2. if $x = x_1\alpha x_2]^{|x_2|}]x_3$, for $\alpha \in N_0 \cup N_1 \cup V'$ and $x_2 \in (N_0 \cup N_1 \cup V')^*$, then α is killed by the occurrence of $]$ in front of x_3.

(In rule 2, all symbols in x_2 are dead, killed by the $|x_2|$ occurrences of $]$ in the right hand of x_2.)

The productions of types 1, 2 produce a string in $L(G_0)$, shuffled with dead symbols and killers. The productions of types 3, 4 simulate corresponding rules in τ. The productions of type 5 move alive symbols from the left to the right, cross dead symbols and killers. This is useful both for preparing substrings AB for productions of type 4 and for transporting to the right alive copies of terminals. In order to obtain a string in RV^* we must use exactly once a production of type 6. This production checks whether the pairs $X], X \in N_1$, appear to the left of symbols used when simulating the work of τ and that all symbols to the left of $\#$ are either killers or dead symbols (hence the derivation in τ is terminal), or alive occurrences of b and c.

The equality $L = (R \backslash L_{in}(G)) \cap V^*$ can be proved in a similar way to the proofs of the corresponding equalities in Theorem 3.4 and its corollary. The following claims are useful for such a proof.

1. For every successful derivation in G (that is a derivation leading to a string in RV^*), $X_0 \Longrightarrow_{in}^* z_1 X_f] z_2 \# w$, there is a derivation $X_0 \Longrightarrow_{in}^* z_1 X_f] w_1 \Longrightarrow_{in}^* z_1 X_f] z_2 \# w$, with productions of types 1, 2 used before any production of other types.

2. If in a derivation as above we use further productions of types 1, 2 for deriving the prefix $z_1 X_f]$ in $z_1 X_f] z_2 \# w$, then we obtain a derivation which is no more successful.

3. If in a derivation step $u \Longrightarrow_{in} v$ we use a production of type 3, $(\{A\}, \{\$]\alpha\})$, for A alive in u, then A will be dead in v and α will be alive in v; if A is dead in u, then both A and α are dead in v.

4. If in a derivation step $u \Longrightarrow_{in} v$ we use a production of type 4, $(\{AB\}, \{\$]]AC\})$, for A, B alive in u, then AB are dead in v and AC will be alive in v; if A, B are dead in u, then all corresponding symbols A, B, C are dead in v.

5. If in a derivation step $u \Longrightarrow_{in} v$ we use a production of type 4, $(\{AB\}, x\{\$]]AC\})$, for A alive and B dead in u, then the new occurrence of A will be alive in v but the old occurrence of A and the occurrences of B and C will be dead in v.

6. If in a derivation step $u \Longrightarrow_{in} v$ we use a production of type 5, adjoining $]\alpha$ to some $\alpha x]^{|x|}$, and α is alive in u, then that occurrence of α is dead in v and the newly introduced one is alive; if α is dead in u, then it will remain dead in v and the newly introduced occurrence of α in v will be dead, too.

Consequently, if $u \Longrightarrow_{in} v$ in G by a production π of types 3, 4, then $alive(u) \Longrightarrow alive(v)$ in τ by using the rule in P associated with π. Moreover, for $u \Longrightarrow_{in} v$ using a production of type 5 we have $alive(u) = alive(v)$.

The productions of type 6 can be used only once and to the right of the symbol $\#$ introduced in this way we must have only symbols in V. Because the selector of a production of type 6 is bounded by the prefix $X_f]$ and a suffix $b^i c$, it follows that the string to which such a rule is applied is of the form $z_1 X_f] u_1 b^i c u_2$ such that $alive(u_1) = \lambda$ and $u_2 \in V^*$, hence we obtain $z_1 X_f] u_1 b^i c \# u_2$. According to the previous claims, if we continue to rewrite such a string, then no new alive symbols can be introduced to the left of $\#$. This proves that $(R \backslash L_{in}(G)) \cap V^* \subseteq L$. The inverse inclusion follows in an easier way. $\qquad \Box$

Corollary 4.2. *The family $1ICC(CF)$ is incomparable with each family F such that $LIN \subseteq F \subset RE$, which is closed under left quotient and intersection with regular languages.*

Again MAT^λ and $ET0L$ are important families of languages satisfying the conditions from the statement of the above result, hence $1ICC(CF) - MAT^\lambda \neq \emptyset$, $1ICC(CF) - ET0L \neq \emptyset$.

Also counterparts of results from Section 3.3 hold for one-sided grammars. Let us denote by $\mathcal{L}_1(X, Y, Z)$ the family of languages generated by one-sided grammars associated with families $\mathcal{L}(X, Y, Z)$ as in Section 3.3.

The following relations are either obvious or are particular cases of results in the previous sections.

Lemma 4.1. $\mathcal{L}_1(X, Y, Z) \subseteq REG = \mathcal{L}_1(X, Re, Z)$, for all values of X, Y, Z.

Moreover, also the following lemma holds.

Lemma 4.2. $\mathcal{L}_1(Fu, Pl, Mo) \subseteq \mathcal{L}_1(Fa, Pl, Id)$ and $\mathcal{L}_1(Sa, Pl, Mo) \subseteq \mathcal{L}_1(Sa, Pl, Id)$.

Proof. Take a language $L = \mathcal{L}_1(Fu, Pl, Mo)$, $L = h(\bigcup_{i=1}^n L(G_i))$ where $h : V^* \longrightarrow U^*$ is a morphism and $G_i = (V, \{w_i\}, P_i)$ for each $1 \leq i \leq n$. The language L is regular. We construct the grammar $G = (U, B, P)$ with

$B = \{h(w_1), \ldots, h(w_n)\}$,

$P = \{(L/\{h(v)\}, \{\$h(v)\}) \mid \text{there is } (D, \{\$v\}) \in P_i, \text{ for some } 1 \leq i \leq n\}$.

Obviously, P is a finite set.

By a standard argument we can see that $L = L_{ex}(G)$.   □

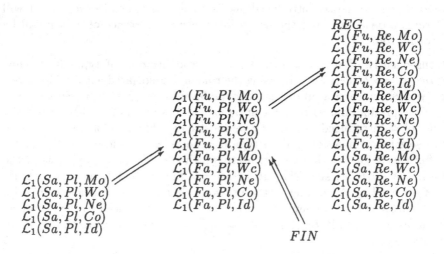

$$
\begin{array}{l}
REG \\
\mathcal{L}_1(Fu, Re, Mo) \\
\mathcal{L}_1(Fu, Re, Wc) \\
\mathcal{L}_1(Fu, Re, Ne) \\
\mathcal{L}_1(Fu, Re, Co) \\
\mathcal{L}_1(Fu, Re, Id) \\
\mathcal{L}_1(Fa, Re, Mo) \\
\mathcal{L}_1(Fa, Re, Wc) \\
\mathcal{L}_1(Fa, Re, Ne) \\
\mathcal{L}_1(Fa, Re, Co) \\
\mathcal{L}_1(Fa, Re, Id) \\
\mathcal{L}_1(Sa, Re, Mo) \\
\mathcal{L}_1(Sa, Re, Wc) \\
\mathcal{L}_1(Sa, Re, Ne) \\
\mathcal{L}_1(Sa, Re, Co) \\
\mathcal{L}_1(Sa, Re, Id)
\end{array}
$$

$$
\begin{array}{l}
\mathcal{L}_1(Fu, Pl, Mo) \\
\mathcal{L}_1(Fu, Pl, Wc) \\
\mathcal{L}_1(Fu, Pl, Ne) \\
\mathcal{L}_1(Fu, Pl, Co) \\
\mathcal{L}_1(Fu, Pl, Id) \\
\mathcal{L}_1(Fa, Pl, Mo) \\
\mathcal{L}_1(Fa, Pl, Wc) \\
\mathcal{L}_1(Fa, Pl, Ne) \\
\mathcal{L}_1(Fa, Pl, Co) \\
\mathcal{L}_1(Fa, Pl, Id)
\end{array}
$$

$$
\begin{array}{l}
\mathcal{L}_1(Sa, Pl, Mo) \\
\mathcal{L}_1(Sa, Pl, Wc) \\
\mathcal{L}_1(Sa, Pl, Ne) \\
\mathcal{L}_1(Sa, Pl, Co) \\
\mathcal{L}_1(Sa, Pl, Id)
\end{array}
$$

FIN

Fig. 9

Consequently, for one-sided grammars we obtain the situation in the diagram from Figure 9, with only three different families. The inclusions are proper, because the regular language $L_1 = ab^+a$ is not in $\mathcal{L}_1(X, Y, Z)$ for any X, Y, Z with $Y \neq Re$, and every finite language is in $\mathcal{L}_1(Fa, Pl, Id)$, implying that $L_2 = \{a, b\}$ considered also at the beginning of Section 3.3 is in $\mathcal{L}_1(Fa, Pl, Id) - \mathcal{L}_1(Sa, Pl, Id)$.

4.3 Leftmost derivation

In the previous two variants of contextual grammars we have imposed restrictions on the grammar components. From now on we consider a series of modifications of the derivation relations defined with respect to a contextual grammar.

For instance, given a contextual grammar $G = (V, B, \{(D_1, C_1), \ldots, (D_n, C_n)\})$, the *leftmost derivation* with respect to G is defined by

$$x \Longrightarrow_{left} y \quad \text{iff} \quad x = x_1 x_2 x_3, y = x_1 u x_2 v x_3,$$

$$\text{for some } x_1, x_2, x_3 \in V^*, x_2 \in D_i, u\$v \in C_i, 1 \le i \le n,$$

$$\text{such that there is no decomposition } x = x_1' x_2' x_3' \text{ with}$$

$$|x_1'| < |x_1| \text{ and } x_2' \in D_j \text{ for some } 1 \le j \le n.$$

Hence one adjoins a context to the leftmost possible place, taking into consideration all possible decompositions of the current string and all components of the grammar.

The language generated in this way is denoted by $L_{left}(G)$ and the family of such languages generated by grammars with F choice is denoted by $ICC_{left}(F)$.

Here are two examples:

$$G_1 = (\{a, b, c, d\}, \{dab\}, \{(\{ab, ac, dc\}, \{a\$b, c\$\})\}),$$
$$G_2 = (\{a, b\}, \{aba\}, \{(a^+ba, \{a\$a\}), (a^+b, \{b\$b\})\}).$$

We have

$$L_{left}(G_1) \cap a^+dcb^+(ac)^+ab^+ = \{a^n dcb^n (ac)^{m-1} ab^m \mid n, m \ge 1\},$$
$$L_{left}(G_2) = \{b^m a^n b^{m+1} a^n \mid n \ge 1, m \ge 0\}.$$

The first language is not linear (LIN is closed under intersection with regular sets), the second one is not context-free.

In fact, we get the following result.

Theorem 4.5. *$ICC_{left}(F)$ is incomparable with LIN and ICC for all F containing the finite languages; $ICC_{left}(F)$ is incomparable with CF for all F containing the regular languages.*

Proof. We have seen that $ICC_{left}(REG) - CF \neq \emptyset$ and $ICC_{left}(FIN) - LIN \neq \emptyset$. Moreover, for the grammar

$$G = (\{a, b\}, \{bb\}, \{(\{bb\}, \{\$a, a\$a\}), (\{ab\}, \{a\$a\})\}),$$

we have

$$L_{left}(G) = \{bba^m, a^{n+1}ba^nba^{m+1} \mid n, m \geq 0\}$$

(starting from bb we can use either the first context in the first production, thus producing $bba^m, m \geq 0$, or the second one; the second one can be added also to bba^m, hence we get $abba^{m+1}$; from now on only the second production can be used, hence we obtain $a^{n+1}ba^nba^{m+1}$). This language is not in ICC: the context $a^i\$a^i \in \varphi(a^jba^k), i \geq 1, j, k \geq 0$, necessary for generating $a^{n+1}ba^nba$ with arbitrarily large n, can be used for $a^{j+2}ba^{j+1}ba^k \Longrightarrow_{in} a^{j+2}ba^{j+i+1}ba^{k+i}$, leading to a parasitic string.

On the other hand, the language $L = \{a^mba^nca^n \mid n, m \geq 1\}$ is in $LIN \cap ICC(FIN)$ but not in ICC_{left}. Assume that $L = L_{left}(G)$ for some $G = (\{a, b, c\}, B, \{(D_1, C_1), \ldots, (D_n, C_n)\})$. In order to generate strings ab^nca^n with arbitrarily large n, we need a context $u\$v = a^r\$a^r \in C_i, r \geq 1$, for some D_i containing $a^gca^h, g, h \geq 0$. In order to generate a^mbaca with arbitrarily large m, we need a context $u'\$v' = a^p\$a^q \in C_j, p + q \geq 1$, for some D_j containing $a^s, s \geq 0$, or a context $u'\$v' = a^p\$ \in C_j, p \geq 1$, for some D_j containing a string $a^sbz, s \geq 0$. Consider such strings in D_j with the smallest s. To every string of the form $a^fba^lca^l, l \geq 1, f \geq s$, respectively $a^fbzy, f \geq s$, only the context $u'\$v'$ can be added, therefore strings $a^mba^nca^n$ with $m \geq s$ and arbitrarily large n cannot be generated. Therefore L cannot be equal to $L_{left}(G)$, hence $L \notin ICC_{left}$ (the type of selectors in G plays no role in this argument).   □

Conjecture. $ICC_{left}(FIN) \subseteq CF$.

4.4 Parallel derivation

For a contextual grammar $G = (V, B, \{(D_1, C_1), \ldots, (D_n, C_n)\})$ we define the *parallel* derivation by

$$x \Longrightarrow_p y \quad \text{iff} \quad x = x_1x_2 \ldots x_r, y = u_1x_1v_1u_2x_2v_2 \ldots u_rx_rv_r,$$

$$\text{for } u_j\$v_j \in C_{i_j}, x_j \in D_{i_j}, 1 \leq j \leq r, 1 \leq i_j \leq n, r \geq 1.$$

Thus, the whole string is splitted into strings in the selectors of G and contexts are added to each such string.

We denote by $L_p(G)$ the language generated in this way and by $ICC_p(F)$ the family of such languages generated by grammars with F choice.

The following examples show that parallel derivation is very powerful.

Let

$$G_1 = (\{a, b, c\}, \{abc, a^2b^2c^2\}, \{(a^+b^+, \{a\$b\}), (bc^+, \{\$c\})\}), \text{ and}$$
$$G_2 = (\{a\}, \{a^k\}, \{(\{a\}, \{\$a^{k-1}\})\}), k \geq 1.$$

Then

$$L_p(G_1) = \{a^n b^n c^n \mid n \geq 1\},$$
$$L_p(G_2) = \{a^{k^n} \mid n \geq 1\}.$$

(Note that both grammars are deterministic and that G_2 is one-sided.)

In fact, we have the following result.

Theorem 4.6. $ICC(F) \subset ICC_p(F)$ for all F containing languages of the form $\{a\}$.

Proof. For a grammar $G = (V, B, P)$, construct $G' = (V, B, P \cup \{(V, \{\$\})\})$. Clearly, $L_{in}(G) = L_p(G')$, hence $ICC(F) \subseteq ICC_p(F)$. The inclusion is proper, because $ICC_p(F)$ contains one-letter non-regular languages (example above) which are not in $ICC(F)$ (Theorem 2.2). $\qquad \square$

Note that in the above proof we have used, for the first time in this chapter, the empty context.

Moreover, the following surprising result (showing that parallel *rewriting* can be simulated by parallel *adjoining*) is true:

Theorem 4.7. $D0L \subset ICC_p(FIN)$.

Proof. There are finite languages not in $D0L$, hence it is enough to prove the inclusion.

Let L be a language in the family $D0L$. Then there exist the alphabet V, the word $w \in V^*$, and the homomorphism $h : V^* \longrightarrow V^*$ such that

$$L = \bigcup_{m=0}^{\infty} \{h^m(w)\}.$$

Consider the sets A_1, A_2, A_3 defined as follows:

$A_1 = \{a \in V \mid a \in alph(h^n(a)) \text{ for some } n \geq 1\}$,
$A_2 = \{b \in V \mid \text{there is } m \geq 1 \text{ such that } b \notin alph(h^n(w)) \text{ for any } n \geq m\}$,
$A_3 = V - (A_1 \cup A_2) = \{c \in V \mid c \notin alph(h^s(c)) \text{ for all } s \geq 1, \text{ but for any } m \geq 1 \text{ there is } n \geq m \text{ such that } c \in alph(h^n(w))\}$.

For each $a \in A_1$ let n_a denote the smallest n from the definition of A_1. Denote also by p the least common multiple of $\{n_a \mid a \in A_1\}$. Clearly, $a \in alph(h^p(a))$ for all $a \in A_1$. This implies that for any $a \in A_1$ there are $u, v \in V^*$ such that $h^p(a) = uav$.

For each $b \in A_2$ let m_b be the smallest m from the definition of A_2 and denote $q = \max\{m_b \mid b \in A_2\}$. Consequently, $b \notin alph(h^n(w))$ for any $n \geq q$ and $b \in A_2$. Let $r = card(A_3)$.

Construct the contextual grammar with finite choice $G = (V, B, C, \varphi)$, where

$$B = \{h^n(w) \mid 0 \le n \le p + q + r - 1\},$$
$$C = \{h^s(u)\$h^s(v) \mid uav = h^p(a), a \in A_1, 0 \le s \le r\},$$

and $\varphi(h^s(a)) = \{h^s(u)\$h^s(v) \mid uav = h^p(a)\},$ for $0 \le s \le r, a \in A_1$.

We can prove that $L_p(G) = L$ making use of the following two claims.

Claim 3. *Let c be an arbitrary symbol from A_3 and let $n \ge q + r$ be such that $h^n(w) = y_1 c y_2$ for some $y_1, y_2 \in V^*$. Then there are $s, 1 \le s \le r$, and $a \in A$ such that*

$$h^{n-s}(w) = z_1 a z_2, \quad \text{for some } z_1, z_2 \in V^*,$$
$$h^s(a) = x_1 c x_2, \quad \text{for some } x_1, x_2 \in V^*,$$
$$y_1 = h^s(z_1) x_1, \quad y_2 = x_2 h^s(z_2).$$

Claim 4.

$$L = \bigcup_{n=0}^{q+r-1} \{h^n(w)\} \cup \bigcup_{i=0}^{p-1} \bigcup_{n=0}^{\infty} \{(h^p)^n(h^{q+r+i}(w))\}. \qquad \square$$

On the other hand, we have the following result.

Theorem 4.8. *All families $P0L, PDT0L, LIN$ contain languages not in ICC_p.*

An example is the language

$$L_0 = \{f\} \cup a^*c \cup b^*d \cup a^+ \cup b^+ \cup \{a^n eb^n \mid n \ge 0\} \cup \{a^n b^n \mid n \ge 1\}.$$

4.5 Maximal/minimal use of selectors

In the basic internal derivation, the string selector to which a context is adjoined can appear anywhere in the current string; in the leftmost derivation the choice is restricted to the first place from the left where a selector occurs in the current string. Further restrictions on the applications of productions are also considered. For instance, we can allow only derivations where the selectors are the largest/smallest possible, in the sense that no superstring/substring of the chosen selector can be also used as a selector.

More formally, for a grammar $G = (V, B, \{(D_1, C_1), \ldots, (D_n, C_n)\})$ we can define the relations

$$x \Longrightarrow_{ml} y \quad \text{iff} \quad x = x_1 x_2 x_3, y = x_1 u x_2 v x_3,$$

for $x_2 \in D_i, (u,v) \in C_i$, for some $1 \le i \le n$,

and there are no $x_1', x_2', x_3' \in V^*$ such that

$x = x_1' x_2' x_3', x_2' \in D_i, |x_1'| \ge |x_1|, |x_3'| \ge |x_3|, |x_2'| < |x_2|;$

$$x \Longrightarrow_{Ml} y \quad \text{iff} \quad x = x_1 x_2 x_3, y = x_1 u x_2 v x_3,$$

for $x_2 \in D_i, (u,v) \in C_i$, for some $1 \le i \le n$,

and there are no $x_1', x_2', x_3' \in V^*$ such that

$x = x_1' x_2' x_3', x_2' \in D_i, |x_1'| \le |x_1|, |x_3'| \le |x_3|, |x_2'| > |x_2|;$

We call $\Longrightarrow_{ml}$ a derivation in the *minimal local* mode and $\Longrightarrow_{Ml}$ a derivation in the *maximal local* mode.

When the condition $x_2' \in D_i$ from the above definitions is replaced by

$$x_2' \in D_j, 1 \le j \le n,$$

then we call the so obtained relations *minimal global* and *maximal global*, and we write $\Longrightarrow_{mg}$ and $\Longrightarrow_{Mg}$, respectively.

For $\alpha \in \{ml, mg, Ml, Mg\}$, we denote by $L_\alpha(G)$ the language generated by G in the mode α. The family of such languages generated by grammars with F choice is denoted by $ICC_\alpha(F)$.

We give now, without proofs, a summary of the results from [33], [32] about the families $ICC_\alpha(F)$ with $F \in \{FIN, REG\}$.

Here is a simple example illustrating the effect of using the selectors of a contextual grammar in the maximal/minimal modes. For the grammar $G = (\{a, b\}, \{a, ab\}, \{((\{a\}, \{\$a\}), (\{a^+b\}, \{a\$b\}))\})$ we obtain

$$L_{Ml}(G) = L_{ml}(G) = L_{in}(G) = a^+ \cup \{a^n b^m \mid n \ge m \ge 1\},$$
$$L_{Mg}(G) = a^+ \cup \{a^n b^n \mid n \ge 1\},$$
$$L_{mg}(G) = a^+ \cup a^+ b.$$

Remember that $a^+ \cup \{a^n b^n \mid n \ge 1\} \notin ICC$.

Theorem 4.9. *The relations in Figure 10 hold; the arrows indicate strict inclusions, the dotted arrow indicates an inclusion not known to be proper. Families not linked by a path in this diagram are not necessarily incomparable, but $ICC_{Mg}(REG)$ is incomparable with all $ICC(F), ICC_{Ml}(F), F \in \{FIN, REG\}$.*

5. Bibliographical notes

The contextual grammars (with arbitrary choice and without choice, with the language generated in the external mode) were introduced in [29]; a presentation can be found also in [30]. The internal mode of derivation has been

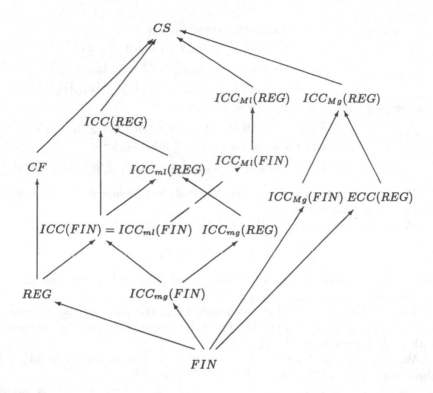

Fig. 10

considered in [69]. Grammars with choice restricted to families in Chomsky
hierarchy (with external derivation) were introduced in [22]. Early papers
devoted to the study of these basic classes of contextual grammars are [21],
[45], [50], [52], etc. The grammars called *total* here are special cases of the
n-contextual grammars considered in [40], [41], [42], [44]. Such a grammar
uses contexts of the form $u_1\$u_2\$\ldots\$u_n$ and a choice mapping defined on
$(V^*)^{n+1}$ (n substrings are inserted, depending on the whole current string,
splitted into $n + 1$ parts). Hence total grammars are 3-contextual grammars
in this sense.

The monograph [60] contains almost all the results up to 1981.

The material included in Sections 2.2–2.7, 3.1 and 4.6 is based mainly on
[60].

The compact presentation of a grammar was used first in [70] (in earlier
papers the contexts are written in the form (u, v) and the choice is given
only by a mapping). Example 9 in Section 2.3 is from [11] (the first internal
contextual grammar with finite choice generating a non-context-free language
appears in [65]). Lemma 2.10 is from [11]; a longer proof is given in [79].

Further (non)closure and decidability results can be found in [60] and its
references. Not covered here is the area of descriptional complexity of contex-

tual languages. Three measures were introduced in [51] for grammars $G = (V, B, C, \varphi)$ working in the external mode: $Baz(G) = card(B), Con(G) = card(C), Fi(G) = card\{D \subseteq V^* \mid \varphi(x) = \varphi(y) \text{ for all } x, y \in D\}$. Many further measures are considered in [14], [67]. In general, the usual decidability problems for such measures (see [60]) are shown to be undecidable for contextual grammars with choice, and all of them induce infinite hierarchies of languages.

The proof of Lemma 3.6 is from [70]; in [65] a proof is given of the inclusion $ECC(REG) \subseteq CF$. The matching normal form in [9] also yields the inclusion $ECC(REG) \subseteq LIN$ (two productions $(D_1, C_1), (D_2, C_2)$ are *matching* if $uxv \in D_2$ for all $x \in D_1, u\$v \in C_1$; a derivation is matching when any two consecutive productions used in it are matching; a grammar G is in the matching normal form if each word in $L_{ex}(G)$ has a matching derivation; therefore the notion has a dynamic character). For every grammar G, irrespective of the type of choice, if it is in the matching normal form, then $L_{ex}(G) \in LIN$. Moreover, if G is a grammar with regular choice, then an externally equivalent grammar G' can be constructed, such that G' is in the matching normal form. Also a static normal form is given in [9] (a grammar is in the *disjoint-equal normal form* if any two selectors of it are either equal or disjoint).

Grammars with external bounded choice in the sense of Definition 6 are called semi-conditional in [64]. A variant of them are the branching grammars, which are usual or one-sided contextual grammars with contexts of the form $u\$v, u, v \in V \cup \{\lambda\}$. They are considered in [56], in relation to [15].

Theorem 3.4 is from [10], where also related forms and consequences of this representation are discussed. The one-sided variant of it (Theorem 4.4) is from [8]. The content of Section 3.3 is based on [10].

The determinism (Section 4.1) is introduced in [70]. Theorem 4.2 is from [7], where it is also proved that the family $DICC(REG)$ contains non-semilinear languages. The result is improved in [20], where one proves that also the family $DICC(FIN)$ contains non-semilinear languages. The study of one-sided contextual grammars is started in [70]. The leftmost derivation is introduced in [70], where also the so-called modular contextual grammars are discussed. They are usual grammars $G = (V, B, (D_1, C_1), \ldots, (D_n, C_n))$ with the "modules" used in the same way as the components of a cooperating grammar system [4]: in an internal or external derivation step according to a component (D_i, C_i), exactly k, at least k, at most k, any number, or the largest possible number of steps are performed. A number of results about such modular grammars can be found in [18], [73]. For instance, modular grammars with regular choice generate by external derivations in the maximal mode only linear languages, a counterpart of Lemma 3.6. The converse is not true: there are linear languages which cannot be generated by modular grammars with regular choice in the external maximal mode. Grammars with linear choice and at least two components generate non-context-free languages. In the $= k$ and $\geq k$ modes of derivation, the type of choice induces hierarchies similar to those from Theorem 3.1.

The parallel derivation, in the *total* mode considered in Section 4.4, is introduced in [72]. The case when not necessarily the whole current string is covered by selectors, but the maximal number of selectors is used (the remaining uncovered substrings contain no further selector string) is considered in [38]. Theorem 4.7, as well as the results from Theorem 4.8 for $P0L$ and $PDT0L$ appear in [80], where also representation theorems for languages in $ICC_p(F)$ are given, for various F. For instance, each language $L \in ICC_p(FIN)$ can be written in the form $L = \cup_{m=0}^{\infty}(h_2(h_1^{-1}))^m(B)$, for a finite language B and morphisms h_1, h_2 such that $|h_2(x)| \geq |h_1(x)|$ for all x.

The language of the strings generated by *blocked derivations* is considered in [72] (a derivation is blocked when no further step is possible) for modular grammars: a change of the module happens only after the derivation is blocked. Again grammars with regular choice and external derivation generate only linear languages.

The results in Section 4.5 are from [33]. Further results about grammars with maximal use of selectors are given in [32]: $ICC_{Mg}(FIN) - CF \neq \emptyset$ and every language $L \in RE$ can be written in the form $L = h_1(h_2^{-1}(L'))$, where $L' \in ICC_{Mg}(FIN)$ and h_1, h_2 are two morphisms. Discussions about the linguistic relevance of contextual grammars with maximal use of selectors can be also found in [32].

Contextual-like grammars based on the shuffle operation appear also in [36], [71]. In [36] they are without choice: $G = (V, B, C)$, $B, C \subseteq V^*$, and the generated language consists of the closure of B under the shuffle with words in C. In [71] the choice is restricted, the grammar is given in the form $G = (V, B, (D_1, C_1), \ldots, (D_n, C_n))$, with all B, D_i, C_i languages over V. The selectors must appear as compact substrings of the string to be derived $(x = x_1 x_2 x_3, x_2 \in D_i)$, but the added string is shuffled in the selector (in x_2 above). Derivations with a leftmost occurrence of the selector, its occurrence in the prefix of the derived string, on an arbitrary position, or equalizing the derived string, as well as parallel derivations are investigated in [71].

In [34] one considers contextual grammars with depth-first derivation: at every step, the selector must contain as a subword at least one of the members of the context used at the previous step. (The λ-context proves to be very important in this framework: using it we can "forget" the last used context.)

The growth of languages generated in a deterministic way (by leftmost derivations with a maximal use of selectors or with selectors appearing as prefixes of the current string) by deterministic grammars is investigated in [74].

Proposals for definitions of the ambiguity in internal contextual grammars can be found in [35]; further results appear in [19].

Several other variants of contextual grammars can be found in the literature. In the ν-contextual grammar in [47] there are two choice mappings, φ and ψ; a derivation is accepted only if it develops using contexts selected by

φ and ends using a context selected by ψ. The generative power of external grammars is significantly increased. The *regular contextual grammars* in [45] are similar to the grammars in the matching normal form: for all $x, y \in V^*$, if $\varphi(x) = \varphi(y)$ and $u\$v \in \varphi(x)$, then $\varphi(uxv) = \varphi(uyv)$. (This is a stronger condition than the matching normal form: the property holds for all strings in V^*, not only for strings which suffice for completing at least one derivation for each string in $L_{ex}(G)$.) It is proved in [45] that such grammars, with arbitrary choice, generate in the external mode linear languages only.

In [60] one also considers internal and external grammars without choice, but with the derivation regulated by matrix, programmed and regular control mechanisms, as those used in the case of Chomsky grammars [6].

Grammars $G = (V, B, C, \varphi)$ with infinite set C of contexts, but with finite $\varphi(x)$ for all $x \in V^*$, are proposed in [66].

In [70] one considers also grammars with both usual contexts and with *erased* contexts (they are not adjoined, but removed from the current string, depending on their selectors), or with *erasing* contexts (their selector is erased from the current string). They increase the generative capacity and also make possible the use of extended symbols, like in Chomsky grammars and in L systems.

The last chapter of [60] presents results about other research topics: the syntactic monoid of contextual languages, the composition of contextual grammars (the starting language, B, of one grammar is the language generated by another one), grammar forms (see further results in [5]), characterizations using equations (following [23]) or automata similar to contraction automata in [81]. Recent characterizations of classes of contextual languages are obtained using restarting automata [25]. An algebraic description of various classes of contextual languages can be found in [48], whereas [49] deals with the grammatical inference problem.

A sort of counterpart of contextual grammars are the *insertion grammars* in [13]: systems $G = (V, B, P)$, with V an alphabet, B a finite set of axioms, P a finite set of rewriting rules of the form $xy \rightarrow xzy$ (the string z is inserted in the context (x, y)). Results about these grammars can be found in [27], [62], [63], [78].

Contextual grammars are also important for [57], [59], [61]. An action system with finite history, as introduced in [57], is an one-sided ν-grammar with bounded external choice: $\Sigma = (A, A_0, A_f, \varphi)$, where A is the alphabet of elementary actions, A_0 is the set of initial actions, A_f is the set of final actions, and $\varphi : (\cup_{i=1}^{k} A^i) \longrightarrow 2^A$ specifies the next action depending on at most k previous actions. The language associated with such a system, $L(\Sigma)$, consists of all strings $w \in A^*$ such that $w = a_0 w' a_f$, $s_0 \in A_0, a_f \in A_f, w' \in A^*$, and for each $i, 2 \le i \le |w|$, we can write $w = w_1 w_2 a w_3$ with $|w_1 w_2 a| = i, a \in A, 1 \le |w_2| \le k, a \in \varphi(w_2)$. Such systems are extensively used in [1] and [59].

Contextual-like array grammars are considered in [12].

References

[1]   T. Bălănescu, M. Gheorghe, Program tracing and languages of action, *Rev. Roum. Ling.*, 32 (1987), 167–170.

[2]   I. Bucurescu, A. Pascu, Fuzzy contextual grammars, *Rev. Roum. Math. Pures Appl.*, 32 (1987), 497–507.

[3]   G. Ciucar, On the syntactic complexity of Galiukschov semicontextual languages, *Cah. Ling. Th. Appl.*, 25 (1988), 23–28.

[4]   E. Csuhaj-Varju, J. Dassow, J. Kelemen, Gh. Păun, *Grammar Systems. A Grammatical Approach to Distribution and Cooperation*, Gordon and Breach, London, 1994.

[5]   J. Dassow, On contextual grammar forms, *Intern. J. Computer Math.*, 17 (1985), 37–52.

[6]   J. Dassow, Gh. Păun, *Regulated Rewriting in Formal Language Theory*, Springer-Verlag, Berlin, Heidelberg, 1989.

[7]   A. Ehrenfeucht, L. Ilie, Gh. Păun, G. Rozenberg, A. Salomaa, On the generative capacity of certain classes of contextual grammars, in *Mathematical Linguistics and Related Topics* (Gh. Păun, ed.), Ed. Academiei, Bucharest, 1995, 105–118.

[8]   A. Ehrenfeucht, A. Mateescu, Gh. Păun, G. Rozenberg, A. Salomaa, On representing RE languages by one-sided internal contextual languages, *Technical Report 95-04*, Dept. of Computer Sci., Leiden Univ., 1995, and *Acta Cybernetica*, to appear.

[9]   A. Ehrenfeucht, Gh. Păun, G. Rozenberg, Normal forms for contextual grammars, in *Mathematical Aspects of Natural and Formal Languages* (Gh. Păun, ed.), World Sci. Publ., Singapore, 1994, 79–96.

[10]   A. Ehrenfeucht, Gh. Păun, G. Rozenberg, The linear landscape of external contextual languages, *Acta Informatica*, 33 (1996), 571–593..

[11]   A. Ehrenfeucht, Gh. Păun, G. Rozenberg, On representing recursively enumerable languages by internal contextual languages, *Th. Computer Sci.*, 1996.

[12]   R. Freund, Gh. Păun, G. Rozenberg, Contextual array grammars, *Technical Report 95-38*, Dept. of Computer Sci., Leiden Univ., 1995.

[13]   B. S. Galiukschov, Semicontextual grammars, *Mat. logica i mat. ling.*, Kalinin Univ., 1981, 38–50 (in Russ.).

[14]   G. Georgescu, Infinite hierarchies of some types of contextual languages, in *Mathematical Aspects of Natural and Formal Languages* (Gh. Păun, ed.), World Sci. Publ., Singapore, 1994, 137–150.

[15]   I. M. Havel, On branching and looping. Part I, *Th. Computer Sci.*, 10 (1980), 187–220.

[16]   T. Head, Formal language theory and DNA: an analysis of the generative capacity of specific recombinant behaviors, *Bull. Math. Biology*, 49 (1987), 737–759.

[17]   T. Head, Gh. Păun, D. Pixton, Language theory and molecular genetics: generative mechanisms suggested by DNA recombination, in this *Handbook*.

[18]   L. Ilie, On contextual grammars with parallel derivation, in vol. *Mathematical Aspects of Natural and Formal Languages* (Gh. Păun, ed.), World Sci. Publ., Singapore, 1994, 165–172.

[19]   L. Ilie, On ambiguity in contextual languages, *Intern. Conf. Math. Ling.*, Tarragona, 1996.

[20]   L. Ilie, A non-semilinear language generated by an internal contextual grammar with finite choice, *Ann. Univ. Buc., Matem.-Inform. Series*, 45, 1 (1996), 63–70.

[21] S. Istrail, A problem about contextual grammars with choice, *Stud. Cerc. Matem.*, 30 (1978), 135–139.

[22] S. Istrail, Contextual grammars with regulated selection, *Stud. Cerc. Matem.*, 30 (1978), 287–294.

[23] S. Istrail, A fixed-point approach to contextual languages, *Rev. Roum. Math. Pures Appl.*, 25 (1980), 861–869.

[24] M. Ito, Generalized contextual languages, *Cah. Ling. Th. Appl.*, 11 (1974), 73–77.

[25] P. Jancar, F. Mraz, M. Platek, M. Prochazka, J. Vogel, Restarting automata, Marcus grammars, and context-free grammars, *Proc. Developments in Language Theory Conf.*, Magdeburg, 1995, 102–111.

[26] L. Kari, On insertion and deletion in formal languages, *PH D Thesis*, Univ. of Turku, 1991.

[27] M. Marcus, Gh. Păun, Regulated Galiukschov semicontextual grammars, *Kybernetika*, 26 (1990), 316–326.

[28] S. Marcus, *Algebraic Linguistics. Analytical Models*, Academic Press, New York, 1967.

[29] S. Marcus, Contextual grammars, *Rev. Roum. Math. Pures Appl.*, 14 (1969), 1525–1534.

[30] S. Marcus, Deux types nouveaux de grammaires génératives, *Cah. Ling. Th. Appl.*, 6 (1969), 69–74.

[31] S. Marcus, Contextual grammars and natural languages, in this *Handbook*.

[32] S. Marcus, C. Martin-Vide, Gh. Păun, On the power of internal contextual grammars with maximal use of selectors, *Fourth meeting on mathematics of language*, Philadelphia, 1995.

[33] C. Martin-Vide, A. Mateescu, J. Miquel-Verges, Gh. Păun, Contextual grammars with maximal, minimal and scattered use of contexts, *BISFAI '95 Conf.* (M. Kappel, E. Shamir, eds.), Jerusalem, 1995, 132–142.

[34] C. Martin-Vide, J. Miquel-Verges, Gh. Păun, Contextual grammars with depth-first derivation, *Tenth Twente Workshop on Language Technology*, Twente, 1995.

[35] C. Martin-Vide, J. Miquel-Verges, Gh. Păun, A. Salomaa, An attempt to define the ambiguity of internal contextual languages, *Intern. Conf. Math. Ling.*, Tarragona, 1996.

[36] A. Mateescu, Marcus contextual grammars with shuffled contexts, in vol. *Mathematical Aspects of Natural and Formal Languages* (Gh. Păun, ed.), World Sci. Publ., Singapore, 1994, 275–284.

[37] H. A. Maurer, A. Salomaa, D. Wood, Pure grammars, *Inform. Control*, 44 (1980), 47–72.

[38] V. Mitrana, Contextual grammars: the strategy of minimal competence, in vol. *Mathematical Aspects of Natural and Formal Languages* (Gh. Păun, ed.), World Sci. Publ., Singapore, 1994, 319–331.

[39] V. Mitrana, Contextual insertion and deletion, in vol. *Mathematical Linguistics and Related Topics* (Gh. Păun, ed.), The Publ. House of the Romanian Academy, Bucharest, 1995, 271–278.

[40] X. M. Nguyen, On the *n*-contextual grammars, *Rev. Roum. Math. Pures Appl.*, 25 (1980), 1395–1406.

[41] X. M. Nguyen, N-contextual grammars with syntactical invariant choice, *Found. Control Engineering*, 5 (1980), 21–29.

[42] X. M. Nguyen, Some classes of *n*-contextual grammars, *Rev. Roum. Math. Pures Appl.*, 26 (1981), 1221–1233.

[43] X. M. Nguyen, *On Some Generalizations of Contextual Grammars*, PhD Thesis, Univ. of Bucharest, 1981.

[44] X. M. Nguyen, Gh. Păun, On the generative capacity of n-contextual grammars, *Bull. Math. Soc. Sci. Math. Roumanie*, 25(74) (1982), 345–354.

[45] M. Novotny, On a class of contextual grammars, *Cah. Ling. Th. Appl.*, 11 (1974), 313–315.

[46] M. Novotny, Each generalized contextual language is context-sensitive, *Rev. Roum. Math. Pures Appl.*, 21 (1976), 353–362.

[47] M. Novotny, On some variants of contextual grammars, *Rev. Roum. Math. Pures Appl.*, 21 (1976), 1053–1062.

[48] M. Novotny, Contextual grammars vs. context-free algebras, *Czech. Math. Journal*, 32 (1982), 529–547.

[49] M. Novotny, On some constructions of grammars for linear languages, *Intern. J. Computer Math.*, 17 (1985), 65–77.

[50] Gh. Păun, On contextual grammars, *Stud. Cerc. Matem.*, 26 (1974), 1111–1129.

[51] Gh. Păun, On the complexity of contextual grammars with choice, *Stud. Cerc. Matem.*, 27 (1975), 559–569.

[52] Gh. Păun, Contextual grammars with restriction in derivation, *Rev. Roum. Math. Pures Appl.*, 22 (1977), 1147–1154.

[53] Gh. Păun, On some classes of contextual grammars, *Bull. Math. Soc. Sci. Math. Roumanie*, 22(70) (1978), 183–189.

[54] Gh. Păun, An infinite hierarchy of contextual languages with choice, *Bull. Math. Soc. Sci. Math. Roumanie*, 22(70) (1978), 425–430.

[55] Gh. Păun, Operations with contextual grammars and languages, *Stud. Cerc. Matem.*, 30 (1978), 425–439.

[56] Gh. Păun, Two infinite hierarchies defined by branching grammars, *Kybernetika*, 14, 5 (1978), 397–407.

[57] Gh. Păun, A formal linguistic model of action systems, *Ars Semeiotica*, 2 (1979), 33–47.

[58] Gh. Păun, Marcus' contextual grammars and languages. A survey, *Rev. Roum. Math. Pures Appl.*, 24 (1979), 1467–1486.

[59] Gh. Păun, *Generative Mechanisms for Some Economic Processes*, The Technical Publ. House, Bucharest, 1980 (in Romanian).

[60] Gh. Păun, *Contextual Grammars*, The Publ. House of the Romanian Academy, Bucharest, 1982 (in Romanian).

[61] Gh. Păun, Modelling economic processes by means of formal grammars: a survey of results at the middle of 1981, *Acta Appl. Math.*, 1 (1983), 79–95.

[62] Gh. Păun, On semicontextual grammars, *Bull. Math. Soc. Sci. Math. Roumanie*, 28(76) (1984), 63–68,

[63] Gh. Păun, Two theorems about Galiukschov semicontextual languages, *Kybernetika*, 21 (1985), 360–365.

[64] Gh. Păun, Semiconditional contextual grammars, *Fundamenta Inform.*, 8 (1985), 151–161.

[65] Gh. Păun, On some open problems about Marcus contextual languages, *Intern. J. Computer Math.*, 17 (1985), 9–23.

[66] Gh. Păun, A new class of contextual grammars, *Rev. Roum. Ling.*, 33 (1988), 167–171.

[67] Gh. Păun, Further remarks on the syntactical complexity of Marcus contextual languages, *Ann. Univ. Buc., Ser. Matem.-Inform.*, 39 - 40 (1990–1991), 72–82.

[68] Gh. Păun, Marcus contextual grammars. After 25 years, *Bulletin EATCS*, 52 (1994), 263–273.

[69] Gh. Păun, X. M. Nguyen, On the inner contextual grammars, *Rev. Roum. Math. Pures Appl.*, 25 (1980), 641–651.

[70] Gh. Păun, G. Rozenberg, A. Salomaa, Contextual grammars: erasing, determinism, one-sided contexts, in *Developments in Language Theory* (G. Rozenberg, A. Salomaa, eds.), World Sci. Publ., Singapore, 1994, 370–388.

[71] Gh. Păun, G. Rozenberg, A. Salomaa, Grammars based on the shuffle operation, *J. Universal Computer Sci.*, 1 (1995), 67–81.

[72] Gh. Păun, G. Rozenberg, A. Salomaa, Contextual grammars: parallelism and blocking of derivation, *Fundamenta Inform.*, 41, 1-2 (1996), 83–108.

[73] Gh. Păun, G. Rozenberg, A. Salomaa, Marcus contextual grammars: modularity and leftmost derivation, in *Mathematical Aspects of Natural and Formal Languages* (Gh. Păun, ed.), World Sci. Publ., Singapore, 1994, 375–392.

[74] Gh. Păun, G. Rozenberg, A. Salomaa, Contextual grammars: deterministic derivations and growth functions, *Rev. Roum. Math. Pures Appl.*, 41, 1-2 (1996), 83–108.

[75] Gh. Păun, A. Salomaa, Thin and slender languages, *Discr. Appl. Math.*, 61 (1995), 257–270.

[76] M. Penttonen, One-sided and two-sided contexts in formal grammars, *Inform. Control*, 25 (1974), 371–392.

[77] A. Salomaa, *Formal Languages*, Academic Press, New York, London, 1973.

[78] C. C. Squier, Semicontextual grammars: an example, *Bull. Math. Soc. Sci. Math. Roumanie*, 32(80) (1988), 167–170.

[79] S. Vicolov, Two theorems about Marcus contextual languages, *Bull. Math. Soc. Sci. Math. Roumanie*, 35(83) (1991), 167–170.

[80] S. Vicolov-Dumitrescu, On the total parallelism in contextual grammars, in vol. *Mathematical Linguistics and Related Topics* (Gh. Păun, ed.), The Publ. House of the Romanian Academy, Bucharest, 1995, 350–360.

[81] S. H. von Solms, The characterization by automata of certain classes of languages in the context-sensitive area, *Inform. Control*, 27 (1975), 262–271.

Language Theory and Molecular Genetics: Generative Mechanisms Suggested by DNA Recombination

Thomas Head, Gheorghe Păun, and Dennis Pixton

1. Introduction

The stimulus for the development of the theory presented in this chapter is the string behaviors exhibited by the group of molecules often referred to collectively as the informational macromolecules. These include the molecules that play central roles in molecular biology and genetics : DNA, RNA, and the polypeptides. The discussion of the motivation for the generative systems is focused here on the recombinant behaviors of double stranded DNA molecules made possible by the presence of specific sets of enzymes. The function of this introduction is to provide richness to the reading of this chapter. It indicates the potential for productive interaction between the systems discussed and molecular biology, biotechnology, and DNA computing. However, the theory developed in this chapter can stand alone. It does not require a concern for its origins in molecular phenomena. Accordingly, only the most central points concerning the molecular connection are given here. An appendix to this chapter is included for those who wish to consider the molecular connection and possible applications in the biosciences. Here we present only enough details to motivate each term in the definition of the concept of a splicing rule that is given in the next section. The splicing rule concept is the foundation for the present chapter.

Here we discuss only the double stranded form of DNA. Such molecules may be regarded as strings over the alphabet consisting of the four compound "two-level" symbols

$$
\begin{array}{cccc}
A & C & G & T \\
T & G & C & A
\end{array}
$$

For illustrating a splicing rule most efficiently in molecular terms we choose two DNA molecules that have critical subwords recognized by specific restriction enzymes as explained below.

Molecule 1:

$$
\begin{array}{l}
AAAAAAA, GA, TC, AAAAAAA \\
TTTTTTT, CT, AG, TTTTTTT
\end{array}
$$

Molecule 2:

$$CCCCCCCCCCC, TGG, CCA, CCCCCCCCCC$$
$$GGGGGGGGGGGG, ACC, GGT, GGGGGGGGGG$$

The commas have been inserted above only for clarity of reference and do not represent physical aspects of the molecules discussed. Suppose the strings above represent molecules in an appropriate aqueous solution in which the restriction enzymes *Dpn*I and *Bal*I are present and a ligase enzyme is present. Then the following molecule may arise:

Molecule 3:

$$AAAAAAA, GA, CCA, CCCCCCCCCC$$
$$TTTTTTT, CT, GGT, GGGGGGGGGG$$

The formation of this molecule is made possible by the three enzymes in the following way: When *Dpn*I encounters a segment of a DNA molecule having the four letter subword

$$GATC$$
$$CTAG$$

it cuts (both strands of) the DNA molecule between the A and T. When *Bal*I encounters a segment of a DNA molecule having the six letter subword

$$TGGCCA$$
$$ACCGGT$$

it cuts (both strands of) the DNA molecule between the G and C. In the case of the first two molecules above, these two actions leave four DNA molecules in the solution:

$$AAAAAAGA \qquad\qquad TCAAAAAAA$$
$$TTTTTTCT \qquad\qquad AGTTTTTTT$$

$$CCCCCCCCCCCTGG \qquad CCACCCCCCCCCC$$
$$GGGGGGGGGGGGACC \qquad GGTGGGGGGGGGG$$

The ligase enzyme has the potential to bind together pairs of these molecules (those which have 5′ phosphates still attached, as will be the case where the fresh cuts have just been made). One of these possibilities allows the first and fourth molecules in the list to be bound producing the recombinant

Molecule 3 (again):

$$AAAAAAAGACCACCCCCCCCCC$$
$$TTTTTTTCTGGTGGGGGGGGGG$$

Of several possible molecules that can be formed here by the ligase, we have chosen to discuss this combination of the first and fourth specifically to allow comparison with the definition of a splicing rule presented in the next section. Note that this recombinant molecule cannot be recut by either of the two enzymes. This is a source of complexity in the class of molecules that may arise from such cut and paste activities.

We now summarize the discussion above abstractly. Let u_1 and u_2 represent the middle two subwords delimited by commas in Molecule 1 above and let u_3 and u_4 represent the middle two subwords delimited by commas in Molecule 2 above. The essence of the modeling capacity of the systems presented in this chapter may now be illustrated: If $Ba\mathit{l}I$, $Dpn\mathrm{I}$, and a ligase are present in an appropriate aqueous solution, then, from any two double stranded DNA molecules of the forms $x_1 u_1 u_2 x_2$ and $y_1 u_3 u_4 y_2$, the molecule $x_1 u_1 u_4 y_2$ may arise.

$Ba\mathit{l}I$ and $Dpn\mathrm{I}$ are examples of restriction endonucleases. Appendix A of [46] provides a reference list of such enzymes that includes the patterns in which they cut DNA. The number of known endonucleases is now in the thousands with the number of distinct cutting patterns numbered in the hundreds. The two endonucleases we have discussed above are said to produce "blunt ends" when they cut DNA. In the appendix to this chapter endonucleases are discussed that produce, not blunt ends, but what are called "sticky overhangs". The discussion there provides an integration of these cases into one generative scheme having splicing rules all of the form suggested in the paragraph above.

This chapter reports developments in formal language theory that provide new generative devices that allow close simulation of molecular recombination processes by corresponding generative processes acting on strings. An initial set of molecules is represented by an initial set of strings. The action of a set of endonucleases and a ligase is represented by a set of splicing rules acting on strings. The language of all possible strings that may be generated serves as a representation of the set of all possible molecules that may be generated by the biochemical recombination processes. Usually we consider that, in the situation modeled, the number of copies of some of the molecules in the initial set is not bounded. This condition allows for the generation of molecules of arbitrarily great length, corresponding to infinite languages of strings. In current experiments it is common to use 10^{12} identical DNA molecules of each given sequence.

Numerous molecular considerations have been ignored in this introduction and many are also ignored in the appendix. DNA molecules occur naturally in both linear and circular form and these forms interact through recombination. Circular and mixed recombination behaviors based on restriction enzymes and a ligase are covered here in Section 5. See [22] and [23] and the biological references listed in them for further discussions of the modeling aspects of the systems treated in this chapter. Splicing concepts have been considered in the context of algorithmic learning theory in [54] and in the context of investigations of proteins in [56] and RNA in [57].

2. Formal language theory prerequisites

We shall first introduce some notation and terminology which we shall use in this chapter. For basic definitions and results in formal language theory, the reader may consult known monographs, such as [21], [25], [49], as well as other chapters of the present handbook. In particular, we refer the reader to [24], [47] for L system theory, to [18] for an overview of AFL area, and to [8] for regulated rewriting.

In what follows, V is always a given alphabet. We denote: $V^* =$ the free monoid generated by V under the operation of concatenation, $\lambda =$ the empty string, $V^+ = V^* - \{\lambda\}$, $|x| =$ the length of the string $x \in V^*$, $Pref(x)$ $(Suf(x), Sub(x)) =$ the set of prefixes (suffixes, subwords, respectively) of $x \in V^*$, $L_2 \backslash L_1 = \{w \in V^* \mid xw \in L_1 \text{ for some } x \in L_2\}$ (the left quotient of L_1 with respect to L_2), $L_1/L_2 = \{w \in V^* \mid wx \in L_1 \text{ for some } x \in L_2\}$ (the right quotient of L_1 with respect to L_2), $\partial_x^l(L) = \{x\} \backslash L$ (the left derivative of $L \subseteq V^*$ with respect to $x \in V^*$), $\partial_x^r(L) = L/\{x\}$ (the right derivative of $L \subseteq V^*$ with respect to $x \in V^*$), $x \,\text{Ш}\, y = \{x_1 y_1 x_2 y_2 \ldots x_n y_n \mid n \geq 1, x = x_1 x_2 \ldots x_n, y = y_1 y_2 \ldots y_n, x_i, y_i \in V^*, 1 \leq i \leq n\}$ (the shuffle operation), $FIN, REG, LIN, CF, CS, RE =$ the families of finite, regular, linear, context-free, context-sensitive, and recursively enumerable languages, respectively, $coding =$ a morphism $h : V_1^* \longrightarrow V_2^*$ such that $h(a) \in V_2$ for each $a \in V_1$, $weak\ coding =$ a morphism $h : V_1^* \longrightarrow V_2^*$ such that $h(a) \in V_2 \cup \{\lambda\}$ for each $a \in V_1$, $restricted\ morphism$ (on a language $L \subseteq (V_1 \cup V_2)^*$) $=$ a morphism $h : (V_1 \cup V_2)^* \longrightarrow V_1^*$ such that $h(a) = \lambda$ for $a \in V_2$, $h(a) \neq \lambda$ for $a \in V_1$, and there is a constant k such that $|x| \leq k$ for each $x \in Sub(L) \cap V_2^*$.

Convention 2.1. In this chapter, two languages are considered equal if they differ by at most the empty string ($L_1 = L_2$ iff $L_1 - \{\lambda\} = L_2 - \{\lambda\}$).

3. The splicing operation

3.1 The uniterated case

Consider an alphabet V and two special symbols, $\#, \$$, not in V. A *splicing rule* (over V) is a string of the form

$$r = u_1 \# u_2 \$ u_3 \# u_4,$$

where $u_i \in V^*, 1 \leq i \leq 4$.

For such a rule r and strings $x, y, z \in V^*$ we write

$$(x, y) \vdash_r z \quad \text{iff} \quad x = x_1 u_1 u_2 x_2, \ y = y_1 u_3 u_4 y_2,$$
$$z = x_1 u_1 u_4 y_2, \text{ for some } x_1, x_2, y_1, y_2 \in V^*.$$

We say that z is obtained by *splicing* x, y, as indicated by the rule r; $u_1 u_2$ and $u_3 u_4$ are called the *sites* of the splicing. We call x the *first term* and y the *second term* of the splicing operation. When understood from the context, we omit the specification of r and we write $\vdash$ instead of $\vdash_r$.

The passing from x, y to z, via $\vdash_r$, can be represented as shown in Fig. 1 below.

An *H scheme* is a pair

$$\sigma = (V, R),$$

where V is an alphabet and $R \subseteq V^* \# V^* \$ V^* \# V^*$ is a set of splicing rules.

Note that R can be infinite, and that we can consider its place in the Chomsky hierarchy, or in another classification of languages. In general, if $R \in FA$, for a given family of languages, FA, then we say that the H scheme σ is of FA type (or an FA scheme).

For a given H scheme $\sigma = (V, R)$ and a language $L \subseteq V^*$, we define

$$\sigma(L) = \{z \in V^* \mid (x, y) \vdash_r z, \text{ for some } x, y \in L, r \in R\}.$$

Some remarks about these definitions are worth mentioning:

– We have defined $\sigma(L)$ only for $L \subseteq V^*$, but this restriction is not important: given $\sigma = (V, R)$, $R \subseteq V^* \# V^* \$ V^* \# V^*$, and $L \subseteq U^*$ such that $V \subset U$ and $L \not\subseteq V^*$, we can consider the H scheme $\sigma' = (U, R)$ and then $\sigma'(L)$ is defined and it equals the language of strings obtained by applying the rules in R to elements of L (on sites being strings over V).
– We have defined σ as a unary operation with languages. It might look more natural to consider the binary operation

$$\sigma(L_1, L_2) = \{z \in V^* \mid (x, y) \vdash_r z, x \in L_1, y \in L_2, r \in R\}.$$

However, the splicing operation seems to be more important from a practical point of view and more interesting from a mathematical point of view in the iterated form; the iteration is based on the unary variant of our operation.

Sometimes, given an H scheme $\sigma = (V, R)$ and two strings $x, y \in V^*$, we also denote

$$\sigma(x, y) = \{z \in V^* \mid (x, y) \vdash_r z, \text{ for some } r \in R\}.$$

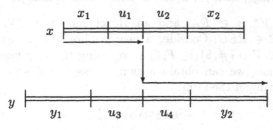

Fig. 1

Notice that $\sigma(x,y)$ is in general different from $\sigma(\{x,y\})$, which is the union of all the four sets $\sigma(x,x), \sigma(x,y), \sigma(y,x), \sigma(y,y)$.

Having defined a new operation with languages, a natural problem is to put it in relation with the known operations in formal language theory. This problem has two parts: to see which operations, together, can simulate the new one and, conversely, to see which operations are implied by the new one in combination with other given operations. We shall approach here these sub-problems in the general framework of abstract families of languages; as consequences, we shall find the closure properties of families in Chomsky hierarchy with respect to the splicing operation.

Given two families of languages, FA_1, FA_2, we define

$$S(FA_1, FA_2) = \{\sigma(L) \mid L \in FA_1, \text{ and } \sigma = (V, R) \text{ with } R \in FA_2\}.$$

Therefore, the family FA_1 is closed under splicing of FA_2 type if $S(FA_1, FA_2) \subseteq FA_1$. In general, the power of the FA_2 splicing is measured by investigating the families $S(FA_1, FA_2)$, for various FA_1.

We shall examine here the families $S(FA_1, FA_2)$ for FA_1, FA_2 in the set $\{FIN, REG, LIN, CF, CS, RE\}$. The results will be collected in a synthesis theorem, after the presentation of a series of general lemmas.

In all these lemmas, all families of languages are supposed to contain at least the finite languages (hence these families are equal or larger than FIN).

Lemma 3.1. *If $FA_1 \subseteq FA_1'$ and $FA_2 \subseteq FA_2'$, then $S(FA_1, FA_2) \subseteq S(FA_1', FA_2')$, for all FA_1, FA_1', FA_2, FA_2'.*

Proof. Obvious, by definitions.   □

Lemma 3.2. *If FA_1 is a family of languages which is closed under concatenation with symbols, then $FA_1 \subseteq S(FA_1, FA_2)$, for all FA_2.*

Proof. Take $L \subseteq V^*, L \in FA_1$, and $c \notin V$. Then $L_0 = Lc \in FA_1$. For the H scheme $\sigma = (V \cup \{c\}, \{\#c\$c\#\})$ we have $L = \sigma(L_0)$, hence $L \in S(FA_1, FA_2)$, for all FA_2.   □

Lemma 3.3. *If FA is a family of languages closed under concatenation and arbitrary gsm mappings, then FA is closed under REG splicing.*

Proof. Take $L \subseteq V^*, L \in FA$, and an H scheme, $\sigma = (V, R)$, $R \subseteq V^*\#V^*\$V^*\#V^*$, $R \in REG$. Consider a new symbol, $c \notin V$, and a finite automaton $A = (K, V \cup \{\#, \$\}, s_0, F, \delta)$ recognizing the language R. By a standard construction, we can obtain a gsm γ, associated with A, which transforms every string of the form

$$w = x_1 u_1 u_2 x_2 c y_1 u_3 u_4 y_2,$$

for $x_1, x_2, y_1, y_2 \in V^*$, $u_1\#u_2\$u_3\#u_4 \in R$, to the string

$$\gamma(w) = x_1 u_1 u_4 y_2.$$

Consequently, $\sigma(L) = \gamma(LcL)$. From the closure properties of FA, we get $\sigma(L) \in FA$. □

Lemma 3.4. *If FA is a family of languages closed under union, concatenation with symbols, and FIN splicing, then FA is closed under concatenation.*

Proof. Take two languages $L_1, L_2 \in FA, L_1, L_2 \subseteq V^*$, consider two new symbols, $c_1, c_2 \notin V$, and the H scheme

$$\sigma = (V \cup \{c_1, c_2\}, \{\#c_1\$c_2\#\}).$$

Obviously,

$$L_1 L_2 = \sigma(L_1 c_1 \cup c_2 L_2).$$

Hence, if FA has the mentioned properties, then $L_1 L_2 \in FA$. □

Lemma 3.5. *If FA is a family of languages closed under concatenation with symbols and FIN splicing, then FA is closed under the operations Pref and Suf.*

Proof. For $L \subseteq V^*, L \in FA$, consider a new symbol, $c \notin V$, and the H schemes

$$\sigma = (V \cup \{c\}, \{\#c\$c\#\} \cup \#V\$c\#),$$
$$\sigma' = (V \cup \{c\}, \{\#c\$c\#\} \cup \#c\$V\#).$$

We have

$$Pref(L) = \sigma(Lc), \quad Suf(L) = \sigma'(cL).$$

It is easy to see that using $r = \#c\$c\#$ we obtain $(xc, yc) \vdash_r x$ and that for each $z \in Pref(x) - \{x\}, x \in L$, there is a rule of the form $r = \#a\$c\#$ in σ such that $(xc, yc) \vdash_r z$. Similarly for σ'. □

Lemma 3.6. *FA is a family of languages which is closed under substitution with λ-free regular sets and arbitrary gsm mappings, then $S(REG, FA) \subseteq FA$.*

Proof. If FA has the above mentioned closure properties, then it is also closed under concatenation with symbols and intersection with regular sets (this follows directly from the closure under gsm mappings). Now, the closure under concatenation with symbols and under substitution with regular sets implies the closure under concatenation with regular sets. We shall use these properties below.

Take $L \subseteq V^*, L \in REG$, and an H scheme $\sigma = (V, R)$ with $R \in FA$. Consider the regular substitution $s : (V \cup \{\#, \$\})^* \longrightarrow 2^{(V \cup \{\#, \$\})^*}$ defined by

$$s(a) = \{a\}, \ a \in V, \ s(\#) = \{\#\}, \ s(\$) = V^*\$V^*,$$

and construct the language

$$L_1 = V^* s(R) V^*.$$

Consider also the language
$$L_2 = (L \, \text{Ш} \, \{\#\})\$(L \, \text{Ш} \, \{\#\}).$$
As $L_1 \in FA$ and $L_2 \in REG$, we have $L_1 \cap L_2 \in FA$. The strings in $L_1 \cap L_2$ are of the form
$$w = x_1 u_1 \# u_2 x_2 \$ y_1 u_3 \# u_4 y_2,$$
for $x_1 u_1 u_2 x_2 \in L, y_1 u_3 u_4 y_2 \in L$ and $u_1 \# u_2 \$ u_3 \# u_4 \in R$.

If γ is a gsm which erases the substring $\# z_2 \$ z_3 \#$ from strings of the form $z_1 \# z_2 \$ z_3 \# z_4, z_i \in V^*, 1 \leq i \leq 4$, then we get $\sigma(L) = \gamma(L_1 \cap L_2)$, hence $\sigma(L) \in FA$. □

Lemma 3.7. *If FA is a family of languages which is closed under concatenation with symbols, then for all $L_1, L_2 \in FA$ we have $L_1/L_2 \in S(FA, FA)$.*

Proof. Take $L_1, L_2 \subseteq V^*, L_1, L_2 \in FA$, and $c \notin V$. For the H scheme
$$\sigma = (V \cup \{c\}, \# L_2 c \$ c \#),$$
we obtain
$$L_1/L_2 = \sigma(L_1 c).$$
Indeed, the only possible splicing of strings in $L_1 c$ is of the form
$$(x_1 x_2 c, yc) \vdash_r x_1, \text{ for } x_1 x_2 \in L_1, x_2 \in L_2, y \in L_1,$$
where $r = \# x_2 c \$ c \#$. □

Lemma 3.8. *If FA is a family of languages closed under concatenation with symbols, then for each $L \in FA, L \subseteq V^*$, and $c \notin V$ we have $cL \in S(REG, FA)$.*

Proof. For L, c as above, consider the H scheme
$$\sigma = (V \cup \{c, c'\}, cL \# c' \$ c' \#),$$
where c' is one further new symbol. Clearly, this is a scheme of FA type. Then,
$$cL = \sigma(cV^* c'),$$
because the only splicings are of the form $(cxc', cyc') \vdash_r cx$, for $r = cx \# c' \$ c' \#, x \in L, y \in V^*$ □

Lemma 3.9. *If FA is a family of languages closed under concatenation with symbols and shuffle with symbols, then for each $L \in FA, L \subseteq V^*$, and $c \notin V$, we have $\{c\}Pref(L) \in S(REG, FA)$.*

Proof. For L, c as above, consider the H scheme of FA type
$$\sigma = (V \cup \{c, c'\}, c(L \, \text{Ш} \, \{\#\})c' \$ c' \#),$$
where c' is one further new symbol. We have
$$\{c\}Pref(L) = \sigma(cV^* c),$$
because the only possible splicings are of the form $(cx_1 x_2 c', cyc') \vdash_r cx_1$, for $r = cx_1 \# x_2 c' \$ c' \#, x_1 x_2 \in L, y \in V^*$. □

We now synthesize the consequences of the previous lemmas for the families in the Chomsky hierarchy.

Theorem 3.1. *The relations in Table 1 hold, where at the intersection of the row marked with FA_1 with the column marked with FA_2 there appear either the family $S(FA_1, FA_2)$, or two families FA_3, FA_4 such that $FA_3 \subset S(FA_1, FA_2) \subset FA_4$. These families FA_3, FA_4 are the best estimations among the six families considered here.*

Table 1

	FIN	*REG*	*LIN*	*CF*	*CS*	*RE*
FIN	*FIN*	*FIN*	*FIN*	*FIN*	*FIN*	*FIN*
REG	*REG*	*REG*	*REG, LIN*	*REG, CF*	*REG, RE*	*REG, RE*
LIN	*LIN, CF*	*LIN, CF*	*RE*	*RE*	*RE*	*RE*
CF	*CF*	*CF*	*RE*	*RE*	*RE*	*RE*
CS	*RE*	*RE*	*RE*	*RE*	*RE*	*RE*
RE	*RE*	*RE*	*RE*	*RE*	*RE*	*RE*

Proof. Clearly, $\sigma(L) \in FIN$ for all $L \in FIN$, whatever σ is. Together with Lemma 3.2, we have $S(FIN, FA) = FIN$ for all families FA.

Lemma 3.3 shows that $S(REG, REG) \subseteq REG$. Together with Lemma 3.2 we have $S(REG, FIN) = S(REG, REG) = REG$.

From Lemma 3.4 we get $S(LIN, FIN) - LIN \neq \emptyset$. From Lemma 3.3 we have $S(CF, REG) \subseteq CF$. Therefore, $LIN \subset S(LIN, FA) \subseteq CF = S(CF, FA)$ for $FA \in \{FIN, REG\}$.

Also the inclusions $S(LIN, FA) \subset CF, FA \in \{FIN, REG\}$, are proper. In order to see this, let us examine again the proof of Lemma 3.3. If $L \subseteq V^*, L \in LIN$, and $\sigma = (V, R)$ is an H scheme of REG type, then $\sigma(L) = \gamma(LcL)$, where $c \notin V$ and γ is a gsm. The language LcL has a context-free index less than or equal to 2.

(The *index* of a language L' is the maximum number of nonterminal occurrences in the sentential forms of derivations of strings in L', taking the most economical grammar from this point of view. Formally, if $G = (N, T, S, P)$ is a grammar and $D : S = w_1 \Longrightarrow w_2 \Longrightarrow \ldots \Longrightarrow w_n = w$ is a terminal derivation in G, then its index is $ind(D, G) = \max\{|w_i|_N \mid 1 \leq i \leq n\}$, the index of $w \in L(G)$ is $ind(w, G) = \min\{ind(D, G) \mid D : S \Longrightarrow^* w\}$, and the index of G is $ind(G) = \sup\{ind(w, G) \mid x \in L(G)\}$. Then, for $L' \in CF$, $ind_{CF}(L') = \inf\{ind(G) \mid L = L(G), G \text{ a context-free grammar}\}$. See details in [1], [19].)

The family of context-free languages of finite index is a full AFL [20], hence it is closed under arbitrary gsm mappings. Consequently, for each $L \in S(LIN, REG)$ we have $ind_{CF}(L) < \infty$. Because there are context-free languages of infinite index [48], it follows that $CF - S(LIN, REG) \neq \emptyset$.

For every language $L \in RE, L \subseteq V^*$, there are $c_1, c_2 \notin V$ and a language $L' \subseteq Lc_1c_2^*$ such that $L' \in CS$ and for each $w \in L$ there is $i \geq 0$ such that $wc_1c_2^i \in L'$ (see Theorem 9.9 in [49]). Take one further new symbol, c_3. The language $L'c_3$ is still in CS. For the H scheme

$$\sigma = (V \cup \{c_1, c_2, c_3\}, \{\#c_1\$c_3\#\}),$$

we have

$$\sigma(L'c_3) = \{w \mid wc_1c_2^ic_3 \in L'c_3 \text{ for some } i \geq 0\} = L.$$

Consequently, $RE \subseteq S(CS, FIN)$. As $S(RE, RE) \subseteq RE$ (by Turing-Church thesis), we get $S(CS, FA) = S(RE, FA) = RE$ for all FA.

According to [28], every language $L \in RE$ can be written as $L = L_1/L_2$, for $L_1, L_2 \in LIN$. By Lemma 3.7, each language L_1/L_2 with linear L_1, L_2 is in $S(LIN, LIN)$. Consequently, $S(LIN, FA) = S(CF, FA) = RE$, too, for all $FA \in \{LIN, CF, CS, RE\}$.

From Lemma 3.6 we have $S(REG, FA) \subseteq FA$ for $FA \in \{LIN, CF, RE\}$. All these inclusions are proper. More exactly, there are linear languages not in $S(REG, RE)$. Such an example is $L = \{a^nb^n \mid n \geq 1\}$.

Assume that $L = \sigma(L_0)$ for some $L_0 \in REG, L_0 \subseteq V^*$, and $\sigma = (V, R)$. Take a finite automaton for $L_0, A = (K, V, s_0, F, \delta)$, let $m = card(K)$, and consider the string $w = a^{m+1}b^{m+1}$ in L. Let $x, y \in L_0$ and $r \in R$ be such that $(x, y) \vdash_r w$, $x = x_1u_1u_2x_2$, $y = y_1u_3u_4y_2$, $w = x_1u_1u_4y_2$, for $r = u_1\#u_2\$u_3\#u_4$. We have either $x_1u_1 = a^{m+1}z$ or $u_4y_2 = z'b^{m+1}$, for some $z, z' \in \{a, b\}^*$. Assume that we have the first case; the second one is similar. Consequently, $x = a^{m+1}zu_2x_2$. When parsing the prefix a^{m+1}, the automaton A uses twice a state in K; the corresponding cycle can be iterated, hence L_0 contains strings of the form $x' = a^{m+1+ti}zu_2x_2$ for $t > 0$ and arbitrary $i \geq 0$. For such a string x' with $i \geq 1$ we have

$$(x', y) \vdash_r a^{m+1+ti}zu_2x_2 = a^{m+1+ti}b^{m+1}.$$

This string is not in L, a contradiction. The argument does not depend on the type of R. (Compare this with Lemma 3.8: $L \notin S(REG, RE)$, but $cL \in S(REG, LIN)$.)

According to Lemma 3.8, $S(REG, LIN) - REG \neq \emptyset$ and $S(REG, CF) - LIN \neq \emptyset$. From Lemma 3.9 we have $S(REG, CS) - CS \neq \emptyset$. (Consequently, $S(REG, CF)$ is incomparable with LIN and $S(REG, CS), S(REG, RE)$ are incomparable with LIN, CF, CS.)

All the assertions represented in the table are proved. □

Some remarks about the results in Table 1 are worth mentioning:

- All families $S(FA_1, FA_2)$ characterize families in Chomsky hierarchy, with the exceptions of $S(REG, FA_2)$, with $FA_2 \in \{LIN, CF, CS, RE\}$, and $S(LIN, FA_2)$ with $FA_2 \in \{FIN, REG\}$, which are strictly intermediate between families in Chomsky hierarchy. These six intermediate families need further investigations concerning their properties (closure under operations and decidability, for instance).

- A series of new characterizations of the family RE are obtained, starting, somewhat surprisingly, from "simple" pairs (FA_1, FA_2); especially interesting is the case (LIN, LIN), in view of the fact that it seems that the actual language of DNA sequences is not regular, or even context-free [3], [52]. Then, according to the previous results, it can be nothing else but recursively enumerable, of the highest complexity (in the Chomsky hierarchy).

We close this section by examining a possible hierarchy between LIN and CF, defined by subfamilies of $S(LIN, FIN)$.

For an H scheme $\sigma = (V, R)$ with a finite R, we define the *radius* of σ as

$$rad(\sigma) = \max\{|x| \mid x = u_i, 1 \le i \le 4, \text{ for some } u_1 \# u_2 \$ u_3 \# u_4 \in R\}.$$

Then, for $p \ge 1$, we denote by $S(FA, p)$ the family of languages $\sigma(L)$, for $L \in FA$ and σ an H scheme of radius less than or equal to p.

Note that in the proofs of Lemma 3.2 (and 3.4, 3.5), as well as in the proof of $RE \subseteq S(CS, FIN)$ in Theorem 3.1, the schemes used are of radius 1. Hence for $FA \in \{FIN, REG, CF, CS, RE\}$ we have

$$S(FA, 1) = S(FA, p), \text{ for all } p \ge 1,$$

that is, these hierarchies collapse. The same is true for $FA = LIN$. This follows from the next lemma.

Lemma 3.10. *If FA is a family of languages closed under λ-free gsm mappings, then $S(FA, FIN) \subseteq S(FA, 1)$.*

Proof. Take an H scheme $\sigma = (V, R)$ with finite R. Assume the rules in R are labelled in an one-to-one manner, $R = \{r_1, \ldots, r_s\}$, $r_i = u_{i,1} \# u_{i,2} \$ u_{i,3} \# u_{i,4}$, $1 \le i \le s$. It is easy to construct a gsm γ associated with R which transforms each string $w = x_1 u_{i,1} u_{i,2} x_2$ in $\gamma(w) = x_1 u_{i,1} c_i u_{i,2} x_2$ and each string $w = y_1 u_{i,3} u_{i,4} y_2$ in $\gamma(w) = y_1 u_{i,3} c'_i u_{i,4} y_2$, for $x_1, x_2, y_1, y_2 \in V^*$ and r_i as above; c_i, c'_i are new symbols, associated with r_i. Consider now the H scheme $\sigma' = (V \cup \{c_i, c'_i \mid 1 \le i \le s\}, \{\# c_i \$ c'_i \# \mid 1 \le i \le s\})$. We have $\gamma(L) \in FA$ for each language $L \subseteq V^*, L \in FA$, and we obviously have $\sigma(L) = \sigma'(\gamma(L))$. (The end of the phrase is not modified.) As $rad(\sigma') = 1$, the proof is complete. □

Theorem 3.2. $LIN \subset S(LIN, p) = S(LIN, FIN), p \ge 1$.

Proof. The inclusions $S(LIN, p) \subseteq S(LIN, p+1) \subseteq S(LIN, FIN), p \ge 1$, follow by the definitions. From Lemma 3.10 we also get $S(LIN, FIN) \subseteq S(LIN, 1)$. The relation $LIN \subset S(LIN, FIN)$ is known from Theorem 3.1. □

3.2 The iterated case

Having an H scheme $\sigma = (V, R)$, as considered in the previous section ($R \subseteq V^* \# V^* \$ V^* \# V^*$), we can also apply σ to a language $L \subseteq V^*$ iteratively. Formally, we define

$$\sigma^0(L) = L,$$
$$\sigma^{i+1}(L) = \sigma^i(L) \cup \sigma(\sigma^i(L)), \ i \geq 0,$$

and

$$\sigma^*(L) = \bigcup_{i \geq 0} \sigma^i(L).$$

In other words, $\sigma^*(L)$ is the smallest language L' which contains L, and is closed under the splicing with respect to $\sigma, \sigma(L') \subseteq L'$.

For two families of languages, FA_1, FA_2, we define

$$H(FA_1, FA_2) = \{\sigma^*(L) \mid L \in FA_1, \text{ and } \sigma = (V, R) \text{ with } R \in FA_2\}.$$

In this way, we have a new operation with languages, the iterated splicing, which must be put in relation with known operations in language theory, and we have new families of languages, $H(FA_1, FA_2)$ corresponding to $S(FA_1, FA_2)$ in the previous section. In the same way, we can consider the hierarchies on the radius of finite H schemes, that is the families $H(FA, p)$ of languages $\sigma^*(L)$, for $L \in FA$ and σ an H scheme with $rad(\sigma) \leq p$.

Lemma 3.11.
(i) If $FA_1 \subseteq FA_1'$ and $FA_2 \subseteq FA_2'$, then $H(FA_1, FA_2) \subseteq H(FA_1', FA_2')$, for all families FA_1, FA_1', FA_2, FA_2'.
(ii) $H(FA, p) \subseteq H(FA, q)$, for all FA and $p \leq q$.

Proof. Obvious from definitions.   □

Lemma 3.12. $FA \subseteq H(FA, 1)$, for all families FA.

Proof. Given $L \subseteq V^*, L \in FA$, consider a symbol $c \notin V$ and the H scheme

$$\sigma = (V \cup \{c\}, \{\#c\$c\#\}).$$

We clearly have $\sigma^i(L) = L$ for all $i \geq 0$, hence $\sigma^*(L) = L$.   □

Lemma 3.13. If FA_1, FA_2, FA_3 are families of languages such that both FA_1 and FA_2 are closed under shuffle with symbols and both FA_2 and FA_3 are closed under intersection with regular sets, then $H(FA_1, FA_2) \subseteq FA_3$ implies $S(FA_1, FA_2) \subseteq FA_3$.

Proof. Take a language $L \subseteq V^*, L \in FA_1$, and an H scheme $\sigma_1 = (V, R)$ with $R \in FA_2$. For $c \notin V$, consider the language

$$L' = L \amalg \{c\}$$

and the H scheme $\sigma_2 = (V \cup \{c\}, R')$ with

$$R' = ((R \amalg \{c\}) \amalg \{c\}) \cap V^* \# c V^* \$ V^* c \# V^*.$$

From the properties of FA_1, FA_2 we have $L' \in FA_1, R' \in FA_2$. Moreover,

$$\sigma_1(L) = \sigma_2^*(L') \cap V^*.$$

Indeed, $\sigma_2^i(L') = \sigma_2(L') \cup L'$ for all $i \geq 1$ (any splicing removes the symbol c from the strings of L', hence no further splicing is possible having as one of its terms the obtained string). Therefore, if $\sigma_2^*(L') \in FA_3$, then $\sigma_1(L) \in FA_3$, too.   □

Corollary 3.1. *In the conditions of Lemma 3.13, each language* $L \in S(FA_1, FA_2)$ *can be written as* $L = L' \cap V^*$, *for* $L' \in H(FA_1, FA_2)$.

Lemma 3.14. *If* FA *is a full AFL then* $H(FA, FIN) \subseteq FA$.

Remark 3.1. We shall only use three properties of the full AFL FA. The first is that FA contains all regular languages, and the second is closure under right and left quotient by regular languages. The third is closure under *substitution into regular languages.* That is, suppose a language $L(a)$ is associated to each symbol a in some alphabet A. To every word $w = a_1 a_2 \ldots a_n$ in A^* we associate the concatenation $L(w) = L(a_1) L(a_2) \ldots L(a_n)$, and to a language R in A^* we associate the union $L(R)$ of all $L(w)$ for $w \in R$. If all the languages $L(a)$ for $a \in A$ are in a full AFL FA and R is regular then $L(R)$ is in FA. See Theorem 11.5 in [25].

Proof. of Lemma 3.14 Let $L \subseteq V^*$ be a language in FA, and let $\sigma = (V, R)$ be a finite H scheme. The proof that $\sigma^*(L)$ is in FA has three main steps, as follows. First we shall construct a language J consisting of strings which record entire splicing derivation trees. This language is not in general in the family FA. Second, we use the language J to construct a language K in the family FA. Third, we verify that $K = \sigma^*(L)$.

The auxiliary language J uses an alphabet $\bar{V}$ consisting of the symbols of V together with 4 additional symbols for each rule $r \in R$. These are *round brackets*, written $(\!(_r,)\!)_r$, and *square brackets*, written $[\![_r,]\!]_r$. Our first requirement on the language J is that these brackets are properly nested, in the following sense. We say a string in $\bar{V}^*$ is a *nest* iff it has the form $(\!(_r x]\!]_r$ or $[\![_r x)\!)_r$ for some $r \in R$ where, recursively, every bracket in x lies in a nest in x. A nest is called a left or right nest depending on whether it begins or ends with a round bracket. We say a string w is *properly nested* iff every bracket in w lies in a nest in w, and we write N for the set of properly nested strings.

This language is context-free but not regular. We define a function E from N to V^* which operates on a string by erasing all nests. This function satisfies $E(zw) = E(z)E(w)$ if z and w are both in N.

In the rest of the proof the following notation will be useful. For a rule $r = u_1\#u_2\$u_3\#u_4$ we write α_r for the left site u_1u_2 and γ_r for the right site u_3u_4. Also, we write β_r for the string u_1u_4.

Now we can describe an analogue of the splicing operation on the strings of N. Suppose that r is a rule in R and x, y are strings in N. Suppose that these strings factor as x_1vx_2 and y_1wy_2 with each factor in N, and suppose that $E(v) = \alpha_r$, $E(w) = \gamma_r$. Then we say the string $z = x_1(_rvx_2]_r\beta_r[_ry_1w)_ry_2$ is obtained from x and y by *justified splicing* using the rule r. Note that z is in N and that $E(z) = E(x_1)\beta_rE(y_2)$ is the result of splicing $E(x) = E(x_1)\alpha_rE(x_2)$ and $E(y) = E(y_1)\gamma_rE(y_2)$ using r. Conversely, suppose x and y are in N and $E(x)$ and $E(y)$ can be spliced using the rule r. Then $E(x) = x_1'\alpha_rx_2'$, $E(y) = y_1'\gamma_ry_2'$, and the result of the splicing is $z' = x_1'\beta_ry_2'$. Since E operates by erasing nests, we can factor x as x_1vx_2 with $E(x_1) = x_1'$, $E(v) = \alpha_r$, $E(x_2) = x_2'$ and with each factor in N. We can similarly factor $y = y_1wy_2$, and then $z' = E(z)$ where z is the result of justified splicing using x, y, and r.

We define J as the smallest language in $\bar{V}^*$ which contains L and which is closed under justified splicing. It is immediate from the remarks above that $E(J) = \sigma^*(L)$.

If w is a string in J then its image $E(w)$ may be described as the concatenation of the substrings which remain when the maximal nests in w are deleted. We shall define K by identifying the possible factors in such a concatenation and the permissible orders in which such factors may be joined. Before giving the details we record a few simple facts about the language J which will be useful. For simplicity we formulate these and later results using only left nests; in each case there is a "dual" statement and proof for right nests.

Fact 3.1. *If $w = x(_r y]_r z$ is in J with $(_r y]_r$ a maximal nest then xy is in J and α_r is a prefix of $E(y)$.*

Proof. This is an induction on the length of w. Since w contains a nest it was obtained by justified splicing, $w = p_1(_s p_2]_s\beta_s[_s q_1)_s q_2$, where p_1p_2 and q_1q_2 are in J, each p_i, q_i is in N, α_s is a prefix of $E(p_2)$ and γ_s is a suffix of $E(q_1)$. Comparing possible locations of maximal nests in w, we reduce to three cases.

First, we may have $(_r y]_r = (_s p_2]_s$, in which case $xy = p_1p_2$ is in J and $\alpha_r = \alpha_s$ is a prefix of $E(y) = E(p_2)$. Second, $(_r y]_r$ may be a substring of p_1, in which case it is a maximal nest in $p_1p_2 = x(_r y]_r z'$ and we finish by induction. Third, $(_r y]_r$ may be a substring of q_2. In this case $x = p_1(_s p_2]_s\beta_s[_s q_1)_s x'$ and $(_r y]_r$ is a maximal nest in $q_1q_2 = q_1x'(_r y]_r z$ so, by induction, $q_1x'y$ is in J and α_r is a prefix of $E(y)$. But then a justified splicing operation on p_1p_2 and $q_1x'y$ using the rule s produces $xy = p_1(_s p_2]_s\beta_s[_s q_1)_s x'y$ in J. $\square$

Fact 3.2. *If $w = x(_r y]_r z$ and $w' = x'(_r y']_r z'$ are in J with the indicated nests maximal then $x(_r y]_r z'$ is in J.*

Proof. We induct on the length of w', so suppose w' was obtained by justified splicing, $w' = p_1(_sp_2]_s\beta_s[_sq_1)_sq_2$, as in the preceding proof. Again there are three cases.

If $(_ry']_r = (_sp_2]_s$ then $r = s$ and $z' = \beta_s[_sq_1)_sq_2$. Hence $\alpha_s = \alpha_r$ is a prefix of $E(y)$ and xy is in J by Fact 1. Thus a justified splicing operation on xy and q_1q_2 demonstrates that $x(_ry]_rz' = x(_sy]_s\beta_s[_sq_1)_sq_2$ is in J.

In the second case $(_ry']_r$ is a substring of q_2, so $q_1q_2 = q_1x''(_ry']_rz'$ and $x(_ry]_rz'$ is in J by an application of the inductive hypothesis to q_1q_2.

In the last case $(_ry']_r$ is a substring of p_1. Hence $p_1p_2 = x'(_ry']_rz''p_2$. Again by induction $x(_ry]_rz''p_2$ is in J. Now justified splicing with q_1q_2 demonstrates that $x(_ry]_rz' = x(_ry]_rz''(_sp_2]_s\beta_s[_sq_1)_sq_2$ is in J.   □

Now suppose we have a string w in J. We start by analyzing the first maximal nest in w and the substring before it. It is clear that this first nest must be a left nest, so $w = y(_rz]_rv$. Then Fact 1 implies that yz is in J so, if z contains no brackets, yz is in L and α_r is a prefix of z. On the other hand, if z contains brackets then the first maximal nest in yz must again be a left nest, so $yz = yz_1(_sz']_sv'$ with z_1 in V^* and we can continue inductively. We eventually arrive at the following form: yz is in L, where $z = z_1z_2\ldots z_n$, corresponding to rules $r_1,\ldots,r_n$. Each z_k, for $k < n$, lies between two consecutive round brackets in w, $(_r$ and $(_s$ with $r = r_k$ and $s = r_{k+1}$, and α_{r_n} is a prefix of z_n. It may be that α_{r_m} is a prefix of z_m for some $m < n$. In this case we choose the first such m and replace n with m and z_m with the concatenation $z_mz_{m+1}\ldots z_n$. Thus we have the representation $z = z_1z_2\ldots z_n$ as above with the further condition that z_k is shorter than α_{r_k} for all $k < n$. The point of this exercise is that all such substrings y lie in the right quotient of L by the set of all such suffixes z. A similar description will apply to all the substrings of w between the maximal nests, and our next task is to give a more systematic account of the suffixes and prefixes which occur this way.

In the rest of the proof we use the following conventions to regularize the notation. First, we delete from R any rules that are not actually used in the construction of J, so we may assume

Fact 3.3. *If $r \in R$ then both α_r and γ_r occur as substrings of strings in $E(J)$.*

Second, we consider the string \$, which cannot be a rule, as a "dummy" rule. We write $L_\$ = L$, and we write $L_r = \{\beta_r\}$ for each rule r in R.

Suppose $r \in R \cup \{\$\}$, and s and t are in R. We define $F_r(s,t)$ as the set of strings y in V^* which are shorter than α_s and for which some string in J contains $(_sy(_t$ as a substring. If $r = \$$ we define $F_r(s,\$) = \alpha_sV^*$. If $r \in R$ we define $F_r(s,\$)$ as the union of α_sV^* and the set of strings y in V^* which are shorter than α_s and for which some string in J contains $(_sy[_r$ as a substring. If $t \neq \$$ then $F_r(s,t)$ is a finite set, and $F_r(s,\$)$ is the union of a regular set and a finite set, so is regular. Now we define the "suffix sets" $S_r(s)$, for

$r \in R \cup \{\$\}$ and $s \in R$, as the union of all concatenations $F_1 F_2 \ldots F_n$ where $n > 0$ and each F_k has the form $F_r(s_k, t_k)$, satisfying the following conditions:

$$s_1 = s, \quad t_n = \$, \quad \text{and for each } k < n, \quad t_k = s_{k+1}.$$

Consider temporarily a finite alphabet $\{f_{st}\}$ indexed by $s \in R$ and $t \in R \cup \{\$\}$. Then the language over this alphabet consisting of all strings $f_{s_1 t_1} f_{s_2 t_2} \ldots f_{s_n t_n}$ which satisfy the conditions displayed above is obviously regular. Since the family of regular languages is closed under substitution into regular languages we conclude that the sets $S_r(s)$ are regular, for $s \in R$. We complete the definition by setting $S_r(\$) = \emptyset$.

In an entirely dual fashion we define "prefix sets" $P_r(s)$ for r, s in $R \cup \{\$\}$, and observe that they are regular.

Now consider again a string w in J. We claim that the strings which occur between maximal nests in w are in the languages $L_r(s, t)$, which are defined as the double quotients $P_r(s) \backslash L_r / S_r(s)$. To see this we consider the various types of such substrings. If w contains no brackets then w itself is in $L = L_\$ = L_\$(\$, \$)$ (remember that each $S_r(\$)$ and $P_r(\$)$ is empty). So suppose that w contains at least one maximal nest. The substring of w before the first maximal nest was analyzed above; it lies in $L_\$ / S_\$(t) = L_\$(\$, t)$ where the first maximal nest in w begins with $(\!(_t$. Dually, the substring following the last maximal nest is in $L_\$(s, \$)$ where the last maximal nest ends with $)\!)_s$. So we consider an internal substring bounded by two maximal nests. We have the following possible configurations.

- $(\!(_s x]\!]_s y[\![_t z)\!)_t$: A simple induction shows that $s = t$ and $y = \beta_s$, so y is in $L_s(\$, \$) = L_s$.
- $(\!(_r x]\!]_r y(\!(_t z]\!]_t$: This is similar to the analysis of $L_\$(\$, t)$, using Fact 1 applied to the nest $(\!(_t z]\!]_t$ to start the induction. We continue "peeling off" such right nests as long as possible, until we are left with a substring of the form $(\!(_r x]\!]_r \beta_r [\![_r z')\!)_r$. Hence y is in $L_r(\$, t)$.
- $[\![_s x)\!)_s y[\![_r z)\!)_r$: This is dual to the preceding case; y is in $L_r(s, \$)$.
- $[\![_s x)\!)_s y(\!(_t z]\!]_t$: This is a combination of the preceding two cases, applying Fact 1 to both the left and right nests. The conclusion is that y is in $L_r(s, t)$, where r is determined by the last square bracket in $[\![_s x)\!)_s$: If this bracket is of the form $[\![_q$ then $r = \$$, and if it is $]\!]_q$ then $r = q$. Of course, r is equivalently determined by the first square bracket in $(\!(_t z]\!]_t$.

Thus each string in $E(w) \in E(J) = \sigma^*(L)$ lies in a concatenation $Y_1 Y_2 \ldots Y_n$ where $n > 0$ and each Y_k has the form $L_{r_k}(s_k, t_k)$. Moreover, we claim that the following conditions are satisfied:

Fact 3.4. $r_1 = s_1 = \$$ and $r_n = t_n = \$$. Also, for $k < n$, either $\$ \neq t_k = r_{k+1}$ and $s_{k+1} = \$$ or $t_k = \$$ and $r_k = s_{k+1} \neq \$$.

Proof. The first condition was noted above. For the second it is enough to see that a maximal nest in w separates any two consecutive segments, say

$y_k \in Y_k$ and $y_{k+1} \in Y_{k+1}$. If this nest is a left nest $(\!(_t z]\!]_t$ then, from the discussion above, we must have $y_k \in L.(\cdot, t)$ and $y_{k+1} \in L_t(\$, \cdot)$, and of course $t \neq \$$. This is the first alternative, and the second corresponds to a right nest between y_k and y_{k+1}. $\square$

We define the language K as the union of all concatenations $Y_1 Y_2 \ldots Y_n$ satisfying the conditions above. Each language $L_r(s, t)$ is in the class FA since it is either finite or the quotient of $L \in FA$ by regular languages, and K is obtained by substituting these members of FA into a regular language (just as in the argument that $S_r(s)$ is regular). Hence K is in FA, and we have already argued that $E(J) = \sigma^*(L) \subseteq K$.

To complete the proof we must show the opposite inclusion, $K \subseteq \sigma^*(L)$. That is, given $y \in K$ we must construct $w \in J$ so that $E(w) = y$. Most of the work is contained in the following:

Fact 3.5. *If $y \in L_r(s, t)$ then there are strings x, z in N so that $xyz \in J$, satisfying:*

> *If $s \neq \$$ then x ends with $)\!)_s$.*
>
> *If $s = \$$ and $r \in R$ then x ends with $]\!]_r$.*
>
> *If $s = \$$ and $r = \$$ then $x = \lambda$.*
>
> *If $t \neq \$$ then z starts with $(\!(_t$.*
>
> *If $t = \$$ and $r \in R$ then z starts with $[\!(_r$.*
>
> *If $t = \$$ and $r = \$$ then $z = \lambda$.*

Proof. We will need the following simple observation:

Fact 3.6. *If $x \in V^*$, b and b' are brackets, and bxb' is a substring of some $w \in J$ then bxb' is a substring of some $w' \in J$ such that either b or b' is an end bracket in a maximal nest.*

Proof. This follows just by choosing w' to contain bxb' and have minimal length. If neither b nor b' is an end bracket of a maximal nest then we can apply Fact 1 to obtain a shorter string in J containing bxb'. $\square$

Now we consider various cases. If $r = s = t = \$$ then $y \in L_{\$}(\$, \$) = L \subseteq J$ and $x = z = \lambda$. If $r \in R$ and $s = t = \$$ then $y = \beta_r$. According to Fact 3 we can find two strings in J for which justified splicing using r is possible, yielding $x_1(\!(_r x_2]\!]_r y[\!(_r z_1)\!)_r z_2$, and we set $x = x_1(\!(_r x_2]\!]_r$ and $z = [\!(_r z_1)\!)_r z_2$.

Now consider $s = \$$, $t \in R$. Then there is $q \in S_r(t)$ so that $yq \in L_r$. We shall induct on the the number of factors of q in the representation $q = q_1 q_2 \ldots q_n \in F_r(t, t_2) F_r(t_2, t_3) \ldots F_r(t_n, \$)$, with $t_1 = t$ variable. By Fact 3 there is a string $w_0 = x_0 y_0 z_0$ in J with all factors in N and $E(y_0) = \gamma_t$. We shall use this in the following. The obvious nests in the following construction are maximal, and we will generally not mention this explicitly.

We first consider the case $n = 1$, so $q \in F_r(t, \$)$. Then $yq \in L_r$, and by the $L_r(\$, \$)$ case we can find x and z' so that $w = xyqz'$ is in J, $x = \lambda$ if $r = \$$, and otherwise x ends with $]_r$. If α_t is a prefix of q then we obtain $y(_t qz']_t z_0$ in J by justified splicing applied to w and w_0, and we set $z = (_t qz']_t z_0$. The alternative is that $r \neq \$$ and q is shorter than α_t. Then z' starts with $[_r$ so we can write $w = xyq[_r z_1')_r z_2'$. By the definition of $F_r(\$, t)$ some string of J contains $(_t q[_r$ as a substring, and by Fact 6 we may find such a string $x'(_t q[_r z_1)_r z_2]_t z_3$ with the nest starting at $(_t$ maximal. An application of Fact 1 shows that $w' = x' q[_r z_1)_r z_2$ is in J and α_t is a prefix of $E(q[_r z_1)_r z_2)$. Applying Fact 2 to w and w' we see that $w'' = xyq[_r z_1)_r z_2$ is in J. Since α_t is a prefix of $E(q[_r z_1)_r z_2)$ we can use justified splicing with w_0 to produce $xy(_t q[_r z_1)_r z_2]_t z_0$ in J, and we set $z = (_t q[_r z_1)_r z_2]_t z_0$.

For the general case, $n > 1$, we have $t = t_1$ and we set $u = t_2$. We can write $q = q_1 q'$ where q_1 is in $F_r(t, u)$ and $q' = q_2 q_3 \ldots q_n$ is in $S_r(u)$. Hence the inductive hypothesis produces x and $(_u z_1']_u z_2'$ so that $w = xyq_1(_u z_1']_u z_2'$ is in J, with $x = \lambda$ or ending in $]_r$. Since q_1 is in $F(t, u)$ we can find a string in J of the form $x'(_t q_1(_u z_1]_u z_2]_t z_3$, and by Fact 6 we can assume that the nest starting with $(_t$ is maximal. By Fact 1 the string $w' = x' q_1(_u z_1]_u z_2$ is also in J, and α_t is a prefix of $E(q_1(_u z_1]_u z_2)$. We can now apply Fact 2 to w and w' to produce $w'' = xyq_1(_u z_1']_u z_2$ in J. Since α_t is a prefix of $E(q_1(_u z_1']_u z_2) = E(q_1(_u z_1]_u z_2)$ we can use justified splicing with w_0 to produce $xy(_t q_1(_u z_1']_u z_2]_t z_0$ in J, and we set $z = (_t q_1(_u z_1']_u z_2]_t z_0$.

The next case, $L_r(s, \$)$ with $s \in R$, is handled by a dual argument. Finally, the case $L_r(s, t)$ with s and t both in R is handled by a double induction, combining the arguments of the previous two cases.   □

Now we are ready to prove that $K \subseteq E(J)$. Select $y \in K$ and represent it as $y_1 y_2 \ldots y_n$ where $y_k \in L_{r_k}(s_k, t_k)$ and the conditions in Fact 4 are satisfied. Let $y_k' = y_1 y_2 \ldots y_k$. We shall show inductively that there is a string $\bar{y}_k z_k$ in J so that both factors are in N, $E(\bar{y}_k) = y_k'$, and either

- $k = n$ and $z_k = \lambda$, or
- $k < n$, $r = r_k \neq \$$ and z_k begins with $[_r$, or
- $k < n$, $t = t_k \neq \$$ and z_k begins with $(_t$.

Note that Fact 4 implies that $r_k = t_k = \$$ if and only if $k = n$.

We start the induction with $k = 1$. Then $L_{r_k}(s_k, t_k) = L_\$(\$, t_1)$. We set $\bar{y}_1 = y_1$ and apply Fact 5 to produce z_1 so that $\bar{y}_1 z_1$ is in J and z_1 satisfies the appropriate alternative above.

For the general case suppose that $1 \leq k < n$, and that $\bar{y}_k$ and z_k have been constructed. We set $r = r_{k+1}$, $s = s_{k+1}$ and $t = t_{k+1}$. Since $y_{k+1} \in L_r(s, t)$ we can find x and z so that $xy_{k+1}z$ is in J and x and z satisfy the conditions of Fact 5. There are now two cases, according to Fact 4.

The first possibility is that $t_k = r \neq \$$ and $s = \$$. Then z_k starts with $(_r$, so $z_k = (_r z']_r z''$, and x ends with $]_r$, so $x = x'(_r x'']_r$. Now Fact 2 applies to $\bar{y}_k z_k = \bar{y}_k(_r z']_r z''$ and $xy_{k+1}z = x'(_r x'']_r y_{k+1}z$ to produce

$\bar{y}_k(_r z')_r y_{k+1} z$ in J. We set $\bar{y}_{k+1} = \bar{y}_k(_r z')_r y_{k+1}$ and $z_{k+1} = z$, and check that $E(\bar{y}_{k+1}) = E(\bar{y}_k)y_{k+1} = y'_k y_{k+1} = y'_{k+1}$ and that z begins with $[_r$ since $r \neq \$$.

The second possibility is that $r_k = s \neq \$$ and $t_k = \$$. Reasoning as above, we see that $\bar{y}_k z_k = \bar{y}_k [_s z'_k)_s z''_k$ and $xy_{k+1}z = x'[_s x'')_s y_{k+1}z$. Again Fact 2 applies to produce $\bar{y}_k [_s x'')_s y_{k+1} z$ in J. We set $\bar{y}_{k+1} = y_k [_s x'')_s y_{k+1}$ and $z_{k+1} = z$, and check that they suffice as above.

The last thing to notice is that if $k + 1 = n$ then $r = \$$, $t = \$$, and the second possibility occurs, so $z = \lambda$.

This finishes the induction. Applying the $k = n$ case now provides $\bar{y}_n \in J$ so that $E(y_n) = y'_n = y$, demonstrating that $K \subseteq E(J)$.

This completes the proof that $K = E(J) = \sigma^*(L)$, thus completing the proof of Lemma 3.14. $\quad\square$

Lemma 3.15. *Every language* $L \in RE, L \subseteq V^*$, *can be written as* $L = L' \cap V^*$ *for some* $L' \in H(FIN, REG)$.

Proof. Consider a type-0 grammar $G = (N, T, S, P)$ and construct the H scheme

$$\sigma = (V, R),$$

where

$$
\begin{aligned}
V &= N \cup T \cup \{X, X', B, Y, Z\} \cup \\
&\quad \cup \{Y_\alpha \mid \alpha \in N \cup T \cup \{B\}\},
\end{aligned}
$$

and R contains the following groups of rules:

1. $Xw\#uY\$Z\#vY$,   for $u \to v \in P, w \in (N \cup T \cup \{B\})^*$,
2. $Xw\#\alpha Y\$Z\#Y_\alpha$,   for $\alpha \in N \cup T \cup \{B\}, w \in (N \cup T \cup \{B\})^*$,
3. $X'\alpha\#Z\$X\#wY_\alpha$,   for $\alpha \in N \cup T \cup \{B\}, w \in (N \cup T \cup \{B\})^*$,
4. $X'w\#Y_\alpha\$Z\#Y$,   for $\alpha \in N \cup T \cup \{B\}, w \in (N \cup T \cup \{B\})^*$,
5. $X\#Z\$X'\#wY$,   for $w \in (N \cup T \cup \{B\})^*$,
6. $\#ZY\$XB\#wY$,   for $w \in T^*$,
7. $\#Y\$XZ\#$.

Consider also the language

$$
\begin{aligned}
L_0 &= \{XBSY, ZY, XZ\} \cup \\
&\quad \cup \{ZvY \mid u \to v \in P\} \cup \\
&\quad \cup \{ZY_\alpha, X'\alpha Z \mid \alpha \in N \cup T \cup \{B\}\}.
\end{aligned}
$$

We obtain $L = \sigma^*(L_0) \cap T^*$.

Indeed, let us examine the work of σ, namely the possibilities to obtain a string in T^*.

No string in L_0 is in T^*. All rules in R involve a string containing the symbol Z, but this symbol will not appear in the string produced by splicing. Therefore, at each step we have to use a string in L_0 and, excepting the case of using the string $XBSY$ in L_0, a string produced at a previous step.

The symbol B is a marker for the beginning of the sentential forms of G simulated by σ.

By rules in group 1 we can simulate the rules in P. Rules in groups 2–5 move symbols from the right hand end of the current string to the left hand end, thus making possible the simulation of rules in P at the right hand end of the string produced by σ. However, because B is always present and marks the place where the string of G begins, we know in each moment which is that string. Namely, if the current string in σ is of the form $\beta_1 w_1 B w_2 \beta_2$, for some β_1, β_2 markers of types $X, X', Y, Y_\alpha, \alpha \in N \cup T \cup \{B\}$, $w_1, w_2 \in (N \cup T)^*$, then $w_2 w_1$ is a sentential form of G.

We start from $XBSY$, hence from the axiom of G, marked to the left hand with B and bracketed by X, Y.

Let us see how the rules 2–5 work. Take a string $Xw\alpha Y$, for some $\alpha \in N \cup T \cup \{B\}$, $w \in (N \cup T \cup \{B\})^*$. By a rule of type 2 we get

$$(Xw\alpha Y, ZY_\alpha) \vdash Xw Y_\alpha.$$

The symbol Y_α memorizes the fact that α has been erased from the right hand end of $w\alpha$. No rule in R can be applied to $Xw Y_\alpha$, except for rules of type 3:

$$(X'\alpha Z, Xw Y_\alpha) \vdash X'\alpha w Y_\alpha.$$

Remark that exactly the symbol α removed at the previous step is now added in the front of w. Again we have only one possibility to continue, namely by using a rule of type 4. We get

$$(X'\alpha w Y_\alpha, ZY) \vdash X'\alpha w Y.$$

If we use now a rule of type 7, removing Y, then X' (and B) can never be removed, the string cannot be turned to a terminal one. We have to use a rule of type 5:

$$(XZ, X'\alpha w Y) \vdash X\alpha w Y.$$

We have started from $Xw\alpha Y$ and we have obtained $X\alpha w Y$, a string with the same end markers. We can iterate these steps as long as we want, hence each circular permutation of the string between X and Y can be produced. Moreover, what we obtain are exactly the circular permutations and nothing more (for instance, at every step we still have one and only one occurrence of B).

To every string XwY we can also apply a rule of type 1, providing w ends with the left hand member of a rule in P. Any rule of P can be simulated in this way, at any place we want in the corresponding sentential form of G, by preparing the string as above, using rules in groups 2–5.

Consequently, for every sentential form w of G there is a string $XBwY$, produced by σ, and, conversely, if $Xw_1 Bw_2 Y$ is produced by σ, then $w_2 w_1$ is a sentential form of G.

The only possibility to remove the symbols not in T from the strings produced by σ is by using rules in groups 6, 7. More precisely, the symbols XB can be removed only in the following conditions: (1) Y is present (hence the work is blocked if we use first rule 7, removing Y: the string cannot participate to any further splicing, and it is not terminal), (2) the current string bracketed by X, Y consists of terminal symbols only, and (3) the symbol B is in the left hand position. After removing X and B we can remove Y, too, and what we obtain is a string in T^*. From the previous discussion, it is clear that such a string is in $L(G)$, hence $\sigma^*(L_0) \cap T^* \subseteq L(G)$. Conversely, each string in $L(G)$ can be produced in this way, hence $L(G) \subseteq \sigma^*(L_0) \cap T^*$. We have the equality $L(G) = \sigma^*(L_0) \cap T^*$, which completes the proof.   □

Lemma 3.16. *Let FA be a family of languages closed under intersection with regular sets and restricted morphisms. For every $L \subseteq V^*, L \notin FA$, and $c, d \notin V$, we have $L' \notin H(FA, RE)$, for*

$$L' = (dc)^* L(dc)^* \cup c(dc)^* L(dc)^* d.$$

Proof. For L, c, d as above, denote

$$L_1 = (dc)^* L(dc)^*,$$
$$L_2 = c(dc)^* L(dc)^* d.$$

Because $L = L_1 \cap V^* = L' \cap V^*$ and $L = h(L_2 \cap cV^*d) = h(L' \cap cV^*d)$, where h is the morphism defined by $h(a) = a, a \in V$, and $h(c) = h(d) = \lambda$, it follows that $L_1 \notin FA, L_2 \notin FA$, and $L' = L_1 \cup L_2 \notin FA$.

Assume that $L' = \sigma^*(L_0)$, for some $L_0 \in FA$ and $\sigma = (V, R)$ with arbitrary R. Because $L' \notin FA$, we need effective splicing in order to produce L' from L_0. That is, splicings $(x, y) \vdash_r z$ with $x \neq z, y \neq z, x, y \in L_0$ are necessary. Write $x = x_1 u_1 u_2 x_2$, $y = y_1 u_3 u_4 y_2$, for some $x_1, x_2, y_1, y_2 \in (V \cup \{c, d\}^*)$, and $r = u_1 \# u_2 \$ u_3 \# u_4 \in R$.

If $x \in L_1$, then $x' = cxd \in L_2, x' = cx_1 u_1 u_2 x_2 d$, hence we can perform $(x', y) \vdash_r z' = cx_1 u_1 u_4 y_2 = cz$. If $z \in L'$, then $cz \notin L'$, a contradiction.

Therefore, x must be from L_2. Then $x' = dxc \in L_1$, $x' = dx_1 u_1 u_2 x_2 c$, hence we can perform $(x', y) \vdash_r z' = dx_1 u_1 u_4 y_2 = dz$. Again we obtain a parasitic string. Since the splicing is not possible, we must have $\sigma^*(L_0) = L_0$, which contradicts $L' \neq L_0$.

As the type of R plays no role in the previous argument, we have $L' \notin H(FA, RE)$.   □

Theorem 3.3. *The relations in Table 2 hold, where at the intersection of the row marked with FA_1 with the column marked with FA_2 there appear either the family $H(FA_1, FA_2)$, or two families FA_3, FA_4 such that $FA_3 \subset H(FA_1, FA_2) \subset FA_4$. These families FA_3, FA_4 are the best estimations among the six families considered here.*

Table 2

	FIN	REG	LIN	CF	CS	RE
FIN	FIN, REG	FIN, RE	FIN, RE	FIN, RE	FIN, RE	FIN, RE
REG	REG	REG, RE	REG, RE	REG, RE	REG, RE	REG, RE
LIN	LIN, CF	LIN, RE	LIN, RE	LIN, RE	LIN, RE	LIN, RE
CF	CF	CF, RE	CF, RE	CF, RE	CF, RE	CF, RE
CS	CS, RE	CS, RE	CS, RE	CS, RE	CS, RE	CS, RE
RE	RE	RE	RE	RE	RE	RE

Proof. From Lemma 3.12 we have the inclusions $FA_1 \subseteq H(FA_1, FA_2)$, for all values of FA_1, FA_2. On the other hand, $H(FA_1, FA_2) \subseteq RE$ for all FA_1, FA_2. With the exception of the families $H(RE, FA_2)$, which are equal to RE, all inclusions $H(FA_1, FA_2) \subseteq RE$ are proper: From Lemma 3.16, we see that all the following differences are non-empty $REG - H(FIN, RE)$, $LIN - H(REG, RE)$, $CF - H(LIN, RE)$, $CS - H(CF, RE)$, $RE - H(CS, RE)$.

Lemma 3.14 and Lemma 3.12 together imply that $H(REG, FIN) = REG$. Hence we have $H(FIN, FIN) \subseteq REG$. This inclusion is strict by Lemma 3.16.

From Lemma 3.13 we obtain the strictness of the inclusions $LIN \subset H(LIN, FIN)$, and $CS \subset H(CS, FIN)$. The same result is obtained if FIN is replaced by any family FA_2.

Lemma 3.14, together with Lemma 3.12, implies $H(CF, FIN) = CF$. We also have $H(LIN, FIN) \subseteq CF$. The inclusion is proper by Lemma 3.16).

From Lemma 3.15 we see that RE is the best estimation for $H(FA_1, FA_2)$, $FA_2 \neq FIN$ ($H(FIN, REG) - FA \neq \emptyset$ for all families $FA \subset RE$ which are closed under intersection with regular sets).

The only thing which remains to be proved is the assertion that $H(FIN, FIN)$ contains infinite languages. This is even true for $H(FIN, 1)$: for $\sigma = (\{a\}, \{a\#\$\#a\})$ we have $\sigma^*(\{a\}) = a^+$. Thus, the proof is complete.   □

Many of the relations in Table 2 are of interest:

– The iterated splicing with respect to regular sets of rules leads the regular languages (even the finite ones) into non-regular (even non-recursive) languages (this is not true for the "weaker" case of uniterated splicing); therefore, the result in [6], [44] cannot be improved.
– The iterated splicing with respect to (at least) regular sets of rules already leads the finite languages to non-context-sensitive languages. In fact, for all FA_2 containing the regular languages, the intersections of the languages in $H(FA_1, FA_2)$ with regular languages of the form V^* characterize the family of recursively enumerable languages.
– However, all the families $H(FA_1, FA_2), FA_1 \neq RE$, have surprising limitations. When FA_1' is the smallest family among those considered here

which strictly includes FA_1, there are languages in FA_1' which are not in $H(FA_1, FA_2)$, for all FA_2, including $FA_2 = RE$.

In view of the equalities $H(FA, FIN) = FA$, for $FA \in \{REG, CF, RE\}$, the hierarchies on the radius of H schemes collapse in these cases. The problem is still *open* for $FA \in \{LIN, CS\}$, but for FIN we have

Theorem 3.4. $FIN \subset H(FIN, 1) \subset H(FIN, 2) \subset \ldots \subset H(FIN, FIN) \subset REG$.

Proof. The inclusions follow from the definitions and Lemma 3.12; the strictness of the first and last inclusions is already known.

For $k \geq 1$, consider the language

$$L_k = \{a^{2k}b^{2k}a^n b^{2k}a^{2k} \mid n \geq 2k + 1\}.$$

It belongs to $H(FIN, k + 1)$, because $L_k = \sigma^*(L_k')$, for

$$L_k' = \{a^{2k}b^{2k}a^{2k+2}b^{2k}a^{2k}\},$$
$$\sigma = (\{a, b\}, \{a^{k+1}\#a^k\$a^{k+1}\#a^k\}).$$

(The splicing rule can be used only with the sites $u_1u_2 = a^{2k+1}$ and $u_3u_4 = a^{2k+1}$ in the central substring, $a^{2k+i}, i \geq 1$, of strings in L_k. Hence we can obtain strings with a^{2k+i+1} as a central substring, for all $i \geq 1$.)

Assume that $L_k = \sigma_1^*(L_k'')$, for some finite language L_k'' and an H scheme $\sigma_1 = (V, R)$ with $rad(\sigma_1) \leq k$. Take a rule $r = u_1\#u_2\$u_3\#u_4 \in R$ and two strings $x, y \in L_k$ to which this rule can be applied, $x = a^{2k}b^{2k}a^n b^{2k}a^{2k}, n \geq 2k + 1$, $y = y_1u_3u_4y_2$. Because $|u_1u_2| \leq 2k$, if $u_1u_2 \in a^*$, then u_1u_2 is a substring of both the prefix a^{2k} and of the suffix a^{2k}, as well as of the central subword a^n of x. Similarly, if $u_1u_2 \in b^*$, then u_1u_2 is a substring of both substrings b^{2k} of x. If $u_1u_2 \in a^+b^+$, then u_1u_2 is a substring of both the prefix $a^{2k}b^{2k}$ and of the subword $a^n b^{2k}$ of x; if $u_1u_2 \in b^+a^+$, then u_1u_2 is a substring of both the suffix $b^{2k}a^{2k}$ and of the subword $b^{2k}a^n$ of x. In all cases, splicing x, y according to the rule r we find at least one parasitic string, hence the equality $L_k = \sigma_1^*(L_k'')$ is not possible. Therefore, $H(FIN, k + 1) - H(FIN, k) \neq \emptyset$, for all $k \geq 1$. □

Although the families $H(FIN, 1)$ and REG seem to be so different (situated at the ends of an infinite hierarchy), they are equal modulo a coding.

Theorem 3.5. *Every regular language is a coding of a language in the family* $H(FIN, 1)$.

Proof. Let $L \in REG$ be generated by a regular grammar $G = (N, T, S, P)$; hence the rules in P have the forms $X \to aY, X \to a$, for $X, Y \in N, a \in T$. Consider the alphabet

$$V = \{[X, a, Y] \mid X \to aY \in P, X, Y \in N, a \in T\} \cup$$
$$\cup \{[X, a, *] \mid X \to a \in P, X \in N, a \in T\},$$

the H scheme

$$\sigma = (V, \quad \{[X, a, Y]\#\$\#[Y, b, Z] \mid [X, a, Y], [Y, b, Z] \in V\} \cup$$
$$\cup \{[X, a, Y]\#\$\#[Y, b, *] \mid [X, a, Y], [Y, b, *] \in V\}),$$

and the finite language

$$L_0 = \{[S, a, *] \mid S \to a \in P, a \in T\} \cup$$
$$\cup \{[X_1, a_1, X_2][X_2, a_2, X_3] \ldots [X_k, a_k, X_{k+1}][X_{k+1}, a_{k+1}, *] \mid$$
$$k \geq 1, \ X_1 = S, \ X_i \to a_i X_{i+1} \in P, \ 1 \leq i \leq k, \ X_{k+1} \to a_{k+1},$$
$$\text{and for no } 1 \leq i_1 < i_2 < i_3 \leq k \text{ we have}$$
$$[X_{i_1}, a_{i_1}, X_{i_1+1}] = [X_{i_2}, a_{i_2}, X_{i_2+1}] = [X_{i_3}, a_{i_3}, X_{i_3+1}]\}$$

(we can have at most pairs of equal symbols of V in a string of L_0, but no triples of equal symbols). Consider also the coding $h : V \longrightarrow T$ defined by

$$h([X, a, Y]) = h([X, a, *]) = a, \ X, Y \in N, a \in T.$$

We have the relation

$$L = h(\sigma^*(L_0)).$$

Indeed, each string in L_0 corresponds to a derivation in G and if x, y are strings in $\sigma^*(L_0)$ describing derivations in G, $x = x_1[X, a, Y][Y, a', Z']x_2$ and $y = y_1[X', b', Y] [Y, b, Z]y_2$, then $z = x_1[X, a, Y][Y, b, Z]y_2 \in \sigma(x, y)$ and obviously z corresponds to a derivation in G, too. The coding h associates to such a string w describing a derivation in G the string $h(w)$ generated by this derivation. Consequently, $h(\sigma^*(L_0)) \subseteq L$.

Conversely, consider the strings in V^* describing derivations in G. Such strings w of length less than or equal to two are in L_0 hence in $\sigma^*(L_0)$. Assume that all such strings of length less than or equal to some $n \geq 2$ are in $\sigma^*(L_0)$ and consider a string w of the smallest length greater than n for which a derivation in G can be found. Because $|w| > n \geq 2$, it follows that $w \notin L_0$, hence w contains a symbol $[X, a, Y]$ on three different positions:

$$w = w_1[X, a, Y]w_2[X, a, Y]w_3[X, a, Y]w_4.$$

Then

$$w' = w_1[X, a, Y]w_2[X, a, Y]w_4,$$
$$w'' = w_1[X, a, Y]w_3[X, a, Y]w_4$$

describe correct derivations in G and $|w'| < |w|, |w''| < |w|$, hence $w', w'' \in \sigma^*(L_0)$ by the induction hypothesis. From the form of w', w'' and of splicing rules of σ we have $w \in \sigma(w', w'')$, hence $w \in \sigma^*(L_0)$, too.

For each derivation in G we find a string $w \in \sigma^*(L_0)$ such that $h(w)$ is exactly the string generated by this derivation. In conclusion, $L \subseteq h(\sigma^*(L_0))$.

□

3.3 The case of multisets

In the previous section, when iterating an H scheme $\sigma = (V, R)$ starting from a language $L \subseteq V^*$, we have assumed that each string in L, or obtained by iterated splicing, is present in an unbounded number of copies: after splicing x, y in order to obtain z, the strings x, y are still available for further splicings. The same is true for z.

However, it is natural (also biologically motivated) to imagine that the terms of the splicing are "consumed" when producing new strings, that the new strings are composed of parts of the strings entering the operation, hence the old strings disappear in this way. This is possible for the definition of $\sigma^*(L)$ in the previous section if we assume that we have infinitely many copies of each string.

Restricting this assumption, we are lead to considering *multisets*, sets with multiplicities associated to their elements.

Take an alphabet V. A multiset over V^* is a mapping $M : V^* \longrightarrow \mathbf{N} \cup \{\infty\}$; $M(x)$ is the number of copies of $x \in V^*$ in this multiset. The set $\{w \in V^* \mid M(w) > 0\}$ is called the *support* of M and it is denoted by $supp(M)$. A usual set S is a multiset with $S(x) = 1$ for $x \in S$ and $S(x) = 0$ for $x \notin S$. For two multisets M_1, M_2, we define the *union* by $(M_1 \cup M_2)(x) = M_1(x) + M_2(x)$ and the *difference* by $(M_1 - M_2)(x) = \max\{0, M_1(x) - M_2(x)\}$.

In the previous section we have implicitly worked with multisets M of the form $M(x) = \infty$ for all $x \in supp(M)$; we call them ω *multisets*.

Consider now an H scheme $\sigma = (V, R)$, with $R \subseteq V^* \# V^* \$ V^* \# V^*$. For two multisets M_1, M_2 over V^* we define

$M_1 \Longrightarrow_\sigma M_2$ iff   there are $x, y, z, w \in V^*$ such that

(i) $M_1(x) > 0, (M_1 - \{(x, 1)\})(y) > 0,$

(ii) $x = x_1 u_1 u_2 x_2, y = y_1 u_3 u_4 y_2,$

$z = x_1 u_1 u_4 y_2, w = y_1 u_3 u_2 x_2,$

for some $x_1, x_2, y_1, y_2 \in V^*$ and

$u_1 \# u_2 \$ u_3 \# u_4 \in R$, and

(iii) $M_2 = (((M_1 - \{(x, 1)\}) - \{(y, 1)\}) \cup \{(z, 1)\}) \cup \{(w, 1)\}.$

At point (iii) we have operations with multisets. The writing above covers also the case when $x = y$ (then we must have $M_1(x) \geq 2$) or $z = w$.

When understood, we omit the subscript σ and we write simply $M_1 \Longrightarrow M_2$. The reflexive and transitive closure of $\Longrightarrow$ is denoted by $\Longrightarrow^*$.

An H scheme $\sigma = (V, R)$ associates a language to a multiset M over V^* as follows:

$$\sigma^*(M) = \{w \in V^* \mid \text{there is } M \Longrightarrow_\sigma^* M' \text{ such that } w \in supp(M')\}.$$

For two families of languages, FA_1, FA_2, we denote by $H(mFA_1, FA_2)$ the family of languages of the form $\sigma^*(M)$, for $\sigma = (V, R), R \in FA_2$, and M a multiset over V^* such that $supp(M) \in FA_1$.

The use of multisets has a surprisingly strong influence on the power of the (iterated) splicing.

Theorem 3.6. *Each language $L \in RE, L \subseteq V^*$, can be written in the form $L = L_1 \cap V^*$, where $L_1 \in H(mFIN, FIN)$.*

Proof. Consider a type-0 Chomsky grammar $G = (N, T, S, P)$, with the rules in P of the form $u \to v$ where $1 \leq |u| \leq 2, 0 \leq |v| \leq 2$, and $u \neq v$ (for instance, we can take G in Kuroda normal form). Also assume that the rules in P are labelled in an one-to-one manner with elements of a set K; we write $r : u \to v$, for r being the label of $u \to v$. By U we denote the set $N \cup T$ and we construct the H scheme $\sigma = (V, R)$, where

$$V = N \cup T \cup \{X_1, X_2, Y, Z_1, Z_2\} \cup \{(r), [r] \mid r \in K\},$$

and the set R contains the following splicing rules:

1. $\delta_1 \delta_2 Y u \# \beta_1 \beta_2 \$(r) v \# [r]$,   for $r : u \to v \in P$,
   $\beta_1, \beta_2 \in U \cup \{X_2\}, \ \delta_1, \delta_2 \in U \cup \{X_1\}$,

2. $Y \# u [r] \$(r) \# v \alpha$,   for $r : u \to v \in P, \ \alpha \in U \cup \{X_2\}$,

3. $\delta_1 \delta_2 Y \alpha \# \beta_1 \beta_2 \$ Z_1 \alpha Y \# Z_2$,   for $\alpha \in U, \ \beta_1, \beta_2 \in U \cup \{X_2\}$,
   $\delta_1, \delta_2 \in U \cup \{X_1\}$,

4. $\delta \# Y \alpha Z_2 \$ Z_1 \# \alpha Y \beta$,   for $\alpha \in U, \ \delta \in U \cup \{X_1\}$,
   $\beta \in U \cup \{X_2\}$,

5. $\delta \alpha Y \# \beta_1 \beta_2 \beta_3 \$ Z_1 Y \alpha \# Z_2$,   for $\alpha \in U, \ \beta_1 \in U, \ \beta_2, \beta_3 \in U \cup \{X_2\}$,
   $\delta \in U \cup \{X_1\}$,

6. $\delta \# \alpha Y Z_2 \$ Z_1 \# Y \alpha \beta$,   for $\alpha \in U, \ \delta \in U \cup \{X_1\}$,
   $\beta \in U \cup \{X_2\}$,

7. $\# Y Y \$ X_1^2 Y \# w$,   for $w \in \{X_2^2\} \cup T\{X_2^2\} \cup T^2\{X_2\} \cup T^3$,

8. $\# X_2^2 \$ Y^3 \#$.

Consider also the multiset M containing the string

$$w_0 = X_1^2 Y S X_2^2,$$

with the multiplicity $M(w_0) = 1$, and the following strings with infinite multiplicity:

$$
\begin{aligned}
w_r &= (r)v[r], & \text{for } r : u \to v \in P, \\
w_\alpha &= Z_1 \alpha Y Z_2, & \text{for } \alpha \in U, \\
w'_\alpha &= Z_1 Y \alpha Z_2, & \text{for } \alpha \in U, \\
w_t &= YY.
\end{aligned}
$$

The idea behind this construction is the following. The rules in groups 1 and 2 simulate rules in P, but only in the presence of the symbol Y. The rules in groups 3 and 4 move the symbol Y to the right, the rules in groups 5 and 6 move the symbol Y to the left. The "starting string" in M is w_0. All the rules of types 1–6 involve a string derived from w_0 and containing such a symbol Y introduced by the string w_0; note that each such rule can only use one string in M which is different from w_0. At any splicing step we have two occurrences of X_1 at the beginning of a string and two occurrences of X_2 at the end of a string (maybe the same string). The rules in groups 1, 3, and 5 separate strings of the form $X_1^2 z X_2^2$ into two strings $X_1^2 z_1$, $z_2 X_2^2$, each having multiplicity one; the rules in groups 2, 4, and 6 bring together these strings, leading to a string of the form $X_1^2 z' X_2^2$. The rules in groups 7 and 8 remove the auxiliary symbols X_1, X_2, Y. If the remaining string is in T^*, then it is an element of $L(G)$. The symbols $(r), [r]$ are associated with the labels in K, and Z_1 and Z_2 are associated with *moving* operations.

Using these explanations, the reader can easily verify that each derivation in G can be simulated in σ, that is we have $L(G) \subseteq \sigma^*(L_0) \cap T^*$.

Let us consider in some detail the opposite inclusion. We claim that if $M \Longrightarrow_\sigma^* M'$ and $w \in T^*, M'(w) > 0$, then $w \in L(G)$.

As we have pointed out above, a direct check confirms that we cannot splice two strings of the form $w_r, w_\alpha, w'_\alpha, w_t$ (for instance, the symbols δ, β in rules of type 4 and 6 prevent the splicing of $w_\alpha, w'_\alpha, \alpha \in U$). In the first step, we have to start with $w_0 = X_1^2 Y S X_2^2$, $M(w_0) = 1$. Now assume that we have a string $X_1^2 w_1 Y w_2 X_2^2$ with multiplicity 1 (w_0 is of this form). If $w_2 = u w_3$ for some $r : u \to v \in P$, then we can apply a rule of type 1. Using the string $(r)v[r]$ from M we obtain

$$(X_1^2 w_1 Y u w_3 X_2^2, (r)v[r]) \vdash (X_1^2 w_1 Y u[r], \ (r)v w_3 X_2^2).$$

No rule from groups 1 or 3–8 can be applied to the obtained strings, because so far no string containing Y^3 has been derived. From group 2, the rule $Y\#u[r]\$(r)\#v\alpha$ can be applied involving both these strings, which leads to

$$(X_1^2 w_1 Y u[r], (r)v w_3 X_2^2) \vdash (X_1^2 w_1 Y v w_3 X_2^2, \ (r)u[r]),$$

where the string $(r)u[r]$ can never participate to a new splicing, because in the rule $r : u \to v$ from P we have assumed $u \neq v$. The multiplicities of $X_1^2 w_1 Y u[r]$ and $(r)v w_3 X_1^2$ have been reduced to 0 again (hence these strings are no more available), the multiplicity of $X_1^2 w_1 Y v w_3 X_2^2$ is one. In this way, we have passed from $X_1^2 w_1 Y u w_3 X_2^2$ to $X_1^2 w_1 Y v w_3 X_2^2$, both having the multiplicity one, which corresponds to using the rule $r : u \to v$ in P. Moreover we see that at each moment there is only one string containing X_1^2 and only one string (maybe the same) containing X_2^2 in the current multiset.

If we apply a rule of type 3 to a string $X_1^2 w_1 Y \alpha w_3 X_2^2$, then we get

$$(X_1^2 w_1 Y \alpha w_3 X_2^2, Z_1 \alpha Y Z_2) \vdash (X_1^2 w_1 Y \alpha Z_2, \ Z_1 \alpha Y w_3 X_2^2).$$

No rule from groups 1–3 and 5–8 can be applied to the obtained strings. By using a rule from group 4 we obtain

$$(X_1^2 w_1 Y \alpha Z_2, Z_1 \alpha Y w_3 X_2^2) \vdash (X_1^2 w_1 \alpha Y w_3 X_2^2, \ Z_1 Y \alpha Z_2).$$

The first of the obtained strings has replaced $X_1^2 w_1 Y \alpha w_3 X_2^2$, which now has the multiplicity 0 (hence we have interchanged Y with α) and the second one is a string in L_0.

In the same way, one can see that the use of a rule of type 5 must be followed by the use of the corresponding rule of type 6, which results in interchanging Y with its left hand neighbour.

Consequently, at each step we have a multiset with either one word $X_1^2 w_1 Y w_2 X_2^2$ or two words $X_1^2 z_1$, $z_2 X_2^2$, each one with multiplicity 1. Only in the first case, provided $w_1 = \lambda$, we can remove $X_1^2 Y$ by using a rule from group 7; then we can also remove X_2^2 by using the rule in group 8. This is the only way to remove these symbols not in T. If the obtained string is not in T^*, then it cannot be processed further, because it does not contain the symbol Y. In conclusion, we can only simulate derivations in G and move Y freely in the string of multiplicity one, hence $\sigma^*(M) \cap T^* \subseteq L(G)$. This completes the proof.   □

The fact that we have worked with multisets is essential, otherwise parasitic strings can be obtained, for instance, by illegally combining strings obtained from different starting strings.

Corollary 3.2. $H(mFIN, FIN)$ *contains non-recursive languages.*

A more precise statement will be given below. This corollary suggests that $H(mFIN, FIN)$ is a "very large" family. Just the opposite conclusion follows from the next result:

Theorem 3.7. *There are regular languages which do not belong to $H(mFIN, RE)$.*

Proof. Consider the language

$$L = (ab)^+ \cup (ba)^+.$$

Assume that $L \in H(mFIN, RE)$, let $\sigma = (V, R)$ be an H scheme with arbitrary R and let M be a multiset with finite support such that $L = \sigma^*(M)$. Consider a rule $r = u_1 \# u_2 \$ u_3 \# u_4$ in R effectively used in a splicing (there is such a rule, because L is infinite and $supp(M)$ is finite). Obviously, r can be applied to all strings w in $(ab)^+$ or in $(ba)^+$ with $|w| \geq |u_1 u_2| + 2$, $|w| \geq |u_3 u_4| + 2$.

Each string $w \in supp(M)$ must be in L, hence it is either in $(ab)^+$ or in $(ba)^+$.

In order to produce a string in $(ab)^+$ we have to start from two strings in $(ab)^+$; in order to produce a string in $(ba)^+$ we have to start from two strings in $(ba)^+$ (otherwise the obtained strings are of the form aza or bzb). Therefore, the strings in $(ab)^+$ are produced independently of the strings in $(ba)^+$. This implies that we can pass from a multiset M' such that $M \Longrightarrow^* M'$, to both a multiset M_1 by producing a new string in $(ab)^+$, and to a multiset M_2 by producing a string in $(ba)^+$. When passing from M' to M_1, the elements of $(ba)^+$ remain unchanged, $M_1(x) = M'(x)$ for all $x \in (ba)^+$. Similarly, when passing from M' to M_2: $M_2(x) = M'(x)$ for all $x \in (ab)^+$. Consequently, there is a multiset M'' such that $M \Longrightarrow^* M''$ and there are $w_1 \in supp(M'') \cap (ab)^+$ and $w_2 \in supp(M'') \cap (ba)^+$ such that $|w_1| \geq |u_1 u_2| + 2$, and $|w_2| \geq |u_3 u_4| + 2$. This means that the splicing rule $u_1 \# u_2 \$ u_3 \# u_4$ can be applied to w_1, w_2 and we obtain two strings of the form $aza, bz'b$, a contradiction. In conclusion, the equality $L = \sigma^*(M)$ is impossible. $\qquad\square$

Corollary 3.3. *All families $H(mFIN, FA)$, FA arbitrary, are incomparable with all families FA' such that $REG \subseteq FA' \subset RE$, and FA' is closed under intersection with regular sets.*

Proof. Combine Theorem 3.6 with Theorem 3.7. $\qquad\square$

Open problems. In the proof of Theorem 3.6 we have used a splicing scheme with radius five. Can this bound be improved? What is the optimal result? Which restrictions on the H schemes and/or on the starting multisets can lead to languages $\sigma^*(M)$ which are context-free or context-sensitive? (Then the membership problem for such languages will be decidable.)

If we have a multiset M over V^*, we can define its *weight* by

$$wht(M) = \sum_{x \in V^*} |x| \cdot M(x).$$

If $\sigma = (V, R)$ and $M \Longrightarrow^* M'$, then clearly $wht(M) = wht(M')$. If we start from M such that $wht(M) < \infty$ (this means that $supp(M)$ is finite and that $M(x) < \infty$ for all $x \in V^*$), then there are only finitely many different multisets M' such that $M \Longrightarrow^* M'$.

This (simple) case raises a series of interesting problems. For instance, given M with $wht(M) < \infty$ and an H scheme $\sigma = (V, R)$, we can construct the finite directed graph (describing the *evolution* of M as determined by σ)

$$\Gamma(\sigma, M) = (U, E),$$

where

$$U = \{M' \mid wht(M') = wht(M)\},$$
$$E = \{(M', M'') \mid M', M'' \in U, M' \Longrightarrow_\sigma M''\}.$$

Therefore, by examining this graph we can decide whether or not a given multiset is reachable from M and whether or not there are infinite chains $M \Longrightarrow M_1 \Longrightarrow \dots$. We can also determine the largest *degree of nondeterminism* in the evolution of M, that is the largest outdegree of vertices in $\Gamma(\sigma, M)$, and so on and so forth.

How complex is the problem of deciding of what type is the evolution of an arbitrarily given multiset M with respect to an arbitrarily given H scheme

σ? Are there pairs (M, σ) leading to evolutions of various types and having a prescribed degree of nondeterminism, length of deterministic intervals, or other prescribed features?

We do not consider in details such problems, but we only present some examples.

Take $V = \{a, b\}$, the multiset M_0 defined by

$$M_0(abbb) = M_0(aba) = 1,$$

(for the not mentioned strings we have $M_0(x) = 0$) as well as the splicing scheme

$$\sigma = (V, \{a\#b\$ab\#b\}).$$

We obtain

$$\Gamma(\sigma, M_0) = (\{M_0, M_1, M_2\}, \{(M_0, M_1), (M_1, M_0), (M_1, M_2), (M_2, M_1)\}),$$

where

$$M_1(abba) = M_1(abb) = 1,$$
$$M_2(ab) = M_2(abbba) = 1.$$

Therefore, the evolution is infinite and its degree of nondeterminism is 2. Although the evolution is infinite, it is important to note that it is not necessarily cyclic. Indeed, there are cube-free infinite strings over the two-letter alphabet $\{0, 1\}$ [55]. Take such an infinite sequence z and assume that it starts with 0 (we can remove the initial occurrences of 1, at most two, if z does not start with 0) and consider the morphism h defined by $h(0) = M_0M_1, h(1) = M_2M_1$. It is easy to see that $h(z)$ is a correct evolution of M_0, because from M_1 we can go in $\Gamma(\sigma, M_0)$ both to M_0 and to M_2. The sequence $h(z)$ is cube-free, hence non-cyclic.

Consider now the alphabet $V = \{a_i \mid 1 \le i \le n\} \cup \{b\}$, for some even n, and the multiset M_0 defined by

$M_0(a_1) = 1,$
$M_0(b^m a_n b^m a_{n-1} a_n a_{n-2} a_{n-1} a_{n-3} a_{n-2} \cdots a_5 a_6 a_4 a_5 a_3 a_4 a_2 a_3 a_1 a_2) = 1.$

Take also the H scheme

$$\sigma = (V, \{\#a_i\$a_i\# \mid 1 \le i \le n\} \cup \{\#b\$b\#\}).$$

We obtain (we write directly the unique strings in the support of multisets):

$$(a_1, \; b^m a_n b^m a_{n-1} \cdots a_5 a_6 a_4 a_5 a_3 a_4 a_2 a_3 a_1 a_2) \Longrightarrow$$
$$(a_2, \; b^m a_n b^m a_{n-1} \cdots a_5 a_6 a_4 a_5 a_3 a_4 a_2 a_3 a_1^2) \Longrightarrow$$
$$(a_3 a_1^2, \; b^m a_n b^m a_{n-1} \cdots a_5 a_6 a_4 a_5 a_3 a_4 a_2^2) \Longrightarrow$$
$$(a_4 a_2^2, \; b^m a_n b^m a_{n-1} \cdots a_5 a_6 a_4 a_5 a_3^2 a_1^2) \Longrightarrow$$
$$(a_5 a_3^2 a_1^2, \; b^m a_n b^m a_{n-1} \cdots a_5 a_6 a_4^2 a_2^2) \Longrightarrow$$
$$(a_6 a_4^2 a_2^2, \; b^m a_n b^m a_{n-1} \cdots a_5^2 a_3^2 a_1^2) \Longrightarrow \cdots \Longrightarrow$$
$$(b^m a_{n-1}^2 a_{n-3}^2 \cdots a_3^2 a_1^2, \; b^m a_n^2 a_{n-2}^2 \cdots a_4^2 a_2^2).$$

Therefore, for n steps the evolution is deterministic. Now, the rule $\#b\$b\#$ can be applied for each b in the first string and each b in the second one, hence the degree of nondeterminism is m^2. From now on, the evolution can continue forever, always using the rule $\#b\$b\#$.

We close this section by inviting the reader to examine the following example:

$$V = \{a, b\},$$
$$M_0(aab) = M_0(aba) = 1,$$
$$\sigma = (V, \{\#a\$a\#\}).$$

The graph $\Gamma(\sigma, M_0)$ contains 11 vertices, 4 of which have no successor (they are "dead"); M_0 reproduces itself (there is an edge (M_0, M_0)), but no other node can evolve back to M_0, there are infinite paths, etc.

However, if we replace σ by

$$\sigma' = (V, \{a\#\$a\#\}),$$

then the obtained graph will be quite different: there are only 8 vertices, all of them reproduce themselves, there is no dead node, and from each node we can return to M_0.

If we replace σ' by

$$\sigma'' = (V, \{a\#\$\#a\}),$$

then the changes are dramatic: the associated graph contains 22 vertices, 7 of them are dead, 2 can reproduce themselves, etc.

Very small modifications of the evolution rules give rise to dramatic changes in the evolution. The interpretation is somewhat expected: the evolution can have any conceivable form, it is very sensitive, and it is difficult to anticipate it. . .

4. Generative mechanisms based on splicing

The usual understanding of the notion of a *generative mechanism* (or *grammar*) is that of a finite device able to generate potentially infinite languages, starting from given *axioms*, via a *derivation process*. Having a finite H scheme, $\sigma = (V, R)$, if we associate to it the language $\sigma^*(L)$, for some (finite) language L, then we have a grammar-like device. We can write it in the form $\gamma = (V, L, R)$. Therefore, in this sense, the whole Section 3.1.2 deals with grammars based on splicing, with the remark that we have considered such triples (V, L, R) with the components R, L not necessarily finite. More explicitly designed grammar-like mechanisms will be considered in the next two sections, starting somewhat from the extreme possibilities: one of the simplest (yet interesting) cases is investigated in Section 4.1, whereas in Section 4.2 we allow also infinite components in our devices.

4.1 Simple H systems

A *simple H system* is a triple

$$\gamma = (V, A, M),$$

where V is an alphabet, $A \subseteq V^*$ is a finite language (of *axioms*) and $M \subseteq V$; the symbols of M are called *markers*.

For $x, y, z \in V^*$ and $a \in M$ we write

$$(x, y) \vdash_a z \quad \text{iff} \quad x = x_1 a x_2, y = y_1 a y_2, z = x_1 a y_2,$$
$$\text{for some } x_1, x_2, y_1, y_2 \in V^*.$$

Therefore, the symbols $a \in M$ are used for building splicing rules of the form $a\#\$a\#$. Consider the associated H scheme

$$\sigma_M = (V, \{a\#\$a\# \mid a \in M\}).$$

We define the *language generated* by γ as

$$L(\gamma) = \sigma_M^*(A),$$

where σ_M^* is defined as in the previous sections (in the free mode).

Consequently, we have here a very particular type of iterated H scheme (with radius one), applied to a finite language, and, according to Lemma 3.14, each language generated in this way is regular. We shall give stronger forms of this assertion below.

We denote by SH the family of languages $L(\gamma)$, where γ is a simple H system. From the previous remarks, we have $SH \subseteq H(FIN, 1)$.

The following statements (characterizations and necessary conditions for a language to be in the family SH) can be easily proven.

Lemma 4.1. *If $\gamma = (V, A, M)$ is a simple H system, then $L(\gamma)$ is the smallest language $L \subseteq V^*$ such that (1) $A \subseteq L$, and (2) $\sigma_M(L) \subseteq L$.*

Lemma 4.2. *(i) If $L \in SH$ is an infinite language, then there is $M \subseteq alph(L), M \neq \emptyset$, such that $\sigma_M(L) \subseteq L$.*
(ii) If $L \in SH, L \subseteq V^*$, and $a^+ \subseteq L$ for some $a \in L$, then $\sigma_{\{a\}}(L) \subseteq L$.

Using point (i) in the previous lemma, we can, for instance, prove that the language $L = a^+b^+a^+b^+$ is not in SH (we have $(abab, abab) \vdash_a ababab \notin L$, and $(abab, abab) \vdash_b ababab \notin L$, hence we cannot have $M \neq \emptyset$ as in the lemma).

Using point (ii), we obtain

Lemma 4.3. *A language $L \subseteq a^*$ is in SH if and only if it is either finite or equal to a^*.*

(Remember that "equal to a^*" means a^* or a^+, because we do not distinguish between languages differing by λ.)

Corollary 4.1. $SH \subset H(FIN, 1)$.

Proof. By the previous lemma, the language $L = \{a^n \mid n \geq 2\}$ is not in SH, but $L = \sigma^*(\{a^3\})$ for $\sigma = (\{a\}, \{a\#a\$a\#a\})$, hence $L \in H(FIN, 1)$. $\square$

Also, a characterization of languages in SH of the form w^*, where w is a string, can be found:

Lemma 4.4. *For $w \in V^+$ we have $w^* \in SH$ if and only if there is $a \in V$ such that $|w|_a = 1$.*

Proof. If $w = w_1 a w_2, |w|_a = 1$, then $w^* = L(\gamma)$ for $\gamma = (V, \{w^2\}, \{a\})$: $\sigma_{\{a\}}(w^2, w^2) = \{w, w^2, w^3\}$ and $(w^i, w^2) \vdash_a w^{i+1}, i \geq 3$.

Conversely, if $w^* \in SH$, then $w^* = L(\gamma)$ for some $\gamma = (V, A, M)$, and we must have $M \neq \emptyset$ and $|w|_M > 0$. Take $a \in M$ and assume that $|w|_a \geq 2, w = w_1 a w_2 a w_3$. Then $(w, w) \vdash_a w_1 a w_3$. Because $|w_1 a w_3| < |w|$ and $w_1 a w_3 \neq \lambda$, we have $w_1 a w_3 \notin w^*$, a contradiction. Therefore, $|w|_a = 1$ for each $a \in M$. $\square$

Using these lemmas we can obtain

Theorem 4.1. *The family SH is an anti-AFL.*

Proof. For union and concatenation we take $L_1 = a^+ b, L_2 = b^+ a$, for Kleene + we take $L = aabb$, for the intersection with regular sets it is enough to note that $V^* \in SH$ for each V, for morphisms we consider $L = a^+ b \cup c^+ d$ and h defined by $h(a) = h(d) = a, h(b) = h(c) = b$, whereas for inverse morphisms we take $L = \{a\}$ and h defined by $h(a) = a, h(b) = \lambda$ (hence $h^{-1}(L) = b^+ a b^+$). $\square$

From Theorem 3.5 we know that each regular language is a coding of a language in $H(FIN, 1)$. A similar result is true for the family SH (which, we have pointed out, is a strict subfamily of $H(FIN, 1)$).

Theorem 4.2. *Every regular language is the image of a language in SH through a coding.*

Proof. Let $G = (N, T, S, P)$ be a regular grammar (hence with rules of the forms $X \to aY, X \to a$, for $X, Y \in N, a \in T$). We construct the simple H system $\gamma = (W, A, W)$, where

$$W = \{X_1, a, X_2 \mid X_1 \in N, a \in T, X_2 \in N \cup \{*\}\},$$

and A contains all strings of the form

$$w_\delta = (S, a_1, X_1)(X_1, a_2, X_2)\ldots(X_{k-1}, a_k, X_k)(X_k, a_{k+1}, *)$$

for each derivation in G

$$\delta : S = X_0 \Longrightarrow a_1 X_1 \Longrightarrow a_1 a_2 X_2 \Longrightarrow \ldots \Longrightarrow a_1 \ldots a_k X_k \Longrightarrow a_1 a_2 \ldots a_k a_{k+1},$$

such that $k \geq 0$ and each rule $X_i \rightarrow a_{i+1} X_{i+1}, 0 \leq i \leq k$, appears in δ at most two times.

The set A is finite (each derivation as above has at most $2 \cdot card(P)$ steps).

Consider the coding $h : W^* \longrightarrow T^*$ defined by $h((X_1, a, X_2)) = a$, $X_1 \in N, a \in T, X_2 \in N \cup \{*\}$.

The inclusion $h(L(\gamma)) \subseteq L(G)$ follows from the construction of γ and the definition of h (the splicing continues correctly the beginning of a derivation in G with the end of another derivation in G).

Conversely, consider a derivation δ in G. If each rule used in δ appears at most twice, then $w_\delta \in A$, and $h(w_\delta)$ is the string generated by δ. If there is a rule $X_i \rightarrow a_{i+1} X_{i+1}$ used at least three times in δ, then w_δ is of the form

$$z = z_1(X_i, a_{i+1}, X_{i+1})z_2(X_i, a_{i+1}, X_{i+1})z_3(X_i, a_{i+1}, X_{i+1})z_4.$$

We have $(z', z'') \vdash (X_i, a_{i+1}, X_{i+1})z$ for

$$z' = z_1(X_i, a_{i+1}, X_{i+1})z_2(X_i, a_{i+1}, X_{i+1})z_4,$$
$$z'' = z_1(X_i, a_{i+1}, X_{i+1})z_3(X_i, a_{i+1}, X_{i+1})z_4,$$

and both z and z' describe derivations in G strictly shorter than δ. By induction on the length of the derivations, we obtain that for each derivation δ in G we find a string w_δ in $L(\gamma)$. As $h(w_\delta)$ is the string generated by δ, we obtain $L(G) \subseteq h(L(\gamma))$.   □

Consider now the problem symmetric to that discussed above: can we represent languages in SH starting from simpler languages and using suitable operations? We will now establish a rather strong representation result.

Theorem 4.3. *For every language $L \in SH$ there are five finite languages L_1, L_2, L_3, L_4, L_5 and a projection h such that $L = h(L_1 L_2^* L_3 \cap L_4^*) \cup L_5$.*

Proof. Let $\gamma = (V, A, M)$ be a simple H system. For each $a \in V$ consider a new symbol, a'; denote $V' = \{a' \mid a \in V\}$.

Define

$$L_1 = \{xa \mid xay \in A, x, y \in V^*, a \in M\},$$
$$L_2 = \{a'xb \mid yaxbz \in A, x, y, z \in V^*, a, b \in M\},$$
$$L_3 = \{a'x \mid yax \in A, x, y \in V^*, a \in M\},$$
$$L_4 = V \cup \{aa' \mid a \in V\},$$
$$L_5 = \{x \in A \mid |x|_M = 0\},$$
$$h : (V \cup V')^* \longrightarrow V^*, \ h(a) = a, a \in V, \ \text{and} \ h(a') = \lambda, a \in V.$$

(Clearly, L_1, L_2, L_3, L_4, L_5 are finite, because A is finite.) Then we claim that

$$L(\gamma) = h(L_1 L_2^* L_3 \cap L_4^*) \cup L_5.$$

Let us denote by B the right hand member of this equality.

(a) $L(\gamma) \subseteq B$.   According to Lemma 4.1, it is enough to prove that B has the two properties (1), (2) in that lemma.

(i) If $x \in A$ and $|x|_M = 0$, then $x \in L_5 \subseteq B$. If $x \in A$ and $x = x_1 a x_2, a \in M$, then $x_1 a \in L_1, a' x_2 \in L_3$, hence $x_1 a a' x_2 \in L_1 L_3$. Clearly, $x_1 a a' x_2 \in L_4^*$. As $h(x_1 a a' x_2) = x_1 a x_2 = x$, we have $x \in B$. Consequently, $A \subseteq B$.

(ii) Take two strings $x, y \in B$. If one of them is in L_5, then $\sigma_M(x, y) = \{x, y\} \subseteq B$.

Take $x', y' \in L_1 L_2^* L_3 \cap L_4^*$ such that $x = h(x'), y = h(y')$, and take $z \in \sigma_M(x, y), (x, y) \vdash_a z$ for some $a \in M$. We have

$$x = z_1 a z_2', \; y = z_1' a z_2, \text{ for } z = z_1 a z_2.$$

Write

$$x' = x_1 a_1 a_1' x_2 \ldots x_k a_k a_k' x_{k+1}, \; k \geq 1,$$
$$y' = y_1 b_1 b_1' y_2 \ldots y_s b_s b_s' y_{s+1}, \; s \geq 1,$$

for $a_i, b_i \in M, x_i, y_i \in V^*, a_{i-1}' x_i a_i, b_{i-1}' y_i b_i \in L_2$ for all i and $x_1 a_1, y_1 b_1 \in L_1, a_k' x_{k+1}, b_s' y_{s+1} \in L_2$. Then

$$x = x_1 a_1 x_2 \ldots x_k a_k x_{k+1}, \; y = y_1 b_1 y_2 \ldots y_s b_s y_{s+1}.$$

Identify the marker a in x, respectively in y, as used in $(x, y) \vdash_a z$.
If $a = a_i$, then $z_1 = x_1 a_1 \ldots x_{i-1} a_{i-1} x_i, \; z_2' = x_{i+1} a_{i+1} \ldots a_k x_{k+1}$.
If $a_i \neq a$ for each $1 \leq i \leq k$, then there is $x_i = x_i' a x_i''$. For $i = 1$ we have $x_1' a \in L_1$ and $a' x_1'' a_1 \in L_2$. For $1 < i < k + 1$ we have $a_{i-1}' x_i' a \in L_2, a' x_i'' a_{i+1} \in L_2$. For $i = k + 1$ we have $a_k' x_{k+1}' a \in L_2$ and $a' x_{k+1}'' \in L_3$. In all cases we can write $x' = w_1 a a' w_2 \in L_1 L_2^* L_3 \cap L_4^*$. Similarly, we can write $y' = w_1' a a' w_2' \in L_1 L_2^* L_3 \cap L_4^*$. For the string $z' = w_1 a a' w_2'$ we clearly have $z' \in L_1 L_2^* L_3 \cap L_4^*$ and $z = h(z')$. Consequently, $z \in B$, which completes the proof of the property (ii), hence of the inclusion $L(\gamma) \subseteq B$.

(b) $B \subseteq L(\gamma)$.   Take $x \in B$. If $x \in L_5$, then $x \in A \subseteq L(\gamma)$.
If $x = h(x'), x' = x_1 a_1 a_1' x_2 a_2 a_2' \ldots x_k' a_k a_k' x_{k+1}, k \geq 1$, with $x_1 a_1 \in L_1$, $a_{i-1}' x_i a_i \in L_2, 2 \leq i \leq k, a_k' x_{k+1} \in L_3$, then, from the definitions of L_1, L_2, L_3, there are the strings $x_1 a_1 x_1', y_i a_{i-1} x_i a_i y_i', 2 \leq i \leq k, z_{k+1} a_k x_{k+1}$, all of them in A. Then

$$(x_1 a_1 x_1', y_2 a_1 x_2 a_2 y_2') \vdash_{a_1} x_1 a_1 x_2 a_2 y_2' = w_2,$$
$$(w_2, y_3 a_2 x_3 a_3 y_3') \vdash_{a_2} x_1 a_1 x_2 a_2 x_3 a_3 y_3' = w_3,$$
$$\ldots\ldots\ldots\ldots$$
$$(w_k, z_{k+1} a_k x_{k+1}) \vdash_{a_k} x_1 a_1 x_2 a_2 \ldots x_k a_k x_{k+1} = x.$$

Consequently, $x \in L(\gamma)$. $\square$

This representation is not a characterization of languages in SH. In fact, a similar result holds true for all regular languages: just combine Theorems 4.2 and 4.3. However, this representation has a series of interesting consequences, some of them referring exactly to the regularity of simple splicing languages: for instance, it directly implies, in this simple way, that $SH \subseteq REG$.

The intersection with L_4^* above can be replaced by an inverse morphism.

Corollary 4.2. *For every $L \in SH$ there are the finite languages L_1, L_2, L_3, L_5, a coding h_1 and a morphism h_2 such that $L = h_1(h_2^{-1}(L_1 L_2^* L_3)) \cup L_5$.*

Proof. We construct L_1, L_2, L_3, L_5 as in the proof of Theorem 4.3 and define

$$h_2 : (V \cup V')^* \longrightarrow (V \cup V')^*,$$
$$h_2(a) = a, a \in V, \text{ and } h_2(a') = aa', a \in V,$$
$$h_1 : (V \cup V')^* \longrightarrow V^*,$$
$$h_1(a) = h_1(a') = a, a \in V.$$

Then $h_2^{-1}(x), x \in L_1 L_2^* L_3$, is defined if and only if $x \in L_4^*$. Because $h_1(h_2^{-1}(x)) = h(x)$, for h as in the proof of Theorem 4.3, we have the equality. $\qquad\square$

From the previous theorem, we also obtain the following useful necessary condition for a language to be in SH.

Corollary 4.3. *If $\gamma = (V, A, M)$ is a simple H system, then for every $x \in Sub(L(\gamma)) \cap (V - M)^*$ we have $|x| \le \max\{|w| \mid w \in A\}$.*

Because the proof of Theorem 4.3 is constructive, the proof for the inclusion $SH \subseteq REG$ obtained in this way is effective, hence all decidable properties of REG are decidable also for SH. Moreover, making use of the property in Corollary 4.3, we get the following result.

Theorem 4.4. *It is decidable whether or not a regular language is a simple H language.*

Proof. Let $L \subseteq V^*$ be a regular language, given by a regular grammar or a finite automaton. For any subset M of V, denote

$$R_M = (V - M)^* \cup$$
$$\cup \{x_1 a_1 x_2 a_2 \ldots x_k a_k x_{k+1} \mid 1 \le k \le 2 \cdot card(M),$$
$$x_i \in (V - M)^*, 1 \le i \le k + 1, a_i \in M, 1 \le i \le k, \text{ and}$$
$$\text{there are no } 1 \le i < j < l \le k \text{ such that } a_i = a_j = a_l\}.$$

(Therefore, R_M contains all strings x over V such that each symbol of M appears at most twice in x.)

(1) If $L \cap R_M$ is an infinite set, then there is no H system $\gamma = (V, A, M)$ such that $L = L(\gamma)$.

Indeed, $L \cap R_M$ being infinite means that there is $x \in Sub(L) \cap (V - M)^*$ of arbitrary length, contradicting the previous corollary.

(2) If $L \cap R_M$ is a finite set, then we consider all H systems $\gamma = (V, A, M)$ with $A \subseteq L \cap R_M$. Then there is an H system $\gamma' = (V, A', M)$ such that $L = L(\gamma')$ if and only if $L = L(\gamma)$ for a system γ constructed above.

(*if*): trivial.

(*only if*): Take $\gamma' = (V, A', M)$ such that $L(\gamma') = L$ and A' is not a subset of $L \cap R_M$. This means that A' contains a string of the form

$$z = x_1 a x_2 a x_3 a x_4,$$

for $x_1, x_2, x_3, x_4 \in V^*, a \in M$. Consider the strings

$$z_1 = x_1 a x_2 a x_4, \quad z_2 = x_1 a x_3 a x_4.$$

Both of them are in $\sigma_M(\{z\})$, hence in $L(\gamma')$. Moreover, $(z_1, z_2) \vdash_a z$. Therefore, replacing A' by

$$A'' = (A' - \{z\}) \cup \{z_1, z_2\},$$

we get an H system $\gamma'' = (V, A'', M)$ such that $L(\gamma') = L(\gamma'')$. Continuing this procedure (for a finite number of times, because A' is finite and $|z_1| < |z|, |z_2| < |z|$) we eventually find a system $\gamma''' = (V, A''', M)$ with $A''' \subseteq L \cap R_M$.

Now, $L \in SH$ if and only if $L = L(\gamma)$ for some $\gamma = (V, A, M)$ with $M \subseteq V$. There are only finitely many such sets M. Proceed as above with each of them. We have $L \in SH$ if and only if there is such a set M_0 for which $L \cap R_{M_0}$ is finite and there is $A_0 \subseteq L \cap R_{M_0}$ (finitely many possibilities) such that $L = L(\gamma_0)$ for $\gamma_0 = (V, A_0, M_0)$. The equality $L = L(\gamma_0)$ can be checked algorithmically. In conclusion, the question of whether or not $L \in SH$ can be decided algorithmically. □

This result cannot be extended to context-free (not even to linear) languages.

Theorem 4.5. *The problem whether or not a linear language is a simple H language is not decidable.*

Proof. Take an arbitrary linear language $L_1 \subseteq \{a, b\}^*$, as well as the language $L_2 = c^+ d^+ c^+ d^+$, which is not in the family SH (Lemma 4.2). Construct the language

$$L = L_1 \{c, d\}^* \cup \{a, b\}^* L_2.$$

This is a linear language.

If $L_1 = \{a, b\}^*$, then $L = \{a, b\}^* \{c, d\}^*$ and this is a simple H language: for $\gamma = (\{a, b, c, d\}, \{xy \mid x \in \{a, b\}^*, y \in \{c, d\}^*, |x| \in \{0, 2\}, |y| \in \{0, 2\}\}, \{a, b, c, d\})$ we have $L(\gamma) = \{a, b\}^* \{c, d\}^*$.

If $L_1 \neq \{a, b\}^*$, then $\{a, b\}^* - L_1 \neq \emptyset$. Take $w \in \{a, b\}^* - L_1$ and consider the string $w' = wcdcd$. It is in L, and $(w', w') \vdash_e wcdcdcd$ for each $e \in \{c, d\}$. The obtained string is not in L, therefore no one of c, d can be a marker in an H system for the language L. But L contains all strings in $wc^+d^+c^+d^+$, hence $L \in SH$ would contradict Corollary 4.3 above.

Consequently, $L \in SH$ if and only if $L_1 = \{a, b\}^*$, which is undecidable.

$\square$

Because $SH \subset REG$, it is of interest to investigate the relationships between SH and other subfamilies of REG. We consider only one (important) such sub-regular family, that of strictly locally testable languages [32], [9].

A language $L \subseteq V^*$ is *p-strictly locally testable*, for some $p \geq 1$, if we can write it in the form

$$L = \{x \in L \mid |x| < 2p\} \cup (Pref(L) \cap V^p)V^*(Suf(L) \cap V^p) - V^*(V^p - Sub(L))V^*.$$

A language is strictly locally testable if it is p-strictly locally testable for some $p \geq 1$. We denote by SLT the family of such languages.

Clearly, $SLT \subset REG$. In fact, SLT is contained in the family of *extended star-free languages*, the smallest family of languages containing the finite languages and closed under boolean operations and under concatenation. This family coincides with that of non-counting languages [32] and has been characterized in [50] as the family of languages having aperiodic syntactic monoids.

Theorem 4.6. $SH \subset SLT$.

Proof. We shall use the characterization given in [9] for strictly locally testable languages.

According to [51], a string $x \in V^*$ is called *constant* with respect to a language $L \subseteq V^*$ if whenever $uxv \in L$ and $u'xv' \in L$, then also $uxv' \in L$ and $u'xv \in L$. In [9] it is proved that a language $L \subseteq V^*$ is strictly locally testable if and only if there is an integer k such that all strings in V^k are constants with respect to L.

Consider now a language $L \in SH, L = L(\gamma)$, for some $\gamma = (V, A, M)$. Take the integer

$$k = \max\{|x| \mid x \in A\} + 1.$$

Every string in V^k is a constant with respect to L. Indeed, take such a string x and two strings $uxv, u'xv'$ in L. Because $|x| = k$, according to Corollary 4.3, we have $|x|_M > 0$. Take $a \in M$ such that $x = x_1 a x_2$. Therefore $uxv = ux_1 a x_2 v, u'xv' = u'x_1 a x_2 v'$, hence $(uxv, u'xv') \vdash_a ux_1 a x_2 v' = uxv'$ and $(u'xv', uxv) \vdash_a u'x_1 a x_2 v = u'xv$. In conclusion, $L \in SLT$ and $SH \subseteq SLT$.

The inclusion is proper: for $w = aabb$ we have $w^* \in SLT$ (obvious), but $w^* \notin SH$ (Lemma 4.4).

$\square$

We close this section with the remark that given a simple H system $\gamma = (V, A, M)$ we can associate to it splicing rules of four (nontrivial) forms: $a\#\$a\#$ as above (we say that they are *of type* (1,3)), $\#a\$\#a$ (of type (2,4)), $a\#\$\#a$ (of type (1,4)), and $\#a\$a\#$ (of type (2,3)). If we denote by $\vdash_a^{(i,j)}$ the splicing with respect to a rule of type (i,j) associated to $a \in M$, we have $\vdash_a^{(1,3)} \equiv \vdash_a^{(2,4)}$, but $\vdash_a^{(1,4)}$ and $\vdash_a^{(2,3)}$ are different. Consequently, to a simple H system γ we can associate three distinct languages, $L_{(1,3)}(\gamma) = L_{(2,4)}(\gamma), L_{(1,4)}(\gamma)$ and $L_{(2,3)}(\gamma)$, defined by iterating on A the corresponding operation $\vdash_a^{(i,j)}$. The reader can verify that each two of the associated families $SH = SH_{(1,3)} = SH_{(2,4)}, SH_{(1,4)}, SH_{(2,3)}$ are incomparable. Therefore, a similar study as that done here for SH is necessary for $SH_{(1,4)}$ and $SH_{(2,3)}$. Most of the results in this section hold true also for these two further families.

4.2 Extended H systems

We return now to the splicing operation based on arbitrarily large sets of rules, and still we add one further feature to the generative devices we consider: extended symbols, as in Chomsky grammars and in L systems.

An *extended H system* is a quadruple

$$\gamma = (V, T, A, R),$$

where V is an alphabet, $T \subseteq V$, $A \subseteq V^*$, and $R \subseteq V^*\#V^*\$V^*\#V^*$, where $\#, \$$ are special symbols not in V.

We call V the alphabet of γ, T is the *terminal* alphabet, A is the set of *axioms*, and R the set of splicing rules. Therefore, we have an *underlying H scheme*, $\sigma = (V, R)$, augmented with a given subset of V and a set of axioms.

The *language generated* by γ is defined by

$$L(\gamma) = \sigma^*(A) \cap T^*,$$

where σ is the underlying H scheme of γ.

For two families of languages, FA_1, FA_2, we denote by $EH(FA_1, FA_2)$ the family of languages $L(\gamma)$ generated by extended H systems $\gamma = (V, T, A, R)$, with $A \in FA_1, R \in FA_2$. Part of the results in the preceding sections can be reformulated in terms of extended H systems. Moreover, we have

Lemma 4.5. $REG \subseteq EH(FIN, FIN)$.

Proof. Take a language $L \in REG, L \subseteq T^*$, generated by a regular grammar $G = (N, T, S, P)$.

We construct the H system

$$\gamma = (N \cup T \cup \{Z\}, T, A_1 \cup A_2 \cup A_3, R_1 \cup R_2),$$

with

$$A_1 = \{S\} \cup (L \cap \{\lambda\}),$$
$$A_2 = \{ZaY \mid X \to aY \in P, a \in T\},$$
$$A_3 = \{ZZa \mid X \to a \in P, a \in T\},$$
$$R_1 = \{\#X\$Z\#aY \mid X \to aY \in P, a \in T\},$$
$$R_2 = \{\#X\$ZZ\#a \mid X \to a \in P, a \in T\}.$$

If we splice a string ZxX, possibly from A_2 (for $x = c$ and $U \to cX \in P$) using a rule in R_1, then we get a string of the form $ZxaY$. The symbol Z cannot be eliminated, hence no terminal string can be obtained if we continue to use the resulting string as the first term of a splicing. On the other hand, such a string ZxX with $|x| \geq 2$ cannot be used as the second term of a splicing. Consequently, the only way to obtain a terminal string is to start from S, splicing with rules in R_1 for an arbitrary number of times and ending with a rule in R_2. Always the first term of the splicing is that obtained by a previous splicing and the second one is from A_2 or from A_3 (at the last step). This corresponds to a derivation in G, hence we have $L(\gamma) = L(G) = L$.   □

Theorem 4.7. *The relations in Table 3 hold, where at the intersection of the row marked with FA_1 with the column marked with FA_2 there appear either the family $EH(FA_1, FA_2)$, or two families FA_3, FA_4 such that $FA_3 \subset EH(FA_1, FA_2) \subseteq FA_4$. These families FA_3, FA_4 are the best estimations among the six families considered here.*

Table 3

	FIN	REG	LIN	CF	CS	RE
FIN	REG	RE	RE	RE	RE	RE
REG	REG	RE	RE	RE	RE	RE
LIN	LIN, CF	RE	RE	RE	RE	RE
CF	CF	RE	RE	RE	RE	RE
CS	RE	RE	RE	RE	RE	RE
RE	RE	RE	RE	RE	RE	RE

Proof. Clearly, $FA \subseteq EH(FA, FIN)$ for all FA. From Lemma 4.5 we also have $REG \subseteq EH(FIN, FA)$ for all FA. From Lemma 3.14 and the closure of REG, CF under intersection with regular sets we obtain $EH(REG, FIN) \subseteq REG$, $EH(CF, FIN) \subseteq CF$. If in the proof of the relation $RE \subseteq S(CS, FIN)$ in Theorem 3.1 we take c_1, c_2, c_3 as nonterminal symbols and V as a terminal alphabet, then we obtain $RE \subseteq EH(CS, FIN)$. Thus, the first column of Table 3 is obtained.

From the proof of Lemma 3.15 we obtain $RE \subseteq EH(FIN, REG)$. As $EH(FA_1, FA_2) \subseteq RE$ for all families FA_1, FA_2 (by Turing-Church thesis), this completes the proof.   □

The only family which is not precisely characterized is $EH(LIN, FIN)$.

One can see in Table 3 that the extended H systems have a surprisingly large generative power. In particular, extended H systems with a finite set of axioms and a regular set of rules prove to be computationally complete, they have the same power as Turing machines.

5. Splicing circular words

Since DNA exists in a circular form as well as a linear form, it is important to extend splicing theory to incorporate circular strings. In this section we give the basic definitions and prove some closure results similar to those for linear splicing, under some extra restrictions on the H scheme. These extra restrictions are natural in the applications to DNA splicing. However, in the more general setting, many questions, including the analysis of generative capacity, remain open.

5.1 Circular words

We start with some basic definitions for circular strings. The idea is that a circular string over an alphabet V is a sequence $x_1 x_2 \ldots x_n$ of elements of V, with the understanding that x_1 follows x_n. In other words, $x_1 x_2 \ldots x_n$ should be regarded as the same as $x_k x_{k+1} \ldots x_n x_1 \ldots x_{k-1}$ for $1 < k < n$.

To make this precise we introduce a relation on strings in V^* by declaring $xy \sim yx$ for all strings x, y. This is easily seen to be an equivalence relation, and we define $\hat{}w$, the circular form of w, to be the equivalence class of the string w. The set of all such circular strings over V is denoted $V^{\hat{}}$. A subset of $V^{\hat{}}$ is called a *circular language* over V. Suppose L is a language in V^* and C is a circular language in $V^{\hat{}}$. The set of all circular strings corresponding to elements of L is written $\mathrm{Cir}(L)$, and is called the *circularization* of L. Any language L for which $\mathrm{Cir}(L) = C$ is called a *linearization* of C. The set of *all* strings in V^* corresponding to elements of C is the *full linearization* of C, written $\mathrm{Lin}(C)$. Given a family of languages FA we form a corresponding family of circular languages $FA^{\hat{}}$ consisting of all those circular languages C which have some linearization in FA. This gives a meaning to such notions as "regular circular language."

It is easy to see which of the standard operations of language theory are usable in the circular sense. For example, length of a circular word is well defined. Althouth the concatenation of two circular words is not generally well defined, the powers $(\hat{}w)^n$ of the circular word $\hat{}w$ are well defined as $\hat{}w^n$. Also, any homomorphism h from V^* to W^* induces a map, also written h, from $V^{\hat{}}$ to $W^{\hat{}}$. We will call such an induced map a homomorphism (although there is no underlying algebraic structure to justify the name).

In discussing closure properties the following observations will be useful. The *cyclic closure* $\text{CYCLE}(L)$ of a language L is the set of all cyclic permutations of words of L. In other words, this is the set of all w' such that $w' \sim w$ for some w in L. It is shown in [25] that each of the families REG, CF, and RE is closed under this operation. If $C = \text{Cir}(L)$ then $\text{Lin}(C) = \text{CYCLE}(L)$, from which the following lemma and its corollary are obvious.

Lemma 5.1. *If FA is a family of languages closed under cyclic closure then a circular language C is in $FA\hat{\ }$ if and only if its full linearization $\text{Lin}(C)$ is in FA.*

Corollary 5.1. *Suppose FA is a family of languages closed under cyclic closure. If FA is closed under finite unions, direct homomorphic images, or inverse homomorphic images then so is $FA\hat{\ }$, and if FA is closed under intersection with reglar sets then $FA\hat{\ }$ is closed under intersection with regular circular languages.*

5.2 Circular splicing

Given a splicing rule $r = u_1 \# u_2 \$ u_3 \# u_4$ we define the *reversal* of r to be the rule $u_3 \# u_4 \$ u_1 \# u_2$, and we use the notation $\bar{r}$ to refer to this reversal. We say an H scheme $\sigma = (V, R)$ is *symmetric* iff $\bar{r}$ is in R whenever r is in R. We say σ is *reflexive* iff whenever $u_1 \# u_2 \$ u_3 \# u_4$ is in R then $u_1 \# u_2 \$ u_1 \# u_2$ and $u_3 \# u_4 \$ u_3 \# u_4$ are also in R.

Now we can discuss splicing in the circular context. We start with a splicing rule $r = u_1 \# u_2 \$ u_3 \# u_4$ and two circular strings $\hat{\ } x u_1 u_2$ and $\hat{\ } y u_3 u_4$ containing the splicing sites $u_1 u_2$ and $u_3 u_4$. The idea of splicing is to cut the first string between u_1 and u_2 and the second one between u_3 and u_4, obtaining the linear strings $u_2 x u_1$ and $u_4 y u_3$, and then to join the appropriate ends. However, the rule r only specifies that the u_1 end should be joined to the u_4 end, resulting in the linear string $u_2 x u_1 u_4 y u_3$. In order to obtain a circular string we must join the two ends of this string together, to obtain $\hat{\ } u_2 x u_1 u_4 y u_3$. We may consider this final closing up of the string to be the effect of simultaneously applying the reversed rule $\bar{r} = u_3 \# u_4 \$ u_1 \# u_2$, which also dictates cutting between u_1 and u_2 and between u_3 and u_4, but which then specifies that the u_3 end should be connected to the u_2 end.

Thus we define the result of splicing $\hat{\ } x u_1 u_2$ and $\hat{\ } y u_3 u_4$ using the rule r to be $\hat{\ } u_2 x u_1 u_4 y u_3$. This is the same as the result of splicing $\hat{\ } y u_3 u_4$ and $\hat{\ } x u_1 u_2$ using the reversed rule $\bar{r}$. In light of this and the discussion above it is natural to assume that any H scheme $\sigma = (V, R)$ used for circular splicing is symmetric.

There is another possibility for circular splicing that did not occur in the linear case. Suppose that $\hat{\ } x u_1 u_2 y u_3 u_4$ is a circular string containing disjoint copies of the sites of the splicing rule r. We can consider this string to be cut at *both* sites to yield the linear strings $u_4 x u_1$ and $u_2 y u_3$, which can then be closed to the circular forms $^\frown u_4 x u_1$ and $^\frown u_2 y u_3$ by joining the u_4, u_1 and u_2, u_3 ends as specified by the pair of rules r and $\bar{r}$. This operation, producing the two circular strings $^\frown u_4 x u_1$ and $^\frown u_2 y u_3$ from $^\frown x u_1 u_2 y u_3 u_4$, will be called *self-splicing* using the rule r. In the linear case the analogous operation can be effected by a sequence of ordinary splicing operations on pairs of strings, but in the circular case this is not so, and self-splicing cannot be subsumed under the binary splicing operation.

If $\sigma = (V, R)$ is a symmetric H scheme and C is a circular language then we define the *circular splicing language* $\sigma^*(C)$ generated by σ and C as the smallest circular language that contains C and is closed under circular splicing and self-splicing using the rules of R.

We now consider a simple example due to Siromoney *et al.* [53]. We let $V = \{a, b\}$, $C = \{^\frown ab\}$, $\sigma = (V, R)$ where $R = \{r, \bar{r}\}$ and $r = a\#b\$b\#a$. One can then calculate that $\sigma^*(C) = \{^\frown a^n b^m \mid n = 0, m > 0 \text{ or } n > 0, m = 0 \text{ or } n = m > 0\}$. Thus in this case a finite H scheme and a finite initial language produce a non-regular splicing language. The problem of characterizing the circular languages that arise in this way, even for finite initial data, remains open.

There is one case where the closure results of Section 3 extend in a satisfactory manner. It is natural to assume that whenever a string can be cut at a site then the two cut ends can be rejoined; in particular, this is consistent with the interpretation in terms of DNA splicing. In other words, if $u_1\#u_2\$u_3\#u_4$ is a splicing rule then the u_1 end resulting from a cut may be joined not just to a u_4 end but also to a u_2 end, and similarly for the other combinations. We can assure this condition by assuming that the splicing scheme is reflexive.

If we reconsider the previous example, but add $a\#b\$a\#b$ and $b\#a\$b\#a$ to R, then we have a reflexive H scheme. We can calculate in this case that the splicing language is the set of all circular strings of positive length, which is regular, as we should expect.

The result of this section is the following version of the closure Lemma 3.14:

Theorem 5.1. *Suppose FA is a full abstract family of languages which is closed under cyclic closure. Suppose $\sigma = (V, R)$ is a finite symmetric and reflexive H scheme and C is a circular language in $FA^\frown$. Then $\sigma^*(C)$ is also in $FA^\frown$.*

Remark 5.1. In applications to DNA splicing the rules are determined by the actions of restriction enzymes and ligases. A restriction enzyme is characterized by a site (u_1, u_2), an overlap prefix of u_2, and a polarity (+ or −). A rule $u_1\#u_2\$u_3\#u_4$ is determined from two such descriptions if and only if the overlap prefixes and polarities are the same. It follows easily that any H scheme constructed in this way is both symmetric and reflexive.

Remark 5.2. As noted above, this result is not true without the reflexivity assumption. It is possible to consider circular splicing languages generated *only* by the binary splicing operation, *without* using self-splicing. If we drop self-splicing in the Siromoney examples we obtain the following: Without the reflexivity assumption the splicing language is the set of all circular strings $^\frown a^n b^n$ with $n > 0$, and, if we enforce the reflexivity assumption, it is the set of all circular strings with positive length, where the number of a's equals the number of b's. Thus without the self-splicing operation we can have non-regularity even if we assume reflexivity. The problem of characterizing the circular languages obtained this way remains open.

Before proving Theorem 5.1 we introduce a construction that is generally useful in dealing with splicing (although we present a version adapted to the case at hand). This construction follows closely the original motivation for splicing in terms of DNA fragments. This motivation can be understood very roughly as follows, as a two-step process. Molecules may be cut (by a restriction enzyme), leaving the two new ends "marked" (single-stranded instead of double stranded). Then these fragments may be pasted back together (by a ligase) which can only join ends that are appropriately marked. Our construction is a modified form of splicing which more closely follows this process.

We start with an H scheme $\sigma = (V, R)$. For later convenience we introduce the following notation: If $r = u_1 \# u_2 \$ u_3 \# u_4$ is in R then we write $\alpha_r = u_1 u_2$, $\gamma_r = u_3 u_4$ and $\beta_r = u_1 u_4$. We also introduce new symbols X_r and Y_r for each r in R, and we let $X = \{X_r \mid r \in R\}$ and $Y = \{Y_r \mid r \in R\}$. Using this notation we define two operations on strings:

Cutting: If r is in R then $x\alpha_r y$ may be **cut** to produce xX_r and $x\gamma_r y$ may be **cut** to produce $Y_r y$.

Pasting: If r is in R then xX_r and $Y_r y$ may be **pasted** to produce $x\beta_r y$.

The idea is that X_r and Y_r are the "marked" versions of the u_1 and u_4 ends after a cut, and they are "unmarked" when the cut ends are rejoined, so $X_r Y_r$ is replaced with $\beta_r = u_1 u_4$.

Lemma 5.2. *Suppose that $\sigma = (V, R)$ is a finite H scheme and that X and Y are defined as above. Suppose that L is a language in YV^*X which lies in the full AFL FA. Suppose that for each $r \in R$ both α_r and γ_r are non-empty, and that both X_r and Y_r occur on strings in L. Then the smallest language $\hat{L}$ which contains L and is closed under cutting and pasting lies in YV^*X and is in the family FA.*

Proof. The fact that cutting and pasting preserve the language YV^*X is immediate from the definitions of the operations, using the fact that α_r and γ_r are non-empty strings over V (so that the site for cutting a word $Y_r z X_s$ must be in z.)

For the closure result we first define a related H scheme $\sigma_1 = (V_1, R_1)$. The new alphabet consists of the symbols of V, the new symbols X_r and Y_r, and a second collection of new symbols X'_r and Y'_r. As above we define $X' = \{X'_r : r \in R\}$ and $Y' = \{Y'_r : r \in R\}$. The rules in R_1 are all those of the forms $\#\alpha_r \$\# X_r$, $Y_r\#\$\gamma_r\#$, $\#X_r\$Y'_r\#$ and $\#X'_r\$Y_r\#$, where r is in R. We also define L_1 by adding to L all strings $Y'_r\beta_r X'_r$ for r in R.

We have $\tilde{L} \subseteq \sigma_1^*(L_1)$. To see this we just need to note that any cut or paste operation can be emulated by splicing operations. For example, a cut operation $x\alpha_r y \to xX_r$ can be obtained by splicing $x\alpha_r y$ to any string ending in X_r using the rule $\#\alpha_r\$\#X_r$. Cutting on the other end is handled similarly. Also, a paste operation $xX_r, Y_r y \to x\beta_r y$ is the result of first splicing xX_r and $Y'_r\beta_r X'_r$ using $\#X_r\$Y'_r\#$, and then splicing the result with $Y_r y$ using $\#X'_r\$Y_r\#$.

We claim that $\tilde{L} = \sigma_1^*(L_1) \cap YV^*X$. Together with Lemma 3.14 this finishes the proof that $\tilde{L}$ is in FA.

To prove the claim we must show that $\sigma_1^*(L_1) \cap YV^*X \subseteq \tilde{L}$. It is clear, as in the first paragraph of this proof, that splicing using σ_1 preserves $(Y \cup Y')V^*(X \cup X')$. Hence each string $w \in \sigma_1^*(L_1)$ has the form BzA where $B \in Y \cup Y'$, $A \in X \cup X'$ and $z \in V^*$. We shall show that if $B \in Y$ and $A \in X$ then $w \in \tilde{L}$ by induction on the number of splicing operations needed to produce w. We actually need to prove the following technicality, which reduces to the statement that $w \in \tilde{L}$ if $B \in Y$ and $A \in X$.

Assertion 1. *Write* $w \in \sigma_1^*(L_1)$ *as* BzA, *as above. If* $B = Y_r$ *let* $t = r$ *and* $z_1 = \lambda$, *and if* $B = Y'_r$ *let* $Y_t z_1 X_r$ *be a string in* $\tilde{L}$. *Similarly, if* $A = Y_s$ *let* $z_2 = \lambda$ *and* $u = s$, *and if* $A = Y'_s$ *let* $Y_s z_2 X_u$ *be a string in* $\tilde{L}$. *Then* $Y_t z_1 z z_2 X_u$ *is in* $\tilde{L}$.

Proof. We start the induction with $w \in L_1$. Then either $B = Y_r$, $A = X_s$, and $Y_t z_1 z z_2 X_u = Y_r z X_s = w \in L \subseteq \tilde{L}$, or $B = Y'_r$, $A = X'_r$, and $w = X'_r\beta_r Y'_r$. But then $Y_t z_1 z z_2 X_u = Y_t z_1 \beta_r z_2 X_u$ is the result of pasting $Y_t z_1 X_r$ and $Y_s z_2 X_u$, so it is in $\tilde{L}$.

For the general case we consider w in $\sigma_1^n(L_1)$ but not in $\sigma_1^{n-1}(L_1)$, and we suppose that the Assertion is true for all strings in $\sigma_1^{n-1}(L_1)$. Then w was obtained from two strings in $\sigma_1^{n-1}(L_1)$ by a splicing operation.

Suppose first that this splicing operation used a rule of the form $\#\alpha_s\$\#X_s$. Thus there are two strings $Bz\alpha_s z'C$ and xX_r in $\sigma_1^{n-1}(L_1)$ which splice to produce $w = BzA$ (with $A = X_s$.) If $C = X_p$ then we set $q = p$ and $z'_2 = \lambda$, and otherwise we choose any string $Y_p z'_2 X_q$ in L_1. Applying the Assertion inductively to $Bz\alpha_s z'C$, with t and z_1 on the left and q and z'_2 on the right, we have $Y_t z_1 z\alpha_s z' z'_2 X_q$ in $\tilde{L}$. Now a cut operation at the site α_s produces $Y_t z_1 z X_s = Y_t z_1 z z_2 X_u$ in $\tilde{L}$, as desired. A similar argument applies if the last splicing operation used the rule $Y_r\#\$\gamma_r\#$.

The alternative is that the last splicing operation used $\#X_v\$Y'_v\#$ (or $\#X'_v\$Y_v\#$ with a similar analysis). In this case there are strings $Bx_1 X_v$ and

$Y'_v x_2 A$ in $\sigma_1^*(L_1)$ which splice using $\#X_v\$Y'_v\#$ to form w, with $z = x_1 x_2$. Applying the Assertion inductively to Bx_1X_v, with t and z_1 on the left and v and λ on the right, shows that $Y_t z_1 x_1 X_v$ is in $\tilde{L}$. Since $Y_t z_1 x_1 X_v$ is in $\tilde{L}$ we can apply the Assertion inductively to $Y'_v x_2 A$, with t and $z_1 x_1$ on the left and u and z_2 on the right, to see that $Y_t z_1 x_1 x_2 z_2 X_u = Y_t z_1 z z_2 X_u$ is in $\tilde{L}$.

This concludes the induction proof of the Assertion, and hence concludes the proof of Lemma 5.2.   □

Now we turn to the proof of Theorem 5.1. We start with FA, C, and $\sigma = (V, R)$ as specified. The main idea is to cut the circular words in C to form a linear language L to which we can apply Lemma 5.2. We then recover $\sigma^*(C)$ by rejoining the ends in the strings of $\tilde{L}$. We need some preliminary analysis so we can satisfy the hypotheses of Lemma 5.2.

First we need to eliminate the possibility of empty sites. Suppose R contains one rule with either $\alpha_r = \lambda$ or $\gamma_r = \lambda$. Then by the reflexivity assumption R contains the rule $\#\$\#$. Now if $\hat{}x$ and $\hat{}y$ are in $\sigma^*(C)$ we can splice them using $\#\$\#$ to obtain $\hat{}xy$ in $\sigma^*(C)$. On the other hand, if $\hat{}xy$ is in $\sigma^*(C)$ then we can apply self-splicing using $\#\$\#$ to obtain $\hat{}x$ and $\hat{}y$ in $\sigma^*(C)$. Hence Lin(C) is closed under arbitrary factorization and concatenation. It follows that $\sigma^*(C) = C$ if $C = \emptyset$ or $C = \{\lambda\}$, and otherwise $\sigma^*(C) = V_0\hat{}$ where V_0 is the set of symbols of V which occur on strings in C. Hence in this case $\sigma^*(C)$ is regular, so it is in the family $FA\hat{}$. Thus for the remainder of the proof we may assume that α_r and γ_r are non-empty for all $r \in R$.

We also need to arrange that any site of any rule in R occurs on a string of C. To do this we first delete from R any rule which can not be used in any splicing operation involving strings of $\sigma^*(C)$. This preserves the symmetry and reflexivity of R. We then select, for each rule $r \in R$, a string $\hat{}w_r \in \sigma^*(C)$ which contains α_r as a substring and we let $C' = C \cup \{\hat{}w_r \mid r \in R\}$. This is still in $FA\hat{}$ since it differs from C by a finite set. Moreover, since $C \subseteq C' \subseteq \sigma^*(C)$, it is clear that $\sigma^*(C') = \sigma^*(C)$. Thus for the proof of Theorem 5.1 we may replace C with C', so we may assume that for every rule r in R there is a string in C which contains α_r as a substring. (By the symmetry assumption, there is also a string in C which contains $\gamma_r = \alpha_{\bar{r}}$.)

We define the sets X and Y of extra symbols as before. Suppose $\hat{}z\alpha_r$ is an element of C containing the site α_r. From this we generate a linear string in YV^*X by cutting before α_r and marking the cut end with X_r; this is the same as cutting after $\alpha_r = \gamma_{\bar{r}}$, so we mark the other cut end with $Y_{\bar{r}}$. That is, the associated linear string is $Y_{\bar{r}}zX_r$. We let L be the set of all such strings generated from elements of C, and we let $\tilde{L}$ be the closure of L under cutting and pasting using the rules in R. We now consider a string of the form $Y_r z X_r$ to be suitable for reconnection to form the circular string $\hat{}z\beta_r$. We define $\tilde{C}$ as the set of all such circular strings obtained from elements of $\tilde{L}$, together with the original strings in C.

We shall finish the proof of Theorem 5.1 by showing that $\tilde{C}$ is in $FA\hat{}$ and that $\tilde{C} = \sigma^*(C)$.

For the first part we describe $\tilde{C}$ in terms of language operations. First, $\text{Lin}(C)$ is in FA by Lemma 5.1. Next let Y_r be the set of strings of the form $z\alpha_r$ where $\hat{\ }z\alpha_r$ is in C. Then $Y_r = \text{Lin}(C) \cap V^*\{\alpha_r\}$, so Y_r is in FA since FA is closed under intersection with regular sets. But then L can be written as the union of the sets $\{Y_{\bar{r}}\}(L_r/\{\alpha_r\})\{X_r\}$, which is in FA since FA is closed under concatenation, quotients, and finite unions. Since any site of any rule in R lies on some string in C we see that each X_r and each Y_r occurs on some string of L. So Lemma 5.2 applies to show that $\tilde{L}$ is in FA. Now the set of circular strings resulting from reconnecting the elements of $\tilde{L}$ of the form $Y_r z X_r$ may be written as $\text{Cir}((\{Y_r\}\backslash\tilde{L}/\{X_r\})\{\beta_r\})$. This is in $FA\hat{\ }$ since FA is closed under quotients and concatenation. Therefore $\tilde{C}$ is in $FA\hat{\ }$ since it is the union of these finitely many sets together with C.

For the last part we need the following two observations connecting $\tilde{C}$ with $\tilde{L}$ and $\tilde{L}$ with $\sigma^*(C)$.

Assertion 2. *If* $r \in R$ *and* $\hat{\ }x\alpha_r$ *is in* $\tilde{C}$ *then* $Y_{\bar{r}}xX_r$ *is in* $\tilde{L}$.

Proof. This is true, by the definition of L, if $\hat{\ }x\alpha_r$ is in C. Otherwise we can find $s \in R$ and $Y_s y X_s$ in $\tilde{L}$, so that $\hat{\ }x\alpha_r = \hat{\ }y\beta_s$. Pasting three copies of $Y_s y X_s$ together shows that $Y_s y \beta_s y \beta_s y Y_s$ is in $\tilde{L}$. Since $y\beta_s$ is a cyclic permutation of $x\alpha_r$ we can factor this as $Y_s w \alpha_r x \alpha_r z Y_s$. If we notice that $\gamma_{\bar{r}} = \alpha_r$ we see that we can perform two cut operations, one after the first α_r and the other before the second α_r, to show that $Y_{\bar{r}} x X_r$ is in $\tilde{L}$. □

Assertion 3. *If* $Y_s x X_r$ *is in* $\tilde{L}$ *then there is a string* y *so that* $\hat{\ }\gamma_s x \alpha_r y$ *is in* $\sigma^*(C)$.

Proof. We prove this by induction on the number of cutting and pasting operations necessary to produce $Y_s x X_r$. First suppose that $Y_s x X_r$ is in L. Then $s = \bar{r}$ and $\hat{\ }x\alpha_r$ is in C. By reflexivity there is a rule $t \in R$ with α_t, β_t, and γ_t all equal to α_r. If we splice together two copies of $\hat{\ }x\alpha_r$ using this rule we obtain $\hat{\ }x\alpha_r x \alpha_r = \hat{\ }\alpha_r x \alpha_r x = \hat{\ }\gamma_{\bar{r}} x \alpha_r x$ in $\sigma^*(C)$, as required.

For the inductive step we must consider two cases. First, suppose that $Y_s x X_r$ is the result of cutting a string w in $\tilde{L}$ for which Assertion 2 is true. Either $w = Y_s x \alpha_r z X_t$ or $w = Y_t z \gamma_s x X_r$, from which we deduce that either $\hat{\ }\gamma_s x \alpha_r z \alpha_t z'$ or $\hat{\ }\gamma_t z \gamma_s x \alpha_r z' = \hat{\ }\gamma_s x \alpha_r z' \gamma_t z$ is in $\sigma^*(C)$, as required. The second case is that $Y_s x X_r$ is the result of pasting two strings in $\tilde{L}$ which satisfy Assertion 2. We write these as $Y_s x_1 X_t$ and $Y_t x_2 X_r$, so $Y_s x X_r = Y_s x_1 \beta_t x_2 X_r$, and we can find strings $\hat{\ }\gamma_s x_1 \alpha_t y_1$ and $\hat{\ }\gamma_t x_2 \alpha_r y_2$ in $\sigma^*(C)$. Now the result of splicing these strings using the rule t is $\hat{\ }\gamma_s x_1 \beta_t x_2 \alpha_r y_2 \beta_{\bar{t}} y_1$, which has the desired form since $x_1 \beta_t x_2 = x$. □

Now we can argue that $\tilde{C} = \sigma^*(C)$.

First we show that $\tilde{C}$ is closed under splicing. Suppose that $\hat{\ }x\alpha_r$ and $\hat{\ }y\gamma_r$ are in $\tilde{C}$. Remembering that $\gamma_r = \alpha_{\bar{r}}$, we can apply Assertion 1 to see that $Y_{\bar{r}} x X_r$ and $Y_r y X_{\bar{r}}$ are in $\tilde{L}$, and pasting these together shows that $Y_{\bar{r}} x \beta_r y X_{\bar{r}}$ is in $\tilde{L}$. Hence $\hat{\ }x\beta_r y\beta_{\bar{r}}$ is in $\tilde{C}$, and this is the result of splicing $\hat{\ }x\alpha_r$ and $\hat{\ }y\gamma_r$.

Next suppose $\hat{}w = \hat{}x\alpha_r y\gamma_r$ is in $\tilde{C}$. If we rewrite $\hat{}w$ as $\hat{}y\gamma_r x\alpha_r$ and apply Assertion 1 we obtain $Y_{\bar{r}}y\gamma_r x X_r$ in $\tilde{L}$. Now a cut operation after γ_r produces $Y_r x X_r$ in $\tilde{L}$, so $\hat{}x\beta_r$ is in $\tilde{C}$. This is one of the two results of self-splicing $\hat{}w$, and the other is produced similarly.

Since $\tilde{C}$ contains C, we have shown that $\sigma^*(C) \subseteq \tilde{C}$.

Finally we show that $\tilde{C} \subseteq \sigma^*(C)$. Take $\hat{}w \in \tilde{C}$. We may assume that $\hat{}w \notin C$, so $\hat{}w = \hat{}x\beta_r$ where $Y_r x X_r$ is in $\tilde{L}$. By Assertion 2 there is a string y so that $\hat{}\gamma_r x\alpha_r y$ is in $\sigma^*(C)$. But then self-splicing this string using the rule r shows that $\hat{}x\beta_r$ is in $\sigma^*(C)$.

This concludes the proof of Theorem 5.1. □

5.3 Mixed splicing

There is another splicing variation which is important for DNA applications, which arises when both linear and circular words are present. We first ask whether there are new possibilities for the splicing operations. Considerations similar to those for circular splicing lead to two new operations: splicing a linear and a circular string, and self-splicing a linear string. Suppose $r = u_1\#u_2\$u_3\#u_4$ is a splicing rule. If $w = xu_1u_2y$ and $\hat{}v = \hat{}zu_3u_4$ are a linear and a circular string containing the splicing sites then we define the result of splicing w and $\hat{}v$ using r and $\bar{r}$ as the linear string $xu_1u_4zu_3u_2y$. If $w = xu_1u_2yu_3u_4z$ is a linear string containing disjoint copies of the two splicing sites then we define the result of self-splicing w using r and $\bar{r}$ as the circular string $\hat{}u_3yu_2$. (The other plausible result of self-splicing w is the linear string xu_1u_4z, but this is the result of ordinary linear splicing on two copies of w.)

By a *mixed language* over V we shall mean a subset M of the (disjoint) union $V^* \cup V^{\hat{}}$. If we have an H scheme $\sigma = (V, R)$ then we define the *mixed splicing language* $\sigma_m^*(M)$ generated by σ and M as the smallest mixed language containing M and closed under *all* the splicing operations defined so far: linear splicing, circular splicing and self-splicing, and mixed splicing and self-splicing.

The analog of Theorem 5.1 still holds for mixed splicing:

Theorem 5.2. *Suppose FA is a full abstract family of languages which is closed under cyclic closure. Suppose $\sigma = (V, R)$ is a finite symmetric and reflexive H scheme and M is a mixed language with $M \cap V^*$ in FA and $M \cap V^{\hat{}}$ in $FA^{\hat{}}$. Then $\sigma_m^*(L) \cap V^*$ is in FA and $\sigma_m^*(L) \cap V^{\hat{}}$ is in $FA^{\hat{}}$.*

Proof. This is proved by minor adjustments to the proof of Theorem 5.1. As in that proof we need to convert the initial language M to a linear language L to which we can apply cutting and pasting operations. We translate the circular words in M as before. For the linear words we introduce two extra symbols X_0 and Y_0 (where 0 is not in R) and we translate the linear string w in M to Y_0wX_0. We then need to translate the resulting linear strings in $\tilde{L}$ back to mixed strings. Elements of $\tilde{L}$ of the form $Y_r w X_r$ with $r \in R$

translate, as before, to circular words. An element of $\tilde{L}$ of the form $Y_0 w X_0$ simply translates to the linear string w.

There are no other significant changes in the proof.   □

Two special cases of mixed splicing occur when the initial language contains only linear strings, or only circular strings. In these cases there is no difference between mixed splicing and ordinary linear or circular splicing. In the circular case this is obvious, since splicing operations starting with circular strings cannot produce linear strings, but in the linear case it requires some argument:

Theorem 5.3. *Suppose $\sigma = (V, R)$ is a symmetric H scheme. If $L \subseteq V^*$ and $C \subseteq V^\char`\^$ then $\sigma_m^*(L) = \sigma^*(L)$ and $\sigma_m^*(C) = \sigma^*(C)$.*

Proof. We only need to consider the linear case. Let $ML_n = \sigma_m^n(L) \cap V^*$ (the linear strings obtained by n or fewer mixed splicing operations) and $MC_n = \sigma_m^n(L) \cap V^\char`\^$ (the circular strings obtained by n or fewer mixed splicing operations). Since $\sigma^*(L) \subseteq \sigma_m^*(L)$ we only have to show that $ML_n \subseteq \sigma^*(L)$ for all $n > 0$. This is part c) of the following, which we shall prove by induction on n.

Assertion 4.
a) *If $\char`\^w \in MC_n$ then $\char`\^w$ can be obtained from a string in $\sigma^*(L)$ by a single self-splicing operation.*
b) *If $\char`\^w \in MC_n$ then there are u, v such that $uw^k v \in \sigma^*(L)$ for all $k > 0$.*
c) *$ML_n \subseteq \sigma^*(L)$.*

The base case is immediate since $ML_0 = L$ and $MC_0 = \emptyset$. So we suppose that the Assertion is true, and we shall prove that it remains true with n replaced with $n+1$. We use the notation introduced in the previous subsection.

Part a): Suppose that $\char`\^w$ is in $MC_{n+1}$ but not in $MC_n$. Then $\char`\^w$ is obtained from $\sigma_m^n(L) = ML_n \cup MC_n$ by one splicing operation. There are three possibilities: self-splicing a linear string, self-splicing a circular string, and splicing two circular strings. In the first case part a) is immediate. For the second possibility, suppose that $\char`\^w$ is the result of self-splicing a circular string, $\char`\^z \in MC_n$. Thus there is a rule r in R so that $\char`\^z = \char`\^x \alpha_r y \gamma_r$ and $\char`\^w = \char`\^x \beta_r$. Then part b) of the Assertion applied to $\char`\^z$ written as $\char`\^y \gamma_r x \alpha_r$ provides a string $uy\gamma_r x \alpha_r v$ in $\sigma^*(L)$, and self-splicing this using r yields $\char`\^x \beta_r = \char`\^w$.

For the third possibility there are $\char`\^z_1 = \char`\^x_1 \alpha_r$ and $\char`\^z_2 = \char`\^x_2 \gamma_r$ in MC_n which splice using $r \in R$ to give $\char`\^w = \char`\^x_1 \beta_r x_2 \beta_{\bar{r}}$. By part b) of the Assertion we can find strings $z_1' = u_1 x_1 \alpha_r x_1 \underline{\alpha_r} v_1$ and $z_2' = u_2 x_2 \gamma_r x_2 \gamma_r v_2$ in $\sigma^*(L)$. Then splicing z_1' to z_2' using r at the underlined sites produces $u_1 x_1 \underline{\alpha_r} x_1 \beta_r x_2 \gamma_r v_2$, and self-splicing this at the underlined sites yields $\char`\^x_1 \beta_r x_2 \beta_{\bar{r}} = \char`\^w$.

Part b): Suppose that $\hat{w}$ is in MC_{n+1}. Since we have now proved the $n+1$ case of part a) we may assume that $\hat{w}$ is the result of self-splicing a linear string $z \in \sigma^*(L)$. That is, $z = u\gamma_r x\alpha_r v$ and $\hat{w} = \hat{}x\beta_r$. If we splice two copies of z, using r, we obtain $u\gamma_r x\beta_r x\alpha_r v$. Splicing this and another copy of z yields $u\gamma_r x\beta_r x\beta_r x\alpha_r v$. Continuing this process, we see that $u\gamma_r(x\beta_r)^{k+1}x\alpha_r v$ is in $\sigma^*(L)$ for all $k \geq 0$. Since $\hat{w} = \hat{}x\beta_r$ the string $x\beta_r$ is a cyclic permutation of w, so $(x\beta_r)^{k+1}$ contains w^k as a substring. Thus $(x\beta_r)^{k+1} = u'w^k v'$, and we have shown that $u\gamma_r u'w^k v'x\alpha_r v$ is in $\sigma^*(L)$ for all $k > 0$, as desired.

Part c): Suppose that w is in ML_{n+1} but not in ML_n. Then w is obtained from $\sigma_m^n(L) = ML_n \cup MC_n$ by one splicing operation. There are two possibilities. In the first case w is obtained by applying a linear splicing operation to two strings in ML_n. By part c) of the Assertion, $ML_n \subseteq \sigma^*(L)$, so w is in $\sigma^*(L)$. In the second case w is the result of splicing $z \in ML_n \subseteq \sigma^*(L)$ and $\hat{}z' \in MC_n$. Then $z = x_1\alpha_r x_2$, $\hat{}z' = \hat{}y\gamma_r$, and $w = x_1\beta_r y\beta_{\bar{r}}x_2$. By part b) of the Assertion we can find a string $z'' = uy\gamma_r y\gamma_r v$ in L. Then splicing z and z'' using r yields $x_1\beta_r y\gamma_r v$, and this can be spliced with z using $\bar{r}$ to yield $x_1\beta_r y\beta_{\bar{r}}x_2 = w$. So w is in $\sigma^*(L)$.

This concludes the proof of the induction step and hence the proof of Theorem 5.3.   □

6. Computing by splicing

From the results in the previous sections we obtain a lot of situations where, starting from a given language and applying to it, iteratively or not, an H scheme, possibly also using some squeezing mechanisms (an intersection with a regular language – of the form T^* – in the case of extended H systems), we can characterize the recursively enumerable languages. In other words, we reach in these cases the (generative/recognizing) power of Turing machines. Therefore, we can "compute" by splicing (in certain conditions), exactly what we can compute by Turing machines, hence by any known type of algorithms.

For instance, we know from Theorem 3.6 that $RE = EH(mFIN, FIN)$, where $EH(mFIN, FIN)$ denotes the family of languages generated by *extended mH systems*, that is constructs $\gamma = (V, T, A, R)$, with A a multiset with finite support and R a finite set of rules. Let us write $EH(m[k], FIN)$ instead of $EH(mFIN, FIN)$, when A above is a multiset with $card(supp(A)) \leq k$. When using ω multisets, we write $EH(\omega FA_1, FA_2), EH(\omega[k], FA_2)$ for denoting the corresponding families.

A still stronger form of the result in Theorem 3.6 can be obtained, useful below. The following lemma will be used in the proof.

Lemma 6.1. $EH(mFIN, FIN) \subseteq EH(m[2], FIN)$.

Proof. Take an mH system $\gamma = (V, T, A, R)$, with finite $supp(A)$. Let $w_1, w_2, \ldots, w_n$ be the strings of $supp(A)$ such that $A(w_i) < \infty, 0 \leq i \leq n$, and let $z_1, \ldots, z_m$ be the strings in $supp(A)$ with $A(z_i) = \infty, 0 \leq i \leq m$. We construct the mH system

$$\gamma' = (V \cup \{c, d_1, d_2\}, T, A', R'),$$

where A' contains the string

$$w = (w_1 c)^{A(w_1)} (w_2 c)^{A(w_2)} \ldots (w_n c)^{A(w_n)},$$

with multiplicity 1, and the string

$$z = d_1 c z_1 c z_2 c \ldots c z_m c d_2,$$

with infinite multiplicity. If $n = 0$, then w does not appear, if $m = 0$, then $z = d_1 c d_2$. Moreover

$$R' = R \cup \{\#c\$d_2\#, \ \#d_1\$c\#\}.$$

The string z can be used for cutting each w_i and each z_j from w and z, respectively. For instance, in order to obtain z_j we splice z with z using $\#d_1\$c\#$ for the occurrence of c to the left hand of z_j, that is

$$(d_1 c z_1 c \ldots z_{j-1} c z_j c \ldots c z_m c d_2, z) \vdash (d_1 c z_1 c \ldots z_{j-1} c z, \ z_j c \ldots c z_m c d_2),$$

then we splice the second string with z again using $\#c\$d_2\#$, and we get

$$(z_j c \ldots c z_m c d_2, z) \vdash (z_j, \ z c z_{j+1} \ldots c z_m c d_2).$$

Arbitrarily many strings z_j can be produced, because $A'(z) = \infty$.

In order to produce the strings $w_i, 1 \leq i \leq n$, we start from the left hand end of w, applying $\#c\$d_2\#$ to w and z; we get w_1 and $zc(w_1 c)^{A(w_1)-1}$ $(w_2 c)^{A(w_2)} \ldots (w_n c)^{A(w_n)}$, both with multiplicity 1. Using the rule $\#d_1\$c\#$ for z and the second string, we obtain zcz and $(w_1 c)^{A(w_1)-1} (w_2 c)^{A(w_2)} \ldots$ $(w_n c)^{A(w_n)}$, again both with multiplicity 1. From the first string we can separate axioms $z_j, 1 \leq j \leq m$, but this is not important, because these axioms appear with infinite multiplicity in A. From the second string we can continue as above, cutting again a prefix w_1. In this way, exactly $A(w_1)$ copies of w_1 will be produced. We can proceed in a similar fashion with the other axioms $w_2, \ldots, w_n$ in order to obtain exactly the $A(w_i)$ copies of $w_i, i = 2, \ldots, n$.

The use of the nonterminals c, d_1 and d_2 guarantees that only the axioms of γ with multiplicity ∞ can be generated in an arbitrary number by the splicing rules in $R' - R$, whereas for each axiom w_i of γ with finite multiplicity $A(w_i)$ we can only obtain $A(w_i)$ copies of w_i. If a rule of R is used for splicing strings of the form $x_1 c x_2$, i.e. containing the nonterminal c, we will finally have to cut such a string by using the rules in $R' - R$ in order to obtain a terminal string. Since we start with the axioms of γ, separated by c, and with the correct multiplicities (guaranteed by the construction of the strings w and z), this result corresponds to a correct splicing in γ. Consequently, $L(\gamma') = L(\gamma)$. □

Theorem 6.1. $REG = EH(m[1], FIN) \subset EH(m[2], FIN) = RE.$

Proof. From Theorem 3.6 we know that $RE \subseteq EH(mFIN, FIN)$. With Lemma 6.1 we get $RE \subseteq EH(m[2], FIN)$. Combining this with the Turing-Church thesis, we have $RE = EH(m[2], FIN)$.

In Lemma 4.5 we have proved that $REG \subseteq EH(FIN, FIN)$. We can write $REG \subseteq EH(\omega FIN, FIN)$, having in mind that we use multisets M with infinite multiplicity for each $x \in supp(M)$ (by an easy modification in the proof of Lemma 4.5 we can work with ωH systems in the same way as with mH systems, that is producing two strings in each splicing operation). We then apply the construction in the proof of Lemma 6.1 to γ constructed in the proof of Lemma 4.5. We obtain a system γ' with only one axiom, that denoted by z in the proof of Lemma 6.1, with infinite multiplicity. Therefore, $REG \subseteq EH(m[1], FIN)$.

Conversely, $EH(m[1], FIN) \subseteq REG$. Indeed, take $\gamma = (V, T, A, R)$ with $supp(A) = \{w\}$. If $A(w) < \infty$, then $L(\gamma)$ is, obviously, a finite language (every string in $L(\gamma)$ has a length not greater than $|w| \cdot A(w)$).

If $A(w) = \infty$, then $L(\gamma) \in EH(\omega[1], FIN) \subseteq EH(\omega FIN, FIN)$. From Lemma 3.14 we know that $H(\omega FIN, FIN) \subseteq REG$. As REG is closed under intersection and $L(\gamma) = L(\gamma') \cap T^*$, for $\gamma' = (V, V, \{w\}, R)$, we obtain $L(\gamma) \in REG$. Hence we conclude $EH(m[1], FIN) \subseteq REG$.

Therefore, $REG = EH(m[1], FIN)$.   □

The fact that we can fix the number of axioms in an mH system without diminishing the computational completeness suggests that we should look for *universal H systems*, that is systems which have all components but one (the axiom set) fixed, which are able to behave as any given H system γ when a code of γ is introduced in the axiom set of the universal system. The existence of such systems would be a theoretical proof that universal programmable DNA computers can be designed.

The existence of universal mH systems can be derived from the proofs of Theorems 3.6 and 6.1. First, let us define the notion of a universal mH in a precise way.

Given an alphabet T and two families of languages, FA_1, FA_2, a construct

$$\gamma_U = (V_U, T, A_U, R_U),$$

where V_U is an alphabet, $A_U \subseteq V_U^*$, $A_U \in FA_1$, and $R_U \subseteq V_U^* \# V_U^* \$ V_U^* \# V_U^*$, $R_U \in FA_2$, is said to be a *universal* H system of type (FA_1, FA_2), if for every H system $\gamma = (V, T, A, R)$ of any type $(FA_1', FA_2'), FA_1', FA_2' \in RE$, there is a language A_γ such that $A_U \cup A_\gamma \in FA_1$ and $L(\gamma) = L(\gamma_U')$, where $\gamma_U' = (V_U, T, A_U \cup A_\gamma, R_U)$.

Note that the type (FA_1, FA_2) of the universal system is fixed, but the universal system is able to simulate systems *of any type* (FA_1', FA_2').

The restriction to a given terminal alphabet cannot be avoided, but this is anyway imposed by the fact that the DNA alphabet has only four letters. It is perhaps no surprise why this alphabet has been chosen: it is the smallest one by which we can codify two disjoint arbitrarily large alphabets (terminal and nonterminal symbols in our terminology), using two disjoint subsets of it. This is known in language and information theory in general, but this works also in the H systems area.

Lemma 6.2. *Given an extended H system* $\gamma = (V, T, A, R)$ *of type* (FA_1, FA_2), *for* FA_1, FA_2 *families of languages closed under* λ-*free morphisms, we can construct an extended H system* $\gamma' = (\{c_1, c_2\} \cup T, T, A', R')$ *of the same type* (FA_1, FA_2), *such that* $L(\gamma) = L(\gamma')$. *This is also true when* γ *is an mH system.*

Proof. If $V - T = \{Z_1, \ldots, Z_n\}$, then we consider the morphism $h : V^* \longrightarrow (\{c_1, c_2\} \cup T)^*$, defined by

$$h(Z_i) = c_1 c_2^i c_1, \ 1 \leq i \leq n,$$
$$h(a) = a, \ a \in T.$$

Then

$$A' = h(A),$$
$$R' = \{h(u_1) \# h(u_2) \$ h(u_3) \# h(u_4) \mid u_1 \# u_2 \$ u_3 \# u_4 \in R\}.$$

The equality $L(\gamma) = L(\gamma')$ is obvious. Due to the form of the axioms in A' and of the rules in R', the blocks $c_1 c_2^i c_1$, $1 \leq i \leq n$, are never broken by splicing, thus they behave in the same way as the corresponding symbols Z_i do.

For the case when γ is an mH, the definitions of the corresponding components of γ' are obvious. □

Theorem 6.2. *For every given alphabet* T *there exists an mH system of type* $(m[2], FIN)$ *which is universal for the class of mH systems with the terminal alphabet* T.

Proof. Consider an alphabet T and two different symbols c_1, c_2 not in T.

For the class of type-0 Chomsky grammars with a given terminal alphabet, there are universal grammars, i. e. constructs $G_U = (N_U, T, -, P_U)$ such that for any given grammar $G = (N, T, S, P)$ there is a string $w(G) \in (N_U \cup T)^*$ (the "code" of G) such that $L(G'_U) = L(G)$ for $G'_U = (N_U, T, w(G), P_U)$. (The language $L(G'_U)$ consists of all terminal strings z such that $w(G) \Longrightarrow^* z$ using the rules in P_U.) This follows from the existence of universal Turing machines and from the way of passing from Turing machines to type-0 grammars and conversely, or it can be proved directly (an effective construction of a universal type-0 grammar can be found in [2]).

For a given universal type-0 grammar $G_U = (N_U, T, -, P_U)$, we follow the construction in the proof of Theorem 3.6, obtaining an mH system $\gamma_1 = (V_1, T, A_1, R_1)$, where the axiom (with multiplicity 1) $w_0 = X_1^2 Y S X_2^2$ is not considered. Remark that all other axioms in A_1 (all having infinite multiplicity) and the rules in R_1 depend on N_U, T and P_U only, hence they are fixed.

As in the proof of Lemma 6.1, we now pass from γ_1 to $\gamma_2 = (V_2, T, A_2, R_2)$, with at most two axioms in A_2. In fact, as A_1 contains only axioms with infinite multiplicity, A_2 consists of only one string (that one denoted by z in the proof of Lemma 6.1), namely one which has infinite multiplicity.

We now follow the proof of Lemma 6.2, codifying all symbols in $V_2 - T$ by strings over $\{c_1, c_2\}$; the obtained system,

$$\gamma_U = (\{c_1, c_2\} \cup T, T, A_U, R_U)$$

is the universal mH system we are looking for.

Indeed, take an arbitrary mH system $\gamma_0 = (V, T, A, R)$. From the Turing-Church thesis we know that $L(\gamma_0) \in RE$, hence there is a type-0 grammar $G_0 = (N_0, T, S_0, P_0)$ such that $L(\gamma_0) = L(G_0)$ (the grammar G_0 can be constructed directly and in an effective way). Construct the code of G_0, $w(G_0)$, as dictated by the definition of the universal type-0 grammars one uses, and consider the string

$$w_0' = X_1^2 Y w(G_0) X_2^2,$$

corresponding to the axiom w_0 in the proof of Theorem 3.6. Codify w_0' over $\{c_1, c_2\} \cup T$ as we have done above with the axioms of γ_2. By $w(\gamma_0)$ denote the obtained string. Then $L(\gamma_U') = L(\gamma_0)$, for $\gamma_U' = (\{c_1, c_2\} \cup T, T, \{(w(\gamma_0), 1)\} \cup A_U, R_U)$.

This can be easily seen: In the proof of Theorem 3.6, the system γ simulates the work of G, starting from the axiom S of G, bracketed as in $X_1^2 Y S X_2^2$. If we replace S with an arbitrary string x over the alphabet of G, then we obtain in γ exactly the language of terminal strings y such that $x \Longrightarrow^* y$ in G. If we start from a universal grammar G_U and S is replaced by the code $w(G_0)$ of a type-0 grammar G_0 equivalent with γ_0, then the system γ_U, associated as above with G_U, will simulate the work of G_U, starting from $w(G_0)$. Hence $L(\gamma_U') = L(G_U') = L(G_0) = L(\gamma_0)$, for $G_U' = (N_U, T, w(G_0), P_U)$. □

The equality $RE = EH(FIN, REG)$ in Theorem 4.7 can also be used, exactly as above, to obtain universal H systems with regular sets of rules (we bound the number of axioms to one and the number of nonterminals to two by Lemmas 6.1, 6.2).

7. Bibliographical notes

The splicing operation has been introduced in [22], in the following form: a splicing rule over an alphabet V is a pair of triples, $((u_1, v, u_2), (u_3, v, u_4))$. Such a rule, applied to two strings of the form $x_1 u_1 v u_2 x_2, y_1 u_3 v u_4 y_2$ (on the specified sites) produces two new strings, $x_1 u_1 v u_4 y_2$ and $y_1 u_3 v u_2 x_2$. Therefore, a splicing scheme (or system) based on such rules, used as above, corresponds to an H scheme (system) as we have considered here, containing two rules associated to the triple $((u_1, v, u_2), (u_3, v, u_4))$, namely $u_1 v \# u_2 \$ u_3 v \# u_4$ and $u_3 v \# u_4 \$ u_1 v \# u_2$. Consequently, the framework we have used here is slightly more general than that in [22] (and closer to the "contextual thinking" in formal language theory, where, in most cases – see [15], [29], [11], etc. – one controls by pairs of words, called *contexts*, various operations, of rewriting, insertion, deletion, etc.).

In [22] one considers only splicing schemes with finite sets of rules, in the iterated mode of application. One of the main problems raised in [22] concerns the power of such schemes. A first (important) answer to this question is given in [6], where it is proved that iterating a (finite) splicing scheme on a regular language we obtain a regular language, too. The proof uses rather complicated arguments, in terms of the semigroup of dominoes. The proof has been essentially simplified in [44], where the same result is obtained in formal language and automata theory terms. Splicing schemes with infinite sets of rules codified as in the present chapter were considered in [35], where results concerning the relationships between the splicing operations with respect to sets of rules being languages in Chomsky hierarchy and usual operations with languages are given. The above mentioned regularity result in [6] is extended in [45] to the case of arbitrary full AFL's (Lemma 3.16). Lemma 3.15 is from [39].

Most of the results about uniterated splicing are from [35], [41]; a few results are new, or improvements of those in [35], [41]. Morphic characterizations of regular languages also appear in [16], [17].

The case of multisets was first considered in [10], where one proves that the range of any Turing machine can be represented in a precise way using the iterated splicing of a multiset. A representation of recursively enumerable languages (as morphic images of the intersection of a language in $H(mFIN, FIN)$ with a regular language) is obtained in [36]. The result is improved in [14] in the form of Theorem 3.6. The discussion about describing/predicting the evolution of finite multisets is new.

The material from Section 4.1 is from [31], where, furthermore, an algebraic characterization of simple H languages is given, as well as further results (concerning, for instance, the descriptional complexity of such languages). The results in Section 4.2 are from [42], where the notion of an extended H system is introduced.

The problem of splicing circular words is formulated in [22]. Section 5 is mainly based on [45]. Circular words are also considered in [53].

The idea of computing by splicing appears first in [42]. Theorem 6.2 is from [14], where one also finds universal H systems with finite sets of rules and finite sets of axioms, with the work controlled by permitting (or forbidding) contexts: the splicing rules have associated finite sets of symbols; a rule r can be applied for splicing two strings x, y only when the symbols associated to r are present in x, y (respectively, not present, in the case of systems with forbidding contexts). Universality results are also proved in [58] and [12].

Many other notions and results related to the splicing operation, in general, on the recombinant behavior of DNA, appear in the literature. We briefly survey some of them.

Regulated variants of the splicing operation. In the previous section, no restriction is imposed to the using of a splicing rule: the two sites u_1u_2, u_3u_4 defined by $r = u_1\#u_2\$u_3\#u_4$ may appear in x, y in any place, $x = x_1u_1u_2x_2$, $y = y_1u_3u_4y_2$, any rule may be applied to any strings x, y and so on. Both from biological and from mathematical point of view it is however reasonable to restrict this freedom. For instance, one may imagine that the evolution (by DNA recombination) has a purpose, a "favourized direction", hence we may select from the splicing possibilities those which fulfill a given criterion. We can consider such restrictions as: to increase/decrease the length, to produce a string which is a strict continuation of the first term of the splicing, to produce a string from a given "target language", etc. Symmetrically, it is natural to restrict the splicing to strings having a given degree of similarity. This can be achieved by partitioning the set of all strings and allowing splicing to occur only among strings in the same class. In particular, the classes can consist of singleton languages, of all strings of equal length, they can be regular sets, etc. Then, we can impose restrictions on the place where the sites u_1u_2, u_3u_4 appear in the spliced strings (on the leftmost/rightmost possible positions, as a prefix/suffix, covering completely the string). Finally, the splicing rule to be used can be selected in such a way that its sites are maximal/minimal among the set of the applicable rules, or it can be selected according to a *priority* criterion (a partial order relation on the set of rules), or according to *favouring* or *inhibiting* symbols (like in the random context grammars in regulated rewriting).

There are a lot of variants. Formal definitions and results about the non-iterated case of the associated splicing operations can be found in [41], [27]. The iterated case is still a *research topic* to be considered.

Crossovering. An operation with DNA sequences, similar to the splicing, is the crossovering: given two strings x, y, we jump from x to y and back to x, and so on, a number of times, combining subwords of x, y as specified by the jumping positions. (The splicing is a particular case, with only one jump, from x to y.) It is obvious that when the number of jumps is not specified, a crossovering operation can be simulated by an iterated splicing. An interesting case appears when the number of jumps is fixed (and the places where they are done are ordered). Results about such an operation appear in [34], [26] (many problems are still open in this area).

Splicing on graphs. The reader is referred to [13], where a series of detailed (and complex) definitions are given. The subject still remains to be investigated.

Generating strings by replication. A sort of 2-crossovering is considered in [33], based on the following idea. Take some given *insertion contexts*, (u, v), over some alphabet V, and a string $x \in V^*$. We interpret each (u, v) as a "weak link", where x can be cut and where a further segment, also being a substring of x, can be inserted, providing the links are reproduced. That is, if $x = x_1uvx_2$ and $y = vy' = y''u$ is a substring of x, then we produce the string $z = x_1uyvx_2 = x_1uvy'vx_2 = x_1uy''uvx_2$. Remark how the string z contains again the links uv, before y' and after y''. Based on this operation of *replication*, we can define *replicating systems*, triples $\gamma = (V, A, R)$, where V is an alphabet, A is a language over V and R is a set of insertion contexts. As for H schemes, we can codify the contexts in R as strings, hence we can allow infinite sets R, of a given type in Chomsky hierarchy. As for the splicing operation, we can also consider replicating schemes, pairs $\sigma = (V, R)$, which then can be applied uniterately or iterately to languages. The language generated by a replicating system $\gamma = (V, A, R)$ is the language obtained by iterating the underlying replicating scheme $\sigma = (V, R)$ on the language A of axioms.

A particular form of replicating systems was investigated in [33]; the cases of more general replicating systems remain to be considered (and they seem to lead to rather difficult problems).

The *simple replicating systems* discussed in [33] are of the form $\gamma = (V, w, (a, b))$, where $w \in V^+$ and $a, b \in V$ (one axiom and only one context, of "radius" one). Several variants of the replication operation as specified above are investigated in [33], namely with restrictions about the choice of the inserted substring y. For instance, one can impose any of the following restrictions on the inserted string y: y is a prefix of the current string, y is a maximal prefix or a minimal prefix, y is the leftmost, maximal leftmost or minimal leftmost of the form $y = vy' = y''u$, y is arbitrarily maximal or minimal among the substrings of the current string. Ten languages $L_g(\gamma)$ can be associated in this way to a replicating system γ, hence ten families (of "snakes" – as termed in [33]) are obtained. A comparison of the size of these families relative to each other and to the families in Chomsky hierarchy, is investigated in [33]. All the relationships between the ten families are settled: somewhat surprisingly, most of them are incomparable.

Also settled in [33] is the regularity of languages generated by simple replicating systems. However, the proofs in [33] concerning regularity cannot be directly extended from simple systems to more general systems. The problem is *open* for replicating systems having finite but arbitrary sets of insertion contexts.

Appendix: The derivation of the splicing concept from DNA recombination

Although the formal systems treated in this chapter have the potential to refer to various linear polymers such as polypeptides, RNA, and single-stranded DNA, the model has been explicitly devised to express the cut and paste activities carried out in vitro on double-stranded DNA with restriction enzymes and a ligase. Such activities have been carried out routinely in recent decades in the activity called gene splicing. Our model is liberal in considering the potential recombinant activities made possible in an environment containing

at the same time several different restriction enzymes and a ligase. Practical
genetic engineering normally uses only one or two restriction enzymes and a
ligase, possibly made available sequentially. For details not considered here
see [22], [23], and the biological references listed in them.

Single stranded DNA molecules are linear polymers consisting of four
bases denoted A, C, G, and T. Each such base is joined to the next by a
phosphate group that is bonded through the so-called 3' carbon atom of the
ribose portion of one base and the 5' carbon of the ribose portion of the next
base. As a result of this 3' − 5' distinction, each single strand of DNA has a
unique representation exemplified by 5' − $GCTTAC$ − 3' where the 5' and
the 3' each tell us that the molecule is being displayed with a 5' carbon to the
left and a 3' carbon to the right. There is redundance here since in every case
a well formed single strand must have a 5' at one end and a 3' at the other.
Bases pair naturally through the formation of weak, easily broken, hydrogen
bonds. A and T pair with each other and C and G pair with each other. This
provides for DNA strands to associate into double stranded DNA molecules
having forms as illustrated by the following three molecules:

$$5' - CCCCCTCGACCCCC - 3' \qquad 5' - AAAAAGCGCAAAAA - 3'$$
$$3' - GGGGGAGCTGGGGG - 5' \qquad 3' - TTTTTCGCGTTTTT - 5'$$

$$5' - TTTTTGCGCTTTTT - 3'$$
$$3' - AAAAACGCGAAAAA - 5'$$

Notice that complementary strands always have opposite 5' − 3' orienta-
tion.

Restriction enzymes (endonucleases) cut double stranded DNA molecules
in specific ways. The recognition sequences at which the enzymes *Taq*I, *Sci*
NI, and *Hha*I cut are, respectively:

$$TCGA \qquad\qquad GCGC \qquad\qquad GCGC$$
$$AGCT \qquad\qquad CGCG \qquad\qquad CGCG$$

When *Taq*I, *Sci*NI, and *Hha*I act, respectively, on the three DNA molecules
listed above they cut these molecules at the unique sites occurring in these
molecules to yeld the six fragments as illustrated:

$$5' - CCCCCT \qquad CGACCCCC - 3'$$
$$3' - GGGGGAGC \qquad TGGGGG - 5'$$

$$5' - AAAAAG \qquad CGCAAAAA - 3'$$
$$3' - TTTTTCGC \qquad GTTTTT - 5'$$

$$5' - TTTTTGCG \qquad CTTTTT - 3'$$
$$3' - AAAAAC \qquad GCGAAAAA - 5'$$

The action of *Taq*I is to sever one covalent bond in each strand of its recognition sequence. In each case it is the bond between the T and the neighboring G that is cut. With these two covalent bonds severed, the hydrogen bonds between the two C/G pairs are not strong enough to hold the molecule firmly together. The two fragments that result can then drift apart. However, the potentiality for forming hydrogen bonds between the G and C bases makes the "overhangs" (each reading CG when read in the 5′ to 3′ direction) "sticky". Notice that the result of the cut made by *Sci*NI leaves precisely similar CG sticky overhangs. Notice that the cut made by *Hha*I also leaves a CG sticky overhang (when reading in the 5′ to 3′ direction). However, there is an absolutely crucial distinction: The free tips of all four overhangs created by *Taq*I and *Sci*NI are at 5′ ends, but the free tips of the two overhangs created by *Hha*I are at 3′ ends. The result is that the fragments produced by the first two enzymes are compatible in the sense that hydrogen bonds can form between the two-base sticky ends of a fragment produced by one of these enzymes and a fragment produced by the other. Carefully note that there is no such compatibility between the fragments produced by *Hha*I and the fragments produced by either of the other two enzymes. We say that *Taq*I and *Sci*NI produce 5′-overhangs and that *Hha*I produces 3′-overhangs. Various overhangs are provided by various restriction enzymes. For example, *Eco*RI produces the 5′-overhang $AATT$. Overhangs are rarely of length more than six bases. See [46] for a detailed discussion of restriction enzymes.

The effect of a ligase enzyme, when introduced into a set of DNA fragments that have been produced by restriction enzymes, is to bind *compatible pairs of fragments* back into double stranded molecules by reestablishing covalent bonds. This allows restoration of molecules of the same form as those previously cut. But it also allows "recombinant" molecules of new types to be formed. Using as examples the six fragments given above, we point out two molecules that may arise through ligation:

$$5' - CCCCCTCGCAAAAA - 3' \qquad 5' - AAAAAGCGACCCCC - 3'$$
$$3' - GGGGGAGCGTTTTT - 5' \qquad 3' - TTTTTCGCTGGGGG - 5'$$

The formation of these recombinant molecules is only possible because the overhangs match appropriately. The action of the ligase enzyme is to create covalent bonds that restore the integrity of the strands. In the left molecule the ligase has constructed a bond TC in the upper strand and a bond CG in the lower strand. In the right molecule the ligase has constructed a bond GC in the upper strand and a bond CT in the lower strand. The key features of the behaviors that the formal systems of this chapter were initiated to model have here been described. Periferal features that can be considered by those interested include the duplicity of representations of double stranded DNA molecules, appropriate buffers to support the enzymatic activities, and energy sources in the form of ATP. We leave these considerations to our references and develop the formal model.

The phenomenon to be modeled is the process of producing new double stranded DNA molecules from pre-existing double stranded DNA molecules, especially by the cut and paste methods outlined above. Several fundamental modeling decisions must be motivated and made clear. The purpose of the model is to allow a formal treatment of the generative power of recombinant processes to be incorporated smoothly into the well established theory of formal languages. Two fundamental issues arise immediately. Should the model used for a double stranded DNA molecule recognize the fact that it can be analysed into two single strands? Should the model recognize fragments such as those with sticky overhangs as illustrated above? The answer given here to both questions is: No. Positive answers would not provide models that blend smoothly into standard formal language theory. The previous sections of this chapter demonstrate that the negative decisions have allowed the development of a theory that works well with such classical features as the Chomsky hierarchy and AFL theory. It is certainly possible to develop models that are based on affirmative answers to both questions and, in fact, the first treatment of regularity was given in [6] in this way. The reader may wish to compare the treatment in [6] with the corresponding portions of the exposition provided in this chapter.

In order to model each double stranded DNA molecule with a single string we must use as our alphabet the four compound "two-level" symbols

$$
\begin{array}{cccc}
A & C & G & T \\
T & G & C & A
\end{array}
$$

Using two lines for a symbol is cumbersome. We adopt a much simpler notation taking advantage of the fact that *we will no longer treat single stranded DNA*. Since the second row of symbols in a representation of a well-formed double stranded DNA molecule is redundant, *we now represent double stranded DNA molecules by writing only one row*. Thus the two recombinant molecules represented above are now represented more simply by:

$$5' - CCCCCTCGCAAAAA - 3' \qquad 5' - AAAAAGCGACCCCC - 3'$$

The set $D = \{A, C, G, T\}$ has now become the alphabet for representing *double stranded* DNA molecules. Be sure to keep in mind that each of $A, C, G,$ and T now represents a two level pair of bases joined by hydrogen bonds: A is now an abbreviation for A/T, C for C/G, G for G/C, and T for T/A.

In order to model the effect of restriction enzymes and a ligase without formally recognizing the fragments having sticky overhangs we proceed as follows: With the set of restriction enzymes being modeled we construct two sets B and C as follows. Subdivide the restriction enzymes into two classes: those that create 5'-overhangs and those that create 3'-overhangs. Set B will be formed by inspecting the recognition sequences and cut sites of the enzymes that produce 5'-overhangs and set C will be formed by inspecting the recognition sequences and cut sites of those that produce 3'-overhangs.

The construction of the sets B and C will be made clear with an example. Let the situation to be modeled be the one in which the set of enzymes to be used consists of TaqI, SciNI, and HhaI. The information we need concerning the cutting activity of TaqI is expressed completely by writing (T, CG, A) and noting that this enzyme produces $5'$-overhangs. *Since $5'$-overhangs are produced, we place (T, CG, A) in the set B.* That this triple is in B encodes the fact that TaqI acts at the recognition sequence $TCGA$ and cuts leaving sticky $5'$-overhangs CG. For exactly similar reasons we place (G, CG, G) in B to represent the action of SciNI. *Since HhaI produces $3'$-overhangs, we place (C, CG, C) in the set C.* That this triple is in C encodes the fact that HhaI acts at the recognition sequence $GCGC$ and cuts leaving sticky $3'$-overhangs CG (reading always in the $5'$ to $3'$ direction). Finally, suppose that we also wish to use EcoRI. This enzyme operates at the recognition sequence $GAATTC$ and cuts leaving sticky $5'$-overhangs at $AATT$. Consequently we merely add $(G, AATT, C)$ to B in order to add the action of EcoRI into our model.

There are restriction enzymes which cut without sticky overhangs. In the introduction to this chapter two such enzymes are given: DpnI and BalI. When there is no overhang it is convenient to write that the overhang is λ, which denotes the null string. Using this convention we give as patterns for these two enzymes as: (GA, λ, TC) and (TGG, λ, CCA), respectively. Enzymes that cut without leaving overhangs are said to leave "blunt ends". Any two blunt end enzymes are compatible as illustrated in the introduction for DpnI and BalI. *We incorporate the blunt end enzymes into our formalism by adding their patterns to the set B.* For the six enzyme system discussed here we have $B = \{(T, CG, A), (G, CG, C), (G, AATT, C), (GA, \lambda, TC), (TGG, \lambda, CCA)\}$ and $C = \{(G, CG, C)\}$. The biochemical discussion is now adequate to motivate the original definition of the concept of a splicing system.

Definition .1. *A splicing system $S = (V, I, B, C)$ consists of a finite alphabet V, a finite set I of initial strings in V^*, and finite sets B and C of triples (y, x, z) with $y, x, z \in V^*$. Each such triple in B or C is called a pattern. For each such triple the string yxz is called a site and the string x is called a crossing. Patterns in B are called left patterns and patterns in C are called right patterns.*

Definition .2. *The language $L = L(S)$ generated by $S = (V, I, B, C)$ is defined inductively to be the union L of the following sequence of languages: $L_0 = I$, and, for each integer $i \geq 0$, $L_{i+1} = L_i \cup \{w \in V^* \mid w$ is either $x_1 p x v y_2$ or $y_1 u x q x_2$ where the following condition holds: $x_1 p x q x_2$ and $y_1 u x v y_2$ are in L_i with (p, x, q) and (u, x, v) patterns of the same hand$\}$.*

With each experiment in which, into an appropriate aqueous solution, a finite set of double stranded DNA molecules (many copies of each), (many copies of) a ligase, and a finite set of restriction enzymes (many copies of each) are to be added we associate the splicing system $S = (V, I, B, C)$: V is the four letter DNA alphabet; I is determined from the sequence of the bases occurring in the initial set of DNA molecules; and B and C are determined from the cutting patterns of the restriction enzymes. See [22] for the exact details of this process. For this model S, $L = L(S)$ represents the set of all possible well formed double stranded DNA molecules that can arise in the in vitro environment in which the molecules present are only those chosen for the experiment. The biochemical background that motivated the original definition of a splicing system [22] is now complete. Additional details are given in [22] and [23].

Notice that the two definitions above, of $S = (V, I, B, C)$ and $L(S)$, have provided for the modeling of the set of well formed double stranded DNA molecules obtainable from an initial set by splicing without granting any formal recognition to fragments having sticky overhangs. In this way splicing can be discussed in the context of ordinary formal language theory. This is a consequence of our giving the common answer of "No" to the two fundamental questions raised earlier about how the model for splicing should be constructed.

In this chapter splicing operations are not discussed within the context of the splicing systems as just defined, but rather within the newer context of H-systems. The final objective of this appendix is to explain the relation between these two types of systems and the value of each.

The sets B and C in $S = (V, I, B, C)$ are quite natural from a biochemical point of view: An endonuclease cuts leaving either a $5'$-overhang or a $3'$-overhang or blunt ends. There is no other possibility. Mathematically the $B : C$ distinction serves only to express one of the forms of limitation on compatibility for recombination: the patterns (p, x, q) and (u, x, v) can produce new strings only if both lie in B or both lie in C. The intrinsic generality of the mathematical results concerning splicing is not fully expressible when these proofs are made using the biochemically natural $S = (V, I, B, C)$ formalism. Moreover, building the $B : C$ distinction into proofs makes the proofs more awkward and less clear than necessary. All this can be seen from an examination of the proofs appearing in the first few papers dealing with splicing systems.

Splicing is discussed in terms of the concept of an H scheme in this chapter. We consider that this chapter demonstrates that the H scheme concept provides a fully adequate response to the mathematical awkwardness of the original splicing system concept as outlined in the previous paragraph. We next show how to express the information in the splicing *scheme* $SS = (V, B, C)$ in the form of an H scheme $\sigma = (V, R)$, where R is a language of a special form over the augmented alphabet $V \cup \{\#, \$\}$: For each pair of patterns (p, x, q) and (u, y, v) of the same hand and having the same crossing $x = y$, place $px\#q\$ux\#v$ in R. Each rule of R is said to be a splicing rule and R is said to be a language of rules. In an H scheme $\sigma = (V, R)$, R is always a subset of $V^*\#V^*\$V^*\#V^*$. Each H scheme $\sigma = (V, R)$ becomes a unary operation on the set of all languages over V by defining: $\sigma(L) = \{x_1 u_1 u_4 y_2 \mid x_1 u_1 u_2 x_2 \in L, y_1 u_3 u_4 y_2 \in L$ and $u_1 \# u_2 \$ u_3 \# u_4 \in R\}$. Exponents applied to σ provide unary operations by defining iteratively: $\sigma^0(L) = L$ and $\sigma^{i+1}(L) = \sigma^i(L) \cup \sigma(\sigma^i(L))$. (Note that $\sigma^1(L)$ is not $\sigma(L)$. Instead it is $L \cup \sigma(L)$.) Finally, σ^* also becomes a unary operator by defining: $\sigma^*(L) = \bigcup\{\sigma^i(L) \mid i \geq 0\}$. A careful reading of all the relevant definitions confirms that the language $L(S)$ generated by the splicing system $S = (V, I, B, C)$ is identical with the language $\sigma^*(I)$ for $\sigma = (V, R)$ derived above from $SS = (V, B, C)$ as above.

When maximum generality is desired in the mathematics of splicing, the H system formulation is appropriate. However, if DNA computation is to be carried out using restriction enzymes and a ligase then it may be necessary to formulate results in terms of splicing systems as defined in this appendix. A representation theory for H systems in terms of splicing systems may therefore be desired. The freedom allowed in choosing a rule set R does not recognize various closure requirements that hold on any R that is built from a splicing system and therefore also one that is built from a concrete set of restriction enzymes. The simplest example of properties that inevitably hold for any R that is constructed from an $S = (V, I, B, C)$ follow:

(1) If $u_1 \# u_2 \$ u_3 \# u_4$ is in R then both $u_1 \# u_2 \$ u_1 \# u_2$ and $u_3 \# u_4 \$ u_3 \# u_4$ are in R.

(2) If $u_1 \# u_2 \$ u_3 \# u_4$ is in R then $u_3 \# u_4 \$ u_1 \# u_2$ is in R.

The first of these conditions is akin to reflexivity and the second to symmetry. An example of one of the more subtle conditions that must hold is:

(3) If $u_1 \# u_2 \$ u_1 \# u_3$, $u_1 \# u_2 \$ u_1 \# u_4$, and $u_1 \# u_2 \$ u_1 \# u_5$ are all in R then either $u_1 \# u_3 \$ u_1 \# u_4$ is in R or $u_1 \# u_3 \$ u_1 \# u_5$ is in R or $u_1 \# u_4 \$ u_1 \# u_5$ is in R.

The distinction between B (5'-overhangs) and C (3'-overhangs) allows the construction of chemical examples for which a natural analog of transitivity fails, i.e., for which, in the H theory formulation, there are $u_1 \# u_2 \$ u_3 \# u_4$ and $u_3 \# u_4 \$ u_5 \# u_6$ in R with $u_1 \# u_2 \$ u_5 \# u_6$ not in R. Ultimately, of course, any computing system based on splicing must consider in detail the exact recognition sequences and the cutting pattern of any enzymes to be used.

Acknowledgement

Thanks are due to Elizabeth Laun, Hendrik Jan Hoogeboom, and Andreas Weber for carefully reading earlier versions of the manuscript.

References

[1] B. Brainerd, An analog of a theorem about context-free languages, *Inform. Control*, 11 (1968), 561–567.

[2] C. Calude, Gh. Păun, Global syntax and semantics for recursively enumerable languages, *Fundamenta Informaticae*, 4, 2 (1981), 245–254.

[3] J. Collado-Vides, The search for a grammatical theory of gene regulation is formally justified by showing the inadequacy of context-free grammars, *CABIOS*, 7 (1991), 321–326.

[4] E. Csuhaj-Varju, L. Freund, L. Kari, Gh. Păun, DNA computing based on splicing: universality results, *First Annual Pacific Symp. on Biocomputing*, Hawaii, Jan. 1996.

[5] E. Csuhaj-Varju, L. Kari, Gh. Păun, Test tube distributed systems based on splicing, *Computers and AI*, 15, 2-3 (1996), 211–232.

[6] K. Culik II, T. Harju, Splicing semigroups of dominoes and DNA, *Discrete Appl. Math.*, 31 (1991), 261–277.

[7] J. Dassow, V. Mitrana, Splicing grammar systems, *Computers and AI*, 15, 2-3 (1996), 109–121.

[8] J. Dassow, Gh. Păun, *Regulated Rewriting in Formal Language Theory*, Springer-Verlag, Berlin, 1989.

[9] A. De Luca, A. Restivo, A characterization of strictly locally testable languages and its application to subsemigroups of a free semigroup, *Inform. Control*, 44 (1980), 300–319.

[10] K. L. Denninghoff, R. W. Gatterdam, On the undecidability of splicing systems, *Intern. J. Computer Math.*, 27 (1989), 133–145.

[11] A. Ehrenfeucht, Gh. Păun, G. Rozenberg, Contextual grammars, in the present *Handbook*.

[12] C. Ferretti, S. Kobayashi, T. Yokomori, DNA splicing systems and Post systems, *First Annual Pacific Symp. on Biocomputing*, Hawaii, Jan. 1996.

[13] R. Freund, Splicing on graphs,*Proc. 1st Intern. Symp. on Intell. in Neural and Biological Systems*, IEEE, Herndon, 1995, 189–194.

[14] R. Freund, L. Kari, Gh. Păun, DNA computing based on splicing: The existence of universal computers, *Technical Report 185-2/FR-2/95*, TU Wien, 1995.

[15] B. S. Galiukschov, Semicontextual grammars, *Math. Logica i Math. Ling.*, Kalinin Univ., 1981, 38–50 (in Russian)

[16] R. W. Gatterdam, Splicing systems and regularity, *Intern. J. Computer Math.*, 31 (1989), 63–67.

[17] R. W. Gatterdam, DNA and twist free splicing systems, in *Words, Languages and Combinatorics, II* (M. Ito, H. Jürgensen, eds.), World Sci. Publ., Singapore, 1994, 170–178.

[18] S. Ginsburg, *Automata and Language-Theoretic Properties of Formal Languages*, North-Holland, Amsterdam, 1975.

[19] J. Gruska, A few remarks on the index of context-free grammars and languages, *Inform. Control*, 19 (1971), 216–223.

[20] J. Gruska, Descriptional complexity of context-free languages, *Proc. MFCS Symp.*, High Tatras, 1973, 71–83.

[21] M. Harrison, *Introduction to Formal Language Theory*, Addison-Wesley, Reading, Mass., 1978.

[22] T. Head, Formal language theory and DNA: an analysis of the generative capacity of specific recombinant behaviors, *Bull. Math. Biology*, 49 (1987), 737–759.

[23] T. Head, Splicing schemes and DNA, in *Lindenmayer Systems: Impacts on Theoretical Computer Science and Developmental Biology* (G. Rozenberg, A. Salomaa, eds.), Springer-Verlag, Berlin, 1992, 371–383.

[24] G. T. Herman, G. Rozenberg, *Developmental Systems and Languages*, North-Holland, Amsterdam, 1975.

[25] J. E. Hopcroft, J. D. Ullman, *Introduction to Automata Theory, Languages, and Computing*, Addison-Wesley, Reading, Mass., 1979.

[26] L. Ilie, V. Mitrana, Crossing-over on languages. A formal representation of the recombination of genes in a chromosome, submitted, 1995.

[27] L. Kari, Gh. Păun, A. Salomaa, The power of restricted splicing with rules from a regular set, *Journal of Univ. Computer Sci.*, 2, 4 (1996).

[28] M. Latteux, B. Leguy, B. Ratoandromanana, The family of one-counter languages is closed under quotient, *Acta Informatica*, 22 (1985), 579–588.

[29] S. Marcus, Contextual grammars, *Rev. Roum. Math. Pures Appl.*, 14 (1969), 1525–1534.

[30] S. Marcus, Linguistic structures and generative devices in molecular genetics, *Cah. ling. th. appl.*, 11, 2 (1974), 77–104.

[31] A. Mateescu, Gh. Păun, G. Rozenberg, A. Salomaa, Simple splicing systems, *Discrete Applied Math.*, to appear.

[32] R. McNaughton, S. Papert, *Counter-Free Automata*, MIT Press, Cambridge, Mass., 1971.

[33] V. Mihalache, Gh. Păun, G. Rozenberg, A. Salomaa, Generating strings by replication: a simple case, *Acta Informatica*, to appear.

[34] V. Mitrana, Crossover systems: a language-theoretic approach to DNA recombination, submitted, 1995.

[35] Gh. Păun, On the splicing operation, *Discrete Appl. Math.*, 70 (1996), 57–79.

[36] Gh. Păun, On the power of the splicing operation, *Intern. J. Computer Math.*, 59 (1995), 27–35.

[37] Gh. Păun, The splicing as an operation on languages, *Proc. 1st Intern. Symp. on Intell. in Neural and Biological Systems*, IEEE, Herndon, 1995, 176–180.

[38] Gh. Păun, Splicing. A challenge to formal language theorists, *Bulletin EATCS*, 57 (1995), 183–194.

[39] Gh. Păun, Regular extended H systems are computationally universal, *J. Automata, Languages, Combinatorics*, 1, 1 (1996), 27–36.

[40] Gh. Păun, On the power of splicing grammar systems, *Ann. Univ. Bucureşti, Matem.-Inform. Series*, 45, 1 (1996), 93–106.

[41] Gh. Păun, G. Rozenberg, A. Salomaa, Restricted use of the splicing operation, *Intern. J. Computer Math.*, 60 (1996), 17–32.

[42] Gh. Păun, G. Rozenberg, A. Salomaa, Computing by splicing, *Theoretical Computer Sci.*, 168, 2 (1996), 321–336.

[43] Gh. Păun, A. Salomaa, DNA computing based on the splicing operation, *Mathematica Japonica*, 43, 3 (1996), 607–632.

[44] D. Pixton, Regularity of splicing languages, *Discrete Appl. Math.*, 69, (1996), 101–124.

[45] D. Pixton, Linear and circular splicing systems, *Proc. 1st Intern. Symp. on Intell. in Neural and Biological Systems*, IEEE, Herndon, 1995, 38–45.

[46] R. J. Roberts, S. M. Lim, R. S. Lloyd (eds.), *Nucleases*, Second Edition, Cold Spring Harbor Laboratory Press, 1993.

[47] G. Rozenberg, A. Salomaa, *The Mathematical Theory of L Systems*, Academic Press, New York, 1980.

[48] A. Salomaa, On the index of context-free grammars and languages, *Inform. Control*, 14 (1969), 474–477.

[49] A. Salomaa, *Formal Languages*, Academic Press, New York, London, 1973.

[50] M. P. Schützenberger, On finite monoids having only trivial subsemigroups, *Inform. Control*, 8 (1965), 190–194.

[51] M. P. Schützenberger, Sur certaines operations de fermeture dans les langages rationels, *Symposia Math.*, 15 (1975), 245–253.

[52] D. B. Searls, The linguistics of DNA, *American Scientist*, 80 (1992), 579–591.

[53] R. Siromoney, K. B. Subramanian, V. Rajkumar Dare, Circular DNA and splicing systems, *Proc. of Parallel Image Analysis, LNCS 654*, Springer-Verlag, Berlin, 1992, 260–273.

[54] Y. Takada, R. Siromoney, On identifying DNA splicing systems from examples, *Proc. AII'92, LNAI 642* (P. K. Jantke, ed.), Springer-Verlag, Berlin, 1992, 305–319.

[55] A. Thue, Über unendliche Zeichenreichen, *Norsche Vid. Selsk. Skr., I. Mat. Nat. Kl.*, Kristiania, 7 (1906), 1–22.

[56] T. Yokomori, M. Ishida, S. Kobayashi, Learning local languages and its application to protein α-chain identification, *Proc. 27th Hawaii Intern. Conf. System Sci.*, vol. V, 1994, 113–122.

[57] T. Yokomori, S. Kobayashi, DNA evolutionary linguistics and RNA structure modelling: a computational approach, *Proc. 1st Intern. Symp. on Intell. in Neural and Biological Systems*, IEEE, Herndon, 1995, 38–45.

[58] T. Yokomori, S. Kobayashi, C. Ferretti, On the power of circular splicing systems and DNA computability, *Report CSIM 95-01*, Univ. of Electro-Comm., Chofu, Tokyo, 1995.

String Editing and Longest Common Subsequences

Alberto Apostolico

Summary. The string editing problem for input strings x and y consists of transforming x into y by performing a series of weighted edit operations on x of overall minimum cost. An edit operation on x can be the deletion of a symbol from x, the insertion of a symbol in x or the substitution of a symbol of x with another symbol. String editing models a variety of problems arising in such diverse areas as text and speech processing, geology and, last but not least, molecular biology. Special cases of string editing include the longest common subsequence problem, local alignment and similarity searching in DNA and protein sequences, and approximate string searching. We describe serial and parallel algorithmic solutions for the problem and some of its basic variants.

1. Introduction

Let x be a string of $|x| = m$ symbols from some alphabet Σ of cardinality s. We say that Σ is *bounded* when s is a constant independent of m, *unbounded* otherwise. We consider three *edit operations* on x, namely, *deletion* of a symbol from x, *insertion* of a new symbol in x, and *substitution* of one of the symbols of x with another symbol from Σ. We assume that each edit operation has an associated nonnegative real number representing the *cost* of that operation. More precisely, the cost of deleting from x an occurrence of symbol a is denoted by $D(a)$, the cost of inserting some symbol a between any two consecutive positions of x is denoted by $I(a)$ and the cost of substituting some occurrence of a in x with an occurrence of b is denoted by $S(a, b)$. An *edit script* on x is any sequence S of viable edit operations on x, and the cost of S is the sum of all costs of the edit operations in S.

For any given string x, there is always some edit script S that transforms x into any other string of Σ^*, so that the only possible questions of interest revolve around the cost of S. Let x and y be two strings of respective lengths $|x| = m$ and $|y| = n \geq m$. The *string editing problem* for input strings x and y consists of finding an edit script $S\prime$ of minimum cost that transforms x into y. The cost of $S\prime$ is the *edit distance from x to y*. Edit distances where individual operations are assigned integer or unit costs occupy a special place. Such distances are often called Levenshtein distances, since they were introduced by Levenshtein [1966] in connection with error correcting codes. String editing finds applications in a broad variety of contexts, ranging from text and speech processing to geology, from computer vision to molecular biology.

It is not difficult to see that the general (i.e., with unbounded alphabet and unrestricted costs) problem of edit distance computation is solved by a serial algorithm in $\Theta(mn)$ time and space, through dynamic programming. Due to widespread application of the problem, however, such a solution and a few basic variants were discovered and published in literature catering to such diverse disciplines (see, e.g., Sankoff and Kruskal [1983] and references therein). In computer science, the problem was dubbed "the string-to-string correction problem" by Wagner and Fischer [1974]. The CS literature was possibly the last to address the problem, but interest in the CS community increased steadily in subsequent years. By the early 1980s, the problem had proved so pervasive, especially in biology, that a book (Sankoff and Kruskal [1983]) was devoted almost entirely to it. Special issues of the Bulletin of Mathematical Biology and various other books and journals routinely dedicate significant portions to it (see, e.g., Martinez [1984], Waterman [1989]).

An $\Omega(mn)$ lower bound was established for string editing by Wong and Chandra [1976] for the case where the queries on symbols of the string are restricted to tests of equality. For unrestricted tests, a lower bound $\Omega(n \log n)$ was given by Hirschberg [1978]. Algorithms slightly faster than $\Theta(mn)$ were devised by Masek and Paterson [1980], thru resort to the so-called "Four Russians Trick". The "Four Russians" are Arlazarov, Dinic, Kronrod, and Faradzev [1970] (see also Aho, Hopcroft and Ullman [1974] for a discussion of their approach). Along these lines, the total execution time becomes $\Theta(n^2/\log n)$ for bounded alphabets and $O(n^2(\log \log n)/\log n)$ for unbounded alphabets. The method applies only to the classical Levenshtein distance metric, and does not extend to general cost matrices. To this date, the problem of finding either tighter lower bounds or faster algorithms is still open.

The criterion that subtends the computation of edit distances by dynamic programming is readily stated. For this, let $C(i,j)$, $(0 \leq i \leq |x|, 0 \leq j \leq |y|)$ be the minimum cost of transforming the prefix of x of length i into the prefix of y of length j. Let s_k denote the kth symbol of string s. Then $C(0,0) = 0$, $C(i,0) = C(i-1,0)+D(x_i)$ $(i = 1,2,...,m)$, $C(0,j) = C(0,j-1)+I(y_j)$ $(j = 1,2,...,n)$, and

$$C(i,j) = \min\{C(i-1,j-1)+S(x_i,y_j), \ C(i-1,j)+D(x_i), \ C(i,j-1)+I(y_j)\}$$

for all i,j, $(1 \leq i \leq |x|; 1 \leq j \leq |y|)$. Observe that, of all entries of the C-matrix, only the three entries $C(i-1,j-1)$, $C(i-1,j)$, and $C(i,j-1)$ are involved in the computation of the final value of $C(i,j)$. Hence $C(i,j)$ can be evaluated row-by-row or column-by-column in $\Theta(|x||y|) = \Theta(mn)$ time. An optimal edit script can be retrieved at the end by backtracking thru the local decisions that were made by the algorithm.

A few important problems are special cases of string editing, including the *longest common subsequence* problem, *local alignment*, i.e., the detection of local similarities of the kind sought typically in the analysis of molecular sequences such as DNA and proteins, and the problem of *searching for approximate occurrences* of a pattern string in a text string. As highlighted in the following brief discussion, a solution to the general string editing problem implies typically similar bounds for all these special cases.

1.1 Approximate string searching

In this problem, we assume unit cost for all edit operations. Given a *pattern* x and a *text* y, the most general variant of the problem consists of computing, for every position of the text, the best edit distance between the pattern x and some substring w of y ending at that position. It is not difficult to express a solution in terms of a suitable adaptation of our previous recurrence. The first obvious change consists of setting all costs to 1 except that $S(x_i, y_j) = 0$ for $x_i = y_j$. Thus, we have now, for all i, j, $(1 \le i \le |x|; 1 \le j \le |y|)$,

$$C(i,j) = \min\{C(i-1, j-1) + 1, \ C(i-1, j) + 1, \ C(i, j-1) + 1\}.$$

A second change consists of setting the initial conditions so that $C(0, 0) = 0$, $C(i, 0) = i$ $(i = 1, 2, ..., m)$, $C(0, j) = 0$ $(j = 1, 2, ..., n)$. This has the effect of setting to zero the cost of prefixing x by any prefix of y. In other words, any prefix of the text can be skipped free of charge in an optimum edit script.

Clearly, the computation of the final value of $C(i, j)$ may proceed as in the general case, and it will still take $\Theta(|x||y|) = \Theta(mn)$ time. Note, however, that we are interested now in the entire last row of matrix C at the outset.

In practical cases, one would be more interested in methods capable of locating only those segments of y that present a high similarity with x. Formally, given a pattern x, a text y and an integer k, this restricted version of the problem consists of locating all substrings w of y such that the edit distance between w and x is at most k. While the recurrence above obviously incorporates all the distances, there are more efficient methods to deal with this restriction, e.g., with a worst case time complexity $O(kn)$ or even sublinear expected time. Landau and Vishkin [1986, 1988], Sellers [1974], Ukkonen [1985], Galil and Giancarlo [1988], Chang and Lawler [1990], are some good sources for the various notions of approximate string searching and their connection to the string editing problem.

1.2 Local similarity searches in DNA and protein sequences

The basic and widely accepted working hypothesis of molecular biology is that all of the information presiding over the propagation and development of living organisms is encoded in the four bases adenine (A), thymine (T), guanine (G), and cytosine (C) along the DNA (or, in some cases, RNA) molecules that compose their genomes. It is also believed that different pathways in the evolution of species are provoked by elementary perturbations that come to these genomic sequences as one of our edit operations. Once such edit operations are assumed at the basis of molecular evolution, then the cost of an optimum edit script between two genetic sequences may be taken as a measure of their phylogenetic distance. This explains the interest for general string editing in molecular biology applications. In fact, there is a considerable number of motivations and contexts for computing edit distances and alignments of molecular sequences: There is global alignment of pairs of sequences that are globally related by common ancestry, local alignments of related sequences, multiple alignments of members of protein families, self-alignments measuring the autocorrelation in a same sequence, auxiliary alignments performed in data base searches, and so on.

Local alignment is motivated by the fact that not all of the genetic material seems equally crucial to the functionality of organisms, so that the process of evolution tends to preserve some segments of DNA while accepting more liberally changes to others. Notable among the most preserved regions of DNA are certain special segments, that translate into sequences of amino acids that fold up as particular proteins, the building blocks of life. Often in the practice of biological sequence comparison, sequences are compared that have a very poor global similarity score, but embed pairs of highly similar segments. In these cases, methods that detect such local resemblances become of interest.

One way of doing this is by resort to a recurrence formulated by Smith and Waterman [1981]. This assumes as initial conditions $C(i, 0) = C(0, j) = 0$, and then sets:

$$C(i, j) = \max\{C(i-1, j-1) + S(x_i, y_j), \ C(i-1, j) + \delta, \ C(i, j-1) + \delta, \ 0\}$$

for all i, j, $(1 \leq i \leq |x|; 1 \leq j \leq |y|)$. Here δ is the negative weight of deleting or inserting a symbol, and S is also negative on average. At the outset, $C(i, j)$ yields the cost of the best edit script to transforms some suffix of $x_1 x_2 ... x_i$ into some corresponding suffix of $y_1 y_2 ... y_j$. Hence high entries in C denounce the terminal points of segments in the two sequences that have a high similarity score.

1.3 Longest common subsequences

Given a string x over an alphabet $\Sigma = (\sigma_1, \sigma_2, ... \sigma_s)$, a *subsequence* of x is any string w that can be obtained from x by deleting zero or more (not necessarily consecutive) symbols. The *longest common subsequence* (LCS) *problem* for input strings $x = x_1 x_2 ... x_m$ and $y = y_1 y_2 ... y_n$ $(m \leq n)$ consists of finding a third string $w = w_1 w_2 ... w_l$ such that w is a subsequence of x and also a subsequence of y, and w is of maximum possible length. In general, string w is not unique.

Like the string editing problem itself, the LCS problem arises in a number of applications spanning from text editing to molecular sequence comparisons, and has been studied extensively over the past. Its relation to string editing can be understood as follows.

Observe that the effect of a given substitution can be always achieved, alternatively, by an appropriate sequence consisting of one deletion and one insertion. When the cost of a non-vacuous substitution (i.e., a substitution of a symbol with a different one) is higher than the global cost of one deletion followed by one insertion, then an optimum edit script will always avoid substitutions and produce instead y from x solely by insertions and deletions of overall minimum cost. Specifically, assume that insertions and deletions have unit costs, and that a cost higher than 2 is assigned to substitutions. Then, the pairs of matching symbols preserved in an optimal edit script constitute a longest common subsequence of x and y. It is not difficult to see that the cost e of such an optimal edit script, the length l of an LCS and the lengths of the input strings obey the simple relationship: $e = n + m - 2l$. Similar considerations can be developed for the variant where matching pairs are assigned weights and a *heaviest* common subsequence is sought (see, e.g., Jacobson and Vo [1992]).

Lower bounds for the LCS problem are time $\Omega(n \log n)$ or linear time, according to whether the size s of Σ is unbounded or bounded (Hirschberg [1978]). Aho, Hirschberg and Ullman [1976] showed that, for unbounded alphabets, any algorithm using only "equal-unequal" comparisons must take $\Omega(nm)$ time in the worst case. The asymptotically fastest general solution rests on the corresponding solution by Masek and Paterson [1980] to the string editing, hence takes time $O(n^2 \log \log n / \log n)$. Time $\Theta(mn)$ is achieved by the following dynamic programming algorithm from Hirschberg [1975].

Let $L[0...m, 0...n]$ be an integer matrix initially filled with zeroes. The following code transforms L in such a way that $L[i, j]$ $(1 \leq i \leq m, 1 \leq j \leq n)$ contains the length of an LCS between $x_1 x_2 ... x_i$ and $y_1 y_2 ... y_j$.

> **for** $i = 1$ **to** m **do**
>   **for** $j = 1$ **to** n **do if** $x_i = y_j$ **then** $L[i, j] = L[i - 1, j - 1] + 1$
>   **else** $L[i, j] = \text{Max} \{L[i, j - 1], L[i - 1, j]\}$

The correctness of this strategy follows from the obvious relations:

$$
\begin{aligned}
L[i - 1, j] &\leq L[i, j] \leq L[i - 1, j] + 1; \\
L[i, j - 1] &\leq L[i, j] \leq L[i, j - 1] + 1; \\
L[i - 1, j - 1] &\leq L[i, j] \leq L[i - 1, j - 1] + 1.
\end{aligned}
$$

If only the length l of an LCS is desired, then this code can be adapted to use only linear space. The basic observation for this is again that the computation of each row of L only needs the preceding row. If an LCS needs be retrieved at the outset by backtracking, then it would seem necessary to keep track of the decision made at every step by the algorithm. However,

Hirschberg [1975] combined the dynamic programming above and divide-and-conquer in such a way as to compute and output an LCS in time $\Theta(nm)$ and linear space.

The LCS problem will constitute the focus of our discussion through most of this paper, in view of the particularly rich variety of algorithmic solutions that have been devised for this problem over the past two decades or so, which made it susceptible to some degrees of unification and systematization of independent and general interest. Our discussion starts with the exposition of two basic approaches to LCS computation, due respectively to Hirschberg [1978] and Hunt and Szymanski [1977]. We then discuss faster implementations of this second paradigm, and the data strucures that support them. In Section 5 we discuss algorithms that use only linear space to compute an LCS and yet do not necessarily take $\Theta(nm)$ time. One, final, such algorithm is presented in Section 6 where many of the ideas and tools accumulated in the course of our discussion find employment together. In Section 7 we make return to string editing in its general formulation and discuss some of its efficient solutions within a parallel model of computation.

2. Two basic paradigms for the LCS problem

The more recent approaches to the LCS problem achieve time complexities better than $\Theta(nm)$ in favorable cases, though a quadratic time complexity is always touched and sometimes even exceeded in the worst cases. These approaches exploit in various ways the *sparsity* inherent to the LC problem. Sparsity allows us to relate algorithmic performances to parameters other than the lengths of the input. Some such parameters are introduced next.

The ordered pair of *positions* i and j of L, denoted $[i, j]$, is a *match* iff $x_i = y_j$. We use r to denote the total number of matches between x and y. If $[i, j]$ is a match, and an LCS $w_{i,j}$ of $x_1 x_2 ... x_i$ and $y_1 y_2 ... y_j$ has length k, then k is the *rank* of $[i, j]$. The match $[i, j]$ is k-*dominant* if it has rank k and for any other pair $[i\prime, j\prime]$ of rank k either $i\prime > i$ and $j\prime \leq j$ or $i\prime \leq i$ and $j\prime > j$. A little reflection establishes that computing the k-dominant matches $(k = 1, 2, ..., l)$ is all is needed to solve the LCS problem (see, e.g., Apostolico and Guerra [1987], Hirschberg [1977]). Clearly, the LCS of x and y has length l iff the maximum rank attained by a dominant match is l. It is also useful to define, on the set of matches in L, the following partial order relation $\mathcal{R}$: match $[i, j]$ *precedes* match $[i\prime, j\prime]$ in $\mathcal{R}$ if $i < i\prime$ and $j < j\prime$. A set of matches such that in any pair one of the matches always precedes the other in $\mathcal{R}$ constitutes a *chain* relative to the partial order relation $\mathcal{R}$. A set of matches such that in any pair neither element of the pair precedes the other in $\mathcal{R}$ is an *antichain*. Then, the LCS problem translates into the problem of finding a longest *chain* in the *poset* of matches induced by $\mathcal{R}$ (cf. Sankoff and Sellers [1973]). A decomposition of a poset into antichains is *minimal* if it partitions the poset into the minimum possible number of antichains (refer, e.g., to Bogart [1983]).

Theorem 2.1. (Dilworth [1950]) *A maximal chain in a poset P meets all antichains in a minimal antichain decomposition of P.*

In other words, the number of antichains in a minimal decomposition represents also the length of a longest chain. Even though it is never explicitly stated, most known approaches to the LCS problem implicitly compute a minimal antichain decomposition for the poset of matches induced by $\mathcal{R}$. The k-th antichain in this decomposition is represented by the set of all matches having rank k. For general posets, a minimal antichain decomposition is computed by flow techniques (see Bogart [1983]), although not in time linear in the number of elements of the poset. Most LCS algorithms that exploit sparsity have their natural predecessors in either Hunt and Szymanski [1977] or Hirschberg [1977]. In terms of antichain decompositions, the approach of Hirschberg [1977] consists of computing the antichains in succession, while that of Hunt and Szymanski [1977] consists of extending partial antichains relative to all ranks already discovered, one new symbol of y at a time. The respective time complexities are $O(nl + n \log s)$ and $O(r \log n)$. Thus, the algorithm of Hunt and Szymanski is favorable in very sparse cases, but worse than quadratic when r tends to nm. An important specilaization of this algorithm is that to the problem of finding a *longest ascending subsequence* in a permutation of the integers from 1 to n. Here, the total number of matches is n, which results in a total complexity $O(n \log n)$. Resort to the *fat-tree* structures introduced by Van Emde Boas [1975] leads to $O(n \log \log n)$ for this problem, a bound which had been shown to be optimal by Fredman [1975].

Figure 2.1. illustrates the concepts introduced thus far, displaying the final L-matrix for the strings $x = atcgtt$ and $y = ctactaata$. We use circles to represent matches, with bold circles denoting dominant matches. Dotted lines thread antichains relative to $\mathcal{R}$ and also separate regions.

		C	T	A	C	T	A	A	T	A
		1	2	3	4	5	6	7	8	9
A	1	0	0	①	1	1	①	①	1	①
T	2	0	①	1	1	②	2	2	②	2
C	3	①	1	1	②	2	2	2	2	3
G	4	1	1	1	2	2	2	2	2	3
T	5	1	②	2	2	③	3	3	③	3
T	6	1	②	2	2	③	3	3	④	4

Fig. 2.1. Illustrating antichain decompositions

2.1 Hirschberg's paradigm: finding antichains one at a time

We outline a $\Theta(mn)$ time LCS algorithm in which antichains of matches relative to the various ranks are discovered one after the other. Let the *dummy* pair $[0, 0]$ be labeled a 0-dominant match, and assume that all $(k-1)$-dominant matches for some k, $0 \leq k \leq l - 1$, have been discovered at the expense of scanning the part of the L-matrix that would lie above or to the left of the antichain $(k-1)$, inclusive. To find the k-th antichain, scan the unexplored area of the L-matrix from right to left and top-down, until a stream of matches is found occurring in some row i. The leftmost such match is the k-dominant match $[i, j]$ with smallest i-value. The scan continues at next row and to the left of this match, and the process is repeated at successive rows until all of the k-th antichain has been identified. Note that for each k the list of $(k - 1)$-dominant matches is enough to describe the shape of the antichain and also to guide the searches involved at the subsequent stage. Thus, also in this case linear space is sufficient if one wishes to compute only the length of w.

An efficient implementation of this scheme leads to the algorithm by Hirschberg [1977], which takes time $O(nl + n \log s)$ and space $O(d + n)$. Some preprocessing is necessary for that algorithm. Specifically, for each distinct symbol σ in x, we need the count $N(\sigma)$ of all distinct occurrences of σ in y, and also a list σ-OCC of the increasing positions of y that correspond to occurrences of σ. We will find it convenient to make the convention that y is always replaced by $y\$$, where $\$$ is a joker symbol not in Σ, whence the last entry of any σ-OCC list is always $n + 1$. In general, producing the σ-OCC lists charges $O(n \log s)$ time, but the lists require only $O(n)$ space, collectively.

Hirschberg [1977] showed that, as a consequence of the introduction of the σ-OCC lists, one may now identify antichains by traveling on such lists rather than on the rows of matrix L. In this way, each antichain requires only $O(n + m)$ steps. Moreover, the condition that l, the length of an LCS, has been reached is testable in constant time, whence this algorithm requires $O(ln + n \log s)$ in total. We leave it as an exercise for the reader to fill in the details of this construction, and describe below a similar algorithm that requires $O(lm + r + n \log s)$ time, inclusive of preprocessing (Apostolico and Guerra [1987]).

We use an array of integers $PEBBLE[1..m]$, initialized to 1, the role of which shall become apparent later. If $x_i = \sigma_p$, $PEBBLE[i]$ either points to (the location of) an entry $j \leq n$ of σ_p-OCC, and is said to be *active*, or it points to (the location of) $n + 1$ and is *inactive*.

Our algorithm consists of l *stages*, stage k being the set of operations involved in identifying all the k-dominant matches. A match is k-*internal* $(k = 1, 2, ..., l)$ if its rank is larger than k. Stage k, $1 \leq k \leq l$ begins with all active entries of $PEBBLE$ pointing to $(k - 1)$-internal matches and ends with those same entries of $PEBBLE$ pointing to k-internal matches. During stage k the *pebbles*: $PEBBLE[k]$, $PEBBLE[k + 1], \ldots , PEBBLE[m]$, are considered in succession (indeed, no $PEBBLE[i]$ with $i < k$ can be active at stage k). The k-dominant matches detected are appended to the k-th list in the array of lists $RANK$ (through the concatenation operation '$\|$'). The entire process terminates as soon as there are no active pebbles left. We also use an auxiliary table called $SYMB$, which is defined as follows. $SYMB[j] = k$, if $y_j = \sigma_p\text{-}OCC$. Thus $SYMB$ gives constant time access to the entry in the $\sigma\text{-}OCC$ list that corresponds to the symbol of y occurring at any position. We make the convention that, each time $PEBBLE[i]$ is being handled by the algorithm $(i = 1, 2, ..., m)$, then $SYMB[n+1]$ takes the value $N(x_i)$. The table $SYMB$ can be prepared in linear time from the $\sigma\text{-}OCC$ lists, quite easily, and we will not spend time on it. Within *Algorithm LCS1*, $SYMB$ is used to speed up the advancement of some $PEBBLE[i']$, if $x_{i'} = x_i$ for some $i < i'$, and T, the *threshold*, was not changed since row i.

Algorithm LCS1

```
0      for i = 1 to m do PEBBLE[i]  =  1; (initialize pebbles)
1      k = 0
2      while  there are active pebbles do (start stage k + 1)
3             begin  T = n + 1; k = k + 1; RANK[k] = Λ;
4             for i  = k to m  do (advance pebbles)
              begin
5                    t = T;
6                    if xᵢ-OCC[PEBBLE[i]] < T then
                     (record a k-dominant match; update threshold)
7                            begin RANK[k] =
                             RANK[k]||[i, xᵢ-OCC[PEBBLE[i]]];
8                            T = xᵢ-OCC[PEBBLE[i]]
                     end ;
                     (advance pebble, if appropriate)
9                    if xᵢ = yₜ
10                   then PEBBLE[i] = SYMB[t] + 1
11·                  else while xᵢ-OCC[PEBBLE[i]] < t do
                     PEBBLE[i]  =  PEBBLE[i] + 1
              end ;
       end.
```

Figure 2.2. shows the positions occupied by the pebbles at the beginning of each of the stages performed by *Algorithm LCS1* on the input strings of Fig. 2.1.

To illustrate the action of the algorithm, we trace its stage 1, which produces the first boundary (consisting of all the 1-dominant matches). The algorithm starts by assigning the value '10' to both T and t, i.e., the variables which will be used to store the current and previous threshold, respectively (lines 3,5). Next, it compares the first occurrence of symbol $a = x_1$ in y (line 6). Such test is passed (3<10), whence the first 1-dominant match is detected

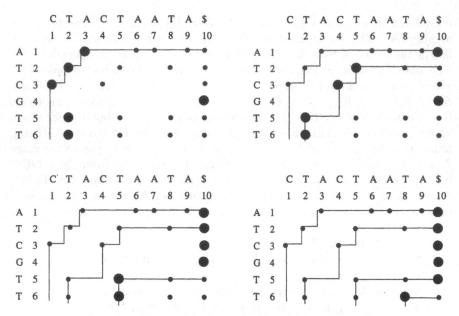

Fig. 2.2. Illustrating the operation of Algorithm LCS1

and appended to the antichain $RANK[1]$ (line 7). Moreover, T is updated to the new value '3' (line 8). At this point, the algorithm tries to advance $PEBBLE[1]$ onto a 1-internal match. By definition of \$, x_i matches \$. Moreover, by our convention, the current value of $SYMB[10]$ is $N(a) = 4$. Thus line 10 is executed following the test of line 9, with the effect of bringing $PEBBLE[1]$ to its rightmost position on $a\text{-}OCC$, and rendering it inactive. As *Algorithm* 1 proceeds to consider $x_2 = b$, the test of line 6 prompts the detection and recording of a new 1-dominant match on column 2 of the L-matrix. This is followed by the advancement of $PEBBLE[2]$ which is thus brought on column 5. $PEBBLE[3]$ is subjected to a similar treatment. In our example, the first three pebbles yield all the dominant matches for the first stage. When $PEBBLE[4]$ is considered, it does not pass the test of line 6; line 11 has no effect and this pebble is left in its inactive status. The last two pebbles are also left in their original position. We encourage the reader to carry out for himself or herself the remainder of this example, with the aid of Fig. 2.2.

As is easy to check, *Algorithm LCS*1 maintains the following invariant condition: if a pebble is considered for the k-th time, then there is no match of rank k on the same row and to the left of that pebble. In other words, if such pebble is active, then it represents either a k-dominant match or a k-internal match. The savings over the algorithm in Hirschberg [1977] is in that in *Algorithm LCS*1 matches whose ranks have been already determined are not reconsidered at subsequent stages.

Theorem 2.2. *Algorithm LCS1 takes time $O(lm + r)$*

Proof. During stage k, $m - k + 1$ pebbles are considered in succession. Each pebble either is advanced some position to the right or it is not moved. The number of moves on one row is bounded by the number of matches on that row, thus the total number of moves is bounded by r. A pebble is considered exactly once during each stage, thus the number of times a pebble can stay put is bounded by l, which yields a total of lm. ☐

If we include preprocessing, the bounds of our strategy become $O(l \times m + r + n \, \log s)$ and $O(d)$ space (linear space if only the length l is sought). If $r < lm$ and m is much smaller than n, this is better than the $O(ln + n \, \log s)$ in Hirschberg [1977]. When r is large compared to $m \times l$, the strongest cause of inefficiency becomes the inner *while* loop of *Algorithm* 1, which generates the $O(r)$ term. Later in our exposition we will see that it is possible to eliminate this term.

2.2 Incremental antichain decompositions and the Hunt-Szymanski paradigm

When the number r of matches is small compared to m^2 (or to the expected value of lm), an algorithm with running time bounded in terms of r may be advantageous. Along these lines, Hunt and Szymanski [1977] set up an algorithm (HS) with a time bound of $O((n + r) \log n)$. This algorithm works by computing, row after row, the ranks of all matches in each row. The treatment of a new row corresponds thus to extending the antichain decomposition relative to all preceding rows. A same match is never considered more than once. On the other hand, the time required by HS degenerates as r gets close to mn. In these cases this algorithm is outperformed by the algorithm of Hirschberg [1977], which exhibits a bound of $O(ln)$ in all situations.

Algorithm HS is reproduced below as our *Algorithm LCS2*. Essentially, it scans the $MATCHLIST$ associated with the i-th row and considers the matches in succession, from right to left. For each match, HS decides whether it is a k-dominant match for some k through a binary search in the array $THRESH$. If this is the case, then the contents of $THRESH[k]$ is suitably updated. Observe that considering the matches in reverse order is crucial to the correct operation of HS. For this, HS needs to preprocess y to obtain the *reverse* of each *sigma-OCC* list: the resulting lists are called $MATCHLIST$s in Hunt and Szymanski [1977].

Algorithm LCS2 "HS": element array $x[1 : m], y[1 : n]$;
integer array $THRESH[0 : m]$; list array $MATCHLIST[1 : m]$;
pointer array $LINK[1 : m]$; pointer PTR;
begin $(PHASE\ 1:$ initializations)
    for $i = 1$ **to** m **do**

$$\text{set } MATCHLIST[i] \ = \ \{ \ j_1, j_2, ..., j_p \}$$
$$\text{such that } j_1 \ > \ j_2... \ > \ j_p$$
$$\text{and } x_i \ = \ y_{j_q} \text{ for } 1 \le q \le p$$
$$\text{set } THRESH[i] \ = \ n+1 \text{ for } 1 \le i \le m;$$
$$THRESH[0] \ = \ 0; \ LINK[0] \ = \ null;$$
($PHASE$ 2 :   find k-dominant matches)
> **for** $i \ = \ 1$ **to** m **do**
>> **for** j **on** $MATCHLIST[i]$ **do**
>>> **begin** *find* k *such that*
>>>> $THRESH[k-1] \ < \ j \ \le \ THRESH[k];$
>>> **if** $j \ < \ THRESH[k]$ **then**
>>>> **begin** $THRESH[k] \ = \ j;$
>>>>> $LINK[k] \ = \ newnode(i, j, \ LINK[k-1])$
>>> **end**
>> **end**

($PHASE$ 3 :   recover LCS w in reverse order)
> $k \ = \ largest \ k \ such \ that \ THRESH[k] \ne n+1;$
> $PTR \ = \ LINK[k];$
> **while** $PTR \ne null$ **do begin**
> *print the match* $[i, j]$ *pointed to by PTR;*
> *advance PTR* **end**

end.

The total time spent by HS is bounded by $O((r+m)\log n \ + \ n \log s)$, where the $n \log s$ term is charged by the preprocessing. The space is bounded by $O(d+n)$. As mentioned, this is good in sparse cases but becomes worse than quadratic for dense r.

3. A speed-up for HS

We can rearrange HS in a way that not only exposes some of its sources of inefficiency but also eliminates them (cf., e.g., Apostolico [1986]).

Instead of considering all the matches in each row, *Algorithm* LCS3 ("HS'") below (cf., e.g., Apostolico [1986]) maintains an *active list* of matches associated with that symbol. A match is in the active list if it is not currently a threshold. The algorithm spots all and only the new dominant matches contributed by any given active list by performing a number of *dictionary primitives* (cf. Aho et al. [1974]) proportional to the number of these new dominant matches, i.e., irrespective of the current size of the active list involved. This might appear counterintuitive at first: one implicit task of the algorithm is to decide which matches are dominant, and yet we are asking that it should never touch a match unless that match will turn out to be dominant.

Algorithm LCS3: "HS/ "
begin
for $i = 1$ **to** m **do**
  begin $\sigma = char(x_i)$;
  $PEBBLE = first(AMATCHLIST[\sigma])$;
  $FLAG = true$;
  while $FLAG$ **do**
    begin
     1) $T = SEARCH(\ PEBBLE\ ,\ THRESH)$;
       $k = rank(SUCC[T])$;
     2) **if** $T = NIL$ **then** $FLAG = false$;
     3) $INSERT(\ PEBBLE\ ,\ THRESH)$;
       $DELETE(\ SUCC[T]\ ,\ THRESH)$;
     4) $LINK[k] = newnode(i, PEBBLE, LINK[k-1])$;
     5) $\sigma\prime = char(y_T)$;
     6) $DELETE(\ PEBBLE\ ,\ AMATCHLIST[\sigma])$;
     7) $PEBBLE = SEARCH(T, AMATCHLIST[\sigma])$;
     8) $INSERT(SUCC[T], AMATCHLIST[\sigma\prime])$;
    end;
  end;
retrieve an LCS as per phase 3 of HS;
end.

Algorithm $LCS3$ uses the same array $THRESH$ of HS. The "active" lists $AMATCHLIST[\sigma_p]$, $p = 1, 2, ...s$ are initialized to coincide with the old $MATCHLIST$s. The primitives $INSERT$ and $DELETE$ have the usual meaning. $SEARCH(\ key\ ,\ LIST)$ returns the largest element in $LIST$ which is not larger than key (NIL, if no such element exists). $SEARCH(NIL, LIST)$ returns NIL without performing any action. Notice that all the searches performed within $HS\prime$ terminate without success (i.e., key is not in $LIST$). The function $char(symbol)$ returns the element of the alphabet Σ which coincides with $symbol$.

To illustrate the operation of $HS\prime$, refer to Fig. 2.1. and assume it resulted from applying $HS\prime$ to $y = CTACTAABA$ and the first six characters of $x = ATCGTTA$.... At this point, $THRESH$ consists of $\{1, 2, 5, 8\}$, since $\sigma_1 = x_7 = A$. Thus, $AMATCHLIST[A] = \{9, 7, 6, 3\}$ coincides with the A-OCC list. The management of x_7 leads to $THRESH = \{1, 2, 3, 6, 9\}$. $AMATCHLIST[A]$ shrinks to just $\{7\}$, while $AMATCHLIST[T]$ is given back the matches '5' and '8'.

The correctness of $HS\prime$ hinges on two easy invariant conditions. Specifically, after the i-th iteration:

1) The k-th entry of $THRESH$ is the smallest position in y such that there is a k-dominant match between x_i and y.

2) $AMATCHLIST[\sigma_t]$ $(t = 1, 2, ..., s)$ contains all and only the occurrences of σ_t in y which are not currently in $THRESH$.

Theorem 3.1. *Throughout its execution, Algorithm $HS\prime$ performs $\Theta(d)$ searches, insertions and deletions.*

Proof. All the searches, insertions and deletions take place in the *while* loop (lines 1–8) controlled by $FLAG$. There is a fixed number of such primitives within these lines, whence it will do to show that $FLAG$ is *true* exactly d times. We can assume w.l.o.g. (as a result of trivial preprocessing) that for $\sigma = char(x_1)$ $AMATCHLIST[\sigma]$ is not empty. Then the last element on this list (i.e., the leftmost match in the form $[1,j]$) is a 1-dominant match, as well as the only dominant match from that list. By initialization, $FLAG$ is *true* the first time it is tested. Since $THRESH$ is empty at this time, lines (3,4) will be executed, whence the first 1-dominant match is recorded. The algorithm updates the other lists so that they are consistent at the next step. Since the $SEARCH$ of line (1) returns NIL, then $FLAG$ is set to the value *false*, which concludes the treatment of the first row. On a generic row, the first match on the $AMATCHLIST$ is certainly a k-dominant match for some k. Assume that a certain number of entries of this $AMATCHLIST$ have been processed and that: (i) the number of times that $FLAG$ was *true* equals the number of dominant matches detected so far, (ii) j is the last dominant match detected, and (iii) j is the only such match which has not been recorded yet. It is easy to see that $HS\prime$: locates the displacement of this match in $THRESH$ (line 1); switches $FLAG$ to *false*, if appropriate (line 2); updates the lists and records this new dominant match in $LINK$ (lines 3–6, 8), and probes into $AMATCHLIST[\sigma]$ seeking the next position to which the $PEBBLE$ should be advanced to mark the next dominant match (line 7, meaningful only if $FLAG$ is *true*). Thus $FLAG$ is *true* as long as conditions (i–iii) hold, that is, exactly for d times.   □

Since dominant matches are a subset of all matches, then it is always $d \leq r$. However, there is no general relationship between d and r. For example, if $x = y$ and x and y are permutations of the first n integers, then we have $d = r$, and also $d = n$. At the other extreme, if $x = y = a^n$, then $r = n^2$, but we still have that $d = n$. hence d is linear both in very dense and very sparse cases. Unfortunately, there are still intermediate cases where $d = \Theta(n^2)$.

A time bound of $HS\prime$ depends on the choice of list implementation. If 2-3 trees or AVL trees (see, e.g., Aho et al. [1974], Mehlhorn [1984]) are used, then $HS\prime$ runs in $O(d \log n + n \log s)$ time, inclusive of preprocessing, and this reduces to $O(d \log \log n + n \log s)$ if one uses a specialized structure for the manipulation of integers [VE]. While this represents an improvement over HS, it is still worse than quadratic in the worst case. Observe, however, that the insertions in each list occur like in merging sorted linear sequences, for which efficient dynamic structures such as *finger-trees* are available (cf., e.g., Brown and Tarjan [1978], Mehlhorn [1984]). Specifically, two lists of sizes k and $f \geq k$, respectively, may be merged in time $O(k \log(2f/k))$. Thus the total time spent by $HS\prime$ for the mergings might be bounded by a form such as

$O(m \log m + d \log(2mn/d))$. In general, this bound no longer applies when deletions are intermixed with insertions in an unpredictable way. Luckily, the special mixture in $HS\prime$ is still susceptible to efficient implementation on finger trees, as we see next.

4. Finger trees

For our purposes, a finger tree is a *level–linked* (a, b)-*tree* $(b \geq 2, a \geq 2)$ with fingers (see Mehlhorn [1984]). A *finger* is simply a pointer to a leaf. A typical finger tree can be obtained, for instance, from a standard 2-3 tree by adding auxiliary links in such a way that it is possible to reach, from each node, its father, children, and neighboring nodes on the same level. Thus, a finger tree can be traversed in any direction.

Let S be a linearly ordered sequence of m elements. The search for an element i in a finger tree T_S storing the sequence S returns the first element j of S not larger than i (NIL, if no such element exists). We assume that any such search originates at the bottom of T_S and from a *finger* leaf f (see, e.g., Brown and Tarjan [1978]). For instance, assume that $i < f$. The search starts by climbing from leaf f toward the root (see Fig. 4.1.), using level-links to inspect left neighboring nodes, until the first node v is found such that the interval of S subtended by v has a left delimiter not larger than i. From this moment on, the search may proceed downwards from node v, driven by the range information stored in each node. Clearly, the effort involved in such a search is proportional to the number of nodes that are traversed in the climb.

In intuitive terms, every new node visited in a climb corresponds to roughly doubling the previous guess for the number of leaves separating j

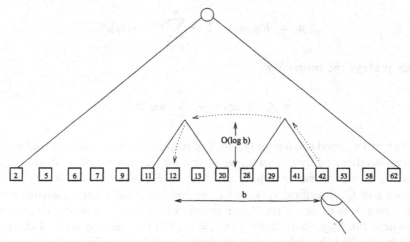

Fig. 4.1. Highlighting a search in a finger tree: searching for an item displaced b positions apart from the finger takes time only $O(\log b)$

from the finger f, much as it happens in an unbounded search (Bentley and Yao [1976]). This observations can be made rigorous, resulting in the following statement (cf., e.g., Mehlhorn [1984]).

Lemma 4.1. *Searching a finger tree for an element that falls b leaves away from a finger takes $O(\log b)$ steps.*

Consider now a sequence of searches in S for an increasing set of keys. We start with the finger pointing to the leftmost leaf of T_S, and advance it, following each search, to the leaf returned by that search. For k consecutive searches, the total effort is bounded by a constant times

$$k + \sum_{j=1}^{k} \log b_j$$

where the b_j's represent the widths of the various *intervals*, and these latter are non-overlapping, i.e., $\sum_{j=1}^{k} b_j \leq 2m$. With this constraint, the above sum is maximum when all the b_j's are equal, which yields a bound of $O(k \log(2m/k))$ for the sequence of searches. When sequences of consecutive insertions or deletions are considered, the problem becomes more complicated, due to the fact that those primitives tamper with the structure of the tree. However, we can still appeal to the following result by Brown and Tarjan [1978].

Lemma 4.2. *Let T be a 2-3 tree with m leaves numbered $1, 2, ..., m$, and let $i_1, i_2, ..., i_k$ be a subsequence of the leaves. Let $i_0 = 0$, and $b_j = i_j - i_{j-1} + 1$ for $j = 1, 2, ..., k$. Furthermore, for i and $i' > i$, let $l(i, i')$ be the number of nodes which are on the path from i' to the root but not on the path from i to the root. Finally, let*

$$p = \log m + 1 + \sum_{j=1}^{k} l(i_{j-1}, i_j).$$

Then p obeys the inequality:

$$p \leq 2(\log m + \sum_{j=1}^{k} \log 2b_j).$$

For two ordered sequences Q and S of respective cardinalities k and m, the expression denoted by p is an upper bound for the process of producing $T_{Q \cup S}$ from T_S or vice versa by orderly insertion in S (deletion from $S \cup Q$) of the elements of Q. Specifically, consider the following three homogeneous series of k operations each: (i) the finger-searches in T_S of each of the elements of Q (where the finger is initially $f = i_0 = 0$), (ii) the insertions of all elements of Q in T_S, (iii) the deletions of all elements of Q from $T_{Q \cup S}$. Then Lemma 4.2 supports the following claim (see, e.g., Mehlhorn [1984]).

Lemma 4.3. *Each one of the series (i)–(iii) takes time* $O(p)$.

Lemma 4.3 does not apply to any hybrid series of dictionary primitives. However, we shall use it to show that it works for the peculiar hybrid series which are involved in $HS\prime$ at each row. Thus, we assume henceforth that all lists in $HS\prime$ are implemented as finger trees. The collective initialization of all trees takes trivially $\Theta(n)$ time.

Theorem 4.1. *Algorithm* $HS\prime$ *can be implemented in* $O(n \log s)$ *preprocessing time and* $O(m \log n + d \log(2mn/d))$ *processing time.*

Proof. It is easy to check that the preprocessing required by $HS\prime$ is basically the same as that required by HS, whence we can concentrate on the second time-bound. Let d_i denote the number of dominant matches which $HS\prime$ introduces handling row i. As seen in the discussion of Theorem 3.1, d_i searches are performed on $THRESH$ at row i. Observe that the arguments of successive searches constitute a strictly decreasing sequence of integers, and that the same can be said of the values returned by those searches. Thus, by Lemma 4.3, the cost of all searches on this row is bounded, up to a multiplicative constant, by

$$\log n + \sum_{k=1}^{d_i} \log b_k,$$

where the intervals b_k are such that

$$\sum_{k=1}^{d_i} b_k \leq 2n,$$

since the finger tree of $THRESH$ contains n leaves. (Note, incidentally, that a $\log n$ term is not necessary for searches.) It follows that, up to a multiplicative constant, the total cost on all rows is bounded by

$$m \log n + \sum_{k=1}^{d} \log b_k,$$

where now $\sum_{k=1}^{d} b_k \leq 2mn$. With this constraint, the previous sum is maximized by choosing all b_i equal, i.e., $b_i = 2mn/d$. The claimed bound then follows.

It is not difficult to show that the same bound holds for the insertions and deletions performed on $THRESH$. We observe the following. First, the two lists of arguments for the insertions and deletions, respectively, represent increasing subsequences of the integers in $[1, n]$. Moreover, the set of items inserted into $THRESH$ is disjoint from the set of items deleted from $THRESH$. The second observation enables us to deal with each one of the two series separately. In other words, the total work involved in the insertions and deletions affecting $THRESH$ at some row is not larger than the work which would be required if one performed all the deletions first, and then performed all the insertions. Thus, through an argument analogous to that used for the searches, the bound follows from Lemma 4.3, and from the fact that, on each row, $THRESH$ is affected by d_i insertions and by a number of deletions which is at least $d_i - 1$ and at most d_i.

We now consider the collection of primitives performed, during the management of a single row, on all the $AMATCHLIST$s involved. The key observation here is that the sum of the cardinalities of all such lists never exceeds n. In fact, there will be exactly n leaves in the forest of finger trees which implement such lists. If the trees corresponding to the various $AMATCHLISTS$s are visualized as aligned one after the other, it is easy to adapt the same argument which was used for $THRESH$ to the primitives affecting the collection of these lists. Indeed, the special conditions on the searches, insertions and deletions still hold locally, on each individual list. This leads to our claimed bound, since the d_i insertions in $THRESH$ correspond in fact to d_i searches with deletions on $AMATCHLIST[\sigma]$, and an equivalent number of insertions take place in the collection of all lists.   □

In conclusion, there is a variation on the Hunt-Szymanzki paradigm in which time depends only on dominant matches, and also is never worse than quadratic in the length of the input. With little extra effort, it is possible to set up simpler *ad hoc* variants of finger trees supporting the same performance (Apostolico and Guerra [1987]). A bound $O(m \log n + d \log(2mn/d))$ was also claimed by Hsu and Du [1984] for one of their constructions, but the claim turned out to be flawed (Apostolico [1987]). Eppstein et al. [1990] observed that using Johnson's [1982] variant of the flat trees of van Emde Boas [1975] would reduce the log factor to $\log \log$.

Before leaving this section, it is useful to mention also some other approaches to the LCS problem. One notable line of research concentrated on cases where the length of an LCS is expected to be close to m, the length of the shorter input string. One early construction in Hirschberg [1977] achieves time $O((m - l)l \log n)$ for this case. (An additional $\Theta(n \log s)$ term, charged by preprocessing, should be added to all time bounds mentioned here.) A subsequent construction requiring $O((m - l)n)$ was proposed by Nakatsu, Kambayashi and Yajima [1982], along with another $O((m - l)l \log n)$ algorithm. The latter bound can be reduced to $O(m(m - l) \min \{\log s, \log m, \log 2n/l\})$ using the techniques we are discussing. In terms of antichain constructions, these approaches may befit the Hunt-Szymanski paradigm. An algorithm taking time $O(ne)$ in terms of the edit distance $e = m + n - 2l$ was proposed by Myers [1986]. Note that since $l \le m$, then $e^2 = \Theta(n^2)$ for $n \ge 2m$. Thus, this bound is comparable to those by Hirschberg [1977] and Nakatsu et al [1982] only for two input strings of nearly equal length. However, this algorithm has expected time $O(n + e^2)$ and a nice, though admittedly impractical, $O(n \log n + e^2)$ variation. Wu, Manber, Myers, and Miller [1990] obtained a slightly faster $O(nP)$ algorithm, where P is the number of deletions in the shortest edit script. Finally, Wu, Manber, and Myers [1991] presented an $O(n + ne/\log n)$ solution by combining features from the algorithms of Masek and Paterson [1980], Ukkonen [1985], and Myers [1986].

5. Linear space

Space saving techniques in computations of string editing and longest common subsequences are useful in many applications and often crucial in molecular sequence alignment. For some time, the early $\Theta(nm)$ algorithm by Hirschberg [1975] was the only one to require linear space. Stated more accurately, most subsequent algorithms could be adapted to run in linear space, but this would be accompanied typically by a degradation of the time performance to $\Theta(nm)$. Apostolico and Guerra [1985] showed that one of their variants of Hirschberg's paradigm could be implemented in linear space at the expense of only an additive term $O(m \log m)$ in time. Also Myers' [1986] strategies can be implemented in linear space. Linear space implementation of the $O((m - l)n)$ by Nakatsu et al. was obtained subsequently in Kumar and Rangan [1987], through a divide and conquer scheme not quite identical to that of Hirschberg. Additional linear space algorithms are found, e.g., in Myers and Miller [1988], Apostolico, Browne and Guerra [1992], Chao [1994], and others. It is not obvious in general that an LCS algorithm can be adapted to run in linear space without substantial alteration of its time complexity. For instance, it is not clear that the $O(m \log n + d \log(2mn/d))$ implementation of the Hunt-Szymanski paradigm, seen in the preceding section, could be adapted to linear space without having to change its time bound.

In this section, we examine a few linear space algorithms that do not always take $\Theta(nm)$ time. Along the way, we shall see that for $m = n$ an $O(n(n - l))$ algorithm of great conceptual simplicity results from introducing some kind of dualization in the structure of HS. Equally simple extensions will enable us to handle the case $m \le n$, in time $O(n(m - l))$ and linear space. At the heart of these algorithms, there is the following divide-and-conquer scheme (cf. Kumar and Rangan [1987]). A recursive procedure takes as input: (1) two strings ϵ and δ such that ϵ is always a substring, say, of y and δ is always a substring of x; (2) the length l of an LCS of ϵ and δ. The task of the procedure is to produce an LCS of ϵ and δ. This is achieved by first computing a suitable *cut* for an LCS of ϵ and δ and then by applying the scheme recursively to the two subdomains of the problem induced by the cut. A cut is any pair $[u, v]$ such that an LCS of ϵ and δ can be formed by concatenating an LCS of the prefixes of length u and v of ϵ and δ, respectively, with an LCS of the corresponding suffixes of those two strings. The detailed description of our scheme is as follows.

Procedure $lcsls$ (ϵ, δ, $i1$, $i2$, $j1$, $j2$, l, LCS)
  begin
    if $l = c$ *or* $min[|\epsilon|, |\delta|] - l = c$
    for some constant c
    then
      determine an LCS in time $O(|\epsilon||\delta|)$
      and space $O(min[|\epsilon|, |\delta|])$
      else begin
        (split the problem into subproblems)
        choose a cut $[u, v]$, $1 \leq u \leq |\epsilon|$, $1 \leq v \leq |\delta|$
        $lcsls(\epsilon, \delta, i1, i1 + u - 1, j1, j1 + v - 1, l_1, LCS1)$;
        $lcsls(\epsilon, \delta, i1 + u, i2, j1 + v, j2, l_2, LCS2)$;
        $LCS = LCS1 \parallel LCS2$;
    end
  end.

The major difference with respect to the scheme Hirschberg [1975] is in that here l is computed prior to running $lcsls$. In what follows, we discuss first some ways of computing l, and then how to choose and compute a suitable cut inside $lcsls$. Obviously, we seek to choose such a cut so as to achieve an optimal balance (cf. Kumar and Rangan [1987]), in the sense that the total time required to solve both induced subproblems should be about one half the time required to solve the original problem.

5.1 Computing the length of a solution

We assume first $n = m$ and outline a simple $O(n(n - l))$ time strategy which is dual to that used in Hunt and Szymanski [1977]. The case $n = m$ arises in the row-wise comparison of digitized pictures and thus has special interest. As seen, the Hunt–Szymanski approach consists of detecting the dominant matches of all available ranks by processing the matches in the L matrix row by row. For this purpose, a list of thresholds which we will now call $row\text{-}THRESH$ is used. After the processing of a row, the k-th entry in $row\text{-}THRESH$ contains the column of the leftmost k-dominant match found so far (refer to Fig. 5.1.). In our example, After processing the sixth row, the final set of row thresholds after processing the sixth row is $\{1,2,5\}$. Note that $m - l = 3$ positions are missing from the final set of thresholds, namely positions 3, 4, and 6. We call each such missing position a *gap*, and we call the sorted list of gaps $row\text{-}COTHRESH$.

We can define similarly a list $colu\text{-}THRESH$ such that its k-th entry contains the row index of the rightmost k-dominant match found so far. For our example, the final set of column thresholds is $\{1,2,5\}$. The corresponding set $colu\text{-}COTHRESH$ of gaps would be $\{3,4,6\}$. Clearly, the $COTHRESH$ lists can be deduced from the $THRESH$ lists, and vice versa. If $m - l < l$, then the $COTHRESH$ lists give a more compact encoding of

		C	T	A	C	T	A	A	T	A
		1	2	3	4	5	6	7	8	9
A	1	0	0	1	1	1	1	1	1	1
T	2	0	1	1	1	2	2	2	2	2
C	3	1	1	1	2	2	2	2	2	3
G	4	1	1	1	2	2	2	2	2	3
T	5	1	2	2	2	3	3	3	3	3
T	6	1	2	2	2	3	3	3	4	4

Fig. 5.1. The trace of row-TRESH

the final set of thresholds. Unfortunately, this is not always true at any stage of the row-by-row computation, since $THRESH$ can be initially more sparse and $COTHRESH$ correspondingly denser. However, if we consider only the upper-left square submatrices of the L-matrix, then we can obtain an useful bound on the size of the $COTHRESH$ lists.

Lemma 5.1. *The total number of gaps falling within the first i positions of either the i-th row or the i-th column of the L-matrix is at most $m - l$.*

Proof. There must be an equal number of gaps, say q, in the i-th row and in the i-th column. Assume that $q > m - l$. Clearly, the number of matches contributed to any LCS by the upper left $i \cdot i$ submatrix of the L-matrix cannot exceed $i - q$. Since the remaining portion of the L-matrix cannot contribute more than $m - i$ matches, it must be the case that $l \leq (m - i) + (i - q) = m - q$. But then $m - l \geq q$, a contradiction. $\square$

Lemma 1 suggests that the length of an LCS of x and y with $|x| = |y|$ can be computed by extending, one row and one column at a time, submatrices of the L-matrix. This is done by a procedure *length1* which we now describe informally. During its i-th iteration, the procedure scans from left to right the $O(m - l)$ cells of the two $COTHRESH$ lists. If in the $row - COTHRESH$ list we find a cell containing position $p < i$ such that $x_i = y_p$, then $[i, p]$ is a dominant match. Continuing the scan, the first cell (if any) is located with an entry larger than $1 + p\prime$, where $p\prime$ is the value stored in the immediately preceding cell. This jump in the list of gaps represents a threshold, namely, the first threshold to the right of p. If such a cell is found, then for some $i\prime < i$, $[i\prime, p\prime + 1]$ is a dominant match having the same rank as $[i, p]$. Hence, gap $p\prime + 1$ is inserted into $row - COTHRESH$. If no such cell is found, then $[i, p]$ is the first dominant match found of its rank, and the cell containing i is removed from $col - COTHRESH$. The processing of the $col - COTHRESH$ list is similar. It is easy to keep track, during these scans, of the number of positions not currently occupied by a gap, whence the rank of any newly detected dominant match is readily computed. The highest rank detected by the algorithm is the length of an LCS for the two input strings. We could add some extra bookkeeping to provide for the retrieval of an LCS at the end. This would, however, havoc the linearity of space. We are interested here mainly in the computation of $|w|$, so that these tedious details can be omitted. The following claim suffices for our discussion.

Lemma 5.2. *The length l of an LCS of two strings x and y such that $|x| = |y| = n$ can be computed in time $O(n(n-l))$ and linear space.*

Note that $O(n(n-l))$ is the same as $O(ne)$, where the edit distance $e = n + m - 2l = 2(n-l)$. When $n > m$ the condition of Lemma 5.1 is no longer met. Even so, it will not be difficult to adapt our technique. Our basic tool is a procedure that tests, for any integer p in the range $[0, m]$, whether x and y have an LCS of length $m - p$. Such a procedure, which we call *length2* is based on the following observation. Suppose strings x and y have an LCS of length l. Then there is at least one such LCS, say, w, that uses only dominant matches. Let $[i,j]$ be one such match. Then, $[i,j]$ appears in the j-th *colu-THRESH* list and, implicitly, in the j-th *colu-COTHRESH* list. Let f be the number of gaps preceding $[i,j]$ in column j of the L matrix. Then the prefix of w that is an LCS for x_i and y_j uses precisely $i - f$ rows among the first i rows of the L matrix. By an argument similar to that of Lemma 5.1, it must be that $f \leq m - l$, since the remaining $m - i$ rows cannot contribute more than $m - i$ matches to w. In other words, no dominant match in an LCS can be preceded by more than $m - l$ gaps in the cothresh list relative to the column where that match occurs.

Therefore, to test whether there is a solution of length $m - p$, it is sufficient to produce the n successive updates of the first p entries of *colu-COTHRESH*. By our preceding discussion, this takes time $O(np)$ and linear space. At the end, either we find a match of rank $m - p$ or higher in this list, or we know that no LCS of length at least $m - p$ exists. Our final procedure *length*, which computes the length of an LCS of x and y in $O(n(m-l))$ time descends now naturally from *length2*. It consists of running the $O(pn)$ procedure *length2* with $p = 0, 1, 2, 4, 8, \ldots$ until it succeeds. Procedure *length2* will succeed when p is at most $2(m - l)$. Thus the total time spent by *length* is proportional to $2n(m-l) + n(m-l) + 1/2n(m-l) + \ldots + 2n + n + n = 4n(m-l) + n$, which is $O(n(m-l))$. This establishes the following claim.

Theorem 5.1. *The length l of an LCS of two strings x and y, of m and n symbols respectively, can be computed in $O(n(m-l))$ time and linear space.*

5.2 Computing an LCS in $O(n(m-l))$ time and linear space

We show here that *length2* and *length* (or *length1*, if $n = m$) can be combined with *lcsls* to produce an LCS of the two input strings x and y. We call the resulting algorithm $lcsls1$. In what follows, we describe the structure of $lcsls1$ and maintain the following bounds.

Theorem 5.2. *Algorithm $lcsls1$ computes an LCS of x and y in time $O(n(m - l))$ and linear space.*

Proof. The proof will be a consequence of the discussion given below.   □

The two issues to be addressed are the computation of l that precedes the execution of $lcsls$ and the choice and computation of a cut inside the body of $lcsls$. We use $length$ to compute l. From this, we know $p = m - l$. This takes time $O(np)$ and linear space. We now call $lcsls$ on $\epsilon = y$ and $\delta = x$. Inside $lcsls$, we will maintain that the value $t = |\delta| - l$ (i.e., the value of p relative to the current subdomain of the problem) is always known. More precisely, we maintain that at the k-th level of recursion, $t \leq \lceil p/2^k \rceil$. This is achieved by computing cuts that always divide t into halves. We call these cuts *balanced cuts*. We will show how the computation of all balanced cuts needed at the k-th level of recursion can be carried out in time $O(np/2^k)$ and linear space. Before describing how this is done, we observe that this condition establishes, for the time bound $T(n, p)$ of $lcsls$, a recurrence of the form: $T(n, p) = cnp + T(n_1, np/2) + T(n_2, np/2)$, with $n_1 + n_2 = n$ and c a constant. With initial conditions of the type $T(h, 0) \leq bph$, where b is another constant, this recurrence has a solution $O(np)$.

Let n and $m \leq n$ be the lengths of ϵ and δ, respectively, and let $l = m - p$ be the length of an LCS for the two strings. The following lemma will be used to find a balanced cut for ϵ and δ (see Fig. 5.2.).

Lemma 5.3. *Assume $m > p \geq 2$ and set $p = p_1 + p_2 + p_3$ with $p_1 \neq 0$, $p_2 = 0$, and $p_3 \neq 0$. Then, there is an LCS $\gamma = \gamma^1 \gamma^2 \gamma^3$ of ϵ and δ for which it is possible to write $\epsilon = \epsilon^1 \epsilon^2 \epsilon^3$ and $\delta = \delta^1 d \delta^2 d\prime \delta^3$ with d and $d\prime$ symbols of Σ, in such a way that: (1) γ consists only of dominant matches; (2) for $i = 1, 2, 3$, γ^i is an LCS of ϵ^i and δ^i and $|\delta^i| - |\gamma^i| = p_i$; (3) let e and $e\prime$ be, respectively, the last symbol of ϵ^1 and the first symbol of ϵ^3, then e and d do not form a dominant match in L and $e\prime \neq d\prime$.*

Proof. In the L-matrix, consider in succession the columns relative to the positions of δ. We start with a counter initialized to zero and update it according to the following. Consider column 1. As is easy to check, if there is any match in column 1, then the one such match occupying the row of lowest index is also the unique dominant match in column 1. If there is a solution γ that uses a match in this column, then we pick the only dominant match in this column and initialize with it a string $\gamma\prime$. If this is not the case, we increment the counter by one. Assume we have handled all columns up to $h - 1$ updating the counter or extending the prefix $\gamma\prime$ of an optimal solution γ, according to the cases met. Considering column h, we increment the counter if and only if no match in that column could be used to extend the length of $\gamma\prime$ by one unit in such a way that the extended string would still be the

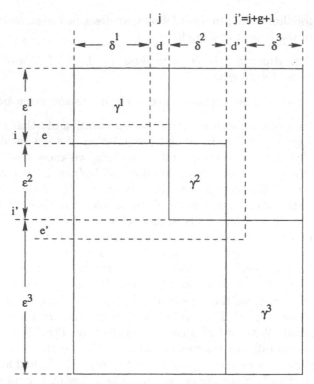

Fig. 5.2. Illustrating Lemma 5.3

prefix of an optimal solution. If some such matches exist, we append to $\gamma\prime$ the one such match contained in the row of smallest possible index (observe that the match thus selected is a dominant match). In conclusion, each column at which the counter is not incremented extends the subsequence $\gamma\prime$ by one new dominant match, while the fact that the counter is incremented at some column h signals that $\gamma\prime$ could not have been continued into an optimal solution γ had we picked a match in column h.

Let now j be the leftmost column at which the counter reaches the value p_1, and let i be the row containing the last one among the matches appended to $\gamma\prime$. We claim that entry $[i, j]$ cannot be a dominant match. In fact, if $[i, j]$ is a match, then clearly its rank is at least $|\gamma\prime|$. Assuming the rank of $[i, j]$ higher than $|\gamma\prime|$ leads to a contradiction. In fact, in this case we can find a string η such that $\eta\gamma\prime\prime$ is an LCS of ϵ and δ, $\gamma = \gamma\prime\gamma\prime\prime$ is also an LCS of ϵ and δ and yet $|\eta\gamma| > |\gamma|$. Thus, either $[i, j]$ is not a match or it is a non dominant match of rank equal to the last match of $\gamma\prime$ used so far. We set δ^1 equal to the prefix of δ of length $j - 1$, ϵ^1 equal to the prefix of ϵ of length i, $\gamma^1 = \gamma\prime$, $e = \epsilon[i]$ and $d = \delta[j]$. These choices are consistent with the properties listed in the lemma for the objects involved.

To continue with the columns of L that fall past column j, we distinguish two cases, according to whether or not $\gamma\prime$ can be extended with a match in column $j + 1$. If $\gamma\prime$ can be extended with a match in column $j + 1$, let $j + 1$, $j + 2, ...$, $j + g$ be the longest run of consecutive columns such that each column contributes a new match to $\gamma\prime$. By the hypothesis $p_1 \; p$, we have $j + gm$ (i.e., we must be forced to skip at least one more column). Let $i\prime$ be the row such that $[i\prime, j+g]$ is a match of $\gamma\prime$. Then, by our choice of g the entry $[i\prime+1, j+g+1]$ cannot be a match. We set ϵ^2 equal to the substring of ϵ that starts at position $i + 1$ and ends at position $i\prime$, δ^2 equal to the substring of δ that starts at $j+1$ and ends at $j+g$, and $e\prime = \epsilon[i\prime+1]$ and $d\prime = \delta[j+g+1]$. Finally, we take the suffix of length g of $\gamma\prime$ as γ^2. Clearly, these assignments satisfy the conditions in the claim. The choices performed so far induce a unique choice of ϵ^3, δ^3, and γ^3. By our construction of $\gamma\prime$, there is an optimal solution γ which has $\gamma\prime = \gamma^1\gamma^2$ as a prefix. In any such solution, $\gamma\prime$ must be followed by an LCS of ϵ^3 and δ^3 of length $|\delta^3| - (p - p_1 - p_2)$, i.e., of length $|\delta^3| - p_3$. Thus the remaining conditions of the claim are also met. If $\gamma\prime$ cannot be extended with a match in column $j + 1$, then the claim still holds by simply taking δ_2 and γ_2 both empty. $\square$

With the choice $p_1 = \lceil p/2 \rceil$ Lemma 5.3 can be used in the computation of a balanced cut for ϵ and δ, as follows. We restrict ourselves to the case where p is even, the case of odd p is quite similar and will be left for an exercise. Let j and $j\prime = j+g+1$ be the positions in δ of d and $d\prime$, respectively, and let i be the position in ϵ of the last symbol of ϵ^1. Clearly, $[i\prime, j\prime - 1]$ is a balanced cut. Observe that this cut coincides with $[i, j]$ if γ^2 is empty.

We now run $length2$ on the ordered pair (δ, ϵ) and with parameter $p/2 + 1$. We use this run to prepare an array $REACH$ with the property that $REACH[i]$ contains the column index relative to the $p/2 + 1$-st gap in the $COTHRESH$ list at row i. Observe that, by condition 3 of the lemma, if $i\prime + 1$ is the position in ϵ of the first symbol of ϵ_3, then $REACH[i\prime + 1]$ equals precisely the position $j\prime$ of $d\prime$ in δ.

Next, we run a copy of $length2$ on the ordered pair (δ^R, ϵ^R) of the *reverse* strings of the two input strings, this time with parameter $p/2$. An array $REVREACH$ similar to $REACH$ is built in this way. Since $[i\prime + 1, j\prime]$ is not a match and we know that $|\delta_3| - |\gamma_3| = p/2$, then $REVREACH[i\prime+1] = j\prime$.

Clearly, any index i^* for which $REACH[i^*] = REVREACH[i^*]$ yields a corresponding balanced cut $[i^* - 1, REACH[i^*] - 1]$. By Lemma 5.3 and the above discussion, at least one such index is guaranteed to exist. In conclusion, we only need to scan the two arrays $REACH$ and $REVREACH$ looking for the first index k such that $REACH[k] = REVREACH[k]$. Having found such an index, we can set, for our balanced cut $[u, v]$, $u = k - 1$ and $v = REACH[k] - 1 = REVREACH[k] - 1$.

As mentioned, the case of odd p is dealt with similarly. At the top level of the recursion, this process takes $O(np)$ time and linear space. Since the parameter p is halved at each level, the overall time taken by the computation of cuts is still $O(np)$. The recursion can stop whenever the current partition of L has an associated value of either the l or p not larger than some preassigned constant. For any such partition, an LCS can be found by known methods in linear space.

This concludes our discussion of Theorem 5.2.

6. Combining few and diverse tools: Hirschberg's paradigm in linear space

In this section, we pick up again Hirschberg's paradigm having in mind a twofold goal: on one hand, we plan to improve the time complexity seen when we last parted from this paradigm; on the other, we want to implement the new version in linear space without having to pay a penalty in time.

We start by presenting a procedure *newlength* that computes the length of an LCS of x and y in time $O(lm\log(\min[s, m, 2n/m]))$. Since symbols not appearing in x cannot contribute to an LCS, we can eliminate such symbols from y and assume henceforth $s \leq m$, which eliminates the $\log m$ from the bound.

The alert reader shall realize quickly that *newlength* is nothing but an offspring of *Algorithm LCS1*. In analogy with that algorithm, the procedure consists of *lsub* stages which identify in succession the *lsub* antichains within a suitable partition of our matrix L. At the beginning, $PEBBLE[i]$ ($i = i_1, .., i_2$) points to the the entry j of $x_i - OCC$, which corresponds to the leftmost occurrence of x_i in the interval $[j_1, .., j_2]$, if any. $PEBBLE[i]$ is then said to be *active*. The procedure advances an active pebble until it becomes *inactive*, i.e., reaches an entry larger than j_2, or the last entry of $x_i - OCC$. By the end of the execution of *newlength*, each pebble is set to point to the rightmost position that it can occupy in the interval $[j_1 \ldots j_2]$.

A significant innovation brought about by *newlength* is in its auxiliary function *closest*, which is defined as follows. For any character σ, $closest(\sigma, y_t)$ returns a pointer to the entry in the $\sigma - OCC$ list corresponding to leftmost occurrence of σ in y which falls past y_t.

One more place where *newlength* departs from Algorithm *LCS1* is that even though *newlength* still detects all dominant matches, it records only the leftmost dominant match incurred for each k. This achieves the linear space bound.

> **Procedure** *newlength (i1, i2, j1, j2, RANK, lsub)*
>   0 $RANK[k] = 0$, $k = 1, 2, ..., (i2 - i1)$;
>   1 $k = 0$
>   2 **while** *there are active pebbles* **do** (start stage $k + 1$)
>   3 **begin** $T = j2 + 1$; $k = k + 1$;
>   4 **for** $i = i1 - 1 + k$ **to** $i2$ **do** (advance pebbles)

```
        begin
 5      t = T;
 6      if PEBBLE[i] is active
        and x_i - OCC[PEBBLE[i]] < T then
        (update threshold, update leftmost k-dominant match )
 7      begin T = x_i - OCC[PEBBLE[i]]; RANK[k] = T end ;
        (advance pebble, or make it inactive)
 8      PEBBLE[i] = closest[x_i, t];
 9      if PEBBLE[i] is active
        and x_i - OCC[PEBBLE[i]] > j2 then
10      begin PEBBLE[i] = PEBBLE[i] - 1;
        make PEBBLE[i] inactive end;
        end ;
     end (lsub = k).
```

All the elementary steps of $newlength$, with the exception of the executions of $closest$, take constant time. On an input of size $n + m$ the procedure handles at most m pebbles during each of the $lsub$ stages. Thus the total time spent by $newlength$ is $O(m \times lsub +$ total time required by $closest)$. The second term depends on implementation. One efficient implementation of $closest$ rests on two auxiliary structures which we now proceed to describe.

The first structure is a table $CLOSE[1...n + 1]$ which is subdivided into consecutive blocks of size s and defined as follows. Letting $p = j \bmod s$ ($j = 1, ..., n$), $CLOSE[j]$ contains the leftmost position not smaller than j where σ_p occurs in y. Table $CLOSE$ is prepared in time $\Theta(n)$, and it allows us to implement $closest$ in time $O(\log s)$ (Apostolico and Guerra [1987]).

Next, we assume that each $\sigma - OCC$ list is assigned a $finger\ tree$. In order to keep track of the fingers we institute a new global variable, namely, the array of integers $FINGER[1...m]$. At its inception, the procedure $newlength$ moves all the fingers $FINGER[i1]$, $FINGER[i1 + 1]$, ..., $FINGER[i2]$, originally coincident with the pebbles, onto the rightmost position in the interval $[j1...j2]$ that they can occupy on their corresponding $\sigma - OCC$ lists. This positioning of each finger is accomplished in $O(\min[\log s, \log(j2 - j1)])$ time through an application of $closest$. Fingers set from different rows on the same $\sigma - OCC$ list merge into one single $representative$ finger.

During the execution of each stage of $newlength$, the (representative) finger associated with each symbol in $[i1...i2]$ is reconsidered immediately following a $closest$ query and the possible consequent update of the pebble (cf. lines 8-10 of $newlength$). At that point, we simply set: $FINGER[i] = PEBBLE[i]$. Thus, through each individual stage, the finger associated with each symbol moves from right to left. Each of the manipulations just described takes constant time. Finally, both fingers and pebbles are taken back to their initial (leftmost) position soon after the last stage of $newlength$ has been completed. Overall, this takes time $O(i2 - i1)$. In the light of our discussion of Section 4. we may list the following:

Theorem 6.1. *By the combined use of FINGER and CLOSE, the procedure* newlength *computes the length lsub of an LCS of $x_{i1}...x_{i2}$ and $y_{j1}....y_{j2}$ in time $O(lsub \cdot (i2 - i1) \cdot min[\log s, \log(2n/(i2 - i1))])$ and linear space.*

We now show that the procedure *newlength* can be cast in the divide-and-conquer scheme of the previous section resulting in an algorithm *lcsls2* that has time bound $O(ml \log(min[s, 2n/l]))$.

We need to remove the previous assumption that *newlength*, called with j-parameters $j1, j2$, always finds pebbles and fingers pointing to the leftmost positions in the interval $[j1...j2]$. We replace that with the new assumption that either all pebbles and fingers occupy the rightmost positions in the interval $[j1...j2]$, or else they all occupy the leftmost one. Procedure *newlength* checks at its inception which case applies, and brings all pebbles to their leftmost positions, if necessary. This does not affect the time bound of the procedure. Algorithm *lcsls2* uses *newlength* both to compute l prior to executing *lcsls* and to compute cuts inside the body of *lcsls*. For this latter task we use an approach similar to that of *lcsls1*. We outline the method for the case of even l, the case of odd l being handled similarly. We run two copies of *newlength*, on the two mirror images of the problem, with the proviso that computation in each row is stopped as soon as a dominant match of rank $l/2$ is detected. All matches of rank $l/2$ so detected by each version of the procedure are stored in one of two dedicated lists. Observe that the number of such matches cannot exceed the total number of dominant matches detected, and this latter number cannot be larger than ml, the maximum number of matches handled by the procedure. At the end, we scan the two lists looking for the first pair of matches, one from one list and one from the other, that form a chain. From the positions in L of these two matches, we can infer a balanced cut. In the present context, a cut is balanced if it identifies two submatrices L' and L'' of L with the property that an optimal solution γ can be formed by concatenating two optimal solutions γ' and γ'' entirely contained, respectively, in L' and L'' and both of length $l/2$. Leaving the details for an exercise, we concentrate on the following claim.

Theorem 6.2. *The procedure* lcsls2 *finds an LCS in linear space and in time $O(ml \log(min[s, 2n/l]))$.*

Proof. Each execution of *newlength* at the k-th level of the recursion can be bounded in terms of $m_f \cdot l/2^k \log(min[s, 2n/m_f])$, where m_f denotes the number of rows assigned to the f-th subproblem. By the preceding discussion, the time needed to scan each pair of antichains of maximum rank in order to find a balanced cut for that pair can be absorbed in this bound. There are 2^k calls at level k, yielding a total time:

$$\sum_{f=1}^{2^k} m_f \frac{l}{2^k} \log(min[s, \frac{2n}{m_f}]),$$

up to a multiplicative constant. Now it is

$$\sum_{f=1}^{2^k} m_f = m.$$

Since $m_f \geq l/2^k$, we have that the total work at this level of recursion can be bounded by the quantity:

$$m \cdot \frac{l}{2^k} \cdot \log(\min[s, \frac{2n}{l}2^k]) \leq m \cdot \frac{l}{2^k} \cdot \log(\min[s2^k, \frac{2n}{l}2^k]).$$

The right term can be rewritten as:

$$m \cdot \frac{l}{2^k} \log(2^k \cdot \min[s, \frac{2n}{l}]) = m \cdot k \cdot \frac{l}{2^k} + m \cdot \frac{l}{2^k} \log(\min[s, \frac{2n}{l}]).$$

Adding up through $k = 1, 2, ..., \log l$ yields:

$$ml \sum_{k=1}^{\log l} \frac{k}{2^k} + ml \ \log(\min[s, \frac{2n}{l}]) \sum_{k=1}^{\log l} \frac{1}{2^k}.$$

whence the $O(ml \log(\min[s, 2n/l])$ time bound follows.   □

7. Parallel algorithms

It is an exercise of moderate difficulty to set up a parallel solution for the string editing problem in the so-called *systolic array* of processors architecture (see, e.g., Lipton and Lopresti [1985], Leighton [1992]). Such an algorithm would deploy n processors to solve the problem in linear time. A *hypercube* algorithm for the case $m = n$ that runs in $O(\sqrt{n}\log n)$ time with n^2 processors can be found in Ranka and Sahni [1988]. Here we concentrate on the model of computation represented by the synchronous, shared-memory machine, usually referred to as the PRAM (see, e.g., Já Já [1992] for these notions). The currently best algorithms are based on the two strongest variants of the PRAM, namely, the CREW and CRCW, respectively. Recall that in the CREW-PRAM model of parallel computation concurrent reads are allowed but no two processors can simultaneously attempt to write in the same memory location (even if they are trying to write the same thing). The CRCW-PRAM differs from the CREW-PRAM in that it allows many processors to write simultaneously in the same memory location: in any such common-write contest, only one processor succeeds, but it is not known in advance which one.

The primary objective of PRAM algorithmic design is to devise algorithms that are both *fast* and *efficient* for problems in the class NC, i.e., in the class of problems that are solvable in $O(\log^{O(1)} n)$ parallel time by a PRAM using a polynomial number of processors. In order for an algorithm to be both fast and efficient, the product of its time and processor complexities must fall within a polylog factor of the time complexity of the best sequential algorithm for the problem it solves. This goal has been elusive for many simple problems, such as topological sorting of a directed acyclic graph and finding a breadth-first search tree of a graph, which are trivially in NC. For some other problems in NC, it seems somewhat counter-intuitive that any fast and efficient algorithm may exist until one is finally found. Such is the case of the string-editing problem.

For the sake of generality, we reason in terms of the cost matrix C introduced in Section 1. The interdependencies among the entries of that matrix in the string editing problem find useful representation in an $(|x|+1) \times (|y|+1)$ *grid* directed acyclic graph. An $l_1 \times l_2$ grid DAG is a directed acyclic graph whose vertices are the $l_1 l_2$ points of an $l_1 \times l_2$ grid, and such that the only edges from grid point (i, j) are to grid points $(i, j+1)$, $(i+1, j)$, and $(i+1, j+1)$. A grid DAG is obtained by drawing the points such that point (i, j) is at the ith row from the top and jth column from the left. Thus, the top-left point is $(0, 0)$ and has no edge entering it (i.e., is a *source*), and the bottom-right point is (m, n) and has no edge leaving it (i.e., is a *sink*). An example of a grid DAG is given in Figure 7.1.

It is quite natural to associate an $(|x|+1) \times (|y|+1)$ grid DAG G with the string editing problem in the natural way: the $(|x|+1)(|y|+1)$ vertices of G are in one-to-one correspondence with the $(|x|+1)(|y|+1)$ entries of the C-matrix, and the *cost* of an edge from vertex (k, l) to vertex (i, j) is equal to $I(y_j)$ if $k = i$ and $l = j - 1$, to $D(x_i)$ if $k = i - 1$ and $l = j$, to $S(x_i, y_j)$ if $k = i - 1$ and $l = j - 1$. Clearly, the edit scripts that transform x into y or vice versa without obviously wasteful operations are in one-to-one correspondence with the weighted paths in G that originate at the source (which corresponds to $C(0, 0)$) and end on the sink (which corresponds to $C(|x|, |y|)$). In conclusion, the string editing problem may be regarded as the

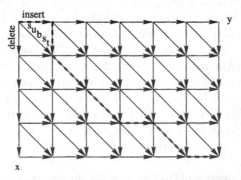

Fig. 7.1. String editing as a path in a grid graph

problem of finding a shortest (i.e., least-cost) source-to-sink path in an $m \times n$ grid DAG G.

This characterization provides a convenient framework for *divide and conquer*, and thus is especially useful in our present context. To see this, assume for simplicity $m = n$, i.e., G is an $m \times m$ grid DAG. Let $DIST_G$ be a $(2m) \times (2m)$ matrix containing the lengths of all shortest paths that begin at the top or left boundary of G, and end at the right or bottom boundary of G. Divide now the $m \times m$ grid into four $(m/2) \times (m/2)$ grids A, B, C, D. Assume that we have (recursively) solved the problem for each of the four grids A, B, C, D, obtaining the analogous distance matrices $DIST_A$, $DIST_B$, $DIST_C$, $DIST_D$. Then we only need to combine these four matrices into the global matrix $DIST_G$.

Let $DIST_{A \cup B}$ be the $(3m/2) \times (3m/2)$ matrix containing the lengths of shortest paths that begin on the top or left boundary of $A \cup B$ and end on its right or bottom boundary. Let $DIST_{C \cup D}$ be analogously defined for $C \cup D$. We may produce $DIST_G$ in three main steps, as follows. First, we use $DIST_A$ and $DIST_B$ to obtain $DIST_{A \cup B}$. We then similarly combine $DIST_C$ and $DIST_D$ into $DIST_{C \cup D}$, and finally obtain $DIST_G$ from $DIST_{A \cup B}$ and $DIST_{C \cup D}$. Clearly, the main problem we face is how to perform such a "conquer" phase efficiently, in parallel.

Since the three steps above are similar in essence, we may restrict consideration to the task of obtaining $DIST_{A \cup B}$ from $DIST_A$ and $DIST_B$. Note that we only need to compute the entries of $DIST_{A \cup B}$ that correspond to paths that begin at a point v on the top or left boundary of A and end at a point w on the right or bottom boundary of B, since all other costs in $DIST_{A \cup B}$ need only to be copied from $DIST_A$ and $DIST_B$.

Observe that $DIST_{A \cup B}(v, w) = \min\{Dist_A(v, p) + Dist_B(p, w) \mid p$ is on the boundary common to A and $B\}$. This shows that we can trivially compute $DIST_{A \cup B}(v, w)$ for a given v, w pair in time $O(q + \log(m/q))$ if we are allowed to use $\Theta(m/q)$ processors for each such pair. This, however, would require an unacceptable $\Theta(m^3/q)$ processors. We would like to perform this task by having only m/q processors assigned to v for computing $DIST_{A \cup B}(v, w)$ for *all* w on the bottom or right boundary of B. It turns out that these m/q processors are enough for doing the job in time $O((q + \log(m/q)) \log m)$. The handle is a *monotonicity* property that allows us to save a linear factor in the number of processor at the cost of a logarithmic factor in time. In order to illustrate this property we need to introduce some additional concepts.

Consider a pair of points (v, w) such that v falls on the left or top boundary of A, and w falls on the bottom or right boundary of B. Let $\theta(v, w)$ denote the *leftmost* p which minimizes $DIST_{A \cup B}(v, w)$. Thus, $\theta(v, w)$ is the leftmost point that can be encountered on the common boundary of A and B in a shortest path from v to w. (see Fig. 7.2.).

Define a linear ordering $<_B$ on the m points at the bottom and right boundaries of B, such that they are encountered in increasing order of $<_B$

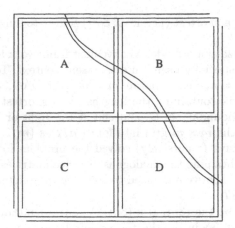

Fig. 7.2. Divide-and-conquer of Matrix $DIST$

by a walk that starts at the leftmost point of the lower boundary of B and ends at the top of the right boundary of B. Let L_B be the list of m points on the lower and right boundaries of B, sorted by increasing order according to the $<_B$ relationship. For any $w_1, w_2 \in L_B$, we have the following:

Lemma 7.1. *If $w_1 <_B w_2$ then $\theta(v, w_1)$ is not to the right of $\theta(v, w_2)$.*

Proof. The proof is by contradiction. Suppose that, for some $w_1, w_2 \in L_B$, we have $w_1 <_B w_2$ and $\theta(v, w_1)$ is to the right of $\theta(v, w_2)$, as shown in Fig. 7.4. By definition of the function θ there is a shortest path from v to w_1 going through $\theta(v, w_1)$ (call this path α), and one from v to w_2 going through $\theta(v, w_2)$ (call it β). Since $w_1 <_B w_2$ and $\theta(v, w_1)$ is to the right of $\theta(v, w_2)$, the two paths α and β must cross at least once somewhere in B: let z be such an intersection point. See Fig. 7.4. Let $prefix(\alpha)$ (respectively, $prefix(\beta)$) be the portion of α (respectively, β) that goes from v to z. Consider the two possible cases:

Case 1. The length of $prefix(\alpha)$ differs from that of $prefix(\beta)$. Without loss of generality, assume it is the length of $prefix(\beta)$ that is the smaller of the two. But then, the v-to-w_1 path obtained from α by replacing $prefix(\alpha)$ by $prefix(\beta)$ is shorter than α, a contradiction.

Case 2. The length of $prefix(\alpha)$ is same as that of $prefix(\beta)$. In α, replacing $prefix(\alpha)$ by $prefix(\beta)$ yields another shortest path between v and w_1, one that crosses the boundary common to A and B at a point to the left of $\theta(v, w_1)$, contradicting the definition of the function θ. $\quad\square$

Lemma 7.1 may be used to obtain an $O((q + \log(m/q))\log m)$ time and $O(m/q)$ processor algorithm for computing $DIST_{A \cup B}(v, w)$ for all $w \in L_B$. We henceforth use $\theta(w)$ as a shorthand for $\theta(v, w)$, with v being understood. It suffices to compute $\theta(w)$ for all $w \in L_B$. The procedure for doing this is recursive, and takes as input:

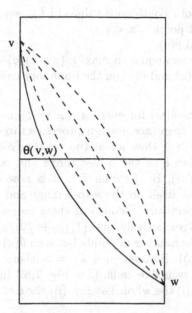

Fig. 7.3. The function θ

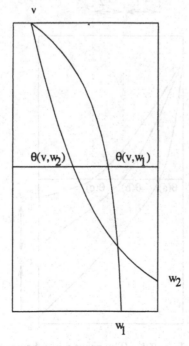

Fig. 7.4. The monotonicity Lemma

– A particular range of r contiguous values in L_B, say a range that begins at point a and ends at point c, $a <_B c$,
– The points $\theta(a)$ and $\theta(c)$,
– A number of processors equal to $\max\{1, (\rho + r)/q\}$ where ρ is the number of points between $\theta(a)$ and $\theta(c)$ on the boundary common to A and B. (See Fig. 7.5.)

The procedure returns $\theta(w)$ for every $a <_B w <_B c$. If $r = 1$ then there is only one such w and there are enough processors to compute $\theta(w)$ in time $O(q + \log(\rho/q))$. If $r > 1$ then all of the $\max\{1, (\rho + r)/q\}$ processors get assigned to the median b of the a-to-c range and compute the value $\theta(b)$ in time $O(q + \log(\rho/q))$. By Lemma 7.1, it is now enough for the procedure to recursively call itself on the a-to-b range and (in parallel) the b-to-c range. The first (respectively, second) of these recursive calls gets assigned $\max\{1, (\rho_1 + r/2)/q\}$ (respectively, $\max\{1, (\rho_2 + r/2)/q\}$) processors, where ρ_1 (respectively, ρ_2) is the number of points between $\theta(a)$ and $\theta(b)$ (respectively, between $\theta(b)$ and $\theta(c)$). Because $\rho_1 + \rho_2 = \rho$, there are enough processors available for the two recursive calls. (See Fig. 7.5.) In the initial call to the procedure, it is given (i) the whole list L_B, (ii) the θ of the first and last point

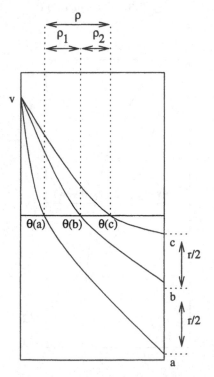

Fig. 7.5. The recursion scheme

of L_B, and (iii) $3m/2q$ processors. The depth of the recursion is $\log m$, at each level of which the time taken is no more than $O(q + \log(m/q))$. Therefore the procedure takes time $O((q + \log(m/q)) \log m)$ with $O(m/q)$ processors.

Theorem 7.1. *The matrix $DIST_G$ can be computed in $O(\log^3 m)$ time, $O(m^2)$ space, and with $O(m^2/\log m)$ processors by a CREW-PRAM.*

Proof. It follows from our discussion that $DIST_G$ can be obtained from $DIST_A$, $DIST_B$, $DIST_C$, $DIST_D$ in parallel in time $O((q + \log m) \log m)$ and with $O(m^2/q)$ processors, where $q \leq m$ is an integer of our choice. Moreover, the whole problem can be solved sequentially in $O(m^2 \log m)$ time. Then, the time and processor complexities of the overall algorithm obey then the following recurrences:

$$T(m) \leq T(m/2) + c_1(q + \log m) \log m,$$

$$P(m) \leq \max(4P(m/2), c_2 m^2/q),$$

with boundary conditions $T(\sqrt{q}) = c_3 q \log q$ and $P(\sqrt{q}) = 1$, where c_1, c_2, c_3 are constants. The solutions are $T(m) = O((q + \log m) \log^2 m)$ and $P(m) = O(m^2/q)$. Choosing $q = \log m$ then establishes the desired result.   □

Parallel algorithms more efficient than the one presented here can be based on the monotonicity property subtending Lemma 7.1 (see, e.g., Apostolico et al. [1988, 1990], Aggarwal and Park [1988], Atallah [1993]). The Lemma itself has a rich history of re-discoveries, which in modern times begin perhaps with Fuchs et al. [1977], but Aggarval and Park [1988] traced back to G. Monge in 1781.

Acknowledgement

This work was supported in parts by NSF Grants CCR-8900305 and CCR-9201078, by NATO GRANT CRG 900293, by the National Research Council of Italy, and by the ESPRIT III Basic Research Programme of the EC under contract No. 9072 (Project GEPPCOM).

References

[1]   AHO, A. V. [1990], Algorithms for finding patterns in strings, Handbook of Theoretical Computer Science, J. VAN LEEUWEN, ED., Elsevier, Amsterdam, 255–300.

[2]   AHO, A. V., D. S. HIRSCHBERG AND J. D. ULLMAN [1976], Bounds on the complexity of the longest common subsequence problem, *J. Assoc. Comput. Mach.*, **23**, 1–12.

[3]   AHO, A. V., J. E. HOPCROFT AND J. D. ULLMAN [1974], *The Design and Analysis of Computer Algorithms*, Addison-Wesley, Reading, MA.

[4]   AGGARWAL, A. AND J. PARK [1988], Notes on searching in multidimensional monotone arrays, in Proc. 29th Annual IEEE Symposium on Foundations of Computer Science, 1988, IEEE Computer Society, Washington, DC, 497–512.

[5]   APOSTOLICO, A. [1986], Improving the worst case performance of the Hunt-Szymanski strategy for the longest common subsequence of two strings, *Information Processing Letters* **23**, 63–69.

[6]   APOSTOLICO, A. [1987], Remark on HSU-DU New Algorithm for the LCS Problem. *Information Processing Letters* **25**, 235–236.

[7]   APOSTOLICO, A., ED. [1994], *Algorithmica* **4/5**, Special Issue on String Algorithmics and Its Applications.

[8]   APOSTOLICO, A., M. J. ATALLAH, L. L. LARMORE AND S. MCFADDIN [1990], Efficient parallel algorithms for string editing and related problems, *SIAM Journal on Computing* **19**, 968–988. Also: *Proceedings of the 26th Allerton Conf. on Comm., Control and Comp.*, Monticello, IL, Sept. 1988, 253–263.

[9]   APOSTOLICO, A., S. BROWNE AND C. GUERRA [1992], Fast linear space computations of longest common subsequences, *Theoretical Computer Science*, **92**, 3–17.

[10]   APOSTOLICO, A. AND Z. GALIL, EDS. [1985], *Combinatorial Algorithms on Words*, Springer-Verlag, Berlin.

[11]   APOSTOLICO, A. AND C. GUERRA [1985], A fast linear space algorithm for computing longest common subsequences, *Proceedings of the 23rd Allerton Conference*, Monticello, IL (1985).

[12]   APOSTOLICO, A. AND C. GUERRA [1987], The longest common subsequence problem revisited, *Algorithmica*, **2**, 315–336.

[13]   ARLAZAROV, V. L., E. A. DINIC, M. A. KRONROD, AND I. A. FARADZEV[1970]. On economical construction of the transitive closure of a directed graph, *Dokl. Akad. Nauk SSSR* **194**, 487–488 (in Russian). English translation in *Soviet Math. Dokl.* 11:5, 1209–1210.

[14]   ATALLAH, M. J. [1993] A Faster Parallel Algorithm for a Matrix Searching Problem, *Algorithmica*, **9**, 156–167.

[15]   BENTLEY, J. L. AND A. C-C. YAO [1976], An almost optimal algorithm for unbounded searching, *Inform. Process. Letters* **5**, 82–87.

[16]   BISHOP, M. J. AND C. J RAWLINGS, EDS. [1987], *Nucleic Acids and Protein Sequence Analysis*, IRL Press, Oxford.

[17]   BOGART, K. P. [1983], *Introductory Combinatorics*, Pitman, N.Y.

[18]   BROWN, M. R. AND R. E. TARJAN [1978], A representation of linear lists with movable fingers. *Proceedings of the 10-th STOC*, San Diego, CA, 19–29.

[19]   CHANG, W. I. AND E. L. LAWLER [1990], Approximate string matching in sublinear expected time, in *Proc. 31st Annual IEEE Symp. on Foundations of Computer Science*, St. Louis, MO, 116–124

[20]   CHAO, K. M. [1994], Computing all suboptimal alignments in linear space, in *Combinatorial Pattern Matching 1994*, M. CROCHEMORE AND D. GUSFIELD, EDS., Proceedings of the 5th Annual Symposium, Asilomar, CA, June 1994, Springer-Verlag Lecture Notes in Computer Science Vol. 807 (1994).

[21]   CROCHEMORE, M. AND W. RYTTER [1994], Text Algorithms, Oxford University Press, N.Y.

[22]   DILWORTH, R. P. [1950], A decomposition theorem for partially ordered sets, *Ann. Math.* **51**, 161–165.

[23]   DOOLITTLE, R. F., ED. [1990], *Molecular Evolution: Computer Analysis of Protein and Nucleic Acid Sequences*, Methods of Enzymology **183**, Academic Press, San Diego, CA.

[24]   VAN EMDE BOAS, P. [1975], Preserving order in a forest in less than logarithmic time, *Proc. 16th FOCS*, 75–84.

[25] EPPSTEIN, D. AND Z. GALIL [1988], Parallel algorithmic techniques for combinatorial computation, *Ann. Rev. Comput. Sci.*, **3**, 233–283.

[26] EPPSTEIN, D., Z. GALIL, R. GIANCARLO, AND G. ITALIANO [1990]. Sparse dynamic programming, *Proc. Symp. on Discrete Algorithms*, San Francisco, CA, 513–522.

[27] FREDMAN, M. L. [1975], On Computing the Length of Longest Increasing Subsequences, *Discrete Mathematics* **11**, 29–35.

[28] FUCHS, H., Z. M. KEDEM, AND S. P. USELTON [1977], Optimal surface reconstruction from planar contours, *Communications of the Assoc. Comput. Mach.*, **20**, 693–702.

[29] GALIL Z. AND R. GIANCARLO [1988], Data structures and algorithms for approximate string matching, *J. Complexity* **4**, 33–72.

[30] GALIL, Z. AND K. PARK [1990], An improved algorithm for approximate string matching, *SIAM Jour. Computing* **19**, 989–999.

[31] GOTOH, O. [1982]. An improved algorithm for matching biological sequences, *J. Mol. Biol.* **162**, 705–708.

[32] VON HEIJNE, G. [1987], *Sequence Analysis in Molecular Biology*, Academic Press, San Diego.

[33] HIRSCHBERG, D.S. [1975], A linear space algorithm for computing maximal common subsequences, *CACM* **18**, 6, 341–343.

[34] HIRSCHBERG, D. S. [1977], Algorithms for the longest common subsequence problem, *JACM* **24**, 4, 664–675.

[35] HIRSCHBERG, D. S. [1978], An information theoretic lower bound for the longest common subsequence problem, *Inform. Process. Lett.* **7**:1, 40–41.

[36] HSU, W. J., AND M. W.DU [1984], New algorithms for the LCS Problem, *J. Comput. System Sci.*, **29**, 133–152.

[37] HUNT, J. W. AND T. G. SZYMANSKI [1977], A fast algorithm for computing longest common subsequences, *CACM* **20**, 5, 350–353.

[38] JÁ JÁ, J. [1992], *An Introduction to Parallel Algorithms*, Addison-Wesley, Reading, MA.

[39] JACOBSON, G. AND K. P. VO [1992], Heaviest increasing/common subsequence problems, in *Combinatorial Pattern Matching, Proceedings of the Third Annual Symposium*, A. APOSTOLICO, M. CROCHEMORE, Z. GALIL AND U. MANBER, EDS., Tucson, Arizona, 1992. Springer-Verlag, Berlin, Lecture Notes in Computer Science 644, 52–66.

[40] JOHNSON, D. B. [1982]. A priority queue in which initialization and queue operations take $O(\log \log D)$ time, *Math. Systems Theory* **15**, 295–309.

[41] IVANOV, A. G. [1985], Recognition of an approximate occurrence of words on a Turing machine in real time, *Math. USSR Izv.*, **24**, 479–522.

[42] KEDEM, Z. M. AND H. FUCHS [1980], On finding several shortest paths in certain graphs, in Proc. 18th Allerton Conference on Communication, Control, and Computing, October 1980, pp. 677–683.

[43] KUMAR, S. K. AND C. P. RANGAN [1987], A linear space algorithm for the LCS problem, *Acta Informatica* **24**, 353–362.

[44] LADNER, R. E., AND M. J. FISCHER [1980], Parallel prefix computation, *J. Assoc. Comput. Mach.*, **27**, 831–838.

[45] LANDAU. G. M. AND U. VISHKIN [1986], Introducing efficient parallelism into approximate string matching and a new serial algorithm, in *Proc. 18th Annual ACM STOC*, New York, 1986, 220–230.

[46] LANDAU, G. M. AND U. VISHKIN [1988], Fast string matching with k differences, *Jour. Comp. and System Sci.* **37**, 63–78.

[47] LEIGHTON, F. T. [1992], *Introduction to Parallel Algorithms and Architectures*, Morgan Kaufmann, San Mateo, CA.

[48]   LEVENSHTEIN, V.I. [1966], Binary codes capable of correcting deletions, in-
       sertions and reversals, Soviet Phys. Dokl., **10**, 707–710.
[49]   LIPTON, R.J. AND D. LOPRESTI [1985], A systolic array for rapid string
       comparison *Proc. Chapel Hill Conf. on Very Large Scale Integration*, H. FUCS,
       ED., Computer Science Press, 363–376.
[50]   H.M. MARTINEZ, ED. [1984], Mathematical and computational problems in
       the analysis of molecular sequences, *Bull. Math. Bio.* **46**, (Special Issue Hon-
       oring M.O. Dayhoff).
[51]   MASEK, W.J. AND M.S. PATERSON [1980], A faster algorithm computing
       string edit distances, *J. Comput. System Sci.*, **20**, 18–31.
[52]   MATHIES, T.R. [1988], A fast parallel algorithm to determine edit distance,
       Tech. Report CMU-CS-88-130, Department of Computer Science, Carnegie
       Mellon University, Pittsburgh, PA, April 1988.
[53]   MEHLHORN, K. [1984], *Data structures and algorithms 1: sorting and search-
       ing*, EATCS Monographs on TCS, Springer-Verlag, Berlin.
[54]   MYERS, E.W. AND W. MILLER [1988], Optimal alignments in linear space,
       Comp. Appl. Biosc. **4**, 1, 11–17.
[55]   MYERS, E.W. [1986], An $O(ND)$ difference algorithm and its variations,
       Algorithmica **1**, 251–266.
[56]   NAKATSU, N. , Y. KAMBAYASHI, AND S. YAJIMA [1982], A longest common
       subsequence algorithm suitable for similar text strings, *Acta Informatica* **18**,
       171–179.
[57]   NEEDLEMAN, R.B. AND C.D. WUNSCH [1973], A general method applicable
       to the search for similarities in the amino-acid sequence of two proteins, *J.
       Molecular Bio.*, **48**, 443–453.
[58]   RANKA, S. AND S. SAHNI [1988], String editing on an SIMD hypercube multi-
       computer, Tech. Report 88-29, Department of Computer Science, University
       of Minnesota, March 1988, *J. Parallel Distributed Comput.*
[59]   SALOMAA, A. [1973] *Formal Languages*, Academic Press, Orlando, Fl.
[60]   SANKOFF, D.[1972], Matching sequences under deletion-insertion constraints,
       Proc. Nat. Acad. Sci. U.S.A., **69**, 4–6.
[61]   SANKOFF, D. AND J.B. KRUSKAL, EDS. [1983], *Time Warps, String Edits and
       Macromolecules: The Theory and Practice of Sequence Comparison*, Addison-
       Wesley, Reading, MA.
[62]   SANKOFF, D. AND P.H. SELLERS [1973], Shortcuts, Diversions and Maximal
       Chains in Partially Ordered Sets, *Discrete Mathematics* , **4**, 287–293.
[63]   SELLERS, P.H. [1980], The theory and computation of evolutionary distance:
       pattern recognition, *J. Algorithms*, **1**, 359–373.
[64]   SMITH, T.F. AND M.S. WATERMAN [1981], Identification of Common Molec-
       ular Subsequences, *Journal of Molecular Biology* **147**, 195–197.
[65]   UKKONEN, E. [1985], Finding approximate patterns in strings, *J. Algorithms*
       6, 132–137.
[67]   WAGNER, R.A. AND M.J. FISCHER [1974], The string to string correction
       problem, *J. Assoc. Comput. Mach.*, **21**, 168–173.
[68]   WATERMAN, M.S. (ED.) [1989], *Mathematical Methods for DNA sequences*,
       CRC Press, Boca Raton.
[69]   WONG, C.K. AND A.K. CHANDRA [1976], Bounds for the string editing
       problem, *J. Assoc. Comput. Mach.*, **23**, 13–16.
[70]   WU, S., U. MANBER, E.W. MYERS, AND W. MILLER [1990]. An $O(NP)$
       sequence comparison algorithm, *Info. Proc. Letters* **35**, 317–323.
[71]   WU, S., U. MANBER, AND E. MYERS [1991]. Improving the running times
       for some string-matching problems.

Automata for Matching Patterns

Maxime Crochemore and Christophe Hancart

1. Pattern matching and automata

This chapter describes several methods of word pattern matching that are based on the use of automata.

Pattern matching (in words) is the problem of locating occurrences of a pattern in a text file. The file is just a string of symbols, but the pattern can be specified in various ways. Here, we only consider patterns described by regular expressions or weaker mechanisms.

Solutions to the problem are basic parts of many text processing tools, such as editors, parsers, and information retrieval systems. They are also widely used in the analysis of biological sequences. The algorithms that solve the problem classically decompose in two steps: a preprocessing phase and a search phase. When the text file is considered to be dynamic (as in editing applications), the preprocessing is applied to the pattern (see Sections 4, 5, and 6). This leads *a posteriori* to a good solution regarding the efficiency of the algorithms of this chapter. When the text file is static (if it is a dictionary, for example) the preprocessing applied to the text builds an index that can later support efficiently several series of queries (see Section 7).

We present solutions in which the search phase is based on automata as opposed to solutions based on combinatorial properties of words. Thus, the algorithms perform on-line searches with a buffer on the text that does not need to store more than one letter at a time. The solutions are adequate for processing sequential-access files or streams of symbols.

The main algorithms of this chapter solve special instances of the determinization or minimization problems of automata. Basically, given an automaton that recognizes the language X on the alphabet A, algorithms build a deterministic, and sometimes minimal, automaton for the language A^*X, which is applied afterwards to search efficiently for words of X.

The time complexity of algorithms is given as a function of the input, and is typically linear in the length of the input. This takes into account the set of letters actually occurring in the input. But the running time may depend on the output as well. So, a careful statement of each problem is necessary, to avoid for example quadratic-size outputs that would obviously imply quadratic-time algorithms.

The complexity of algorithms is analyzed in a model of a machine in which the basic operation on letters is comparison in the form less-equal-greater. The implicit ordering on the alphabet is exploited in several algorithms. The assumption on the model makes it possible to process words over a potentially unbounded alphabet. Some algorithms for the simplest pattern-matching problem (searching for only one word) operates in a weaker model (comparison in the form equal-unequal). We also mention how the running times of most algorithms are affected when branchings in automata are performed by looking up a transition table (see Section 3). This is valid if the alphabet is known in advance and if the letters can be assimilated to indices on a table. Otherwise, a straightforward simulation implies that the running times are multiplied by $O(\log \operatorname{card}(A))$, while in the comparison model the running times of some algorithms are independent of the alphabet.

The regular-expression-matching problem (Section 4) is when the pattern is a general regular expression. The standard solution is certainly by Thompson (1968). The mechanism is one of the basic features of the UNIX operating system and of its tools.

When the language described by the pattern reduces to a finite set of words (Section 5), called a dictionary, the pattern-matching algorithm runs in linear time (on a fixed alphabet) instead of quadratic time for the general solution. Moreover, when the pattern is only one word (Section 6), the same running time holds, independently of the alphabet.

The suffix automata presented in Section 7 serve as indexes. They provide a solution to the pattern-matching instance where the searched text has to be preprocessed. The main point of the section is the linear-time construction of suffix automata (on a fixed alphabet), which results partially from their linear size.

The efficiency of pattern-matching algorithms based on automata strongly relies on particular representations of these automata. This is why a review of several techniques is given in Section 3. The regular-expression-matching problem, the dictionary-matching problem, and the string-matching problem are treated respectively in Sections 4, 5, and 6. Section 7 deals with suffix automata and their applications.

2. Notations

This section is devoted to a review of the material used in this chapter: alphabet, words, languages, regular expressions, finite automata, algorithms for matching patterns.

2.1 Alphabet and words

Let A be a finite set, called the *alphabet*. Its elements are called *letters*, and, for convenience, we denote them by a, b, c, and so on. Furthermore, we assume that there is an ordering on the alphabet.

A *word* is a finite-length sequence of letters. The *length* of a word u is denoted by $|u|$, and its j-th letter by u_j. The set of all words is denoted by A^*, the empty word by ε, and A^+ stands for $A^* \backslash \{\varepsilon\}$.

The *product* of two words u and v, denoted by $u \cdot v$ or uv, is the word obtained by writing sequentially the letters of u then the letters of v. Given a word u, the product of k words identical with u is denoted by u^k, setting $u^0 = \varepsilon$. Denoted respectively by uw^{-1} and $v^{-1}u$ are the words v and w when $u = vw$.

A word v is said to be a *factor* of a word u if $u = u'vu''$ for some words u' and u''; it is a *proper* factor of u if $v \neq u$, a *prefix* of u if $u' = \varepsilon$, and a *suffix* of u if $u'' = \varepsilon$.

2.2 Languages

A *language* is any subset of A^*. The *product* of two languages U and V, denoted by $U \cdot V$ or UV, is the language $\{uv \mid (u,v) \in U \times V\}$. Denoted by U^k is the set of words obtained by making products of k words of U. The *star* of U, denoted by U^*, is the language $\bigcup_{k \geq 0} U^k$. By convention, the order of decreasing precedence for language operations in expressions denoting languages is star or power, product, union. By misuse, a language reduced to only one word u may be denoted by u itself if no confusion arises (with further notations).

The sets of prefixes, of factors, and of suffixes of a language U are denoted respectively by $Pref(U)$, $Fact(U)$, and $Suff(U)$. If U is finite, $|U|$ stands for $\sum_{u \in U} |u|$ (therefore, note that $card(A) = |A|$).

The *right context* of a word u according to a language W is the language $\{u^{-1}w \mid w \in W\}$. The equivalence generated over A^* by the relations

$$u^{-1}W = v^{-1}W, \quad u, v \in A^*$$

is denoted by $\equiv_W$; it is the *right syntactic congruence* associated with the language W.

2.3 Regular expressions

Regular expressions and the languages they describe, the *regular languages*, are defined inductively as follows:

- **0, 1**, and a are regular expressions and describes respectively $\emptyset$ (the empty set), $\{\varepsilon\}$, and $\{a\}$, for each $a \in A$;

– if u and v are regular expressions describing respectively the regular languages U and V, then $u+v$, $u \cdot v$, and u^* are regular expressions describing respectively the regular languages $U \cup V$, $U \cdot V$, and U^*.

By convention, the order of decreasing precedence for operations in regular expressions is star (*), product ($\cdot$), sum ($+$). The dot $\cdot$ is often omitted. Parenthesizing can be used to change the precedence order of operators.

The language described by a regular expression u is denoted by $Lang(u)$. The *length* $|u|$ of a regular expression u is the length of u reckoned on the alphabet $A \cup \{0, 1, +, ^*\}$ (parentheses and product operator $\cdot$ are not reckoned).

2.4 Finite automata

A *(finite) automaton (with one initial state)* is given by a finite set Q, whose elements are called *states*, an *initial* state i, a subset $T \subseteq Q$ of *terminal* states, and a set $E \subseteq Q \times A \times Q$ of *edges*.

An edge (p, a, q) of the automaton (Q, i, T, E) is an *outgoing* edge for state p and an *ingoing* edge for state q; state p is the *source* of this edge, letter a its *label*, and state q its *target*. The number of edges outgoing a state p is called the *(outgoing) degree* of p. We say that there is a *path* labeled by u from state p to state q if there is a finite sequence $(r_{j-1}, a_j, r_j)_{1 \le j \le n}$ of edges such that $n = |u|$, $a_j = u_j$ for each $j \in \{1, \ldots, n\}$, $r_0 = p$ and $r_n = q$. One agrees to define a unique path labeled by ε from each state p to itself.

A word u is *recognized* by the automaton $\mathcal{A} = (Q, i, T, E)$ if there exists a path labeled by u from i to some state in T. The set of all words recognized by $\mathcal{A}$ is denoted by $Lang(\mathcal{A})$. A language X is *recognizable* if there exists an automaton $\mathcal{A}$ such that $X = Lang(\mathcal{A})$.

As an example, the automaton depicted in Figure 2.1 recognizes the language $\{a, b\}^*abaaab$. Its initial state is 0, and its only terminal state is 6.

The automaton (Q, i, T, E) is *deterministic* if for each $(p, a) \in Q \times A$ there is at most one state q such that $(p, a, q) \in E$. It is *complete* if for each $(p, a) \in Q \times A$ there is at least one state q such that $(p, a, q) \in E$. It

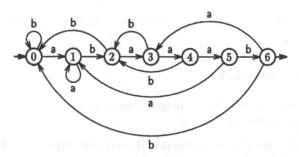

Fig. 2.1. An automaton recognizing the language $\{a, b\}^*abaaab$.

is *normalized* if card$(T) = 1$, the initial state has no ingoing edge, and the terminal state has no outgoing edge. It is *minimal* if it is deterministic and if each deterministic automaton recognizing the same language maps onto it; it has the minimal number of states. The minimal automaton recognizing the language U is denoted by $\mathcal{M}(U)$. It can be defined with the help of right contexts by:

$$\Big(\{u^{-1}U \mid u \in A^*\}, \ \{U\}, \ \{u^{-1}U \mid u \in U\},$$
$$\{(u^{-1}U, a, (ua)^{-1}U) \mid u \in A^*, a \in A\}\Big)$$

In case $\mathcal{A} = (Q, i, T, E)$ is a deterministic automaton, it is convenient to consider the *transition function* $\delta: Q \times A \to Q$ of $\mathcal{A}$ defined for each $(p, a) \in Q \times A$ such that there is an outgoing edge labeled by a for p by

$$\delta(p, a) = q \iff (p, a, q) \in E$$

(notice that δ is a partial function). Equivalently, the quadruple (Q, i, T, δ) denotes the automaton $\mathcal{A}$. In a natural way, the transition function extends to a function mapping from $Q \times A^*$ to Q and also denoted by δ setting

$$\delta(p, u) = \begin{cases} p, & \text{if } u = \varepsilon, \\ \delta(\delta(p, a), v), & \text{if } \delta(p, a) \text{ is defined and } u = av \\ & \qquad \text{for some } (a, v) \in A \times A^*, \\ \text{undefined}, & \text{otherwise}, \end{cases}$$

for each $(p, u) \in Q \times A^*$.

In algorithms that manipulate automata, we constantly use the function STATE-CREATION described in Figure 2.2 (+ stands for the union of sets). This avoids going into details of the implementation of automata that is precisely the subject of Section 3.

> STATE-CREATION
> 1 chose a state q out of Q
> 2 $Q \leftarrow Q + \{q\}$
> 3 **return** q

Fig. 2.2. Creation of a new state and adjunction to the set of states Q.

2.5 Algorithms for matching patterns

The pattern matching problem is to search and locate occurrences of patterns in words (or textual data, less formally speaking). A *pattern* represents a language and is described either by a word, by a finite set of words, or more generally, by a regular expression. We do not consider patterns described by other mechanisms.

Let y be the searched word. An *occurrence* in y of a pattern represented by the language X is a triple (u, x, v) where $u, v \in A^*$, $x \in X$, and such that $y = uxv$. The *position* of the occurrence (u, x, v) of x in y is the length $|u|$; it is sometimes more convenient to consider the *end-position* of the same occurrence, which is defined as the length $|ux|$. Observe that searching y for words in a language X is equivalent to search for prefixes of y that belong to the language $A^* X$; the language of most automata considered in this chapter is of this form.

According to a specific matching problem, the input of an algorithm is a language X described by a word, by a finite set of words, or by a regular expression, and a word y. The output can have several forms. To implement an algorithm that tests whether the pattern occurs in the word or not, the output is just the boolean value TRUE or FALSE respectively. In an on-line search, what is desired is the word, say z, on the alphabet $\{0, 1\}$ that encodes the existence of end-positions of the pattern; the length of z is $|y| + 1$, and its $j + 1$-th letter is 1 exactly when an occurrence of the pattern ends at position j in y. The output can also be the set, say P, of positions (or end-positions) in y of the pattern. To avoid presenting several variants of algorithms, we introduce the statement

occurrenceif e

where e is an appropriate predicate. It can be translated by

if e
then return TRUE

in the first case,

if e
then $z \leftarrow z \cdot 1$
else $z \leftarrow z \cdot 0$

in the second case, and

if e
then $P \leftarrow P + \{$the current position in $y\}$

in the third case. In the first case, the "**return** FALSE" statement has to be included correspondingly at the end of the algorithm; and in the other cases, word z and set P should be initialized at the beginning of the algorithm and returned at the end of the algorithm. From now on, the standard algorithm for matching patterns in words can be written as in Figure 2.3.

The asymptotic time and space complexities of algorithm MATCHER depend on the representation of the automaton, and more specifically, on the representation of the transition function δ (see Section 3). More generally, the complexities of algorithms, functions or procedures developed in this chapter are expressions of the size of the input. They include the size of the language, the length of the searched word, and the size of the alphabet. We assume that

MATCHER(X, y)
 Preprocessing phase
1 built an automaton (Q, i, T, E) recognizing A^*X
 Search phase
 let δ be the transition function of (Q, i, T, E)
2 $p \leftarrow i$
3 **occurrenceif** $p \in T$
4 **for** letter a from first to last letter of y
5 **loop** $p \leftarrow \delta(p, a)$
6   **occurrenceif** $p \in T$

Fig. 2.3. Given a regular language X and a word y, locate all occurrences of words in X that are factors of y.

the "**occurrenceif** e" statement is performed in constant time. Nevertheless, an *ad hoc* output often underlies the complexity result.

3. Representations of deterministic automata

Several pattern matching algorithms rely on a particular representation of the deterministic automaton underlying the method. Implementing a deterministic automaton (Q, i, T, E) remains to implement the transition function δ of the automaton, which is the general problem of realizing partial functions. Five methods are described in this section: transition matrix, adjacency lists, transition list, failure function, and table-compression.

The choice of the representation of the automaton influence the time needed to compute a transition, *i.e.* the time to evaluate $\delta(p, a)$, for any state p and any letter a. This time is called the *delay*, in that it is also the time spent on letter a before moving to the next letter of the input word. Basically: on the one hand, the time to evaluate $\delta(p, a)$ is constant in a model where branchings are allowed and a transition matrix implements δ; on the other hand, if comparison of letters is the only operation allowed on them, the time to evaluate $\delta(p, a)$ is $O(\log \operatorname{card}(A))$, assuming that any two letters can be compared in one unit of time (using binary operations $=$, $\neq$, $<$ or $>$). In the following, we give the memory space and the delay associated to each type of representation. There is an obvious trade-off between these two quantities.

In the chapter, having a representation R of the transition function δ, the automaton is indifferently denoted by (Q, i, T, E), (Q, i, T, δ), and (Q, i, T, R).

3.1 Transition matrix

The simplest method to implement the transition function δ is to store its values in a $Q \times A$-matrix. This is a method of choice for a complete deterministic automaton on a small alphabet and when letters can be assimilated to indices on an array. The space required is $O(\text{card}(Q) \times \text{card}(A))$ and the delay is $O(1)$.

When the automaton is not complete, the representation still works except that the searching procedure can stop on an undefined transition. The matrix can even be initialized in time proportional to the number of edges of the automaton with the help of a sparse matrix representation technique. The above complexities are still valid in this situation.

This kind of representation implicitly assumes that the working alphabet is fixed and known by advance. This contrasts with the representations of Sections 3.2 and 3.4 for which the basic operation on the alphabet is comparing letters.

3.2 Adjacency lists

A traditional way of implementing graphs is to use adjacency lists. This applies to automata as well. Doing so, the set of couples $(a, \delta(p, a))$, whenever $\delta(p, a)$ is defined, is associated with each state $p \in Q$. The space required to represent the $\text{card}(Q)$ adjacency lists of the automaton is $O(\text{card}(Q) + \text{card}(E))$. Contrary to the previous method, this one works even if the only possible elementary operation on letters is comparison. Denoting by d the maximum degree of states of the automaton, the delay is $O(\log d)$, which is also $O(\log \min\{\text{card}(Q), \text{card}(A)\})$, using an efficient implementation of sets based for instance on balanced trees.

The space complexity may be further reduced by considering a *default (target) state* associated to each adjacency list (the most frequently occurring target of a given adjacency list is an obvious choice as default for this adjacency list). The delay can even be improved at the same time because adjacency lists become smaller.

When implementing the automaton, each adjacency list is stored in an array G indexed by Q. If the deterministic automaton is complete and if the initial state i is the uniform default state (i as default state fits in perfectly with pattern matching applications), the computation of a transition from any state p by any letter a, that is, the computation of $\delta(p, a)$, is done by the function of Figure 3.1.

AdjacencyLists-Transition(p, a)
1   $p \leftarrow$ state of the couple of label a in $G[p]$
2   **if** $p = \text{NIL}$
3   **then** $p \leftarrow i$
4   **return** p

Fig. 3.1. Computation of the transition from a state p by a letter a when an array G of adjacency lists represents the transition function.

3.3 Transition list

The transition list method consists in implementing the list of triples (p, a, q) of edges of the automaton. The space required by the implementation is only $O(\text{card}(E))$. Doing so, it is assumed that the list is stored in a hashing table to provide fast computations of transitions. The corresponding hashing function is defined on couples (p, a) from $Q \times A$. Then, given a couple (p, a), the access to the transition (p, a, q), if it appears in the list, is performed in constant time on the average under usual hypotheses on the technique.

3.4 Failure function

The main idea of the failure function method is to reduce the space needed by δ, by deferring, in most possible cases, the computation of the transition from the current state to the computation of the transition from an other given state with the same input letter. It serves to implement deterministic automata in the comparison model. Its main advantage is that, in general, it provides a linear-space representation, and, simultaneously, gives a linear-time cost for a series of transitions, though the time to compute one transition is not necessarily constant.

We only consider the case where the deterministic automata is complete and where i is the default state (extensions of the following statement are not needed in the chapter).

Let γ be a function from $Q \times A$ to Q, and let f be a function from Q into itself. We say that the couple (γ, f) represents the transition function δ if γ is a subfunction of δ and if

$$\delta(p, a) = \begin{cases} \gamma(p, a), & \text{if } \gamma(p, a) \text{ is defined,} \\ \delta(f(p), a), & \text{if } \gamma(p, a) \text{ is undefined and } f(p) \text{ is defined,} \\ i, & \text{otherwise,} \end{cases}$$

for each $(p, a) \in Q \times A$. In this situation, the state $f(p)$ is a stand-in of state p. The functions γ and f are respectively said to be a *subtransition function* and a *failure function*, according to δ. However, this representation is correct if we assume that f defines an order on Q.

Assuming a representation of γ by adjacency lists, the space needed to represent δ by the couple (γ, f) is $O(\text{card}(Q) + \text{card}(E'))$, where

$$E' = \{(p, a, q) \mid (p, a, q) \in E \text{ and } \gamma(p, a) \text{ is defined}\}),$$

which is of course $O(\text{card}(Q) + \text{card}(E))$ since $E' \subseteq E$. (Notice that γ is the transition function of the automaton (Q, i, T, E').) If d is the maximum degree of states of the automaton (Q, i, T, E'), the delay is typically $O(\text{card}(Q) \times \log d)$, that is also $O(\text{card}(Q) \times \log \text{card}(A))$.

When implementing the automaton, the values of the failure function f are stored in an array F indexed by Q. The computation of a transition is done by the function of Figure 3.2. The function always stops if we assume that f defines an order on Q.

FAILUREFUNCTION-TRANSITION(p, a)
1   **while** $p \neq$ NIL and $\gamma(p, a) =$ NIL
2   **loop** $p \leftarrow F[p]$
3   **if** $p \neq$ NIL
4   **then** $p \leftarrow \gamma(p, a)$
5   **else** $p \leftarrow i$
6   **return** p

Fig. 3.2. Computation of the transition from a state p by a letter a when a subtransition γ and an array F corresponding to a failure function represent the transition function.

3.5 Table-compression

The latest method is a mix of the previous ones that provides fast computations of transitions via arrays and a compact representation of edges via failure function.

Four arrays, denoted here by *fail*, *base*, *target*, and *check*, are used. The *fail* and *base* arrays are indexed by Q, and, for each $(p, a) \in Q \times A$, $base[p] + a$ is an index on *target* and *check* arrays, assimilating letters to integers.

The computation of the transition from some state p with some input letter a proceeds as follows: let $k = base[p] + a$; then, if $check[k] = p$, $target[k]$ is the target of the edge of source p and label a; otherwise, this statement is repeated recursively with state $fail[p]$ and letter a. (Notice that it is correct if *fail* defines an order on Q, as in Section 3.4, and if the targets from the smallest element are all defined.) The corresponding function is given in Figure 3.3.

TABLECOMPRESSION-TRANSITION(p, a)
1   **while** $check[base[p] + a] \neq p$
2   **loop** $p \leftarrow fail[p]$
3   **return** $target[base[p] + a]$

Fig. 3.3. Computation of the transition from a state p by a letter a in the table-compression method with suitable arrays *fail*, *base*, *target*, and *check*.

In the worst case, the space needed is $O(\mathrm{card}(Q) \times \mathrm{card}(A))$ and the delay is $O(\mathrm{card}(Q))$. However the method can reduce the space to $O(\mathrm{card}(Q) + \mathrm{card}(A))$ with an $O(1)$ delay in best possible situation.

4. Matching regular expressions

4.1 Outline

Problem 4.1. (Regular-expression-matching problem.) Given a regular expression x, preprocess it in order to locate all occurrences of words of *Lang*(x) that occur in any given word y.

A well-known solution to the above problem is composed of two phases. First, transform the regular expression x into a nondeterministic automaton that recognizes the language described by x, following a construction due to Thompson. Second, simulate the obtained automaton with input word y in such a way that it recognizes each prefix of y that belongs to $A^* Lang(x)$.

Main Theorem 4.1. *The regular expression-matching problem for x and y can be achieved in the following terms:*

- *a preprocessing phase on x building an automaton of size $O(|x|)$, performed in time $O(|x|)$ and $O(|x|)$ extra-space;*
- *a search phase executing the automaton on y performed in time $O(|x| |y|)$ and $O(|x|)$ space, the time spent on each letter of y being $O(|x|)$.*

The construction of the automaton is given in Section 4.2. In Section 4.3, we first show how to solve the membership test, namely, "does y belongs to $Lang(x)$?"; we then present the solution to the search phase of the regular-expression-matching problem as a mere transformation of the previous test. Finally, in Section 4.4, we discuss a possible use of a deterministic automaton to solve the problem.

In the whole section, we assume that the regular expression contains no redundant parentheses, because otherwise the parsing of the expression would not be necessarily asymptotically linear in the length of the expression.

4.2 Regular-expression-matching automata

In order to solve the problem in space linear in the length of the regular expression, we consider special nondeterministic automata.

We say that an automaton is *extended* if it is defined on the extended alphabet $A \cup \{\varepsilon\}$. Observe that a transition from a state to another in an extended automaton may either result of the reading of a letter from the input word, or not (ε-transition).

Theorem 4.2. *Let x be a regular expression. There exists a normalized extended automaton recognizing $Lang(x)$ satisfying the following conditions:*

(i)   *the number of states is bounded by $2|x|$;*

(ii) *the number of edges labeled by letters of A is bounded by $|x|$, and the number of edges labeled by ε is bounded by $4|x|$;*

(iii) *for each state the number of ingoing or outgoing edges is at most 2, and it is exactly 2 only when the edges are labeled by ε.*

Proof. The proof is by induction on the length of regular expressions.

The regular expressions of length equal to 1 are $\mathbf{0}$, $\mathbf{1}$, and a, for each $a \in A$. They are respectively recognized by normalized extended automata in the form

$$\Big(\{i,t\},\ i,\ \{t\},\ \emptyset\Big),$$

$$\big(\{i,t\},\ i,\ \{t\},\ \{(i,\varepsilon,t)\}\big),$$

and

$$\big(\{i,t\},\ i,\ \{t\},\ \{(i,a,t)\}\big),$$

where i and t are two distinct states. The automata are depicted in Figure 4.1.

Now, let $(Q',i',\{t'\},E')$ and $(Q'',i'',\{t''\},E'')$ be normalized extended automata recognizing respectively the regular expressions u and v, assuming that $Q' \cap Q'' = \emptyset$. Then the regular expressions $u + v$, $u \cdot v$, and u^* are respectively recognized by normalized extended automata in the form

$$\Big(Q' \cup Q'' \cup \{i,t\},\ i,\ \{t\},\ E' \cup E'' \cup \{(i,\varepsilon,i'),(i,\varepsilon,i''),(t',\varepsilon,t),(t'',\varepsilon,t)\}\Big)$$

where i and t are two distinct states chosen out of $Q' \cup Q''$,

$$\Big(Q' \cup Q'',\ i',\ \{t''\},\ E' \cup E'' \cup \{(t',\varepsilon,i'')\}\Big),$$

and

$$\Big(Q' \cup \{i,t\},\ i,\ \{t\},\ E' \cup E'' \cup \{(i,\varepsilon,i'),(i,\varepsilon,t),(t',\varepsilon,i'),(t',\varepsilon,t)\}\Big)$$

where i and t are two distinct states chosen out of Q'. The automata are depicted in Figure 4.2.

The above construction clearly proves the existence of a normalized extended automaton recognizing the language described by any given regular

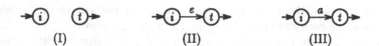

Fig. 4.1. Normalized extended automata recognizing the regular expressions $\mathbf{0}$ (I), $\mathbf{1}$ (II), and a (III) for some $a \in A$.

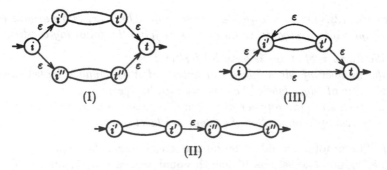

Fig. 4.2. Normalized extended automata recognizing the regular expressions $u + v$ (I), $u \cdot v$ (II), and u^* (III), obtained from normalized extended automata $(Q',i',\{t'\},E')$ and $(Q'',i'',\{t''\},E'')$ recognizing respectively the regular expressions u and v.

expression. It remains to check that the automaton satisfies conditions (i) to (iii). Condition (i) holds since exactly two nodes are created for each letter of a regular expression accounting for its length. Condition (ii) is easy to establish, using similar arguments. And the last condition follows from construction.   □

The previous result proves one half of the theorem of Kleene (the second half of the proof may be found in any standard textbook on automata or formal language theory).

Theorem 4.3 (Kleene, 1956). *A language is recognizable if and only if it is regular.*

We denote by $\mathcal{E}(x)$ the normalized extended automaton constructed in the proof of Theorem 4.2 from the regular expression x, and we call it the *regular-expression-matching automaton* of x.

To evaluate the time complexity of the above construction, it is necessary to give some hints about the data structures involved in the representation of regular-expression-matching automata. Due to the conditions stated in Theorem 4.2, a special representation of regular-expression-matching automata is possible providing an efficient implementation of the construction function. States are simply indices on an array that store edges; each cell of the array has to store at most two edges whose ingoing state is its index. Indices of the initial state and of the terminal state are stored separately. This shows that the space required to store $\mathcal{E}(x)$ is linear in its number of states, which is linear in the length of x according to Theorem 4.2.

Hence, each of the operations induced by $+$, $\cdot$ or $*$ can be implemented to work in constant time. This proves that the time spent on each letter of x is constant. In addition, the function which builds the regular-expression-matching automaton corresponding to a given regular expression is driven by a parser of regular expressions. Then, if parenthesizing in x is not redundant, the time and the space needed for the construction of $\mathcal{E}(x)$ is linear in the length of x.

We have established the following result:

Theorem 4.4. *Let x be a regular expression. The space needed to represent $\mathcal{E}(x)$ is $O(|x|)$. The computation of the automaton is performed in time and space $O(|x|)$.*

4.3 Searching with regular-expression-matching automata

The search for end-positions of words in $Lang(x)$ is performed with a simulation of a deterministic automaton recognizing $A^*Lang(x)$. Indeed, the determinization is avoided because it may lead to an automaton with a number of states which is exponential in the length of the regular expression (see

Section 4.4). But the determinization via the subset construction is just simulated: at a given time, the automaton is not in a given state, but in a set of states. This subset is recomputed whenever necessary in the execution of the search.

As for the determinization of automata with ε-transitions, the searching procedure needs the notion of *closure* of a set of states: if S is a set of states, its closure is the set of states q such that there exists a path labeled by ε from a state of S to q. From the closure of a set of states, it is possible to compute effectively the transitions induced by any input letter.

The simulation of a regular-expression-matching automaton consists in repeating the two operations closure and computation of transitions on a set of states. These two operations are respectively performed by functions CLOSURE and TRANSITIONS of Figures 4.3 and 4.4. With careful implementation, based on standard manipulation of sets and queues, the time and the space required to compute a closure or the transitions from a closure are linear in the size of involved sets of states.

A basic use of an automaton consists in testing whether it recognizes some given word. Testing whether y is in the language described by x is implemented by the algorithm of Figure 4.5. The next proposition states the complexity of such a test.

```
CLOSURE(E, S)
 1  R ← S
 2  ϑ ← EMPTYQUEUE
 3  for each state p in S
 4      loop ENQUEUE(ϑ, p)
 5  while not QUEUEISEMPTY(ϑ)
 6      loop p ← DEQUEUE(ϑ)
 7          for each state q such that (p, ε, q) is in E
 8              loop if q is not in R
 9                  then R ← R + {q}
10                       ENQUEUE(ϑ, q)
11  return R
```

Fig. 4.3. Computation of the closure of a set S of states, with respect to a set E of edges.

```
TRANSITIONS(E, S, a)
 1  R ← ∅
 2  for each state p in S
 3      loop for each state q such that (p, a, q) is in E
 4          loop R ← R + {q}
 5  return R
```

Fig. 4.4. Computation of the transitions by a letter a from states of a set S, with respect to a set E of edges.

REGULAREXPRESSIONTESTER(x, y)
1   built the regular-expression-matching automaton $(Q, i, \{t\}, E)$ of x
2    $C \leftarrow$ CLOSURE$(E, \{i\})$
3   **for** letter a from first to last letter of y
4   **loop** $C \leftarrow$ CLOSURE$(E, \text{TRANSITIONS}(E, C, a))$
5   **return** $t \in C$

Fig. 4.5. Algorithm for testing whether a word y belongs to $Lang(x)$, x being a regular expression.

Proposition 4.1. *Given a regular expression x, testing whether a word y belongs to $Lang(x)$ can be performed in time $O(|x| |y|)$ and space $O(|x|)$.*

Proof. The proof is given by algorithm TESTER of Figure 4.5 for which we analyze the complexity.

According to Theorem 4.4, the regular-expression-matching automaton $(Q, i, \{t\}, E)$ of x can be built in time and space $O(|x|)$.

Each computation of functions CLOSURE and of TRANSITIONS requires time and space $O(\text{card}(Q))$, which is $O(|x|)$ from Theorem 4.2. This is repeated $|y|$ times. This gives $O(|x| |y|)$ time.   □

We now come back to our main problem. It is slightly different than the previous one, because the answer to the test has to be reported for each factor of y, and not only on y itself. But no transformation of $\mathcal{E}(x)$ is necessary. A mere transformation of the search phase of the algorithm is sufficient: at each iteration of the closure computation, the initial state is integrated to the current set of states. Doing so, each factor of y is tested. Moreover, the "**occurrenceif** $t \in C$" instruction is done at each stage. The entire algorithm is given in Figure 4.6. The following theorem established the complexity of the search phase of the algorithm.

Theorem 4.5. *Let x be a regular expression and y be a word. Finding all end-positions of factors of y that are recognized by $\mathcal{E}(x)$ can be performed in time $O(|x| |y|)$ and space $O(|x|)$. The time spent on each letter of y is $O(|x|)$.*

Proof. See the proof of Proposition 4.1. The second part of the statement comes from that fact: the time spent on each letter of y is linear in the time required by the computations of functions CLOSURE and TRANSITIONS.   □

REGULAREXPRESSIONMATCHER(x, y)
1   built the regular-expression-matching automaton $(Q, i, \{t\}, E)$ of x
2    $C \leftarrow$ CLOSURE$(E, \{i\})$
3   **occurrenceif** $t \in C$
4   **for** letter a from first to last letter of y
5   **loop** $C \leftarrow$ CLOSURE$(E, \text{TRANSITIONS}(E, C, a) + \{i\})$
6   **occurrenceif** $t \in C$

Fig. 4.6. Algorithm for computing prefixes of a word y that belong to $A^* Lang(x)$, x being a regular expression.

4.4 Time-space trade-off

The regular-expression-matching problem for a regular expression x and a word y admits a solution based on deterministic automata. It proceeds as follow: build the automaton $\mathcal{E}(x)$; built an equivalent deterministic automaton; search with the deterministic automaton. The drawback of this approach is that the deterministic automaton can have a number of states exponential in the length of x. This is the situation, for example, when

$$x = \mathsf{a} \overbrace{(\mathsf{a} + \mathsf{b}) \cdots (\mathsf{a} + \mathsf{b})}^{m-1 \text{ times}}$$

for some $m \geq 1$; here, the minimal deterministic automaton recognizing $A^* Lang(x)$ has exactly 2^m states since the recognition process has to memorize the last m letters read from the input word y. However, all states of the deterministic automaton for $A^* Lang(x)$ are not necessarily met during the search phase. So, a lazy construction of the deterministic automaton during the search is a possible compromise for practical purposes.

5. Matching finite sets of words

5.1 Outline

Problem 5.1. (Dictionary-matching problem.) Given a finite set of words X, the dictionary, preprocess it in order to locate words of X that occur in any given word y.

The classical solution to this problem is due to Aho and Corasick. It essentially consists in a linear-space implementation of a complete deterministic automaton recognizing the language $A^* X$. The implementation uses both adjacency lists and an appropriate failure function.

Main Theorem 5.1 (Aho and Corasick, 1975). *The dictionary-matching problem for X and y can be achieved in the following terms:*

- *a preprocessing phase on X building an implementation of size $O(|X|)$ of an automaton recognizing $A^* X$, performed in time $O(|X| \times \log \mathrm{card}(A))$ and $O(\mathrm{card}(X))$ extra-space;*
- *a search phase executing the automaton on y performed in time $O(|y| \times \log \mathrm{card}(A))$ and constant extra-space, the delay being $O(|X| \times \log \mathrm{card}(A))$.*

If we allow more extra-space, the asymptotic time complexities can be reduced. This is achieved, for instance, by using techniques of Section 3 for representing deterministic automata with a sparse matrix, and assuming that $O(|X| \times \mathrm{card}(A))$ space is available. The time complexities of the preprocessing and search phases are respectively reduced to $O(|X|)$ and $O(|y|)$, and the delay to $O(|X|)$. Nevertheless, notice that the times complexities given in the above theorem are still linear in $|X|$ or $|y|$ if we consider fixed alphabets.

The method behind Theorem 5.1 is based on a specific automaton recognizing A^*X: its states are the prefixes of words in X (their number is finite as X is). The automaton is not minimal in the general case. It is presented in Section 5.2, and its implementation with a failure function is given in Section 5.3. Section 5.4 is devoted to the search for X with the automaton.

5.2 Dictionary-matching automata

We give a complete deterministic automaton that recognizes A^*X. In order to formalize this automaton, we introduce for each language U the mapping $h_U : A^* \to Pref(U)$ defined for each word v by

$$h_U(v) = \text{the longest suffix of } v \text{ that belongs to } Pref(U).$$

(In the whole Section 5, U refers to an ordinary language, and X refers to a finite language.)

Proposition 5.1. *Let X be a finite language. Then the automaton*

$$\Big(Pref(X), \ \varepsilon, \ Pref(X) \cap A^*X, \ \{(p, a, h_X(pa)) \mid p \in Pref(X), a \in A\} \Big)$$

recognizes the language A^*X. This automaton is deterministic and complete.

In the following, we denote by $\mathcal{D}(X)$ the automaton of Proposition 5.1 applied to X, and we call it the *dictionary-matching automaton* of X.

The proof of Proposition 5.1 relies on the following result.

Lemma 5.1. *Let $U \subseteq A^*$. Then*

(i) $v \in A^*U$ *iff* $h_U(v) \in A^*U$, *for each $v \in A^*$.*

Furthermore, h_U satisfies the relations:

(ii) $h_U(\varepsilon) = \varepsilon$;

(iii) $h_U(va) = h_U(h_U(v)a)$, *for each $(v, a) \in A^* \times A$.*

Proof. If $v \in A^*U$, then v is in the form wu where $w \in A^*$ and $u \in U$; by definition of h_U, u is necessarily a suffix of $h_U(v)$; therefore $h_U(v) \in A^*U$. Conversely, if $h_U(v) \in A^*U$, we have also $v \in A^*U$, because $h_U(v)$ is a suffix of v. Which proves (i).

Property (ii) clearly holds.

It remains to prove (iii). Both words $h_U(va)$ and $h_U(v)a$ are suffixes of va, and therefore one of them is a suffix of the other. Then we distinguish two cases according to which word is a suffix of the other.

First case: $h_U(v)a$ is a proper suffix of $h_U(va)$ (hence $h_U(va) \neq \varepsilon$). Consider the word w defined by $w = h_U(va)a^{-1}$. Thus we have: $h_U(v)$ is a proper suffix of w, w is a suffix of v, and since $h_U(va) \in Pref(U)$, $w \in Pref(U)$. Whence w is a suffix of v that belongs to $Pref(U)$, but strictly longest than $h_U(v)$. This contradicts the maximality of $|h_U(v)|$. So this case is impossible.

Second case: $h_U(va)$ is a suffix of $h_U(v)a$. Then, $h_U(va)$ is a suffix of $h_U(h_U(v)a)$. Now, since $h_U(v)a$ is a suffix of va, $h_U(h_U(v)a)$ is a suffix of $h_U(va)$. Both properties implies $h_U(va) = h_U(h_U(v)a)$, and the expected result follows. $\square$

Proof of Proposition 5.1. Let $v \in A^*$. It follows from properties (ii) and (iii) of Lemma 5.1 that

$$\bigl(h_X(v_1 v_2 \cdots v_{j-1}),\ v_i,\ h_X(v_1 v_2 \cdots v_j)\bigr)_{1 \le j \le |v|}$$

is a path labeled by v from the initial state ε to the state $h_X(v)$.

If $v \in A^*X$, we get $h_X(v) \in A^*X$ from (i) of Lemma 5.1; which shows that $h_X(v)$ is a terminal state, and finally that v is recognized by the automaton.

Conversely, if v is recognized by the automaton, we have $h_X(v) \in A^*X$ by definition of the automaton. This implies that $v \in A^*X$ from (i) of Lemma 5.1 again. $\square$

We show how to implement the automaton $\mathcal{D}(X)$ in the next section.

5.3 Linear dictionary-matching automata

The automaton $\mathcal{D}(X)$ is implemented with a failure function. The aim is to get a representation that does not depend on the size of the alphabet.

For each language U, let $f_U : Pref(U) \to Pref(U)$ be the function defined for each nonempty word u in $Pref(U)$ by

$$f_U(u) = \text{the longest proper suffix of } u \text{ that belongs to } Pref(U).$$

Lemma 5.2. *Let $U \subseteq A^*$. For each $(u, a) \in Pref(U) \times A$, we have:*

$$h_U(ua) = \begin{cases} ua, & \text{if } ua \in Pref(U), \\ h_U(f_U(u)a), & \text{if } u \ne \varepsilon \text{ and } ua \notin Pref(U), \\ \varepsilon, & \text{otherwise.} \end{cases}$$

Proof. The identity clearly holds when $ua \in Pref(U)$ or when $ua \notin Pref(U)$ but $u = \varepsilon$.

It remains to examine the case where $ua \notin Pref(U)$ and $u \ne \varepsilon$. Here, $f_U(u)a$ is a proper suffix of ua. What is more, $h_U(f_U(u)a)$ is the longest suffix of ua that belongs to $Pref(U)$. Indeed, if we assume the existence of a suffix v of ua satisfying $v \in Pref(U)$ and $|v| \ge |f_U(v)a|$, we get that va^{-1} is a proper suffix of u belonging to $Pref(U)$; then $va^{-1} = f_U(u)$ because of the maximality of $|f_U(u)|$. Which achieves the proof. $\square$

We introduce for each language U the function $\gamma_U : Pref(U) \times A \to Pref(U)$ associating with each $(u, a) \in Pref(U) \times A$ such that $ua \in Pref(U)$ the word ua. Thus, with conventions of Section 3.4, we have:

Proposition 5.2. *For each finite language X, the couple (γ_X, f_X) represents the transition function of $\mathcal{D}(X)$; function γ_X is a subtransition function and function f_x a failure function, according to the transition function of $\mathcal{D}(X)$.*

Proof. Follows from Lemma 5.2.   □

Now, let us observe that function γ_X is exactly the transition function of the deterministic automaton

$$\Big(Pref(X), \ \varepsilon, \ X, \ \{(p, a, pa) \mid p \in Pref(X), a \in A, pa \in Pref(X)\} \Big).$$

This automaton recognizes the language X, and is classically called the *trie* of X, as a reference to "information re*trie*val". It is built by function TRIE of Figure 5.1.

Proposition 5.3. *Function TRIE applied to any finite language X builds the trie of X. If the edges of the automaton are implemented via adjacency lists, the size of the trie is $O(|X|)$, and the construction is performed in time $O(|X| \times \log d)$ within constant extra-space, d being the maximum degree of states.*

When $X = \{\mathsf{ab}, \mathsf{babb}, \mathsf{bb}\}$, the trie of X is as depicted in Figure 5.2. This example shall be considered twice in the following.

To achieve the goal of implementing $\mathcal{D}(X)$ in linear size, we use Proposition 5.2. Then, it remains to give methods for computing f_X and for marking the set of terminal states. This can be done by a breadth first search on the graph underlying the trie starting at the initial state, as shown by the two following lemmas.

```
TRIE(X)
        let δ be the transition function of (Q, i, T, E)
 1   (Q, T, E) ← (∅, ∅, ∅)
 2   i ← STATE-CREATION
 3   for word x from first to last word of X
 4       loop t ← i
 5           for letter a from first to last letter of x
 6               loop q ← δ(t, a)
 7                   if q = NIL
 8                       then q ← STATE-CREATION
 9                           E ← E + {(t, a, q)}
10                   t ← q
11           T ← T + {t}
12   return (Q, i, T, E)
```

Fig. 5.1. Construction of the trie of a finite set of words X.

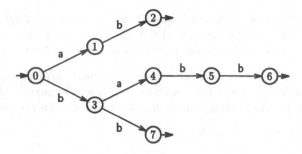

Fig. 5.2. The trie of $\{ab, babb, bb\}$.

Lemma 5.3. *Let $U \subseteq A^*$. For each $(u, a) \in Pref(U) \times A$, we have:*

$$f_U(ua) = \begin{cases} h_U(f_U(u)a), & \text{if } u \neq \varepsilon, \\ \varepsilon, & \text{otherwise.} \end{cases}$$

Proof. Similar to the proof of Lemma 5.2.   □

Lemma 5.4. *Let $U \subseteq A^*$. For each $u \in Pref(U)$, we have:*

$$u \in A^*U \quad \Longleftrightarrow \quad (u \in U) \text{ or } (u \neq \varepsilon \text{ and } f_U(u) \in A^*U).$$

Proof. It is clearly sufficient to prove that

$$u \in (A^*U)\backslash U \quad \Longrightarrow \quad f_U(u) \in A^*U.$$

So, let $u \in (A^*U)\backslash U$. The word u is in the form vw where $v \in A^*$ and w is a proper suffix of u belonging to U. Then, by definition of f_U, w is a suffix of $f_U(u)$. Therefore $f_U(u) \in A^*U$. Which ends the proof.   □

The complete function constructing the representation of $\mathcal{D}(X)$ with the subtransition γ_X and the failure function f_X is given in Figure 5.3. Let us recall that the transition function δ of $\mathcal{D}(X)$ is assumed to be computed by function FAILUREFUNCTION-TRANSITION of Section 3.4. The next theorem states the correctness of the construction and its time and space complexities. We call this representation of $\mathcal{D}(X)$ the *linear dictionary-matching automaton* of X. The term "linear" (in $|X|$ is understood) is suitable if we work with a fixed alphabet, since degrees are upper-bounded by card(A).

Theorem 5.2. *The linear dictionary-matching automaton of any finite language X is built by function* LINEARDICTIONARYMATCHINGAUTOMATON. *The size of this representation of $\mathcal{D}(X)$ is $O(|X|)$. The construction is performed in time $O(|X| \times \log d)$ within $O(\text{card}(X))$ extra-space, d being the maximum degree of states of the trie of X.*

LinearDictionaryMatchingAutomaton(X)
    let γ be the transition function of (Q, i, T, E')
    let δ be the transition function of $(Q, i, T, (\gamma, F))$
1    $(Q, i, T, E') \leftarrow$ Trie(X)
2    $F[i] \leftarrow$ nil
3    $\vartheta \leftarrow$ EmptyQueue
4   Enqueue(ϑ, i)
5   **while** not QueueIsEmpty(ϑ)
6   **loop** $p \leftarrow$ Dequeue(ϑ)
7   **for** each letter a such that $\gamma(p, a) \neq$ nil
8   **loop** $q \leftarrow \gamma(p, a)$
9            $F[q] \leftarrow \delta(F[p], a)$
10   **if** $F[q]$ is in T
11   **then** $T \leftarrow T + \{q\}$
12   Enqueue(ϑ, q)
13 **return** $(Q, i, T, (\gamma, F))$

Fig. 5.3. Construction of the linear dictionary-matching automaton of a finite set of words X.

Proof. The correctness of the function and the order of the size of the representation is consecutive to Propositions 5.1, 5.2, and 5.3, and Lemmas 5.2, 5.3, and 5.4. The extra-space is linear in the size of the queue ϑ, which has always less than card(X) elements.

In order to prove the announced time complexity, we shall see that the last test of the loop of function FailureFunction-Transition (for computing $\delta(F[p], a)$; see Section 3.4) is executed less than $2|X|$ times. To avoid ambiguity, the state variable p of function FailureFunction-Transition is renamed r.

First. We remark that less tests are executed on the trie than if the words of X were considered separately.

Second. Considering separately each word x of X, and assimilating variables p and r with the prefixes of x they represent, the quantity $2|p| - |r|$ grows of at least one unity between two consecutive tests "$\gamma(r, a) =$ nil". When $|x| \leq 1$, no test is performed. But when $|x| \geq 2$, this quantity is equal to 2 before the execution of the first test ($|p| = 1$, $|r| = 0$), and is less than $2|x| - 2$ after the execution of the last test ($|p| = |x| - 1$, $|r| \geq 0$); which shows that less than $2|x| - 3$ tests are executed in this case.

This proves the expected result on the number of tests.

Now, since each of these tests is performed in time $O(\log d)$, the loop of lines 5–12 of function LinearDictionaryMatchingAutomaton is performed in time $O(|X| \times \log d)$. This is also the time complexity of the whole function, since line 1 is also performed in time $O(|X| \times \log d)$ according to Proposition 5.3. □

Figure 5.4 displays the linear dictionary-matching automaton of X when $X = \{\mathsf{ab}, \mathsf{babb}, \mathsf{bb}\}$. The failure function f_X is depicted with non-labeled discontinuous edges.

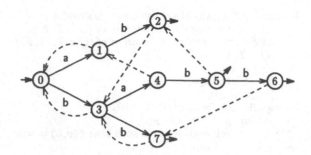

Fig. 5.4. The linear dictionary-matching automaton of $\{ab, babb, bb\}$.

To be complete, we add that f_U can be expressed independently of h_U for any language U.

Lemma 5.5. *Let $U \in A^*$. For each $(u, a) \in Pref(U) \times A$, we have:*

$$f_U(ua) = \begin{cases} f_U(u)a, & \text{if } u \neq \varepsilon \text{ and } ua \in Pref(U), \\ f_U(f_U(u)a), & \text{if } u \neq \varepsilon \text{ and } ua \notin Pref(U), \\ \varepsilon, & \text{if } u = \varepsilon. \end{cases}$$

Proof. This follows from Lemmas 5.2 and 5.3. □

However interesting this result is, it does not lead to another computation of linear dictionary-matching automata than the computation performed by the function of Figure 5.3.

5.4 Searching with linear dictionary-matching automata

We prove in this section that matching a finite set of words can be performed in linear time on fixed alphabets. This is stated in the following theorem.

Theorem 5.3. *Let X be a finite set of words and y be a word. Let ℓ be the maximum length of words of X and d be the maximum degree of states of the trie of X. Using the linear dictionary-matching automaton of X, searching for all occurrences of words of X as factors of y (search phase of algorithm* MATCHER*) is performed in time $O(|y| \times \log d)$, constant extra-space, within a delay of $O(\ell \times \log d)$.*

Proof. The proof is similar to the proof of Theorem 5.2.

Here, instead of the quantity $2|p| - |r|$, we consider the quantity $2|y'| - |p|$ where y' is the already read prefix of y. We obtain that less than $2|y| - 1$ tests "$\gamma(p, a) = $ NIL" are executed. This proves that the total time is $O(|y| \times \log d)$. For the delay, the test "$\gamma(p, a) = $ NIL" cannot be executed strictly more than ℓ times on each input letter a, which gives a time $O(\ell \times \log d)$. □

The search phase can be improved to prevent unnecessary calls to the failure function as far as it is possible.

Assume for instance that during the search state 5 of Figure 5.4 has been reached, and that the next letter of the input word, say c, is not b. The failure function has to be iterated at least twice since neither $\gamma_X(5, c)$ nor $\gamma_X(2, c)$ are defined. It is clear that the test on state 2 is useless, whatever c is. The next attempt is to compute $\gamma_X(3, c)$. Here, state 3 plays its role because c might be equal to a. But now, if $\gamma_X(3, c)$ is undefined, it is needless to iterate again the failure function on state 0, since c is then neither a nor b.

Following a similar reasoning for each states of the linear dictionary-matching automaton of X when $X = \{$ab, babb, bb$\}$ leads to consider the representation depicted in Figure 5.5.

More generally, given a finite language X and the failure function f_X, the representation of $\mathcal{D}(X)$ can be optimized by considering another failure function, denoted here by $\hat{f}_X$. Introducing the notation $Follow_U(u)$ to denote the set defined for each language U and for each word u in $Pref(U)$ by

$$Follow_U(u) = \{a \mid a \in A, ua \in Pref(U)\},$$

it is set that

$$\hat{f}_X(p) = \begin{cases} f_X(p), & \text{if } p \neq \varepsilon \text{ and } Follow_X(f_X(p)) \nsubseteq Follow_X(p), \\ \hat{f}_X(f_X(p)), & \text{if } p \neq \varepsilon \text{ and } Follow_X(f_X(p)) \subseteq Follow_X(p), \\ \text{undefined}, & \text{otherwise}, \end{cases}$$

for each $p \in Pref(X)$. The couple $(\gamma_X, \hat{f}_X)$ represents clearly the transition function of $\mathcal{D}(X)$. New failure states can be computed during a second breadth first search, and this can be done directly on array F_X.

However, substituting $\hat{f}_X$ to f_X does not affect the maximum delay of the searching algorithm that still remains $O(l \times \log d)$. To show this point, we give a worst case example. Let $\varphi(m)$ be the language defined for each $m \geq 1$ by:

$$\varphi(m) = \{a^{m-1}b\} \cup \{a^{2j-1}ba \mid 1 \leq j < \lceil m/2 \rceil\} \cup \{a^{2j}bb \mid 0 \leq j < \lfloor m/2 \rfloor\}.$$

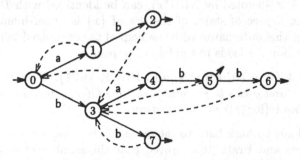

Fig. 5.5. The optimized representation of $\mathcal{D}(\{$ab, babb, bb$\})$.

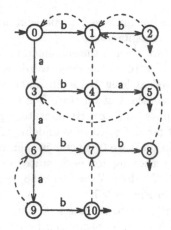

Fig. 5.6. The optimized representation of $\mathcal{D}(\varphi(4))$.

If $X = \varphi(m)$ for some $m \geq 1$, and if $\mathsf{a}^{m-1}\mathsf{bc}$ is the already read prefix of the input, m accesses to the failure function of the linear dictionary-matching automaton of X are made when reading letter c, whatever function f_X or $\hat{f}_X$ is chosen. (See the example given in Figure 5.6.)

6. Matching words

6.1 Outline

Problem 6.1. (String-matching problem.) Given a word x, preprocess it in order to locate all its occurrences in any given word y.

Let us first observe that this problem can be viewed as a particular case of the dictionary-matching problem (see Section 5). Here, the dictionary has only one element. Moreover, the dictionary-matching automaton $\mathcal{D}(\{x\})$, which recognizes the language A^*x, has the minimum number of states required to recognize A^*x, i.e. $|x|+1$ states. Therefore, the minimal automaton recognizing A^*x, denoted by $\mathcal{M}(A^*x)$, can be identified with $\mathcal{D}(\{x\})$. Since the maximum degree of states of the trie of $\{x\}$ is upper-bounded by one, implementing this automaton with the help of the optimized failure function described Section 5.4 leads to the following results.

Theorem 6.1 (Knuth, Morris, and Pratt, 1977). *The string-matching problem for x and y can be performed in time $O(|x| + |y|)$ and space $O(|x|)$, the delay being $\Theta(\log|x|)$ in the worst-case.*

We just have to hark back to the order of the delay for the algorithm of Knuth, Morris and Pratt. It is proved that the number of times the transition function of the trie of $\{x\}$ is performed on any input letter cannot exceed $\lfloor \log_{\Phi}(|x|+1) \rfloor$ where $\Phi = (1 + \sqrt{5})/2$ is the golden ratio. This upper bound is a consequence of a combinatorial property of words due to Fine and Wilf (known as the "periodicity lemma"). But it is closed to the worst-case bound, obtained when x is a prefix of the infinite Fibonacci word (see Chapter "Combinatorics of words").

However, as we shall see, implementing $\mathcal{M}(A^*x)$ with adjacency lists solves the string-matching problem with the additional feature of having a real-time search phase on fixed alphabets, *i.e.* with a delay bounded by a constant.

Main Theorem 6.2. *The string-matching problem for x and y can be achieved in the following terms:*

- *a preprocessing phase on x building an implementation of $\mathcal{M}(A^*x)$ of size $O(|x|)$, performed in time $O(|x|)$ and constant extra-space;*
- *a search phase executing the automaton on y performed in time $O(|y|)$ and constant extra-space, the delay being $O(\log \min\{1 + \lfloor \log_2 |x| \rfloor, \text{card}(A)\})$.*

Underlying the above result are indeed optimal bounds on the complexity of string-matching algorithms for which the search phase is on-line with a one-letter buffer. Relaxing the on-line condition leads to another theorem stated below. But its proof is based on combinatorial properties of words unrelated to automata and not considered in this chapter.

Theorem 6.3 (Galil and Seiferas, 1983). *The string-matching problem for x and y previously stored in memory can be performed in time $O(|x|+|y|)$ and constant extra-space.*

In Section 6.2, we describe an on-line construction of $\mathcal{M}(A^*x)$. The linear implementation via adjacency lists is discussed in Section 6.3. We establish in Section 6.4 properties of $\mathcal{M}(A^*x)$ that are used in Section 6.5 to prove the asymptotic bounds of the search phase claimed in Theorem 6.2.

6.2 String-matching automata

We give a method to build the automaton $\mathcal{M}(A^*x)$. The feature of this method is that it is based on an on-line construction and that it does not use the usual procedures of determinization and minimization of automata.

In the remainder of Section 6 we identify $\mathcal{M}(A^*x)$ with $\mathcal{D}(\{x\})$, which is the automaton

$$\Big(Pref(x), \ \varepsilon, \ \{x\}, \ \{(p, a, h_x(pa)) \mid p \in Pref(x), a \in A\} \Big),$$

$h_x(v)$ being the longest suffix of v which is a prefix of x, for each $v \in A^*$. We call this automaton the *string-matching automaton* of x.

An example of string-matching automaton is given in Figure 2.1: the depicted automaton is $\mathcal{M}(A^*\text{abaaab})$ assuming that $A = \{a, b\}$.

We introduce the notions of "border" as follows. A word v is said to be a *border* of a word u if v is both a prefix and a suffix of u. The longest proper border of a nonempty word u is said to be *the border* of u and is denoted by $Bord(u)$. As a consequence of definitions, we have:

$$h_x(pa) = \begin{cases} pa, & \text{if } pa \text{ is a prefix of } x, \\ Bord(pa), & \text{otherwise,} \end{cases}$$

for each $(p, a) \in Pref(x) \times A$.

In order to build $\mathcal{M}(A^*x)$, the construction of the set of edges of the string-matching automaton of x is to be settled. The construction is on-line, as suggested by the following lemma.

Lemma 6.1. *Let us denote by E_u the set of edges of $\mathcal{M}(A^*u)$ for any $u \in A^*$. We have:*

$$E_\varepsilon = \{(\varepsilon, b, \varepsilon) \mid b \in A\}.$$

Furthermore, for each $(u, a) \in A^* \times A$ we have:

$$E_{ua} = E'_{ua} \cup E''_{ua}$$

with

$$E'_{ua} = \left(E_u \backslash \{(u, a, h_u(ua))\}\right) \cup \{(u, a, ua)\}$$

and

$$E''_{ua} = \{(ua, b, w) \mid (h_u(ua), b, w) \in E'_{ua}\}.$$

Proof. The property for E_ε clearly holds.

Now, let $u \in A^*$ and $a \in A$, let E'_{ua} and E''_{ua} be as in the lemma, and set $v = h_u(ua)$.

Each edge in E_{ua} outgoing a state no longer than $|u|$ belongs to E'_{ua}. The converse is also true.

It remains to prove that each edge in E_{ua} outgoing state ua belongs to E''_{ua}, and that the converse holds. This is to prove that for each $b \in A$, the targets w and w' of the edges (v, b, w) and (ua, b, w'), both in E_{ua}, are identical.

Since v is a border of ua, w is both a suffix of uab and a prefix of ua. Which implies that w is shorter than w'.

Conversely. We have that $|w'| \leq |vb|$. (Assuming the contrary leads to consider that $w'b^{-1}$ is a border of ua contradicting the maximality of v.) Since w' and vb are both suffixes of uab, w' is a suffix of vb. Now w' is also a prefix of ua. This shows that w' is shorter than w, and ends the proof. $\square$

The construction of $\mathcal{M}(A^*ua)$ from $\mathcal{M}(A^*u)$ can be interpreted in a visual point of view as the "unfolding" of the edge $(u, a, h_u(ua))$ of the automaton $\mathcal{M}(A^*u)$. An example is given in Figure 6.1 that depicts four steps related to the construction of $\mathcal{M}(A^*\text{abaaab})$.

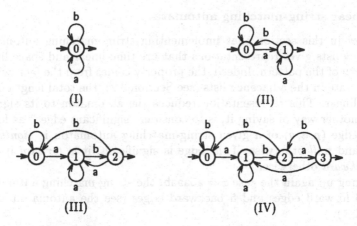

Fig. 6.1. During the construction of the string-matching automaton of abaaab, unfolding of the edge $(\varepsilon, a, \varepsilon)$ from step "ε" (I) to step "a" (II), of the edge (a, b, ε) from step "a" to step "ab" (III), and of the edge (ab, a, a) from step "ab" to step "aba" (IV). It is assumed that $A = \{a, b\}$.

A function that builds the string-matching automaton of x following the method suggested by Lemma 6.1 is given in Figure 6.2. This can be used straightforwardly to implement the automaton via its transition matrix. Following the same scheme, we describe in the next section an implementation of the string-matching automaton of x which size is both linear in $|x|$ and independent of the alphabet.

STRINGMATCHINGAUTOMATON(x)
  let δ be the transition function of $(Q, i, \emptyset, E)$
1 $(Q, E) \leftarrow (\emptyset, \emptyset)$
2 $i \leftarrow$ STATE-CREATION
3 for each letter b in A
4   loop $E \leftarrow E + \{(i, b, i)\}$
5 $t \leftarrow i$
6 for letter a from first to last letter of x
7   loop $r \leftarrow \delta(t, a)$
8    $q \leftarrow$ STATE-CREATION
9    $E \leftarrow E - \{(t, a, r)\} + \{(t, a, q)\}$
10   for each letter b in A
11   loop $E \leftarrow E + \{(q, b, \delta(r, b))\}$
12   $t \leftarrow q$
13 return $(Q, i, \{t\}, E)$

Fig. 6.2. Construction of the string-matching automaton of a word x.

6.3 Linear string-matching automata

We show in this section that implementing string-matching automata via adjacency lists gives representations that are time-linear and space-linear in the length of the pattern. Indeed, the property comes from the fact: with ε as default state in the adjacency lists (see Section 3.2), the total length of these lists is linear. This representation reduces the automaton to its significant part. Another way of saying it, is to consider "significant edges" as follows.

An edge (p, a, q) of a given string-matching automaton is *significant* if $q \neq \varepsilon$, and *null* otherwise; if the edge is significant, it is *forward* if $q = pa$ and *backward* otherwise.

Picking up again the case $x = \mathsf{abaaab}$, the string-matching automaton of x has 6 forward edges and 5 backward edges (see the automaton given in Figure 2.1).

Proposition 6.1. *The number of significant edges of the string-matching automaton of any word x is upper-bounded by $2|x|$; more precisely, its number of forward edges is exactly $|x|$, and its number of backward edges is upper-bounded by $|x|$. The bounds are reached for instance when the first letter of x occurs only at the first position in x.*

In order to prove Proposition 6.1, we shall establish the following result.

Lemma 6.2. *Let (p, a, q) and (p', a', q') be two distinct backward edges of the string-matching automaton of some word u. Then $|p| - |q| \neq |p'| - |q'|$.*

Proof. Suppose for a contradiction the existence of two distinct backward edges (p, a, q) and (p', a', q') of $\mathcal{M}(A^*u)$ satisfying $|p| - |q| = |p'| - |q'|$.

In case $p = p'$, we have that $q = q'$. Since the two edges are significant, this implies that $a = a'$. Which is impossible.

Thus, we can assume without loss of generality that $|p| > |p'|$, thus, $|q| > |q'|$. Since qa^{-1} is a border of p ($|p| - |qa^{-1}|$ is a period of p) and since q' is a proper prefix of q, we have

$$a' = p_{|q'|} = p_{|q'|+|p|-|qa^{-1}|} = p_{|q'|+|p'|-|q'|+1} = p_{|p'|+1}.$$

Which contradicts the fact that (p', a', q') is a backward edge.   □

Proof of Proposition 6.1. The number of forward edges of the automaton is obviously $|x|$.

Let us prove the upper bound on the number of backward edges. Since the number $|p| - |q|$ associated to the backward edge (p, a, q) ranges from 0 to $|x| - 1$, Lemma 6.2 implies that the total number of backward edges is bounded by $|x|$.

We show that the upper bound on the number of backward edges is optimal. Consider that the first letter of x occurs only at the first position in x. The edge (p, x_1, x_1) is an outgoing edge for each state p of non-zero length of the automaton, and this edge is a backward edge. So, the total number of backward edges is $|x|$ in this case.   □

Figure 6.3 displays a string-matching automaton which number of significant edges is maximum for a word of length 7.

From the previous proposition, an implementation of $\mathcal{M}(A^*x)$ via adjacency lists with the initial state ε as uniform default state has a size linear in $|x|$, since the edges represented in the adjacency lists are the significant edges of the automaton. We call this representation of the string-matching automaton of x the *linear string-matching automaton* of x. It is constructed by the function given in Figure 6.4. This function is a mere adaptation of the general function given in Figure 6.2. Recall that the transition function of the linear string-matching automaton of x is assumed to be computed by function ADJACENCYLISTS-TRANSITION of Section 3.2.

Theorem 6.4. *Function* LINEARSTRINGMATCHINGAUTOMATON *builds the linear string-matching automaton of any given word x. The size of this representation of $\mathcal{M}(A^*x)$ is $O(|x|)$. The construction is performed in time $O(|x|)$ and constant extra-space.*

Proof. The correctness of the function is consecutive to Lemma 6.1. The order of the size of the representation follows from Proposition 6.1.

The time required to build the set of all significant edges outgoing a given state is linear in their number (the operations executed on the adjacency list associated to a given state $p \neq x$ are the operations occurring in Figure 6.4 at line 3 if $p = i$ and at line 11 otherwise, then at line 9 if necessary, then finally at line 10; the corresponding operations for state x are at line 3 if $x = \varepsilon$ and at line 11 otherwise). Hence, the total time is $O(|x|)$ from Proposition 6.1. □

We show in Section 6.5 that the linear representation of the string-matching automaton of x described above yields a search for occurrences of x in y that runs in time linear in $|y|$. Before that, we establish combinatorial properties of string-matching automata in the next section.

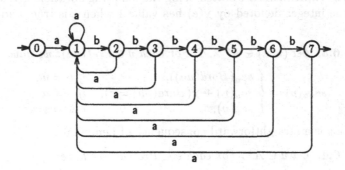

Fig. 6.3. A string-matching automaton with the maximum number of significant edges. The significant edges are the only depicted edges; the target of other edges is 0.

LINEARSTRINGMATCHINGAUTOMATON(x)
    let δ be the transition function of $(Q, i, \emptyset, G)$
1   $Q \leftarrow \emptyset$
2   $i \leftarrow$ STATE-CREATION
3   $G[i] \leftarrow \emptyset$
4   $t \leftarrow i$
5   **for** letter a from first to last letter of x
6   **loop** $r \leftarrow \delta(t, a)$
7           $q \leftarrow$ STATE-CREATION
8   **if** $r \neq i$
9   **then** $G[t] \leftarrow G[t] - \{(a, r)\}$
10          $G[t] \leftarrow G[t] + \{(a, q)\}$
11          $G[q] \leftarrow G[r]$
12          $t \leftarrow q$
13 **return** $(Q, i, \{t\}, G)$

Fig. 6.4. Construction of the linear string-matching automaton of a word x.

6.4 Properties of string-matching automata

We establish in this section some upper bounds for the number of significant edges of string-matching automata. These bounds complete the global bound given in Proposition 6.1, by focusing on the number of outgoing significant edges. The two main results, namely Propositions 6.2 and 6.3, are intensively used in Section 6.5.

Given a word u, we denote by $se_u(p)$ the number of significant edges outgoing the state p of the string-matching automaton of u; if p is a prefix of u and q a prefix of p, the notation $se_u(p, q)$ stands for the number of significant edges which sources range in the set of prefixes of u from q to p, i.e. the number

$$se_u(q) + se_u(q \cdot p_{|q|+1}) + \cdots + se_u(q \cdot p_{|q|+1} \cdots p_{|p|-1}) + se_u(p).$$

The next two lemmas provide recurrence relations satisfied by the numbers $se_u(p)$. The expressions are stated using the following notation: given a predicate e, the integer denoted by $\chi(e)$ has value 1 when e is true, and value 0 otherwise.

Lemma 6.3. *Let* $(u, a) \in A^* \times A$. *For each* $v \in \mathit{Pref}(ua)$, *we have:*

$$se_{ua}(v) = \begin{cases} se_u(Bord(ua)), & \text{if } v = ua, \\ se_u(u) + \chi(Bord(ua) = \varepsilon), & \text{if } v = u, \\ se_u(v), & \text{otherwise.} \end{cases}$$

Proof. This is a straightforward consequence of Lemma 6.1.   □

Lemma 6.4. *Let* $u \in A^+$. *For each* $v \in \mathit{Pref}(u)$, *we have:*

$$se_u(v) = \begin{cases} se_u(Bord(u)), & \text{if } v = u, \\ se_u(Bord(v)) + \chi(Bord(va) = \varepsilon), & \text{if } va \in \mathit{Pref}(u) \\ & \qquad \text{for some } a \in A, \\ 1, & \text{if } v = \varepsilon. \end{cases}$$

Proof. Follows from Lemma 6.3. □

The next lemma is the "cornerstone" of the proof of the logarithmic bound given in Proposition 6.2 stated afterwards.

Lemma 6.5. *Let $u \in A^+$. For each $v \in Pref(u)\backslash\{\varepsilon\}$, we have:*

$$2|Bord(v)| \geq |v| \implies se_u(Bord(v)) = se_u(Bord^2(v)).$$

Proof. Set $k = 2|Bord(v)| - |v|$, $w = v_1 v_2 \cdots v_k$, and $a = v_{k+1}$. Since wa is a proper border of $Bord(v)a$, the border of $Bord(v)a$ is nonempty. Then we apply Lemma 6.4 to the proper prefix $Bord(v)$ of u. □

Proposition 6.2. *Let $u \in A^*$. For each state p of $\mathcal{M}(A^*u)$, we have:*

$$se_u(p) \leq 1 + \lfloor \log_2(|p| + 1) \rfloor.$$

Proof. We prove the result by induction on $|p|$. From Lemma 6.4, this is true if $|p| = 0$. Next, suppose $|p| \geq 1$.

Let j be the integer such that

$$2^j \leq |p| + 1 < 2^{j+1},$$

then let k be the integer such that

$$|Bord^{k+1}(p)| + 1 < 2^j \leq |Bord^k(p)| + 1.$$

Let $\ell \in \{0, \dots, k-1\}$; we have $2|Bord^{\ell+1}(p)| \geq 2^{j+1} - 2 \geq |p| \geq |Bord^\ell(p)|$; which implies $se_u(Bord^{\ell+1}(p)) = se_u(Bord^{\ell+2}(p))$ from Lemma 6.5. Hence we get the equality

$$se_u(Bord(p)) = se_u(Bord^{k+1}(p)).$$

From the induction hypothesis applied to the state $Bord^{k+1}(p)$, we get

$$se_u(Bord^{k+1}(p)) \leq 1 + \lfloor \log_2(|Bord^{k+1}(p)| + 1) \rfloor.$$

Now Lemma 6.4 implies

$$se_u(p) \leq se_u(Bord(p)) + 1.$$

This shows that

$$se_u(p) \leq j + 1 = 1 + \lfloor \log_2(|p| + 1) \rfloor,$$

and ends the proof. □

By way of illustration, we consider the case where $x = \mathsf{abacabad}$. Given a state p of $\mathcal{M}(A^*x)$ (see Figure 6.5), the $1 + \lfloor \log_2(|p| + 1) \rfloor$ bound for the number of significative edges outgoing p is reached when $|p| = 0, 1, 3$, or 7.

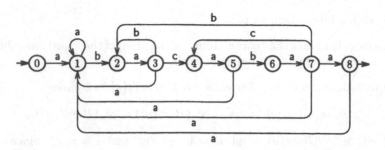

Fig. 6.5. The string-matching automaton of abacabad without its null edges.

Proposition 6.3. *Let $u \in A^*$. For each backward or null edge (p, a, q) of $\mathcal{M}(A^*u)$, we have:*

$$se_u(p, q) \leq 2|p| - 2|q| + 2 - \chi(p = u) - \chi(q = \varepsilon).$$

Proof. The property clearly holds when u is the power of some letter. The remainder of the proof is by induction.

So, let $u \in A^*$, $b \in A$, and let (p, a, q) be a backward or null edge of $\mathcal{M}(A^*ub)$.

If $|p| \leq |u|$, (p, a, q) is also an edge of $\mathcal{M}(A^*u)$. By application of Lemma 6.3, we obtain that

$$se_{ub}(p, q) \leq se_u(p, q) + \chi(p = u).$$

Otherwise $p = ub$. Let r be the border of ub. We only have to examine the case where r is a proper prefix of u (if $r = u$, then $u \in b^*$). Thus (r, a, q) is an edge of $\mathcal{M}(A^*u)$ and of $\mathcal{M}(A^*ub)$. If it is a forward edge, *i.e.* if $q = ra$, we obtain from Lemma 6.3 that

$$se_{ub}(p, q) = se_u(u, r) + \chi(r = \varepsilon),$$

and if it is a backward edge we obtain that

$$se_{ub}(p, q) = se_u(r, q) + se_u(u, r) + \chi(r = \varepsilon).$$

The result now follows by applying of the induction hypothesis to u. $\square$

The previous result is illustrated by the example given in Figure 6.3, *i.e.* when $x = $ abbbbbb. In this case, the $2|p| - 2|q| + 2 - \chi(p = x) - \chi(q = \varepsilon)$ bound is reached for any backward or null edge of $\mathcal{M}(A^*x)$.

Observe that Proposition 6.3 provides another proof of the $2|x|$ bound given in Proposition 6.1 as follows. We consider a null edge outgoing state x (possibly extending the alphabet by one letter). For this edge, with the notation of Proposition 6.3, we have $p = u = x$ and $q = \varepsilon$. Thus, $se_x(x, \varepsilon)$, which is the total number of significant edges of $\mathcal{M}(A^*x)$, is not greater than $2|x| - 2|\varepsilon| + 2 - 1 - 1 = 2|x|$.

6.5 Searching with linear string-matching automata

Our proof of Theorem 6.2 consists in considering linear string-matching automata for matching words. We then consider the model of computation where none ordering on the alphabet is assumed, and give some optimal bounds for string-matching algorithms for which the search phase is on-line with a one-letter buffer.

Consider the search phase of algorithm MATCHER using the linear string-matching automaton of a given word x. For each backward or null edge (p, a, q) of the string-matching automaton of x, let us denote by $c_x(p, q)$ the maximum time for executing the series of transitions from q to q via p, $i.e.$ for reading the word $x_{|q|+1}x_{|q|+2}\cdots x_{|p|}a$ starting in state q. Let us also denote by $C_x(y)$ the time for executing the search phase when y is the searched word, $i.e.$ the time for executing the automaton on y.

Lemma 6.6. Let $x, y \in A^*$. There exists a finite sequence of backward or null edges of $\mathcal{M}(A^*x)$, say $((p_j, a_j, q_j))_{1 \leq j \leq k}$, satisfying the three following conditions:

(i) $q_k = \varepsilon$;

(ii) $\sum_{j=1}^{k}(|p_j| - |q_j| + 1) = |y|$;

(iii) $C_x(y) \leq \sum_{j=1}^{k} c_x(p_j, q_j)$.

Proof. The proof is by induction on $|y|$. Since the property trivially holds when $|y| = 0$, we assume that $|y| \geq 1$.

Observe first that since an upper bound is expected for $C_x(y)$, we can assume, even if the alphabet has to be extended by one letter, that the last letter of y is not a letter occurring actually in x. Hence, we can assume that the lastly performed transition corresponds to a null edge.

Now, let $(p_\ell)_{0 \leq \ell \leq |y|}$ be the sequence of successive values of the current state p of algorithm MATCHER (in other words $p_\ell = h_x(y_1y_2\cdots y_\ell)$). Let m, $0 \leq m \leq |y| - 1$, be the minimal integer satisfying $p_{m+1} = p_{m'}$ for some $m' \leq m$, then let m' be the integer in $\{0, \ldots, m\}$ such that $p_{m'} = p_{m+1}$. Thus, the triple $(p_m, y_{m+1}, p_{m'})$ is a backward or null edge of $\mathcal{M}(A^*x)$, and the $m - m' + 1$ successive letters $y_{m'+1}, y_{m'+2}, \ldots, y_{m+1}$ of y have been read during the computation of the transitions from $p_{m'}$ to p_{m+1} via p_m. Consider the word y' defined by $y' = y_1y_2\cdots y_{m'} \cdot y_{m+2}y_{m+3}\cdots y_{|y|}$. Following the definition of m and m' we have that $C_x(y) \leq c_x(p_m, p_{m'}) + C_x(y')$. Applying the induction hypothesis to y' gives the existence of a finite sequence e' as depicted in the statement. The expected sequence related to y can then be obtained by adding the edge $((p_m, y_{m+1}, p_{m'}))$ in front of the sequence e'. It clearly satisfies conditions (i) to (iii). This ends the proof.   □

Theorem 6.5. Let x and y be words. Using the linear string-matching automaton of x, searching for all occurrences of x as factors of y (search phase of algorithm MATCHER) is performed in time $O(|y|)$, constant extra-space, within a delay $O(\log \min\{1 + \lfloor \log_2 |x| \rfloor, \text{card}(A)\})$.

Proof. Whatever efficient is the implementation of adjacency lists, we may assume that the time for executing the transition from the current state by the current input letter is asymptotically linear in the number of significant edges outgoing the involved state. For each backward or null edge (p, a, q) of $\mathcal{M}(A^*x)$, this assumption implies that

$$c_x(p, q) = O(se_x(p, q));$$

which leads to

$$c_x(p, q) = O(|p| - |q| + 1),$$

by application of Proposition 6.3. We finally apply Lemma 6.6, and get the $O(|y|)$ bound.

We now turn to the proof of the delay. The cardinality of each adjacency list is both upper-bounded by card(A), and, from Lemma 6.3 and Proposition 6.2, by $1 + \lfloor \log_2 |x| \rfloor$. Now, observe that each adjacency list can be arranged in a balanced tree when computing it, without loosing the linear-time complexity of the construction. This provides a logarithmic time for computing a transition. Which proves the asymptotic bound of the delay. $\square$

In the remainder of the section, no ordering on the alphabet is assumed, contrary to what is assumed for the previous statement. The model of computation is the *comparison model* in which algorithms have access to the input words by comparing pairs of letters to test whether they are equal or not. Within this model, given a word x, we denote by $\mathcal{S}(x)$ the family of the string-matching algorithms for which the search phase is on-line with a one-letter buffer.

String-matching algorithms based on the linear string-matching automaton of x can be classified according to the way the adjacency lists are ordered or scanned. For example, the adjacency lists can be ordered by decreasing length of target, or by the frequency of labels as letters of the prefix already read; each adjacency list can also be scanned according to a random processing. (Let us observe that any of these variations preserves the linear time of the search). We denote by $\mathcal{L}(x)$ the subfamily of algorithms in $\mathcal{S}(x)$ which use the linear string-matching automaton of x to search a given word for occurrences of x.

Theorem 6.6. *Given $x \in A^+$, consider an algorithm λ in $\mathcal{L}(x)$, and an input of non-zero length n. In the comparison model, λ performs no more than $2n - 1$ letter comparisons, and compares each of the n letter of the input less than $\min\{1 + \lfloor \log_2 |x| \rfloor, \text{card}(A)\}$ times.*

Proof. This is similar to the proof of Theorem 6.5. The term "-1" of the $2n - 1$ bound results from the fact that we can assume that at least one transition by a null edge of $\mathcal{M}(A^*x)$ is simulated (the edge (q_{k-1}, a_k, q_k) of Lemma 6.6). $\square$

The $2n - 1$ bound of Theorem 6.6 is also the bound reached by the algorithm of Section 5 when ab is a prefix of the only word of the dictionary and the input is in a^*. However, this worst-case bound can be lowered in $\mathcal{L}(x)$, using the special strategy described in the following statement for computing transitions.

Theorem 6.7. *Given $x \in A^+$, consider an algorithm λ in $\mathcal{L}(x)$ that applies the following strategy: to compute a transition from any state p, scan the edges outgoing p in such a way that the forward edge (if any) is scanned last. Then, in the comparison model, λ executes no more than $\lfloor (2 - 1/|x|) \times n \rfloor$ letter comparisons on any input of length n.*

Proof. Let (p, a, q) be a backward or null edge of $\mathcal{M}(A^*x)$. The number of comparisons while executing the series of transitions from q to q via p is bounded by $se_x(p, q) - \chi(p \neq x)$. By application of Proposition 6.3, we obtain that at most $(2 - 1/|x|) \times (|p| - |q| + 1)$ comparisons are performed during this series of transitions. Then we apply Lemma 6.6 and obtain the expected bound.   □

For a given length m of patterns, the delay $\min\{1 + \lfloor \log_2 m \rfloor, \mathrm{card}(A)\}$ (Theorem 6.6) and the coefficient $2 - 1/m$ (Theorem 6.7) are optimal quantities. This is proved by the next two propositions.

Proposition 6.4. *Consider the comparison model. Then, for each $m \geq 1$, for each $n \geq m$, there exist $x \in A^m$ and $y \in A^n$ such that any algorithm in $\mathcal{S}(x)$ performs at least $\min\{1 + \lfloor \log_2 m \rfloor, \mathrm{card}(A)\}$ letter comparisons in the worst-case on some letter of the searched word y.*

Proof. Define recursively the mapping $\xi \colon A^* \to A^*$ by $\xi(ua) = \xi(u) \cdot a \cdot \xi(u)$ for each $(u, a) \in A^* \times A$ and $\xi(\varepsilon) = \varepsilon$. (For instance, we have $\xi(\mathsf{abcd}) = \mathsf{abacabadabacaba}$.)

Set $k = \min\{1 + \lfloor \log_2 m \rfloor, \mathrm{card}(A)\}$, choose k pairwise distinct letters, say $a_1, a_2, \ldots, a_k$, and assume that $\xi(a_1 a_2 \cdots a_{k-1})a_k$ is a prefix of x. If $\xi(a_1 a_2 \cdots a_{k-1})$ is the already read prefix of y, the algorithm can suppose that an occurrence of x starts at one of the positions in the form $2^{k-\ell}$, $1 \leq \ell \leq k$. Hence, the algorithm performs never less than k letter comparisons at position 2^{k-1} in the worst-case.   □

Proposition 6.5. *Consider the comparison model. Then, for each $m \geq 1$, for each $n \geq 0$, there exist $x \in A^m$ and $y \in A^n$ such that any algorithm in $\mathcal{S}(x)$ performs at least $\lfloor (2 - 1/m) \times n \rfloor$ letter comparisons in the worst-case when searching y.*

Proof. Assume that $x = \mathsf{ab}^{m-1}$ and $y \in Pref((\mathsf{a\{a, b\}}^{m-1})^*)$. Then let j, $1 \leq j \leq n$, be the current position on y, and let v be the longest suffix of $y_1 y_2 \cdots y_{j-1}$ that is also a proper prefix of x.

If $v \neq \varepsilon$, the algorithm has to query both if $y_j = \mathsf{a}$ and if $y_j = \mathsf{b}$ in the worst-case, in order to be able to report later an occurrence of x at position either j or $j-|v|$. Otherwise $v = \varepsilon$, and the algorithm can just query if $y_j = $ a. But, according to the definition of y, the second case, namely $v = \varepsilon$, may occur only when $j \equiv 1 \pmod{m}$. Therefore, the number of letter comparisons performed on y is then never less than $2n - \lceil n/m \rceil = \lfloor (2 - 1/m) \times n \rfloor$ in the worst-case.   □

7. Suffix automata

7.1 Outline

The *suffix automaton* of a word x is defined as the minimal deterministic (non necessarily complete) automaton that recognizes the (finite) set of suffixes of x. It is denoted by $\mathcal{M}(\mathit{Suff}(x))$ according to notations of Section 2.

An example of suffix automaton is displayed in Figure 7.1.

The automaton $\mathcal{M}(\mathit{Suff}(x))$ can be used as an index on x to solve the following problem.

Problem 7.1. (Index problem.) Given a word x, preprocess it in order to locate all occurrences of any word y in x.

An alternative solution to the problem is to implement data structures techniques based on a representation of the set of suffixes of x by compact tries. This structure is known as the suffix tree of x.

The suffix automaton provides another solution to the string-matching problem (see Section 6) since it can also be used to search a word y for factors of x. This yields a space efficient solution to the search for rotations (Section 7.5) of a given word.

The surprising property of suffix automata is that their size is linear although the number of factors of a word can be quadratic in the length of the word. The construction of suffix automata is also linear on fixed alphabets.

Main Theorem 7.1. *The size of the suffix automaton of a word x is $O(|x|)$. The automaton can be implemented in time $O(|x| \times \log \operatorname{card}(A))$ and $O(|x|)$ extra-space.*

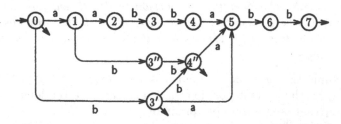

Fig. 7.1. The minimal deterministic automaton recognizing the suffixes of **aabbabb**.

The implementation which is referred to in the theorem is based on adjacency lists. As in Section 5, if we allow more extra-space, the time complexity reduces to $O(|x|)$. This is valid also for Theorems 7.5 and 7.6, Propositions 7.2, 7.3, 7.4, 7.5, and 7.8.

We first review in Section 7.2 properties of suffix automata that are useful to design a construction method. At the same time, we provide exact bounds on the size of the automata. These results have consequences on the running time of the method. Section 7.3 is devoted to the construction of suffix automata itself. The same approach is presented in Section 7.6 for factor automata. Sections 7.4 and 7.5 show how these automata can be used either as indexes or as string-matching automata.

7.2 Sizes and properties

7.2.1 End-positions

Right contexts according to $Suff(x)$ satisfy a few properties stated in the next lemmas and used later in the chapter. The first remark concerns the context of a suffix of a word.

Lemma 7.1. *Let* $u, v \in A^*$. *If* $u \in Suff(v)$, *then* $v^{-1} Suff(x) \subseteq u^{-1} Suff(x)$.

Proof. If $v^{-1} Suff(x) = \emptyset$ the inclusion trivially holds. Otherwise, let $z \in v^{-1} Suff(x)$. Then, $vz \in Suff(x)$ and, since $u \in Suff(v)$, $uz \in Suff(x)$. So, $z \in u^{-1} Suff(x)$. $\square$

Right contexts satisfy a kind of converse statement. To formalize it, we introduce the function $endpos_x : Fact(x) \to \mathbb{N}$ defined for each word u by

$$endpos_x(u) = \min\{|w| \mid w \text{ is a prefix of } x \text{ and } u \text{ is a suffix of } w\}.$$

The value $endpos_x(u)$ marks the ending position of the first (or leftmost) occurrence of u in x.

Lemma 7.2. *Let* $u, v \in Fact(x)$. *If* $u \equiv_{Suff(x)} v$, *we have the equality* $endpos_x(u) = endpos_x(v)$, *which is equivalent to say that one of the words* u *and* v *is a suffix of the other.*

Proof. Let $y, z \in A^*$ be such that $x = yz$ and $u \in Suff(y)$. We assume in addition that $|y| = endpos_x(u)$. Then z is the longest word of $u^{-1} Suff(x)$. The hypothesis implies that z is also the longest word of $v^{-1} Suff(x)$, which shows that $|y| = endpos_x(v)$. In this situation, u and v are both suffixes of y, which proves that one of them is a suffix of the other. $\square$

Another often used property of the syntactic congruence associated with $Suff(x)$ is that it partitions the suffixes of factors into intervals (with respect to the lengths of suffixes).

Lemma 7.3. *Let $u, v, w \in Fact(x)$. Then, if $u \in Suff(v)$, $v \in Suff(w)$, and $u \equiv_{Suff(x)} w$, we have $u \equiv_{Suff(x)} v \equiv_{Suff(x)} w$.*

Proof. By Lemma 7.1, we have the inclusions $w^{-1}Suff(x) \subseteq v^{-1}Suff(x) \subseteq u^{-1}Suff(x)$. But then, the equality $u^{-1}Suff(x) = w^{-1}Suff(x)$ implies the conclusion of the statement.   □

A consequence of the next property is that the direct inclusion of right contexts relative to $Suff(x)$ induces a tree structure on them. In the tree, the parent link corresponds to the proper direct inclusion. This link is discussed in Section 7.2.2 where it is called the "suffix function".

Corollary 7.1. *Let $u, v \in A^*$. Then, one of the three following conditions holds:*

(i) $u^{-1}Suff(x) \subseteq v^{-1}Suff(x)$;
(ii) $v^{-1}Suff(x) \subseteq u^{-1}Suff(x)$;
(iii) $u^{-1}Suff(x) \cap v^{-1}Suff(x) = \emptyset$.

Proof. We just have to show that if $u^{-1}Suff(x) \cap v^{-1}Suff(x) \neq \emptyset$, then we have the inclusion $u^{-1}Suff(x) \subseteq v^{-1}Suff(x)$ or the inclusion $v^{-1}Suff(x) \subseteq u^{-1}Suff(x)$. Let $z \in u^{-1}Suff(x) \cap v^{-1}Suff(x)$. Then, uz and vz are suffixes of x. So, u and v are suffixes of xz^{-1}, which implies that one of the words u and v is a suffix of the other. Therefore, the conclusion follows by Lemma 7.1.   □

7.2.2 Suffix function
We consider the function $s_x \colon Fact(x) \to Fact(x)$ defined for each nonempty word v in $Fact(x)$ by

$$s_x(v) = \text{the longest } u \in Suff(v) \text{ such that } u \not\equiv_{Suff(x)} v.$$

Regarding Lemma 7.1, this is equivalent to

$$s_x(v) = \text{the longest } u \in Suff(v) \text{ such that } v^{-1}Suff(x) \subset u^{-1}Suff(x).$$

The function s_x is called the *suffix function* relative to x. An obvious consequence of the definition is that $s_x(v)$ is a proper suffix of v. The next lemma shows that the suffix function s_x induces what we call a "suffix link" on states of $\mathcal{M}(Suff(x))$.

Lemma 7.4. *Assuming $x \neq \varepsilon$, let $u, v \in Fact(x) \backslash \{\varepsilon\}$. If $u \equiv_{Suff(x)} v$, then $s_x(u) = s_x(v)$.*

Proof. From Lemma 7.2 we can assume without loss of generality that $u \in Suff(v)$. The word u cannot be a suffix of $s_x(v)$, because Lemma 7.3 would then imply $s_x(v)^{-1}Suff(x) = v^{-1}Suff(x)$, which contradicts the definition of $s_x(v)$. Therefore, $s_x(v)$ is a suffix of u. Since, by definition, it is the longest suffix of v non equivalent to it, it is equal to $s_x(u)$.   □

Lemma 7.5. *If $x \neq \varepsilon$, $s_x(x)$ is the longest suffix of x that occurs at least twice in x.*

Proof. The set $x^{-1}Suff(x)$ is equal to $\{\varepsilon\}$. Since x and $s_x(x)$ are not equivalent, the set $s_x(x)^{-1}Suff(x)$ contains some nonempty word z. Therefore, $s_x(x)z$ and $s_x(x)$ are suffixes of x, which proves that $s_x(x)$ occurs at least twice in x. Any suffix w of x, longer than $s_x(x)$, satisfies $w^{-1}Suff(x) = x^{-1}Suff(x) = \{\varepsilon\}$ by definition of $s_x(x)$. Thus, w occurs only as a suffix of x, which ends the proof. □

The next lemma shows that the image of a factor of x by the suffix function is a word of maximum length in its own congruence class. This fact is needed in Section 7.5 where the suffix automaton is used as a matching automaton.

Lemma 7.6. *Assuming $x \neq \varepsilon$, let $u \in Fact(x) \backslash \{\varepsilon\}$. Then, any word equivalent to $s_x(u)$ is a suffix of $s_x(u)$.*

Proof. Let $w = s_x(u)$ and $v \equiv_{Suff(x)} s_x(u)$. The word w is a proper suffix of u. If the conclusion of the statement is false, Lemma 7.2 insures that w is a proper suffix of v. Let $z \in u^{-1}Suff(x)$. Since w is a suffix of u and is equivalent to v, we have $z \in w^{-1}Suff(x) = v^{-1}Suff(x)$. Therefore, u and v are both suffixes of xz^{-1}, which implies that one of them is a suffix of the other. But this contradicts either the definition of w, or the conclusion of Lemma 7.3. This proves that v is necessarily a suffix of w. □

7.2.3 State splitting

In this section we present the properties that yield to the on-line construction of suffix automata described in Section 7.3. This is achieved by deriving relations between the congruences $\equiv_{Suff(w)}$ and $\equiv_{Suff(wa)}$ for any couple $(w, a) \in A^* \times A$. The first property, stated in Lemma 7.8, is that $\equiv_{Suff(wa)}$ is a refinement of $\equiv_{Suff(w)}$. The next lemma shows how right contexts evolves.

Lemma 7.7. *Let $w \in A^*$ and $a \in A$. For each $u \in A^*$, we have:*

$$u^{-1}Suff(wa) = \begin{cases} u^{-1}Suff(w)a \cup \{\varepsilon\}, & \text{if } u \in Suff(wa), \\ u^{-1}Suff(w)a, & \text{otherwise.} \end{cases}$$

Proof. Note first that $\varepsilon \in u^{-1}Suff(wa)$ is equivalent to $u \in Suff(wa)$. So, it remains to prove $u^{-1}Suff(wa) \backslash \{\varepsilon\} = u^{-1}Suff(w)a$.

Let z be a nonempty word in $u^{-1}Suff(wa)$. This means $uz \in Suff(wa)$. The word uz can then be written $uz'a$ with $uz' \in Suff(w)$. Thus, $z' \in u^{-1}Suff(w)$, and $z \in u^{-1}Suff(w)a$.

Conversely. Let z be a (nonempty) word in $u^{-1}Suff(w)a$. It can be written $z'a$ for some $z' \in u^{-1}Suff(w)$. Therefore, $uz' \in Suff(w)$, which implies $uz = uz'a \in Suff(wa)$, that is $z \in u^{-1}Suff(wa)$. □

Lemma 7.8. *Let $w \in A^*$ and $a \in A$. The congruence $\equiv_{Suff(wa)}$ is a refinement of the congruence $\equiv_{Suff(w)}$, that is, for each $u, v \in A^*$, $u \equiv_{Suff(wa)} v$ implies $u \equiv_{Suff(w)} v$.*

Proof. We assume $u \equiv_{Suff(wa)} v$, that is, $u^{-1} Suff(wa) = v^{-1} Suff(wa)$, and prove $u \equiv_{Suff(w)} v$, that is, $u^{-1} Suff(w) = v^{-1} Suff(w)$. We only show that $u^{-1} Suff(w) \subseteq v^{-1} Suff(w)$ because the reverse inclusion follows by symmetry.

If $u^{-1} Suff(w)$ is empty, the inclusion trivially holds. Otherwise, let $z \in u^{-1} Suff(w)$. This is equivalent to $uz \in Suff(w)$, which implies $uza \in Suff(wa)$. The hypothesis gives $vza \in Suff(wa)$, and thus $vz \in Suff(w)$ or $z \in v^{-1} Suff(w)$, which achieves the proof.   □

Given a word w, the congruence $\equiv_{Suff(w)}$ partitions A^* into classes. And Lemma 7.8 remains to say that these classes are union of classes according to $\equiv_{Suff(wa)}$, $a \in A$. It turns out that only one or two classes according to $\equiv_{Suff(w)}$ split into two sub-classes to get the partition induced by $\equiv_{Suff(wa)}$. One of the class that splits is the class of words not occurring in w. It contains the word wa itself that gives rise to a new class and a new state of the suffix automaton (see Lemma 7.9). Theorem 7.2 and its corollaries exhibit conditions under which another class also splits and how it splits.

Lemma 7.9. *Let $w \in A^*$ and $a \in A$. Let z be the longest suffix of wa occurring in w. If u is a suffix of wa such that $|u| > |z|$, $u \equiv_{Suff(wa)} wa$.*

Proof. This is a straightforward consequence of Lemma 7.5.   □

Theorem 7.2. *Let $w \in A^*$ and $a \in A$. Let z be the longest suffix of wa occurring in w. Let z' be the longest factor of w such that $z' \equiv_{Suff(w)} z$. For each $u, v \in Fact(w)$, we have:*

$$u \equiv_{Suff(w)} v \quad and \quad u \not\equiv_{Suff(w)} z \quad \Longrightarrow \quad u \equiv_{Suff(wa)} v.$$

Furthermore, for each $u \in A^*$, we have:

$$u \equiv_{Suff(w)} z \quad \Longrightarrow \quad \begin{cases} u \equiv_{Suff(wa)} z, & if \ |u| \le |z|, \\ u \equiv_{Suff(wa)} z', & otherwise. \end{cases}$$

Proof. Let $u, v \in Fact(w)$ be such that $u \equiv_{Suff(w)} v$, that is, $u^{-1} Suff(w) = v^{-1} Suff(w)$. We first assume $u \not\equiv_{Suff(w)} z$ and prove $u \equiv_{Suff(wa)} v$, that is $u^{-1} Suff(wa) = v^{-1} Suff(wa)$.

By Lemma 7.7, we just have to prove that $u \in Suff(wa)$ is equivalent to $v \in Suff(wa)$. Indeed, it is even sufficient to prove that $u \in Suff(wa)$ implies $v \in Suff(wa)$ because the reverse implication comes by symmetry.

Assume then that $u \in Suff(wa)$. Since $u \in Fact(w)$, u is a suffix of z, by definition of z. So, we can consider the largest integer $k \ge 0$ such that $|u| \le |s_w^k(z)|$. Note that $s_w^k(z)$ is a suffix of wa (as z is), and that Lemma 7.3 insures that $u \equiv_{Suff(w)} s_w^k(z)$. So, $v \equiv_{Suff(w)} s_w^k(z)$ by transitivity.

Since $u \not\equiv_{Suff(w)} z$, we have that $k > 0$. Thus, Lemma 7.6 implies that v is a suffix of $s_w^k(z)$, and then that v is a suffix of wa as expected. This proves the first part of the statement.

Consider now a word u such that $u \equiv_{Suff(w)} z$.

If $|u| \leq |z|$, to prove $u \equiv_{Suff(wa)} z$, using the above argument, we just have to show that $u \in Suff(wa)$ because $z \in Suff(wa)$. Indeed, this is a simple consequence of Lemma 7.2.

Conversely, assume that $|u| > |z|$. When such a word u exists, $z' \neq z$ and $|z'| > |z|$ (z is a proper suffix of z'). Therefore, by the definition of z, u and z' are not suffixes of wa. Using again the above argument, this shows that $u \equiv_{Suff(wa)} z'$.

This proves the second part of the statement and ends the proof.   □

Corollary 7.2. *Let $w \in A^*$ and $a \in A$. Let z be the longest suffix of wa occurring in w. Let z' be the longest word such that $z' \equiv_{Suff(w)} z$. If $z' = z$, then, for each $u, v \in Fact(w)$, $u \equiv_{Suff(w)} v$ implies $u \equiv_{Suff(wa)} v$.*

Proof. The conclusion follows directly from Theorem 7.2 if $u \not\equiv_{Suff(w)} z$. Otherwise, $u \equiv_{Suff(w)} z$, and by the hypothesis on z and Lemma 7.2, we get $|u| \leq |z|$. Thus, Theorem 7.2 again gives the same conclusion.   □

Corollary 7.3. *Let $w \in A^*$ and $a \in A$. Assume that letter a does not occur in w. Then, for each $u, v \in Fact(w)$, $u \equiv_{Suff(w)} v$ implies $u \equiv_{Suff(wa)} v$.*

Proof. Since a does not occur in w, the word z of Corollary 7.2 is the empty word. This word is the longest word in its own congruence class. So, the hypothesis of Corollary 7.2 holds. Therefore, the same conclusion follows.   □

7.2.4 Sizes of suffix automata

We discuss the size of suffix automata both in term of number of states and number of edges. We show that the global size of $\mathcal{M}(Suff(x))$ is $O(|x|)$. The set of states and the set of edges of $\mathcal{M}(Suff(x))$ are respectively denoted by Q and E (without mention of x that is implicit in statements).

Corollary 7.4. *If $|x| = 0$, card$(Q) = 1$; and if $|x| = 1$, card$(Q) = 2$. Otherwise $|x| \geq 2$; then, $|x| + 1 \leq$ card$(Q) \leq 2|x| - 1$ and the upper bound is reached only when x is in the form $ab^{|x|-1}$ for two distinct letters a and b.*

Proof. The minimum number of states is obviously $|x| + 1$, and is reached when x is in the form $a^{|x|}$ for some $a \in A$. Moreover, we have exactly card$(Q) = |x| + 1$ when $|x| \leq 2$.

Assume now that $|x| \geq 3$. By Theorem 7.2, each symbol x_k, $3 \leq k \leq |x|$, increases by at most 2 the number of states of $\mathcal{M}(Suff(x_1 x_2 \cdots x_{k-1}))$. Since the number of states for a word of length 2 is 3, we get that

$$\text{card}(Q) \leq 3 + 2(|x| - 2) = 2|x| - 1,$$

as announced in the statement.

The construction of a word x reaching the upper bound for the number of states of $\mathcal{M}(Suff(x))$ is a mere application of Theorem 7.2 considering that each letter x_k, $3 \leq k \leq |x|$, should effectively increase by 2 the number of states of $\mathcal{M}(Suff(x_1 x_2 \cdots x_{k-1}))$.   □

Figure 7.2 displays a suffix automaton whose number of states is maximum for a word of length 7.

Let $length_x: Q \rightarrow \mathbb{N}$ be the function associating to each state q of $\mathcal{M}(\mathit{Suff}(x))$ the length of the longest word u in the congruence class q. It is also the length of the longest path from the initial state to q. (This path is labeled by u.) Longest paths form a spanning tree on $\mathcal{M}(\mathit{Suff}(x))$ (a consequence of Lemma 7.2). Transitions that belong to that tree are called *solid* edges. Equivalently, for each edge (p, a, q) of $\mathcal{M}(\mathit{Suff}(x))$, we have that:

$$(p, a, q) \text{ is solid} \quad \Longleftrightarrow \quad length_x(q) = length_x(p) + 1.$$

This notion is used in the construction of suffix automata to test the condition stated in Theorem 7.2. We use it here to derive exact bounds on the number of edges of suffix automata.

Lemma 7.10. *Assuming* $|x| \geq 1$, $\mathrm{card}(E) \leq \mathrm{card}(Q) + |x| - 2$.

Proof. Consider the spanning tree of longest paths from the initial state in $\mathcal{M}(\mathit{Suff}(x))$. The tree contains $\mathrm{card}(E) - 1$ edges of $\mathcal{M}(\mathit{Suff}(x))$, which are the solid edges.

To each non-solid edge (p, a, q) we associate the suffix uav of x defined as follows: u is the label of the longest path from the initial state to p, and v is the label of the longest path from q to a terminal state. Note that, doing so, two different non-solid edges are associated with two different suffixes of x. Since suffixes x and ε are labels of paths in the tree, they are not considered in the correspondence. Thus, the number of non-solid edges is at most $|x| - 1$.

Counting together the number of both kinds of edges gives the expected upper bound. $\qquad\square$

Corollary 7.5. *If* $|x| = 0$, $\mathrm{card}(E) = 0$; *if* $|x| = 1$, $\mathrm{card}(E) = 1$; *and if* $|x| = 2$, $2 \leq \mathrm{card}(E) \leq 3$. *Otherwise* $|x| \geq 3$; *then* $|x| \leq \mathrm{card}(E) \leq 3|x| - 4$, *and the upper bound is reached when* x *is in the form* $ab^{|x|-2}c$, *for three pairwise distinct letters* a, b, *and* c.

Proof. The lower bound is obvious, and reached when x is in the form $a^{|x|}$ for some $a \in A$. The upper bound can be checked by hand for the cases where $|x| \leq 2$.

Assume now that $|x| \geq 3$. By Corollary 7.4 and Lemma 7.10 we have $\mathrm{card}(E) \leq 2|x| - 1 + |x| - 2 = 3|x| - 3$. The quantity $2|x| - 1$ is the maximum

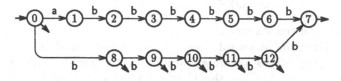

Fig. 7.2. A suffix automaton with the maximum number of states.

number of states obtained only when x is in the form $ab^{|x|-1}$ for two distinct letters a and b. But the number of edges in $\mathcal{M}(\textit{Suff}(ab^{|x|-1}))$ is only $2|x| - 1$. So, card$(E) \leq 3|x| - 4$.

The automaton $\mathcal{M}(\textit{Suff}(ab^{|x|-2}c))$, for three pairwise distinct letters a, b and c, has $2|x| - 2$ states and exactly $3|x| - 4$ edges composed of $2|x| - 3$ solid edges and $|x| - 1$ non-solid edges.   □

Figure 7.3 displays a suffix automaton whose number of edges is maximum for a word of length 7.

As a conclusion of Section 7.2, we get the following statement, direct consequence of Corollaries 7.4 and 7.5.

Theorem 7.3. *The total size of the suffix automaton of a word is linear in the length of the word.*

7.3 Construction

We describe in Sections 7.3.1, 7.3.2 and 7.3.3 an on-line construction of the suffix automaton $\mathcal{M}(\textit{Suff}(x))$.

7.3.1 Suffix links and suffix paths

The construction of $\mathcal{M}(\textit{Suff}(x))$ follows Theorem 7.2 and its corollaries stated in Section 7.2. Conditions that appear in these statements are checked on the automaton with the help of a function defined on its states and called the "suffix link". It is a failure function in the sense of Section 3.4, and is used with this purpose in Section 7.5.

Let $(Q, i, T, E) = \mathcal{M}(\textit{Suff}(x))$ and δ be the corresponding transition function. Let $p \in Q \backslash \{i\}$. State p is a class of factors of x congruent with respect to $\equiv_{\textit{Suff}(x)}$. Let u be any word in the class of p ($u \neq \varepsilon$ because $p \neq i$). Then, the *suffix link* of p is the congruence class of $s_x(u)$. By Lemma 7.4 the value $s_x(u)$ is independent of the word u chosen in the class p, which makes the definition coherent. We denote by f_x the function assignating to each state p its congruence class $s_x(u)$.

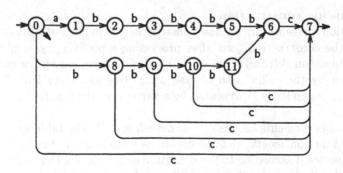

Fig. 7.3. A suffix automaton with the maximum number of edges.

Suffix links induce by iteration "suffix paths" in $\mathcal{M}(\mathit{Suff}(x))$. Note that if $q = f_x(p)$, then $\mathit{length}_x(q) < \mathit{length}_x(p)$. Therefore, the sequence

$$(p, \ f_x(p), \ f_x{}^2(p), \ \ldots)$$

is finite and ends with the initial state i. It is called the *suffix path* of p.

We denote by last_x the state of $\mathcal{M}(\mathit{Suff}(x))$ that is the class of x itself. State last_x has no outgoing edge (otherwise $\mathcal{M}(\mathit{Suff}(x))$ would recognize words longer than x). The suffix path of last_x, i.e.

$$(\mathit{last}_x, \ f_x(\mathit{last}_x), \ f_x{}^2(\mathit{last}_x), \ \ldots),$$

plays an important role in the on-line construction. It is used to test efficiently conditions appearing in statements of the previous section.

Proposition 7.1. *Let $u \in \mathit{Fact}(x)\backslash\{\varepsilon\}$ and set $p = \delta(i, u)$. Then, for any integer $k \geq 0$ for which $s_x{}^k(u)$ is defined, $f_x{}^k(p) = \delta(i, s_x{}^k(u))$.*

Proof. The proof is by induction on k.

For $k = 0$, the equality holds by hypothesis.

Next, let $k \geq 1$ such that $s_x{}^k(u)$ is defined and assume that $f_x{}^{k-1}(p) = \delta(i, s_x{}^{k-1}(u))$. By definition of f_x, $f_x(f_x{}^{k-1}(p))$ is the congruence class of the word $s_x(s_x{}^{k-1}(u))$. Therefore, $f_x{}^k(p) = \delta(i, s_x{}^k(u))$ as expected.   □

Corollary 7.6. *Terminal states of $\mathcal{M}(\mathit{Suff}(x))$, the states in T, are exactly the states of the suffix path of state last_x.*

Proof. Let p be a state of the suffix path of last_x. Then, $p = f_x{}^k(\mathit{last}_x)$ for some integer $k \geq 0$. By Proposition 7.1, since $\mathit{last}_x = \delta(i, x)$, we have $p = \delta(i, s_x{}^k(x))$. Since $s_x{}^k(x)$ is a suffix of x, $p \in T$.

Conversely, let $p \in T$. So, for some $u \in \mathit{Suff}(x)$, $p = \delta(i, u)$. Since $u \in \mathit{Suff}(x)$, we can consider the largest integer $k \geq 0$ such that $|u| \leq |s_x{}^k(x)|$. By Lemma 7.3 we get $u \equiv_{\mathit{Suff}(x)} s_x{}^k(x)$. Thus, $p = \delta(i, s_x{}^k(x))$ by definition of $\mathcal{M}(\mathit{Suff}(x))$. Then, Proposition 7.1 applied to x shows that $p = f_x{}^k(\mathit{last}_x)$, which proves that p belongs to the suffix path of last_x.   □

7.3.2 On-line construction

This section presents an on-line construction of suffix automata. At each stage of the construction, just after processing a prefix $x_1 x_2 \cdots x_\ell$ of x, the suffix automaton $\mathcal{M}(\mathit{Suff}(x_1 x_2 \cdots x_\ell))$ is built. Terminal states are implicitly known by the suffix path of $\mathit{last}_{x_1 x_2 \cdots x_\ell}$ (see Corollary 7.6). The state $\mathit{last}_{x_1 x_2 \cdots x_\ell}$ is explicitly represented by a variable in the function building the automaton.

Two other elements are also used: *Length* and *F*. The table *Length* represents the function length_x defined on states of the automaton. All edges are solid or non-solid according to the definition of Section 7.2 that relies on function length_x. Suffix links of states (different from the initial state) are stored

SUFFIXAUTOMATON(x)
    let δ be the transition function of (Q, i, T, E)
1   $(Q, E) \leftarrow (\emptyset, \emptyset)$
2   $i \leftarrow$ STATE-CREATION
3   $Length[i] \leftarrow 0$
4   $F[i] \leftarrow$ NIL
5   $last \leftarrow i$
6   **for** ℓ from 1 up to $|x|$
7   **loop** SA-EXTEND(ℓ)
8   $T \leftarrow \emptyset$
9   $p \leftarrow last$
10 **loop** $T \leftarrow T + \{p\}$
11       $p \leftarrow F[p]$
12   **while** $p \neq$ NIL
13 **return** $((Q, i, T, E), Length, F)$

Fig. 7.4. On-line construction of the suffix automaton of a word x.

in a table denoted by F that stands for the function f_x. The implementation of $\mathcal{M}(\mathit{Suff}(x))$ with these extra features is discussed in the next section.

The on-line construction in Figure 7.4 is based on procedure SA-EXTEND given in Figure 7.5. The latter procedure processes the next letter, say x_ℓ, of the word x. It transforms the suffix automaton $\mathcal{M}(\mathit{Suff}(x_1 x_2 \cdots x_{\ell-1}))$ already built into the suffix automaton $\mathcal{M}(\mathit{Suff}(x_1 x_2 \cdots x_\ell))$.

We illustrate how procedure SA-EXTEND processes the current automaton through three examples. Let us consider that $x_1 x_2 \cdots x_{\ell-1} =$ ccccbbccc, and let us examine three possible cases according to $a = x_\ell$, namely $a =$ d, $a =$ c, and $a =$ b. The suffix automaton of $x_1 x_2 \cdots x_{\ell-1}$ is depicted in Figure 7.6. Figures 7.7, 7.8, and 7.9 display respectively $\mathcal{M}(\mathit{Suff}(\text{ccccbbcccd}))$, $\mathcal{M}(\mathit{Suff}(\text{ccccbbcccc}))$, and $\mathcal{M}(\mathit{Suff}(\text{ccccbbcccb}))$.

During the execution of the first loop of the procedure, state p runs through a part of the suffix path of $last$. At the same time, edges labeled by a are created from p to the newly created state, unless such an edge already exists in which case the loop stops.

If $a =$ d, the execution of the loop stops at the initial state. The edges labeled by d start at terminal states of $\mathcal{M}(\mathit{Suff}(\text{ccccbbccc}))$. This case corresponds to Corollary 7.3. The resulting automaton is given in Figure 7.7.

If $a =$ c, the loop stops on state $3 = F[last]$ (of the automaton depicted in Figure 7.6) because an edge labeled by c is defined on it. Moreover, the edge is solid, so, we get the suffix link of the new state. Nothing else should be done according to Corollary 7.2. This gives the automaton of Figure 7.8.

Finally, when $a =$ b, the loop stops on state $3 = F[last]$ for the same reason, but the edge labeled by b from 3 is non-solid. The word cccb is a suffix of the new word ccccbbcccb but ccccb is not. Since these two words reach state 5, this state is duplicated into a new state that becomes a terminal state. Suffixes ccb and cb are re-directed to this new state, according to Theorem 7.2. We get the automaton of Figure 7.9.

SA-EXTEND(ℓ)
1   $a \leftarrow x_\ell$
2   $newlast \leftarrow$ STATE-CREATION
3   $Length[newlast] \leftarrow Length[last] + 1$
4   $p \leftarrow last$
5   **loop** $E \leftarrow E + \{(p, a, newlast)\}$
6         $p \leftarrow F[p]$
7   **while** $p \neq$ NIL **and** $\delta(p, a) =$ NIL
8   **if** $p =$ NIL
9   **then** $F[newlast] \leftarrow i$
10   **else** $q \leftarrow \delta(p, a)$
11   **if** $Length[q] = Length[p] + 1$
12   **then** $F[newlast] \leftarrow q$
13   **else** $q' \leftarrow$ STATE-CREATION
14   **for each** letter b such that $\delta(q, b) \neq$ NIL
15   **loop** $E \leftarrow E + \{(q', b, \delta(q, b))\}$
16              $Length[q'] \leftarrow Length[p] + 1$
17              $F[newlast] \leftarrow q'$
18              $F[q'] \leftarrow F[q]$
19              $F[q] \leftarrow q'$
20   **loop** $E \leftarrow E - \{(p, a, q)\} + \{(p, a, q')\}$
21                    $p \leftarrow F[p]$
22   **while** $p \neq$ NIL **and** $\delta(p, a) = q$
23   $last \leftarrow newlast$

Fig. 7.5. From $\mathcal{M}(\mathit{Suff}(x_1 x_2 \cdots x_{\ell-1}))$ to $\mathcal{M}(\mathit{Suff}(x_1 x_2 \cdots x_\ell))$.

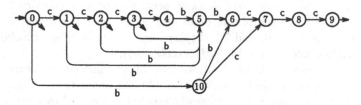

Fig. 7.6. $\mathcal{M}(\mathit{Suff}(\text{ccccbbccc}))$.

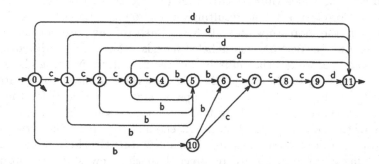

Fig. 7.7. $\mathcal{M}(\mathit{Suff}(\text{ccccbbcccd}))$.

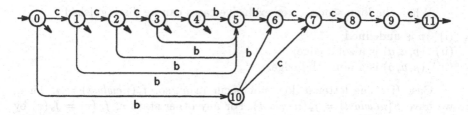

Fig. 7.8. $\mathcal{M}(\textit{Suff}(\text{ccccbbcccc}))$.

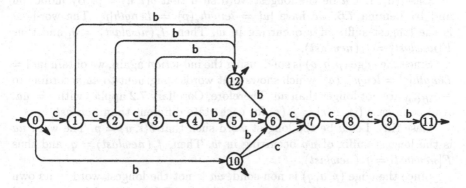

Fig. 7.9. $\mathcal{M}(\textit{Suff}(\text{ccccbbccccb}))$.

Theorem 7.4. *Function* SUFFIXAUTOMATON *builds the suffix automaton of any given word* x.

Proof. The proof is by induction on the length of x. It heavily relies on the properties stated previously.

If $x = \varepsilon$, the function builds an automaton with only one state that is both initial and terminal. No edge is defined. So, the automaton recognizes the language $\{\varepsilon\}$, which is $\textit{Suff}(x)$.

Otherwise $x \neq \varepsilon$. Let $w \in A^*$ and $a \in A$ be such that $x = wa$. We assume, after preprocessing w, that the current values of Q and E are respectively the set of states and of edges of $\mathcal{M}(\textit{Suff}(w))$, that $\textit{last}$ is the state $\delta(i, w)$, that $\textit{Length}[r] = \textit{length}_w(r)$ for each $r \in Q$, and that $F[r] = f_w(r)$ for each $r \in Q\backslash\{i\}$. We prove first that procedure SA-EXTEND correctly updates sets Q and E, variable $\textit{last}$, and tables $\textit{Length}$ and F. Then, we show that terminal states are eventually correctly marked by function SUFFIXAUTOMATON.

The variable p of procedure SA-EXTEND runs through the states of the suffix path of $\textit{last}$ of $\mathcal{M}(\textit{Suff}(w))$. The first loop creates edges by letter a onto the new created state $\textit{newstate}$ according to Lemma 7.9, and we have the equality $\textit{Length}[\textit{newlast}] = \textit{length}_x(\textit{newlast})$.

When the loop stops, three exclusive cases can be distinguished:

(i) p is undefined;

(ii) (p, a, q) is a solid edge;

(iii) (p, a, q) is a non-solid edge.

Case (i). The letter a does not occur in w, so, $f_x(newlast) = i$. Then, we have $F[newlast] = f_x(newlast)$. For any other state r, $f_w(r) = f_x(r)$ by Corollary 7.3. Then, again $F[r] = f_x(r)$ at the end of execution of procedure SA-EXTEND.

Case (ii). Let u be the longest word such that $\delta(i, u) = p$. By induction and by Lemma 7.6, we have $|u| = length_x(p) = Length[p]$. The word ua is the longest suffix of x occurring in w. Then, $f_x(newlast) = q$, and thus $F[newlast] = f_x(newlast)$.

Since the edge (p, a, q) is solid, using the induction again, we obtain $|ua| = Length[q] = length_x(q)$, which shows that words congruent to ua according to $\equiv_{Suff(w)}$ are not longer than ua. Therefore, Corollary 7.2 applies with $z = ua$. And as in case (i), $F[r] = f_x(r)$ for each state different than $newlast$.

Case (iii). Let u be the longest word such that $\delta(i, u) = p$. The word ua is the longest suffix of wa occurring in w. Then, $f_x(newlast) = q$, and thus $F[newlast] = f_x(newlast)$.

Since the edge (p, a, q) is non-solid, ua is not the longest word in its own congruence class according to $\equiv_{Suff(w)}$. Theorem 7.2 applies with $z = ua$, and z' the longest word, label of the path from i to q. The class of ua according to $\equiv_{Suff(w)}$ splits into two classes according to $\equiv_{Suff(x)}$. They are represented by states q and q'.

Words v shorter than ua and such that $v \equiv_{Suff(w)} ua$ are in the form $v'a$ with $v' \in Suff(u)$ (consequence of Lemma 7.2). Before the execution of the last loop, all these words v satisfy $q = \delta(i, v)$. Therefore, after the execution of the loop, they satisfy $q' = \delta(i, v)$, as expected from Theorem 7.2. Words v longer than ua and such that $v \equiv_{Suff(w)} ua$ satisfy $q = \delta(i, v)$ after the execution of the loop, as expected from Theorem 7.2 again. It is easy to check that suffix links are correctly updated.

Finally, in the three cases (i), (ii), and (iii), the value of $last$ is correctly updated at the end of procedure SA-EXTEND.

Thus, the induction proves that the sets Q and E, variable $last$, tables $Length$ and F are correct after the execution of procedure SA-EXTEND.

That terminal states are correctly marked during the last loop of function SUFFIXAUTOMATON is a consequence of Corollary 7.6.   □

7.3.3 Complexity

In order to analyze the complexity of the above construction, we first describe a possible implementation of elements required by the construction. We assume that the automaton is represented by adjacency lists. Doing so, the operations of adding, updating, and accessing a transition (computing

$\delta(p, a))$ take $O(\log \operatorname{card}(A))$ time with an efficient implementation of adjacency lists (see Section 3.2). Function f_x is implemented by the array F that gives access to $f_x(p)$ in constant time.

For the implementation of the solid/non-solid quality of edges, we have chosen to use an array, namely *Length*, representing function *length*$_x$, as suggested by the description of procedure SA-EXTEND. Another possible implementation is to tie a boolean value to edges themselves. Doing so, the first edges created at steps 5 and 20 should be marked as solid. The other edges should be defined as non-solid. This type of implementation do not require the array *Length* that can be eliminated. But the array can be used in applications like the one presented in Section 7.5. Both types of implementation provide a constant-time access to the quality of edges.

Theorem 7.5. *Function* SUFFIXAUTOMATON *can be implemented to work in time* $O(|x| \times \log \operatorname{card}(A))$ *within* $O(|x|)$ *space on each given word* x.

Proof. The set of states of $\mathcal{M}(Suff(x))$ and arrays *Length* and F require $O(\operatorname{card}(Q))$ space. The set of adjacency lists require $O(\operatorname{card}(E))$ space. Thus, the implementation takes $O(|x|)$ space by Corollaries 7.4 and 7.5.

Another consequence of these corollaries is that all operations executed once for each state or each edge take $O(|x| \times \log \operatorname{card}(A))$ on the overall. The same result holds for operations executed once for each letter of x. So, it remains to prove that the total running time of the two loops of lines 5–6 and lines 20–21 inside procedure SA-EXTEND is also $O(|x| \times \log \operatorname{card}(A))$.

Assume that procedure SA-EXTEND is going to update $\mathcal{M}(Suff(w))$, w being a prefix of x. Let u be the longest word reaching state p during the test of the loop of lines 5–6. The initial value of u is $s_w(w)$, and its final value satisfies $ua = s_{wa}(wa)$ (if p is defined). Let k be the quantity $|w| - |u|$, which is the position of the suffix occurrence of u in w. Then, each test strictly increases the value of k during a single run of the procedure. Moreover, the final value of k after a run of the procedure is not greater than its initial value at the beginning of the next run. Therefore, tests and instructions of that loop are executed at most $|x|$ times.

We use a similar argument for the loop of lines 20–21 of procedure SA-EXTEND. Let v be the longest word reaching state p during the test of this loop. The initial value of v equals $s_w{}^k(w)$ for some integer $k \geq 2$, and its final value satisfies $va = s_{wa}{}^2(wa)$ (if p is defined). Then, the position of v as a suffix of w strictly increases at each test over all runs of the procedure. Again, tests and instructions of that loop are executed at most $|x|$ times.

Therefore, the accumulated running time of the two loops of lines 5–6 and lines 20–21 altogether is $O(|x| \times \log \operatorname{card}(A))$. Which ends the proof.   □

7.4 As indexes

The suffix automaton of a word naturally provides an index on its factors. We consider four basic operations on indexes: membership, first position, number of occurrences, and list of positions. The suffix automaton also helps computing efficiently the number of factors in a word, as well as the longest factor occurring at least twice in a word.

7.4.1 Membership
Problem 7.2. (Membership problem for $Fact(x)$.) Given $w \in A^*$, find its longest prefix that belongs to $Fact(x)$.

Proposition 7.2. *With $\mathcal{M}(Suff(x))$, computing the longest prefix u of a word w such that $u \in Fact(x)$ can be performed in time $O(|u| \times \log card(A))$.*

Proof. Just spell the word w in $\mathcal{M}(Suff(x))$ considering the two implementations described in Section 7.3. Stopping the search on the first undefined transition gives the longest prefix u of w for which $\delta(i, u)$ is defined, which means that it is a factor of x.   □

7.4.2 First position
Problem 7.3. (First (respectively last) position of w in x.) Given $w \in Fact(x)$, find its first (respectively last) position in x.

We assume that $w \in Fact(x)$. This test ("does w belong to $Fact(x)$?") can be performed separately as in Section 7.4.1, or can be merged with the solution of the present problem.

The problem of finding the first position $fp_x(w)$ of w in x is equivalent to computing $endpos_x(w)$ because

$$fp_x(w) = endpos_x(w) - |w|.$$

Moreover, this is also equivalent to computing the maximum length of right contexts of w in x,

$$lc_x(w) = \max\{|z| \mid z \in w^{-1}Fact(x)\},$$

because

$$fp_x(w) = |x| - lc_x(w) - |w|.$$

Symmetrically, finding the last position $lp_x(w)$ of w in x remains to computing the smallest length $sc_x(w)$ of its right contexts because

$$lp_x(w) = |x| - sc_x(w) - |w|.$$

To be able to answer efficiently requests on the first or last positions of factors of x, we precompute arrays indexed by states of $\mathcal{M}(Suff(x))$ representing functions lc_x and sc_x. We get the next result.

Proposition 7.3. *The automaton $\mathcal{M}(Suff(x))$ can be preprocessed in time $O(|x|)$ so that the first (or last) position in x of any word $w \in Fact(x)$ can be computed in time $O(|w| \times \log card(A))$ within $O(|x|)$ space.*

Proof. We consider an array LC defined on states of $\mathcal{M}(Suff(x))$ as follows. Let p be a state and u be such that $p = \delta(i, u)$; then, we define $LC[p] = lc_x(u)$. Note that the value of $LC[p]$ does not depend on the word u because for an equivalent word v, $lc_x(u) = lc_x(v)$ (by Lemma 7.2). The array LC satisfies the induction relation:

$$LC[p] = \begin{cases} 0, & \text{if } p = last_x, \\ 1 + \max\{LC[q] \mid q = \delta(p, a), a \in A\}, & \text{otherwise.} \end{cases}$$

So, the computation of LC can be done during a depth-first traversal of the graph of $\mathcal{M}(Suff(x))$. Since the total size of the graph is $O(|x|)$ (Theorem 7.3), this takes time $O(|x|)$.

To compute $fp_x(w)$, we first locate the state $p = \delta(i, w)$, and then return $|x| - |w| - LC[p]$. This takes the same time as for the membership problem.

To find the last occurrence of w in x we consider the array SC that represents the function sc_x. If $p = \delta(i, u)$, we set $SC[p] = sc_x(u)$, which is a coherent definition. We then use the next relation to compute the array during a depth-first traversal of $\mathcal{M}(Suff(x))$:

$$SC[p] = \begin{cases} 0, & \text{if } p \in T, \\ 1 + \min\{SC[q] \mid q = \delta(p, a), a \in A\}, & \text{otherwise.} \end{cases}$$

After the preprocessing, we get the same complexity as above. This ends the proof. $\qquad\square$

7.4.3 Occurrence number
Problem 7.4. (Number of occurrences of w in x.) Given $w \in Fact(x)$, find how many times w occurs in x.

Proposition 7.4. *The automaton $\mathcal{M}(Suff(x))$ can be preprocessed in time $O(|x|)$ so that the number of occurrences in x of any word $w \in Fact(x)$ can be computed in time $O(|w| \times \log card(A))$ within $O(|x|)$ space.*

Proof. The number of occurrences of w in x is

$$card\{z \mid z \in A^* \text{ and } wz \in Suff(x)\}.$$

If $\delta(i, w) = p$, this is also

$$card\{z \mid z \in A^* \text{ and } \delta(p, z) \in T\}.$$

Let $NB[p]$ be this quantity, for any state of $\mathcal{M}(Suff(x))$.

The array NB satisfies the recurrence relation:

$$NB[p] = \begin{cases} 1 + \sum_{q=\delta(p,a), a \in A} NB[q], & \text{if } p \in T, \\ \sum_{q=\delta(p,a), a \in A} NB[q], & \text{otherwise,} \end{cases}$$

which shows that the array NB can be computed in time proportional to the size of the automaton during a depth-first traversal of the graph. This takes $O(|x|)$ time.

Afterwards, the problem remains to access $NB[p]$ for $p = \delta(i, w)$. (If p is undefined, w does not occur in x.) Computing p takes the time announced in the statement.   □

An argument similar to that of the previous proof gives the computation of the number of factors occurring in x, *i.e.* the size of $Fact(x)$. Indeed, $Fact(x)$ is the particular right context associated with the initial state of $\mathcal{M}(Suff(x))$. And to compute its size, we evaluate contexts sizes $CS[p]$ of all states of the automaton using the relation:

$$CS[p] = \begin{cases} 1, & \text{if } p = last_x, \\ 1 + \sum_{q = \delta(p,a), a \in A} CS[q], & \text{otherwise.} \end{cases}$$

This provides a linear-time computation of $\mathrm{card}(Fact(x)) = CS[i]$.

7.4.4 List of positions
Problem 7.5. (Positions of w in x.) Given $w \in Fact(x)$, produce the list of positions of w in x.

Proposition 7.5. *The automaton $\mathcal{M}(Suff(x))$ can be preprocessed in time $O(|x|)$ so that the list L of positions in x of any $w \in Fact(x)$ can be computed in time $O(|w| \times \log \mathrm{card}(A) + \mathrm{card}(L))$ within $O(|x|)$ space.*

Proof. We just sketch the proof of the statement. The automaton is preprocessed in order to create shortcuts over states on which exactly one edge is defined and that are not terminal states. To do so, we create a graph structure superimposed on the automaton. The nodes of the graph are either terminal states or states whose degree is at least two. Arcs of the graph are labeled by the labels of the corresponding path in the automaton. From a given state, labels of outgoing arcs start with pairwise distinct letters (because the automaton is deterministic).

Once the node q associated with w (or an extension of it) is found in the graph, the list of positions of w in x is computed by traversing the subgraph rooted at q. Consider the tree of the traversal. Its internal nodes have at least two children, and its leaves are associated with distinct positions (some positions can correspond to internal nodes). Therefore, the number of nodes of the tree is less than $2 \times \mathrm{card}(L)$, which proves that the time of the traversal is $O(\mathrm{card}(L))$. The extra running time is used to find q.   □

7.4.5 Longest repeated factor
There are two dual problems efficiently solvable with the suffix automaton of x:

− find a longest factor repeated in x;
− find a shortest factor occurring only once in x.

Problem 7.6. (Longest repeated factor in x.) Produce a longest word $u \in Fact(x)$ that occurs twice in x.

If the table NB used to compute the number of occurrences of a factor is already computed, the problem is equivalent to find the deepest state p in $\mathcal{M}(Suff(x))$ for which $NB[p] > 1$. The label of the path from the initial state to p is a solution to the problem. In fact, the problem can be solved without any use of the table NB. We just consider the deepest state p which satisfies one of the two conditions:

(i)   the degree of p is at least two;
(ii) p is a terminal state and the degree of p is at least one.

Doing so, no preprocessing on $\mathcal{M}(Suff(x))$ is even needed, which gives the following result.

Proposition 7.6. *With $\mathcal{M}(Suff(x))$, computing a longest repeated factor of x can be performed in time $O(|x|)$.*

Given a longest repeated factor u of x, ua is a factor of x for some letter a. It is clear that this word is a shortest factor occurring once only in x, *i.e.*, this word is a solution to the dual problem. Hence, the proposition also holds for the second problem.

7.5 As string-matching automata

The suffix automaton $\mathcal{M}(Suff(x))$ of x can be used to solve the string-matching problem, to locate the occurrences of x in a word y. The search procedure behaves like the search phase of algorithm MATCHER (see Section 2) that processes y in an on-line manner. The existence of failure links in $\mathcal{M}(Suff(x))$ is essential for this application, which gives them their name. The search procedure is a consequence of a generic procedure, given Figure 7.10, that can be used for other purposes.

7.5.1 Ending factors
Procedure ENDINGFACTORS of Figure 7.10 computes the longest factor of x ending at each position in y, or more exactly the length of this factor. More precisely, we define for each $k \in \{0, \ldots, |y|\}$ the number

$$\ell_k = \max\{|w| \mid w \in Suff(y_1 y_2 \cdots y_k) \cap Fact(x)\}.$$

The procedure ENDINGFACTORS performs an on-line computation of the sequence $(\ell_k)_{0 \le k \le |y|}$ of lengths of longest ending factors. The output is given as a word on the alphabet $\{0, \ldots, |x|\}$. Function $length_x$ of Section 7.2 (implemented via table $Length$) is used to reset properly the current length just after a suffix link has been traversed.

The core of procedure ENDINGFACTORS is the computation of transitions with the failure table F (implementing the suffix link f_x), similarly as in the general method described in Sections 3.4 and 5.4.

```
ENDINGFACTORS((Q, i, T, E), Length, F, y)
     let δ be the transition function of (Q, i, T, E)
 1   (ℓ, p) ← (0, i)
 2   L ← 0
 3   for letter a from first to last letter of y
 4       loop if δ(p, a) ≠ NIL
 5               then (ℓ, p) ← (ℓ + 1, δ(p, a))
 6               else loop    p ← F[p]
 7                       while p ≠ NIL and δ(p, a) = NIL
 8                       if p ≠ NIL
 9                           then (ℓ, p) ← (Length[p] + 1, δ(p, a))
10                           else (ℓ, p) ← (0, i)
11               L ← L · ℓ
12   return L
```

Fig. 7.10. Computing lengths of factors of a word x ending at all positions in a word y, with $(Q, i, T, E) = \mathcal{M}(\mathit{Suff}(x))$.

Theorem 7.6. *Procedure* ENDINGFACTORS *computes the lengths of longest ending factors of x in y in time $O(|y| \times \log \mathrm{card}(A))$. It executes less than $2|y|$ transitions in $\mathcal{M}(\mathit{Suff}(x))$, and requires $O(|x|)$ space.*

Proof. See the proof of Theorem 5.3.   □

7.5.2 Optimization of suffix links

Indeed, instead of the suffix link f_x, we rather use another link, denoted by $\hat{f}_x$, that optimizes the delay of searches. Its definition is based on transitions defined on states of the automaton, and parallels what is done in Section 5.4.

The "follow set" of a state q of $\mathcal{M}(\mathit{Suff}(x))$ is

$$\mathit{Follow}_x(q) = \{a \mid a \in A, \delta(q, a) \text{ is defined}\}.$$

Then, $\hat{f}_x(q)$ is defined by the relation:

$$\hat{f}_x(q) = \begin{cases} f_x(q), & \text{if } \mathit{Follow}_x(f_x(q)) \not\subseteq \mathit{Follow}_x(q), \\ \hat{f}_x(f_x(q)), & \text{otherwise.} \end{cases}$$

Note that $\hat{f}_x(q)$ can be left undefined with this definition.

A property of Follow_x sets simplifies the computation of $\hat{f}_x$. In the suffix automaton we always have $\mathit{Follow}_x(q) \subseteq \mathit{Follow}_x(f_x(q))$. This is because $f_x(q)$ corresponds to a suffix v of any word u for which $q = \delta(i, u)$. Then, any letter following u in x also follows v (see Lemma 7.1). And this property transfers to follow sets of q and $f_x(q)$ respectively. With this remark, the definition of the failure function $\hat{f}_x$ can be equivalently stated as:

$$\hat{f}_x(q) = \begin{cases} f_x(q), & \text{if the degrees of } q \text{ and of } f_x(q) \text{ are different,} \\ \hat{f}_x(f_x(q)), & \text{otherwise.} \end{cases}$$

Thus, the computation of $\hat{f}_x$ has only to consider degrees of states of the automaton $\mathcal{M}(\mathit{Suff}(x))$, and can be executed in linear time.

Proposition 7.7. *For procedure* ENDINGFACTORS *using a table, say* $\hat{F}$, *implementing the suffix link* $\hat{f}_x$ *instead of the table* F, *the delay is* $O(\text{card}(A))$.

Proof. This is a consequence of:

$$Follow_x(q) \subset Follow_x(\hat{f}_x(q)) \subseteq A$$

for any state q for which $\hat{f}_x(q)$ is defined.   □

7.5.3 Searching for rotations

The knowledge of the sequence of lengths $(\ell_k)_{0 \le k \le |y|}$ leads to several applications such as searching for x in y, computing $lcf(x,y)$, the maximum length of a factor common to x and y, or computing the *subword distance* between two words:

$$d(x,y) = |x| + |y| - 2 \times lcf(x,y).$$

The computation of positions of x in y relies on the simple observation:

$$\ell_k = |x| \iff x \text{ occurs at position } k - |x| \text{ in } y.$$

The same remark applies as well to design an efficient solution to the next problem. A *rotation* (or a *conjugate*) of a word u is a word in the form wv, $w, v \in A^*$, when $u = vw$.

Problem 7.7. (Searching for rotations.) Given $x \in A^*$, locate all occurrences of rotations of x in any given word y.

A first solution to the problem is to apply the algorithm of Section 5 to the set of rotations of x. However, the space required by this solution can be quadratic in $|x|$ like can be the size of the corresponding trie. A solution based on suffix automata keeps the memory space linear.

Proposition 7.8. *After a preprocessing on* x *in time* $O(|x| \times \log \text{card}(A))$, *positions of occurrences of rotations of* x *occurring in* y *can be computed in time* $O(|y| \times \log \text{card}(A))$ *within* $O(|x|)$ *space.*

Proof. Note that the factors of length $|x|$ of the word xx are all the rotations of x. And that longer factors have a rotation of x as a suffix. (In fact, the word xuA^{-1}, where u is the shortest period of x, satisfies the same property.)

The solution consists in running the procedure ENDINGFACTORS with the automaton $\mathcal{M}(Suff(xx))$ after adding this modification: retain position $k - |x|$ each time $\ell_k \ge |x|$. Indeed, $\ell_k \ge |x|$ if and only if the longest factor w of xx ending at position k is not shorter than x. Thus, the suffix of length $|x|$ of w is a rotation of x. The complexity of the new procedure is the same as that of procedure ENDINGFACTORS.   □

7.6 Factor automata

The *factor automaton* of a word x is the minimal deterministic automaton recognizing $Fact(x)$. It is denoted by $\mathcal{M}(Fact(x))$. It is clear that the suffix automaton $\mathcal{M}(Suff(x))$ recognizes $Fact(x)$ if all its states are transformed into terminal states. But the automaton so obtained is not always minimal. For example, the factor automaton of aabbabb, shown in Figure 7.11, is smaller than the suffix automaton of the same word (Figure 7.1). In this section we briefly review few elements related to factor automata: their relation to suffix automata, their sizes, and their construction.

7.6.1 Relation to suffix automata

The construction of factor automata by an on-line algorithm is slightly more tricky than the construction of suffix automata. The latter can be simply deduced from a procedure that builds factor automata as follows. To get $\mathcal{M}(Suff(x))$, first build $\mathcal{M}(Fact(x\$))$, extending alphabet A by letter $\$$, then set as terminal states only those states from which an edge by letter $\$$ outgoes, and finally remove all edges by letter $\$$ and the state they reach. The correctness of this procedure is straightforward, but is also a consequence of Theorem 7.7 below.

Conversely, the construction of $\mathcal{M}(Fact(x))$ from $\mathcal{M}(Suff(x))$ requires a minimization procedure. This is related to the non-solid path in $\mathcal{M}(Fact(x))$ considered in the on-line construction, and that is presented here.

Let us denote by ff_x the suffix function corresponding to the right syntactic congruence associated with $Fact(x)$ (and denoted by $\equiv_{Fact(x)}$ in this chapter). Let $z = ff_x(x)$ (the longest suffix of x occurring at least twice in it). Let $(p_j)_{0 \le j \le |z|}$ be the sequence of states of $\mathcal{M}(Fact(x))$ defined by $p_0 = i$, and, for $0 < j \le |z|$, $p_j = \delta(p_{j-1}, z_j)$, where δ is the transition function of $\mathcal{M}(Fact(x))$, and i its initial state. Let k, $0 \le k \le |z|$, be the smallest integer for which (p_k, z_{k+1}, p_{k+1}) is a non-solid edge (setting $k = |z|$ if no such edge exists). Then, the non-solid path of $\mathcal{M}(Fact(x))$ is composed of edges

$$\cdot (p_k, z_{k+1}, p_{k+1}), \quad (p_{k+1}, z_{k+2}, p_{k+2}), \quad \ldots, \quad (p_{|z|-1}, z_{|z|}, p_{|z|}).$$

In equivalent terms, the word z is decomposed into uv, where $u = z_1 z_2 \cdots z_k$ and $v = z_{k+1} z_{k+2} \cdots z_{|z|}$. The word u is the longest prefix of z which is the

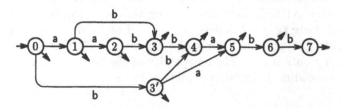

Fig. 7.11. Minimal deterministic automaton recognizing the factors of aabbabb.

longest word in its own congruence class according to $\equiv_{Fact(x)}$. This implies that all shorter prefixes of z satisfy the same condition while longer prefixes do not. The word v labels the non-solid path of $\mathcal{M}(Fact(x))$. It is the empty word if the non-solid path contains no edge.

With the above notion we can describe an alternative method to derive $\mathcal{M}(Suff(x))$ from $\mathcal{M}(Fact(x))$. It consists of first building $\mathcal{M}(Fact(x))$, and then duplicating states p_{k+1}, p_{k+2}, ..., $p_{|z|}$, of the non-solid path of $\mathcal{M}(Fact(x))$ into terminal states while creating edges and suffix links accordingly. This gives also an idea of how, by symmetry, the automaton $\mathcal{M}(Suff(x))$ can be minimized efficiently into $\mathcal{M}(Fact(x))$.

For example, in the automaton $\mathcal{M}(Fact(\text{aabbabb}))$ of Figure 7.11, the non-solid path is composed of the two edges $(1, b, 3)$ and $(3, b, 4)$. The automaton $\mathcal{M}(Suff(\text{aabbabb}))$ is obtained by cloning respectively states 3 and 4 into states $3''$ and $4''$ of Figure 7.1.

The duplication of the non-solid path labeled by v as above is implemented by the procedure of Figure 7.12. The input (r, k), which represents the non-solid path, is defined by $k = |xv^{-1}|$ and $r = \delta(i, xv^{-1})$. For example, with the automaton $\mathcal{M}(Fact(\text{aabbabb}))$ of Figure 7.11 the input is $(5, 5)$.

7.6.2 Size of factor automata

Bounds on the size of factor automata are similar to those of suffix automata. We state the results in this section. We set $(Q, i, T, E) = \mathcal{M}(Fact(x))$.

Proposition 7.9. *If* $|x| \leq 2$, $\text{card}(Q) = |x| + 1$. *Otherwise* $|x| \geq 3$, *and* $|x| + 1 \leq \text{card}(Q) \leq 2|x| - 2$. *If* $|x| \geq 4$, *the upper bound is reached only when* x *is in the form* $ab^{|x|-2}c$ *for three letters* a, b, *and* c, *such that* $a \neq b \neq c$.

Proof. The argument of the proof of Corollary 7.4 works again, except that the last letter of x yields the creation of only one state. Therefore, the upper bound on the number of states is one unit less than that of suffix automata.

```
FA-TO-SA(r, k)
 1    for letter a from k + 1-st to last letter of x
 2        loop t ← δ(r, a)
 3            p ← F[t]
 4            q ← δ(p, a)
 5            q' ← STATE-CREATION
 6            for each letter b such that δ(q, b) ≠ NIL
 7                loop E ← E + {(q', b, δ(q, b))}
 8            Length[q'] ← Length[p] + 1
 9            F[t] ← q'
10            F[q'] ← F[q]
11            F[q] ← q'
12            loop    E ← E − {(p, a, q)} + {(p, a, q')}
13                p ← F[q]
14            while p ≠ NIL and δ(p, a) = q
15            r ← t
```

Fig. 7.12. From $\mathcal{M}(Fact(x))$ to $\mathcal{M}(Suff(x))$. It is assumed that the couple (r, k) is $(\delta(i, xv^{-1}), |xv^{-1}|)$ where v is the label of the non-solid path of $\mathcal{M}(Fact(x))$.

By Theorem 7.2, in order to get the maximum number of states, letters $x_3, x_4, \ldots, x_{|x|-1}$ should eventually lead to the creation of two states (letters x_1, x_2, and $x_{|x|}$ cannot). This happens only when $x_1 \neq x_2$, $x_2 = x_3 = \cdots = x_{|x|-1}$. If $x_{|x|} = x_{|x|-1}$ the word x is in the form ab^{m-1}, with $m \geq 3$ and $a \neq b$, and its factor automaton has exactly $m+1$ states. Therefore, we must have $x_{|x|} \neq x_{|x|-1}$ to get the $2|x| - 2$ bound, and this is also sufficient. $\square$

Figure 7.13 displays a factor automaton having the maximum number of states for a word of length 7.

Proposition 7.10. *If $|x| \leq 2$, $|x| \leq \text{card}(E) \leq 2|x| - 1$. Otherwise $|x| \geq 3$, and $|x| \leq \text{card}(E) \leq 3|x| - 4$. If $|x| \geq 4$, the upper bound is reached only when x is in the form $ab^{|x|-2}c$ for three pairwise distinct letters a, b, and c.*

Proof. Lemma 7.10 is still valid for factor automata, and gives the upper bound $3|x| - 4$ for $|x| \geq 3$. The rest can be checked by hand.

To reach the upper bound, regarding Lemma 7.10 again, $\mathcal{M}(Fact(x))$ should have the maximum number of states. Note that if x is in the form $ab^{|x|-2}a$, with $a, b \in A$ and $a \neq b$, $\text{card}(E) = 2|x| - 1$ only. If x is in the form $ab^{|x|-2}c$ for three pairwise letters a, b, and c, $\text{card}(E) = 3|x| - 4$. Therefore, by Proposition 7.9, this is the only possibility to reach the upper bound. $\square$

Figure 7.14 displays a factor automaton having the maximum number of edges for a word of length 7.

7.6.3 On-line construction

The on-line construction of factor automata is similar to the construction of suffix automata (Section 7.3). The main difference is that the non-solid path of the current automaton is stored and updated after the processing of each letter, and that table F implements ff_x instead of f_x that is related to $\equiv_{Suff(x)}$. The couple of variables (r, k) represents the path as explained previously. The couple gives access to the waiting list of states that are possibly duplicated afterwards.

The function of Figure 7.15 relies on procedure FA-EXTEND (given in Figure 7.16) aimed at transforming the current automaton $\mathcal{M}(Fact(w))$ into $\mathcal{M}(Fact(wa))$, $a \in A$. The correctness of the function is based on a crucial property stated in the next theorem. For each nonempty word w, we denote by $\mathcal{R}(\mathcal{M}(Fact(w)))$ the automaton obtained from $\mathcal{M}(Fact(w))$ by removing the state $last_w$ and all edges reaching it.

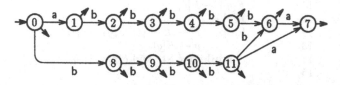

Fig. 7.13. A factor automaton with the maximum number of states.

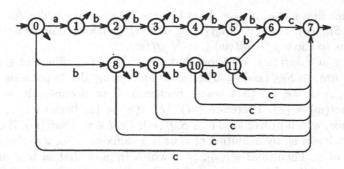

Fig. 7.14. A factor automaton with the maximum number of edges.

FACTORAUTOMATON(x)
1   $(Q, E) \leftarrow (\emptyset, \emptyset)$
2   $i \leftarrow$ STATE-CREATION
3   $Length[i] \leftarrow 0$
4   $F[i] \leftarrow$ NIL
5   $last \leftarrow i$
6   $(r, k) \leftarrow (i, 0)$
7   for ℓ from 1 up to x
8   loop FA-EXTEND(ℓ)
9   return $(Q, i, Q, E), Length, F$

Fig. 7.15. On-line construction of the factor automaton of a word x.

Theorem 7.7. *Let $w \in A^*$, z the longest suffix of w that occurs at least twice in it, and $a \in A$. If $za \notin Fact(w)$, then, disregarding terminal states,*

$$\mathcal{R}(\mathcal{M}(Fact(wa))) = \mathcal{M}(Suff(w)).$$

Before proving the theorem, we prove the following lemma.

Lemma 7.11. *Let $w \in A^*$, z the longest suffix of w that occurs twice in it, and $a \in A$. If $za \notin Fact(w)$, then, for each $s \in A^*$, we have*

$$s \equiv_{Fact(wa)} z \quad \Longrightarrow \quad s \in Suff(z).$$

Proof. The condition on z implies that $(za)^{-1}Fact(wa) = \{\varepsilon\}$. Then, we have $(sa)^{-1}Fact(wa) = \{\varepsilon\}$, which implies that s is a suffix of w. Since z occurs twice in w, this is also the case for s (because some word ta, with $t \neq \varepsilon$, belongs to $z^{-1}Fact(wa) = s^{-1}Fact(wa)$). Therefore, by the definition of z, s is a suffix of z. □

Proof of Theorem 7.7. We prove that, for any words $u, v \in Fact(w)$,

$$u \equiv_{Fact(wa)} v \quad \Longleftrightarrow \quad u \equiv_{Suff(w)} v,$$

which is a re-statement of the conclusion of the theorem.

Assume first $u \equiv_{Fact(wa)} v$. After Lemma 7.2 we can consider, for example, that $u \in Suff(v)$. Then, $v^{-1}Suff(w) \subseteq u^{-1}Suff(w)$, and to prove $u \equiv_{Suff(w)} v$ it remains to show $u^{-1}Suff(w) \subseteq v^{-1}Suff(w)$.

Let $t \in u^{-1}Suff(w)$. We show that $t \in v^{-1}Suff(w)$. Since $ut \in Suff(w)$, we have $uta \in Suff(wa) \subseteq Fact(wa)$, and using the hypothesis, namely $u \equiv_{Fact(wa)} v$, we get that $vta \in Fact(wa)$. If ut occurs only once in w, $(ut)^{-1}Fact(w) = \{\varepsilon\}$. Therefore, $(vt)^{-1}Fact(w) = \{\varepsilon\}$ because $\equiv_{Fact(wa)}$ is a congruence, which proves that $vt \in Suff(w)$, i.e. $t \in v^{-1}Suff(w)$. If ut occurs at least twice in w, by definition of z, ut is a suffix of z. So, $z = z'ut$ for some prefix z' of z. Then, $z'vt \equiv_{Fact(wa)} z$, which implies that vt is a suffix of z and consequently of w by Lemma 7.11. Hence again, $t \in v^{-1}Suff(w)$. This ends the first part of the statement.

Conversely, let us consider that $u \equiv_{Suff(w)} v$, and prove $u \equiv_{Fact(wa)} v$. Without loss of generality, we assume $u \in Suff(v)$, so it remains to prove $u^{-1}Fact(wa) \subseteq v^{-1}Fact(wa)$.

Let $t \in u^{-1}Fact(wa)$. If $t = \varepsilon$, $t \in v^{-1}Fact(wa)$ because $v \in Fact(w)$. We then assume that t is a nonempty word. If $ut \in Fact(w)$, for some $t' \in A^*$, $utt' \in Suff(w)$, that is $tt' \in u^{-1}Suff(w)$. The hypothesis $u \equiv_{Suff(w)} v$ implies $tt' \in v^{-1}Suff(w)$, and consequently $t \in v^{-1}Fact(wa)$. If $vt \notin Fact(w)$, t is a suffix of wa. It can be written $t'a$ for some suffix t' of w. So, $t' \in u^{-1}Suff(w)$, and then $t' \in v^{-1}Suff(w)$ which shows that vt' is a suffix of w. Therefore, $t \in v^{-1}Fact(wa)$. This ends both the second part of the statement and the whole proof.   □

During the construction of $\mathcal{M}(Fact(x))$, the following property is invariant: let $\mathcal{M}(Fact(w))$ be the current automaton and z be as in Theorem 7.7, then $\delta(i,z) = F[last]$. Consequently, the condition on z that appears in Theorem 7.7 translates into the test "$\delta(F[last], a) \neq$ NIL". If its value is TRUE, the automaton and the suffix z extend, and the procedure FA-EXTEND updates the pair (r, k) if necessary in a natural way. Otherwise, Theorem 7.7 applies, which leads to first transform $\mathcal{M}(Fact(w))$ into $\mathcal{M}(Suff(w))$. After that, the automaton is extended by the letter a. In this situation, the new non-solid path is composed of at most one edge, as a consequence of Lemma 7.6 (see also the proof of Theorem 7.7).

Finally, we state the complexity of function FACTORAUTOMATON: the construction of factor automata by this function takes linear time on a fixed alphabet.

Theorem 7.8. *Function* FACTORAUTOMATON *can be implemented to work on the input word x in time $O(|x| \times \log card(A))$ within $O(|x|)$ space.*

Proof. If, for a moment, we do not consider the calls to procedure FA-TO-SA, it is rather simple to see that there is a linear number of instructions executed to built $\mathcal{M}(Fact(x))$. Implementing the automaton with adjacency lists, the cost of computing a transition is $O(\log card(A))$. Which gives the $O(|x| \times \log card(A))$ time for the considered instructions.

FA-EXTEND(ℓ)
1    $a \leftarrow x_\ell$
2   **if** $\delta(F[last], a) \neq$ NIL
3   **then** $newlast \leftarrow$ STATE-CREATION
4          $E \leftarrow E + \{(lasta, newlast)\}$
5          $Length[newlast] \leftarrow Length[last] + 1$
6          $F[newlast] \leftarrow \delta(F[last], a)$
7   **if** $k = \ell$ **and** $Length[F[last]] + 1 = Length[\delta(F[last], a)]$
8   **then** $(r, k) \leftarrow (newlast, \ell + 1)$
9   **else** FA-TO-SA(r, k)
10          $newlast \leftarrow$ STATE-CREATION
11          $p \leftarrow last$
12   **loop** $E \leftarrow E + \{(p, a, newlast)\}$
13             $p \leftarrow F[p]$
14   **while** $p \neq$ NIL **and** $\delta(p, a) =$ NIL
15   **if** $p =$ NIL
16   **then** $F[newlast] \leftarrow i$
17                $(r, k) \leftarrow (newlast, \ell + 1)$
18   **else** $q \leftarrow \delta(p, a)$
19                $F[newlast] \leftarrow q$
20   **if** $Length[p] + 1 = Length[q]$
21   **then** $(r, k) \leftarrow (newlast, \ell + 1)$
22   **else** $(r, k) \leftarrow (last, \ell)$
23    $last \leftarrow newlast$

Fig. 7.16. From $\mathcal{M}(Fact(x_1 \cdots x_{\ell-1}))$ to $\mathcal{M}(Fact(x_1 \cdots x_\ell))$.

The sequence of calls to procedure FA-TO-SA behaves like a construction of $\mathcal{M}(Suff(x))$ (or just of $\mathcal{M}(Suff(x'))$, for some prefix x' of x). Thus, their accumulated running time is as announced by the theorems of Section 7.3, that is, $O(|x| \times \log card(A))$.

This sketches the proof of the result.   □

Bibliographic notes

Only a few books are entirely devoted to pattern matching. One can refer to [19] and [31]. The topic is treated in some books on the design of algorithms such as [3], [7], [23], [15]. An extensive bibliography is also included in [1], and the subject is partially treated in relation to automata in [29].

The notion of a failure function to represent efficiently an automaton is implicit in the work of Morris and Pratt (1970). It is intensively used in [26] and [2]. The table-compression method is explained in [4]. It is the base of some implementations of **lex** (compiler of lexical analyzer) and **yacc** (compiler of compiler) UNIX software tools that involve automata.

The regular-expression-matching problem is considered for the construction of compilers (see [4] for example). The transformation of a regular expression into an automaton is treated in standard textbooks on the subject. The construction described in Section 4 is by Thompson [32]. Many tools under the UNIX operating system use a similar method. For example the **grep** command implements the method with reduced regular expressions. While the command `egrep` operates on complete regular expressions and uses a lazy determinization of the underlying automaton.

The first linear-time solution of the string-matching problem was given by Morris and Pratt, and improved in [26]. The analogue solution to the dictionary-matching problem was designed by Aho and Corasick [2] and is implemented by the `fgrep` UNIX command.

Several authors have considered a variant of the dictionary-matching problem in which the set of words changes during the search. This is called the "dynamic-dictionary-matching problem" (see [5], [6], [25]). A related work based on suffix automata is treated in [27]. A solution to the problem restricted to uniform dictionaries is given in [11] in the comparison model of computation.

The linear size of suffix automata (also called "directed acyclic word graphs" and denoted by the acronym DAWG) and factor automata has been first noticed by Blumer et al., and their efficient construction is in [9] and [16]. An alternative data structure that stores efficiently the factors (subwords) of a word is the suffix tree. It has been first introduced as a position tree by Weiner [35], but the most practical algorithms are by McCreight [28] and Ukkonen [33]. Suffix automata and suffix trees have similar applications to the implementation of indexes (inverted files), to pattern matching, and to data compression.

The solution of the string-matching problem presented in Section 6 is adapted from an original algorithm of Simon [30]. It has been analyzed and improved by Hancart [24]. These algorithms as well as algorithms in [26] solve the string-prefix-matching problem which is the computation of ending prefixes at each position on the searched word. Breslauer et al. [12] gave lower bounds to this problem that meet the upper bounds of Theorem 6.7. Galil showed in [20] how to transform the algorithms of Knuth, Morris and Pratt, so that the search phase works in real time (independently of the alphabet). This applies as well to the algorithm of Simon.

There are many other solutions to the string-matching problem if we relax the condition that the buffer on the searched word is reduced to only one letter. The most practically efficient solutions are based on the algorithm of Boyer and Moore [10]. With the use of automata, it has been extended to the dictionary-matching problem by Commentz-Walter [14] (see also [1]) and by Crochemore et al. (see [19]). The set of configurations possibly met during the search phase of a variant of the algorithm of Boyer and Moore in which all information is retained leads to what is called the "Boyer-Moore automaton". It is still unknown if the number of states of the automaton is polynomial as conjectured by Knuth (see [26, 13, 8]). Theorem 6.3 is by Galil and Seiferas [21]. Like their solution, several other proofs (in [18, 17, 22]) of the same result are based on combinatorial properties of words (see Chapter "Combinatorics of words").

Other kinds of patterns are considered in the approximate string-matching problem (see [19] for references on the subject). The first kind arises when mismatches are allowed between a given word and factors of the searched word (base of the Hamming distance). A second kind of patterns arises when approximate patterns are defined by transformation rules (substitution, insertion, and deletion) that yield the edit distance (see Chapter "String editing and DNA"). This notion is widely used for matching patterns in DNA sequences (see [34]).

References

[1]   A. V. Aho. Algorithms for finding patterns in strings. In J. van Leeuwen, editor, *Handbook of Theoretical Computer Science*, volume A, chapter 5, pages 255–300. Elsevier, Amsterdam, 1990.

[2]   A. V. Aho and M. J. Corasick. Efficient string matching: an aid to bibliographic search. *Communications of the ACM*, 18:333–340, 1975.

[3]   A. V. Aho, J. E. Hopcroft, and J. D. Ullman. *The design and analysis of computer algorithms*. Addison-Wesley, Reading, MA, 1974.

[4]   A. V. Aho, R. Sethi, and J. D. Ullman. *Compilers – Principles, techniques and tools*. Addison-Wesley, Reading, MA, 1986.

[5]   A. Amir and M. Farach. Adaptative dictionary matching. In *Proceedings of the 32th IEEE Annual Symposium on Foundations of Computer Science*, pages 760–766. IEEE Computer Society Press, 1991.

[6]   A. Amir, M. Farach, Z. Galil, R. Giancarlo, and K. Park. Fully dynamic dictionary matching. *Journal of Computer and System Sciences*, 49:208–222, 1994.

[7]   S. Baase. *Computer algorithms – Introduction to design and analysis*. Addison-Wesley, Reading, MA, 1988.

[8]   R. A. Baeza-Yates, C. Choffrut, and G. H. Gonnet. On Boyer-Moore automata. *Algorithmica*, 12:268–292, 1994.

[9]   A. Blumer, J. Blumer, A. Ehrenfeucht, D. Haussler, M. T. Chen, and J. Seiferas. The smallest automaton recognizing the subwords of a text. *Theoretical Computer Science*, 40:31–55, 1985.

[10]   R. S. Boyer and J. S. Moore. A fast string searching algorithm. *Communications of the ACM*, 20:762–772, 1977.

[11]   D. Breslauer. Dictionary-matching on unbounded alphabets: uniform length dictionaries. *Journal of Algorithms*, 18:278–295, 1995.

[12]   D. Breslauer, L. Colussi, and L. Toniolo. Tight comparison bounds for the string prefix-matching problem. *Information Processing Letters*, 47:51–57, 1993.

[13]   V. Bruyère. *Automates de Boyer-Moore*. Thèse annexe, Université de Mons-Hainaut, Belgique, 1991.

[14]   B. Commentz-Walter. A string matching algorithm fast on the average. In *Proceedings of the 6th International Conference on Automata, Languages and Programming*, Lecture Notes in Computer Science, pages 118–132. Springer-Verlag, Berlin, 1979.

[15]   T. H. Cormen, C. E. Leiserson, and R. L. Rivest. *Introduction to algorithms*. MIT Press, 1990.

[16]   M. Crochemore. Transducers and repetitions. *Theoretical Computer Science*, 45:63–86, 1986.

[17]   M. Crochemore. String-matching on ordered alphabets. *Theoretical Computer Science*, 92:33–47, 1992.

[18]   M. Crochemore and D. Perrin. Two-way string-matching. *Journal of the ACM*, 38:651–675, 1991.

[19]   M. Crochemore and W. Rytter. *Text algorithms*. Oxford University Press, 1994.

[20]   Z. Galil. String matching in real time. *Journal of the ACM*, 28:134–149, 1981.

[21]   Z. Galil and J. Seiferas. Time-space optimal string matching. *Journal of Computer and System Sciences*, 26:280–294, 1983.

[22] L. Gąsieniec, W. Plandowski, and W. Rytter. The zooming method: a recursive approach to time-space efficient string-matching. *Theoretical Computer Science*, 147:19–30, 1995.

[23] G. H. Gonnet and R. A. Baeza-Yates. *Handbook of algorithms and data structures*. Addison-Wesley, Reading, MA, 1991.

[24] C. Hancart. On Simon's string searching algorithm. *Information Processing Letters*, 47:95–99, 1993.

[25] R. M. Idury and A. A. Schäffer. Dynamic dictionary matching with failure function. *Theoretical Computer Science*, 131:295–310, 1994.

[26] D. E. Knuth, J. H. Morris, Jr, and V. R. Pratt. Fast pattern matching in strings. *SIAM Journal on Computing*, 6:323–350, 1977.

[27] G. Kucherov and M. Rusinowitch. Matching a set of strings with variable length don't cares. In *Proceedings of the 6th Annual Symposium on Combinatorial Pattern Matching*, Lecture Notes in Computer Science, vol. 937, pages 230–247. Springer-Verlag, Berlin, 1995.

[28] E. M. McCreight. A space-economical suffix tree construction algorithm. *Journal of Algorithms*, 23:262–272, 1976.

[29] D. Perrin. Finite automata. In J. van Leeuwen, editor, *Handbook of Theoretical Computer Science*, volume B, chapter 1, pages 1–57. Elsevier, Amsterdam, 1990.

[30] I. Simon. String matching algorithms and automata. In *Results and Trends in Theoretical Computer Science*, Lecture Notes in Computer Science, vol. 814, pages 386–395. Springer-Verlag, Berlin, 1994.

[31] G. A. Stephen. *String searching algorithms*. World Scientific Press, Singapore, 1994.

[32] K. Thompson. Regular expression search algorithm. *Communications of the ACM*, 11:419–422, 1968.

[33] E. Ukkonen. Constructing suffix trees on-line in linear time. In J. van Leeuwen, editor, *Proceedings of the IFIP 12th World Computer Congress*, Madrid, 1992, pages 484–492, North-Holland.

[34] M. S. Waterman. *Introduction to computational biology*. Chapman and Hall, London, 1995.

[35] P. Weiner. Linear pattern matching algorithm. In *Proceedings of the 14th Annual IEEE Symposium on Switching and Automata Theory*, pages 1–11, 1973.

Symbolic Dynamics and Finite Automata

Marie-Pierre Béal and Dominique Perrin

1. Introduction

Symbolic dynamics is a field which was born with the work in topology of Marston Morse at the beginning of the 1920s [44]. It is, according to Morse, an *"algebra and geometry of recurrence"*. The idea is the following. Divide a surface into regions named by certain symbols. We then study the sequences of symbols obtained by scanning the successive regions while following a trajectory starting from a given point. A further paper by Morse and Hedlund [45] gave the basic results of this theory. Later, the theory was developed by many authors as a branch of ergodic theory (see for example the collected works in [59] or [12]). One of the main directions of research has been the problem of the *isomorphism of shifts of finite type* (see below the definition of these terms). This problem is not yet completely solved although the latest results of Kim and Roush [35] indicate a counterexample to a long-standing conjecture formulated by F. Williams [61].

There are many links between symbolic dynamics and the theory of automata, as pointed out by R. Adler and B. Weiss [60]. Actually a very early reference on this connection can be found in a paper of A. Gleason published many years later [30] after a series of lectures given at the Institute for Defense Analysis in 1960. In this paper, based on notes by R. Beals and M. Spivak, methods of finite semigroups were introduced to obtain some of the results of G. Hedlund.

The idea of considering infinite words also appears, of course, in the framework of automaton theory, independently of symbolic dynamics. This theory was developed initially by R. Büchi and R. McNaughton. Since the beginning it has, however, taken a different direction and is connected with problems of logic rather than with the topological ones raised in symbolic dynamics.

In this chapter, we present some interconnections between automata and symbolic systems and discuss some of the new results that have been obtained in this direction together with some interesting open problems.

The material presented here does not cover all existing connections of this kind. There are, in particular, interesting links between symbolic dynamics and representation of numbers that are not presented here (see [28]). There are also important connections with cellular automata (see, e.g., [16]). The applications of symbolic dynamics to coding are not treated (see [2] or the book of D. Lind and B. Marcus [38]).

The chapter is organized as follows. The first sections (Sections 2, 3, 4) constitute an introduction to symbolic dynamics. It is essentially self-contained although some proofs are only sketched. The concepts introduced include shift spaces, symbolic systems, minimal systems, sofic systems, and systems of finite type.

In the next section (Section 5), we show how the notion of a minimal deterministic automaton translates in the framework of symbolic dynamics to the notion of a Fischer cover. Both notions essentially coincide but for the choice of an initial state.

The following section (Section 6) describes the relation between unambiguous automata and codes with a class of maps called finite-to-one. We show that some results on the completion of codes can be translated into ones on shifts of finite type.

Section 7 is an introduction to the technique of state splitting. We show in particular how this operation is related to automata minimization.

The next section (Section 8) deals with the notion of the isomorphism of shifts of finite type. We define shift equivalence and strong shift equivalence. We show that two shifts of finite type are isomorphic iff their matrices are strong shift equivalent (William's theorem).

Section 9 contains the definition of entropy. We show how this notion is related to well-known concepts in automata and coding theory such as the Kraft inequality. We also state without proof a recent result of D. Handelman that characterizes the entropies of systems generated by finite codes.

The following sections present various topics relating symbolic dynamics and finite automata, covering in particular zeta functions and circular codes.

This chapter is a survey of many results and concepts, not all of them presented with the same degree of detail. As far as the definitions are concerned, it is self-contained for a reader familiar with basic notions of automata theory. Concerning the proofs of the results, the situation is more variable: some of them are given completely, even if sometimes condensed. Some others are only sketched or not even given here, as not being in the scope of this survey.

The material presented is an extended version of a survey by the second author at the conference MFCS in September 1995 in Prague [51].

We would like to thank many people for their help during the preparation of this work and, in particular, Frédérique Bassino, Véronique Bruyère, Aldo De Luca, and Paul Schupp.

2. Symbolic dynamical systems

We present in this section a short introduction to the concepts of symbolic dynamics. For a much more detailed and complete exposition, we refer to the new book of Doug Lind and Brian Marcus [38], which is the first exposition in book form of this theory. Our presentation aims especially at a public of computer scientists already familiar with such concepts as finite automata and transductions.

Let A be a finite alphabet. We denote by A^* the set of finite words on A. The empty word is denoted ϵ and the set of nonempty words is thus $A^+ = A^* - \epsilon$. We consider the set $A^{\mathbf{Z}}$ of two-sided infinite words as a topological space with respect to the usual product topology. An element of $A^{\mathbf{Z}}$ is a sequence

$$
\begin{aligned}
x &= (x_n)_{n \in \mathbf{Z}} \\
&= \ldots x_{-1} x_0 x_1 \ldots
\end{aligned}
$$

The topology is defined by the distance for which words are close if they coincide on a long interval centered at 0. Formally, we may define the distance of x, y as

$$ d(x, y) = 2^{-e(x, y)} $$

where

$$ e(x, y) = \max\{ n \geq 0 \mid x_i = y_i, \ -n \leq i \leq n \} $$

using the convention $e(x, y) = \infty$ if $x = y$, and $e(x, y) = -1$ if $x_0 \neq y_0$.

The *shift* transformation σ acts on $A^{\mathbf{Z}}$ bijectively. It associates to $x \in A^{\mathbf{Z}}$ the element $y = \sigma(x) \in A^{\mathbf{Z}}$ defined for $n \in \mathbf{Z}$ by

$$ y_n = x_{n+1} $$

and obtained by shifting all symbols one place left.

A *symbolic dynamical system* or *subshift* is a subset S of $A^{\mathbf{Z}}$ which is both

1. topologically closed and,
2. shift-invariant, i.e., such that $\sigma(S) = S$.

Thus, and in more intuitive terms, a subshift is a set of bi-infinite words whose definition does not make reference to the origin and allows one passing to the limit.

The system is denoted S or (S, σ) to emphasize the role of the shift σ.

For example, $(A^{\mathbf{Z}}, \sigma)$ itself is a symbolic dynamical system, often called the *full shift* in contrast with the subshifts whose name refer to the embedding in the full shift $A^{\mathbf{Z}}$.

As a less trivial example, the set of sequences on $A = \{a, b\}$ such that a symbol b is always followed by a symbol a is a subshift often called the *golden mean* system. We shall use this example several times in the rest of the chapter.

Let G be a directed graph with E as its set of edges. We actually use multigraphs instead of ordinary graphs in order to be able to have several distinct edges with the same origin and end. Formally, a directed multigraph is given by two sets E (the edges) and V (the vertices) and two functions

$\alpha, \beta : E \to V$. The vertex $\alpha(e)$ is the *origin* of the edge e and $\beta(e)$ is its *end*. We shall always say "graph" for "directed multigraph".

Let S_G be the subset of $E^{\mathbb{Z}}$ formed by all bi-infinite paths in G. It is clear that S_G is a subshift called the *edge shift* on G. Indeed S_G is closed and shift invariant by definition.

Figure 2.1 presents an example of a graph with three edges which defines an edge shift on three symbols.

Let $\mathcal{A} = (Q, E)$ be a *finite automaton* on an alphabet A given by a finite set Q of states and a set $E \subset Q \times A \times Q$ of edges but without initial or final states. The set of all labels $\cdots a_{-1} a_0 a_1 \cdots$ of the bi-infinite paths

$$\cdots p_{-1} \xrightarrow{a_{-1}} p_0 \xrightarrow{a_0} p_1 \cdots$$

is a subshift S. We say that S is the subshift *recognized* by the automaton $\mathcal{A}$.

An automaton can be considered as a graph in which the set of edges is contained in $Q \times A \times Q$. Thus we may consider the subshift formed by all bi-infinite paths in $\mathcal{A}$, which is really the edge shift $S_{\mathcal{A}}$. The subshift recognized by $\mathcal{A}$ is the image of the edge shift $S_{\mathcal{A}}$ under the map assigning to a path its label. This map is continuous. Since the automaton is finite, the set S_G is compact and so is S which is thus closed. A subshift obtained in this way is called a *sofic* shift or a sofic system.

We will use in the sequel finite automata, either in the classical sense as a tuple (Q, E, I, T) with $I \subset Q$ as set of initial states and $T \subset Q$ as set of final states, to recognize a set of finite words, or, as above, without initial and final states, to recognize a subshift.

For a subset X of A^*, we say that X is *recognizable* if it can be recognized by a finite automaton.

A finite word v is said to be a *factor* (or also a *block*) of a finite or infinite word z if its symbols appear consecutively in z.

We may associate with a subshift $S \subset A^{\mathbb{Z}}$ the set

$$F_S = \{x_i \cdots x_j \mid x \in S,\ i \leq j\} \cup \epsilon$$

of factors of its elements.

We may also consider the complement $I_S = A^* - F_S$ of this set, which is the set of *forbidden blocks*.

The set $F = F_S \subset A^*$ satisfies the conditions of being

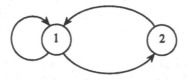

Fig. 2.1. An edge shift

1. factorial: $uvw \in F \Rightarrow v \in F$.
2. extendable: $\forall v \in F, \exists a, b \in A : avb \in F$.

Conversely, such a set of finite words is the set of factors of a subshift as shown by the following proposition.

Proposition 2.1. *Let $F \subset A^*$ be a factorial and extendable set. The set*

$$S = S_F = \{x \in A^{\mathbb{Z}} \mid x_i \cdots x_j \in F \ (i \leq j)\}$$

is a subshift and $F_{S_F} = F$.

Proof. It is clear that S is a subshift and that $F_S \subset F$. To show that $F \subset F_S$, consider $v \in F$. Since F is extendable, we can construct a two-sided infinite word x of the form $\cdots a_n \cdots a_1 v b_1 \cdots b_n \cdots$ such that $a_n \cdots a_1 v b_1 \cdots b_n \in F$ for all n. Then all factors of x are in F and thus $x \in S$, whence $v \in F_S$.   □

One may also define a subshift by a set of forbidden blocks. Indeed, for any set $I \subset A^+$ the set

$$S = \{x \in A^{\mathbb{Z}} \mid x_i \cdots x_j \notin I \ (i \leq j)\}$$

is always a subshift whatever be the set I, but we only have the inclusion $I_S \supset I$ instead of an equality.

For instance, if S is the golden mean system, the set I_S of forbidden blocks is formed by all words containing bb.

The simple relation between a subshift and its set of factors shows that subshifts are closely related to ordinary sets of finite words, in contrast with more general sets of infinite words defined by Büchi automata. The latter are indeed not topologically closed in general and thus have a larger topological complexity in the Borel hierarchy (on Büchi automata, see [24], [58], or the chapter by W. Thomas in this handbook).

We say that an automaton $\mathcal{A} = (Q, E)$ is *trim* if any state is on a bi-infinite path. We shall consider only here trim automata since we are interested in bi-infinite paths. For an automaton $\mathcal{A} = (Q, E, I, T)$ with initial and final states, we say that $\mathcal{A}$ is *trim* if any state is accessible from I and can access T.

Proposition 2.2. *Let $\mathcal{A} = (Q, E)$ be a finite trim automaton. The set F_S of factors of the subshift S recognized by $\mathcal{A}$ is recognized by the automaton (Q, E, Q, Q) in which all states are both initial and final. Conversely, if F_S is recognized by a trim finite automaton $\mathcal{A} = (Q, E, i, T)$, then S is recognized by the automaton (Q, E).*

In particular, S is sofic iff F_S is recognizable.

Proof. The first assertion is clear. For the second one, let S' be the subshift recognized by $\mathcal{A}$. Since $F_S = F_{S'}$ we have $S = S'$ and thus the conclusion. The third statement is a direct consequence of the first ones.   □

The notions introduced above can also be formulated in the context of one-sided infinite words. Indeed the set A^N of one-sided infinite words is also a topological space and the shift transformation is also defined upon it although it is no longer one-to-one. A *one-sided symbolic system*, or one-sided subshift, is a set $S \subset A^N$ which is both closed and invariant. Equivalently, it is the set of right infinite sequences that appear in a shift. We shall usually work with two-sided subshifts because two-sided shifts take into account both the past and the future. An exception will be made in Section 3 concerning the notion of recurrence.

A subshift S is said to be *irreducible* if for any $x, y \in F_S$ there is a word u such that $xuy \in F_S$.

For example, the golden mean system is irreducible. In fact, if x and y avoid bb, then xay also does. On the contrary, the set of infinite words on $\{a, b\}$ avoiding ba is reducible since for all u, the word bua contains a factor ba.

Let S be the edge graph of a graph G. If G is strongly connected, then S is irreducible. The converse is true provided the graph satisfies the condition that each vertex has positive in- and out-degree.

An automaton with a strongly connected graph is said to be *transitive*.

In general, the definition of an irreducible system can be formulated in topological terms and it is related to the possibility of a decomposition into simpler elements.

A subshift S is said to be *primitive* if there is an integer $n > 0$ such that for all $x, y \in F_S$ there exists a word u of length n such that $xuy \in F_S$.

For example, the golden mean system is primitive. On the contrary, the system S formed by all words on $\{a, b\}$ avoiding aa and bb is not primitive. Indeed, for $x = a$, $y = b$ the only adequate u has to be in $(ba)^*$ and thus of even length, whereas for $x = a$, $y = a$ it has to be in $b(ab)^*$ and thus of odd length.

When S is the edge graph of a strongly connected graph G, then S is primitive if and only if the gcd of the cycle lengths is one.

A *morphism* between two subshifts (S, σ) and (T, τ) is a map $f : S \to T$ which is continuous and commutes with the shifts, i.e. such that $f\sigma = \tau f$.

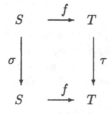

When f is a morphism from S onto T, it is said that T is a *factor* of S.

Let $S \subset A^{\mathbb{Z}}$ and $T \subset B^{\mathbb{Z}}$ be two subshifts and let $k \geq 1$ be an integer. A function $f : S \to T$ is said to be *k-local* or local if there exists a function $\overline{f} : A^k \to B$ and an integer $m \in \mathbb{Z}$ such that for all $x \in S$ the word $y = f(x)$

is defined for $n \in \mathbb{Z}$ by

$$y_{n+m} = \overline{f}(x_{n-(k-1)} \cdots x_{n-1} x_n) \qquad (2.1)$$

Thus the value of a symbol in the image is a function of the symbols contained in a window of length k above it, called a *sliding window* (represented on Figure 2.2 in the case $m = 0$). A local function defined by a formula like Formula (2.1) with $m = 0$ is called *sequential*. Thus a sequential local function is one that writes below the right end of the window.

A local function is also called a *sliding block code* and a k-local function is also called a *k-block map*. A 1-local function or one-block map is nothing else than an alphabetic substitution (or very fine morphism in [24]).

The following result is well-known. In Hedlund's article [33], it is credited to M. L. Curtis, G. Hedlund, and R. C. Lyndon[1].

Theorem 2.1. *Let $S \subset A^{\mathbb{Z}}, T \subset B^{\mathbb{Z}}$ be two symbolic systems defined over finite alphabets A and B. A function $f : S \subset A^{\mathbb{Z}} \to T \subset B^{\mathbb{Z}}$ is a morphism if it is k-local for some $k \geq 0$.*

Proof. It is clear that a local function is a morphism since it is continuous and commutes with the shift by definition. Conversely, let $f : S \to T$ be a continuous map. Since A is finite, $A^{\mathbb{Z}}$ is compact and so is S which is closed in $A^{\mathbb{Z}}$. Thus f is uniformly continuous. This implies that there is an integer k such that the symbol $f(x)_0$ is determined by the window $x_{-k} \cdots x_0 \cdots x_k$. Since f commutes with the shift the other symbols of $f(x)$ are also determined by the corresponding window of length $2k + 1$. $\qquad \square$

An *isomorphism* (also called a *conjugacy*) is a bijective morphism. If f is an isomorphism from S onto T, then S and T are said to be *conjugate*. Since the alphabet A is finite, the space S is compact. The inverse of a continuous function $f : S \to T$ is also continuous when S is compact. Thus the inverse of a conjugacy is a conjugacy.

As a general rule, the concepts studied in symbolic dynamics are invariant under conjugacy and a lot of the attention is given to the search of *complete invariants*, i.e., invariants that characterize a subshift up to conjugacy.

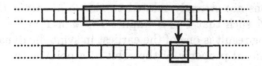

Fig. 2.2. A sequential local function

---

3. Recurrence and minimality

In this section, we concentrate on a special kind of symbolic dynamical systems: the smallest system containing a given infinite word. It is more appropriate to present it in the one-sided case. For a one-sided infinite word $x \in A^{\mathbb{N}}$, we define $F(x)$ to be the set of factors of x. We also define $S(x) = \{y \in A^{\mathbb{N}} \mid F(y) \subset F(x)\}$. The set $S(x)$ is obviously the smallest subshift containing x.

A one-sided infinite word $x \in A^{\mathbb{N}}$ is said to be *recurrent* if any block occurring in x has an infinite number of occurrences. It is obviously enough for x to be recurrent that any prefix of x has a second occurrence.

It is easy to verify that x is recurrent if and only if the subshift $S(x)$ is irreducible. Indeed, if $S(x)$ is irreducible, then for any prefix u of x there is a v such that $uvu \in F(x)$ and thus u has a second occurrence. Conversely, if x is recurrent then for any $u, v \in F(x)$, v has an occurrence following any occurrence of u and thus there is a word w such that $uwv \in F(x)$.

The notion of a recurrent word is linked to the concept of a *sesquipower* (or $(1 + \frac{1}{2})$-power). A word w is called a sesquipower of order n if it can be written $w = uvu$ with u a sesquipower of order $n-1$ and v any word. A sesquipower of order 0 is any nonempty word.

A recurrent word is one that can be written as an infinite *sesquipower*, i.e., as

$$
\begin{aligned}
x &= u_0 \ldots \\
&= u_0 u_1 u_0 \ldots \\
&= u_0 u_1 u_0 u_2 u_0 u_1 u_0 \ldots
\end{aligned}
\tag{3.1}
$$

Indeed, we can choose u_0 to be the first symbol of x. Then, since u_0 has a second occurrence in x, we can find u_1 such that $u_0 u_1 u_0$ is a prefix of x and so on.

A word $x \in A^{\mathbb{N}}$ is said to be *uniformly recurrent* if every block of x appears infinitely often at bounded distance.

A periodic word is obviously uniformly recurrent. A simple way to construct a uniformly recurrent non periodic word is to use words u_i of bounded length in equation (3.1). We shall see examples in more detail below.

These notions are strongly related to that of a *minimal subshift*, i.e., a subshift $S \subset A^{\mathbb{Z}}$ such that $T \subset S$ implies $T = \emptyset$ or $T = S$.

The following result is one of the earliest in symbolic dynamics ([15],[45]). It links a dynamical property of the orbit of a point with a property of the words representing the orbit.

Theorem 3.1. *Let $x \in A^{\mathbb{N}}$ be a one-sided infinite word. The following conditions are equivalent.*

1. x is uniformly recurrent.
2. $S(x)$ is minimal.

Proof. 1 ⇒ 2 Let $S \subset S(x)$ be a subshift and let $y \in S$. Then $S(y) \subset S$. Since $y \in S(x)$, we have $F(y) \subset F(x)$ by the definition of $S(x)$. Let $w \in F(x)$. Since x is uniformly recurrent, w appears in every long enough block of x. Hence w appears in the long enough blocks of y. Whence $w \in F(y)$. This shows that $F(x) = F(y)$ and this implies that $S(y) = S = S(x)$.

2 ⇒ 1 Any block w of x appears in all $y \in S(x)$ since $S(x)$ is minimal. For a given block w of x, we define $i_w(y)$ to be the function assigning to $y \in S(x)$ the least integer i such that $y = uwz$ with $|u| = i$. Since i_w is continuous and $S(x)$ is compact, i_w is bounded. Let w be a block of $x = uwy$. Since $y \in S(x)$, $w \in F(y)$ and thus w has a second occurrence in x at a distance bounded by $i_w(y)$. □

Example 3.1. The word of Thue-Morse is the infinite word obtained by iterating the substitution f defined by $f(a) = ab$, $f(b) = ba$. The word $m = f^\omega(a)$ is uniformly recurrent. Indeed, aaa or bbb are not in $F(m)$. Thus successive occurrences of a or b are separated by at most two symbols. It follows that any block of m appears at bounded distance since it has to appear in some $f^k(a)$ or $f^k(b)$. The system $S(m)$ is known as the *Morse minimal set*.

It is possible to generalize the example of the Morse minimal set as follows. Let $f : A \to A^*$ be a substitution such that for any symbols $a, b \in A$ there is at least an occurrence of b in $f(a)$. Then any infinite word x such that $f(x) = x$ is uniformly recurrent [43].

We used in the proof of Theorem 3.1 a possible variant of the definition of a uniformly recurrent word: for all $n > 0$ there is an $m > n$ such that any factor of length n appears in any factor of length m. This condition can be used as a definition for a uniformly recurrent two-sided infinite word. It also leads to the definition of a function $r_x(n)$ called the *recurrence index* of x. We let $r_x(n) = m$ if m is the smallest possible integer such that any factor of length n appears in any factor of length m. It is well defined for all integers n iff x is uniformly recurrent.

It is worth mentioning yet another equivalent definition of uniformly recurrent words. It uses the notion of a well-quasi-order. Recall (see [39],[37] or [23]) that a partial order on a set X is called a *well-quasi-order* if

1. there are no infinite descending chains
2. Any set of pairwise incomparable elements is finite

Well-quasi-orders are a generalization of well orders which are total orders satisfying condition (1) , or equivalently such that any nonempty subset has a smallest element.

We consider the *factor ordering* on sets of words. It is the partial order defined by $u < v$ if u is a factor of v. The first condition in the above definition is then automatically satisfied since the length of words cannot decrease indefinitely.

We have the following result.

Proposition 3.1. *A recurrent one-sided infinite word x is uniformly recurrent if and only if the factor ordering on the set $F(x)$ is a well-quasi-order.*

Proof. The condition is obviously necessary since the order considered on the set $F(x)$ of factors of a uniformly recurrent word x satisfies the stronger property that the set of elements incomparable with a given $u \in F(x)$ is finite. Conversely, if x is not uniformly recurrent, we can find a sequence $vu_nv \in F(x)$ of words of increasing length such that the only occurrences of v in vu_nv are the two ones as a prefix and a suffix. Then the words vu_nv are incomparable and thus the order is not a well partial order.   □

Note that the factor order on the set A^* itself is not a well-quasi-order (and in fact quite the opposite since it is a maximal set of factors instead of a minimal one). By a classical theorem of Higman (see [39]), it is well-quasi-ordered by a different order, namely the *subword ordering* defined by $u < v$ if $u = u_1u_2\ldots u_n$ and $v = x_0u_1x_1\cdots x_{n-1}u_nx_n$.

In this section, we have only touched the subject of minimal subshifts. There are many other interesting developments in this direction, such as the one of *Sturmian words* (see the Chapter by Aldo De Luca and Stefano Varricchio in this handbook). Minimal subshifts are in a sense at the opposite of the subshifts that we are going to study now. To be more precise, minimal subshifts contain no periodic point unless they are finite, whereas in the sofic subshifts introduced below, the set of periodic points is dense.

4. Sofic systems and shifts of finite type

Recall from Section 2 that a sofic system is a subshift S recognized by a finite automaton. For instance the system given in Figure 4.1 is a sofic system. It consists of all binary sequences such that the length of the blocks of b's between two a's are of even length. This system is sometimes called the *even system*.

A *shift of finite type* is a subshift which is made of all infinite words avoiding a given finite set of blocks. For example, the golden mean system defined in Section 2 as formed by all words on $\{a, b\}$ avoiding bb is a shift of finite type.

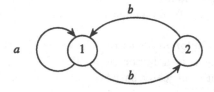

Fig. 4.1. The even system

Shifts of finite type are closely related to a particular and well-known class of finite automata called local automata. We first give their definition.

Proposition 4.1. *Let $\mathcal{A}$ be a finite transitive automaton. The following conditions are equivalent.*

1. *The current state on a path is determined by a bounded number of labels in the past and in the future.*
2. *There is at most one infinite path with a given label.*
3. *There is at most one periodic infinite path with a given label.*

Proof. $1 \Rightarrow 2$. If

$$\cdots p_{n-1} \xrightarrow{a_{n-1}} p_n \xrightarrow{a_n} p_{n+1} \cdots$$

is an infinite path, the current state p_n is determined by (a bounded number of) the symbols $\cdots a_{n-1}a_n \cdots$ and thus the label determines the path.

$2 \Rightarrow 3$ is clear.

$3 \Rightarrow 1$. If condition 1 is not true, there exist by König's lemma two distinct infinite paths $\cdots p_{n-1} \xrightarrow{a_{n-1}} p_n \cdots$ and $\cdots q_{n-1} \xrightarrow{a_{n-1}} q_n \cdots$ with the same label. As the paths are distinct, there is an index n such that $p_n \neq q_n$. Since the automaton is finite, both paths use the same pair of states (p, q) infinitely often before time n, and the same pair of states (r, s) infinitely often after time n. If $p \neq q$ or $r \neq s$, this defines two distinct periodic paths with the same label. If not, we have $p = q$ and $r = s$. As the automaton is transitive, there exists a path from r to q and this again defines two distinct periodic paths with the same label. $\qquad\square$

A finite automaton is said to be *local* if it satisfies condition 1 above (equivalent to 2 and 3 when the automaton is transitive). The bound on the window size corresponding to condition 1 above can actually be shown to be quadratic as a function of the number of states (see, e.g., [7], p. 45).

We recall that an automaton is called *deterministic* if it admits at most one edge leaving a given state and with a given label.

Proposition 4.2. *Let $\mathcal{A}$ be a finite deterministic automaton. The following conditions are equivalent.*

1. *The current state on a path is determined by a bounded number of labels in the past.*
2. *The automaton is local.*

Proof. $1 \Rightarrow 2$ is clear.

$2 \Rightarrow 1$. By definition there exist n and m such that the current state on a path is determined by n symbols in the past and m symbols in the future. Any block of $n + m$ symbols determines the final state. Indeed, on a path

$$p_0 \xrightarrow{a_1} \cdots \xrightarrow{a_n} p_n \xrightarrow{a_{n+1}} \cdots \xrightarrow{a_{n+m}} p_{n+m}$$

the state p_n is determined. But then $p_{n+1}, \ldots, p_{n+m}$ are also determined because the automaton is deterministic. Thus $n + m$ symbols in the past determine the current state.   □

The basic example of a deterministic local automaton is the *standard k-local automaton* or *De Bruijn graph*. Its set of states is the set A^k of words of length k and its edges are the triples (au, b, ub) for $u \in A^{k-1}$ and $a, b \in A$.

The following result shows that shifts of finite type correspond to local automata.

Proposition 4.3. *A shift of finite type is a sofic system. More precisely, a subshift is of finite type if and only if it is recognized by a local automaton.*

Proof. Let S be a shift of finite type defined by a set of forbidden blocks of maximal length k. We use the standard $(k - 1)$-local automaton, erasing all edges (au, b, ub) such that aub contains a forbidden block. In this way, we obtain a finite automaton recognizing S as a sofic system.

Conversely, let $\mathcal{A}$ be a local automaton. By definition, the current state on a path is determined by a bounded number n of symbols in the past and a bounded number m of symbols in the future. Let I be the set of words of length $n + m + 1$ which are not labels of paths of $\mathcal{A}$. The sofic system recognized by $\mathcal{A}$ is then equal to the set of bi-infinite words whose blocks of length $n + m + 1$ avoid I. As a consequence, it is of finite type.   □

We shall see in the next section how this result can be used to check effectively whether a sofic system is of finite type or not.

As an illustration of the above proposition, we may consider again the golden mean system. The set of allowed blocks of length 2 is $\{aa, ab, ba\}$ thus giving the automaton of Figure 4.2.

As a particular class of shifts of finite type, one may define a *Markov shift* as a shift of finite type defined by a set of forbidden blocks of length 2. It is clear that the edge shift S_G on a graph G is a Markov shift. The previous example is also a Markov shift since it can be defined by the set of words on $\{a, b\}$ avoiding bb.

Any shift of finite type S can be obtained, up to conjugacy, as the edge shift of some finite graph. Indeed, in a local automaton the map from edges to labels is a 1-block conjugacy from the edge shift onto S.

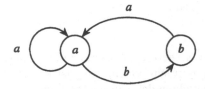

Fig. 4.2. The golden mean system

We will prove that both the notion of a shift of finite type as well as that of a sofic system, are invariant under conjugacy. We begin with shifts of finite type.

Proposition 4.4. *Any conjugate of a shift of finite type is of finite type.*

Proof. Let $S \subset A^{\mathbb{Z}}$ be a shift of finite type conjugate by f to $T \subset B^{\mathbb{Z}}$. By definition, S is the set of words avoiding a finite set of blocks $I \subset A^+$. We may assume that all words of I have the same length s. The map f^{-1} is a k-local map and thus there is a k-local map $g : B^{\mathbb{Z}} \to A^{\mathbb{Z}}$ such that $f^{-1} = g$ on T. Up to some composition with the shift, we may suppose that g is sequential.

Let $J = B^{k+s-1} - F_T$ be the complement of the set of words of length $(k + s - 1)$ that appear as factors of T. We claim that T is the set of words avoiding the finite set J, which proves that T is of finite type.

First it is clear that T avoids the set J. Conversely, let y be an infinite word with all factors of length $(k + s - 1)$ belonging to the complement of J. Let $x = g(y)$. We claim that x belongs to S and therefore that $y \in T$. Let $x_i x_{i+1} \ldots x_{i+s-1}$ be a factor of x of length s. The factor $y_{i-(k-1)} \cdots y_i y_{i+1} \cdots y_{i+s-1}$ of y of length $(k + s - 1)$ is a factor of a word y' of T. By definition of the k-local map f^{-1}, the factor of x equal to $x_i x_{i+1} \ldots x_{i+s-1}$ is also a factor of the word $f^{-1}(y') \in S$. It then does not belong to I. We finally get that x belongs to S and $y = f(x)$ belongs to T. □

Before proceeding to prove further properties of shifts of finite type and sofic systems, we define a useful way to realize a morphism between subshifts.

A (synchronous) *transducer* on $A \times B$ is a finite automaton (Q, E) with edges labeled by $A \times B$. If (p, a, b, q) is an edge, we say that a is the *input label* and b the *output label*. We will only consider here synchronous transducers instead of the more general notion of a transducer in which the edges are labeled by pairs of words on A, B. We shall also only use transducers such that if two distinct edges (p, a, b, q) and (p, a', b', q) have the same origin p and end q then $a \neq a'$ and $b \neq b'$.

The hypothesis made on transducers implies that, by removing the input labels or the output labels, we get an automaton. The automaton we get by removing the input labels is called the *output automaton* of the transducer and the automaton we get by removing the output labels is called the *input automaton* of the transducer.

Let $f : S \to T$ be a morphism from a subshift S into a subshift T. A transducer $\mathcal{T}$ is said to *realize* f if for all $x \in S$, there is a path with input label x and all of them have output label $y = f(x)$ (we admit the possibility of several paths with input label x).

A morphism $f : S \to T$ between two subshifts $S \subset A^{\mathbb{Z}}$ and $T \subset B^{\mathbb{Z}}$ can be realized by a transducer $\mathcal{T} = (Q, E)$ such that the input automaton is local. Let in fact k be such that f is k-local. Up to some composition with a power of the shift, we may assume for simplicity that f is sequential. The set Q of states of T is the set of factors of S of length $(k-1)$, and there is an edge between au and ub labeled $(b, f(aub))$.

If S is moreover of finite type, we may choose k large enough to ensure that S is recognized by the input automaton of T.

For example, the transducer of Figure 4.3 realizes the morphism coding the overlapping blocks of length 2 over the binary alphabet $A = \{a, b\}$ by a symbol from the alphabet $B = \{x, y, z, t\}$.

The transducer that we have associated with a morphism is such that the input automaton is local since it is part of a De Bruijn graph. The following proposition gives a practical method to check whether a morphism between shifts of finite type is a conjugacy.

Proposition 4.5. *Let S be a shift of finite type and let $f : S \to T$ be a morphism from S onto a subshift T. Let T be a transducer realizing f and such that its input automaton is a local automaton recognizing S. Then f is a conjugacy iff the output automaton of T is local.*

Proof. If both the input and the output automaton are local, then f is one-to-one and therefore a conjugacy. Conversely, if the output automaton is not local, there exist two distinct paths with the same output label and thus f is not one-to-one.   □

The following proposition is analogous to the well-known result that the class of rational languages is the closure of local languages under substitutions (Medvedev's theorem, see, e.g., [24], p. 27).

Proposition 4.6. *The factors of shifts of finite type are the sofic systems.*

Proof. A sofic system is by definition recognized by a finite automaton. As such, it is a factor of the edge graph associated with the automaton.

Conversely, let $f : S \to T$ be a morphism from a shift of finite type S onto a subshift T. Up to some composition of f with a power of the shift, we may realize f by a transducer T which may also be chosen such that its input automaton recognizes S. Then the output automaton of T recognizes T which is therefore sofic.   □

We finally obtain the desired result on the invariance under conjugacy of the class of sofic systems as a corollary of the above statement.

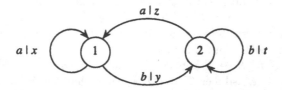

Fig. 4.3. A 2-block map

Proposition 4.7. *Any conjugate of a sofic system is sofic.*

Proof. Let S be a sofic system conjugate to T by f. The sofic system S is a factor of some system of finite type U by an onto morphism g. Then T is the image of U by the morphism $f \circ g$ which proves that it is sofic.   □

5. Minimal automaton of a subshift

The close connection between a subshift S and the set F_S of its finite blocks leads to a possibility of studying the same objects from both the points of view of symbolic dynamics and of finite automata. As a first example of a result with equivalent formulations in terms of symbolic dynamics and in terms of finite automata, the existence of a unique minimal deterministic automaton takes the following form for sofic systems.

We recall that an automaton is said to be transitive if its graph is strongly connected. If $\mathcal{A} = (Q, E)$ is a deterministic automaton, we often denote by $p \cdot a$ the unique state q such that $(p, a, q) \in E$ if it exists. The notation is extended to words. Thus a deterministic automaton is transitive iff for any states p, q there exists a word x such that $p \cdot x = q$.

Proposition 5.1. *Any sofic system can be recognized by a deterministic automaton. The system is irreducible iff the automaton can be chosen transitive.*

Proof. By Proposition 2.2 a sofic system S can be recognized by any automaton recognizing the set F_S. It follows that this automaton can be chosen to be deterministic. If the automaton is transitive, the system is clearly irreducible. Conversely, we consider a deterministic automaton $\mathcal{A} = (Q, E, i, Q)$ recognizing the set F_S of factors of an irreducible sofic system S. Let C be a maximal connected component accessible from i of the automaton (here maximal means that any edge starting in C also ends in C). Then C is a deterministic transitive automaton recognizing S. Indeed, any label of a finite path of C is in F_S. Conversely, let w be a word of F_S and u be the label of a path from i to a state of C. As S is irreducible, there exists a word v such that uvw belongs to F_S. Since the automaton C is transitive and $\mathcal{A}$ is deterministic, we get that w is the label of a path of C. The set of labels of finite paths in C is F_S and thus the automaton C recognizes S.   □

A *reduction* from an automaton $\mathcal{A} = (Q, E)$ onto an automaton $\mathcal{B} = (R, F)$ is a surjective mapping $\rho : Q \to R$ such that $(p, a, q) \in E$ iff $(\rho(p), a, \rho(q)) \in F$. We will show that an irreducible sofic system S has a unique minimal automaton $\mathcal{A}_S$, in the sense that for any transitive automaton $\mathcal{B}$ recognizing S, there is a reduction from $\mathcal{B}$ onto the minimal automaton $\mathcal{A}_S$. In particular, the automaton $\mathcal{A}_S$ has the minimum possible number of states. This result is due to Fischer [25] and the minimal automaton is also called the Fischer cover. It was also obtained independently by D. Beauquier [11].

Proposition 5.2. *Any irreducible sofic system has a unique minimal automaton.*

The proof of this result does not follow immediately from the corresponding well-known statement for ordinary finite automata because of the absence of an initial state. It relies on the notion of a synchronizing word which allows one to fix an initial state. Let S be a non empty irreducible sofic system and let F_S be its set of finite factors. A word x of F_S is a *synchronizing* word of S iff for all words u, v

$$ux, xv \in F_S \Rightarrow uxv \in F_S.$$

Let $\mathcal{A} = (Q, E)$ be a deterministic automaton. For a finite word $x \in A^*$, we define the *rank* of x as the cardinality of the set $Q \cdot x = \{q \cdot x \mid q \in Q\}$.

Proposition 5.3. *Any non-empty irreducible sofic system admits a synchronizing word. In fact any word of minimal nonzero rank is synchronizing.*

Proof. Let x be a word of F_S of minimal nonzero rank. If u is a word such that $ux \in F_S$, then $\emptyset \neq Q \cdot ux \subset Q \cdot x$. By minimality of the rank of x, this implies that $Q \cdot ux = Q \cdot x$. Let $ux, xv \in F_S$. Let $p \xrightarrow{x} q \xrightarrow{v} r$ be a path of label xv. Since $Q \cdot x = Q \cdot ux$, there is a path $s \xrightarrow{ux} q$ and thus a path

$$s \xrightarrow{ux} q \xrightarrow{v} r$$

We conclude that $uxv \in F_S$ and thus x is synchronizing.   □

Proof. of Proposition 5.2. We choose a synchronizing word x of S. Let $\mathcal{A}'$ be the minimal automaton of the set of finite words $x^{-1}F_S = \{y \mid xy \in F_S\}$.

We denote by $\mathcal{A}$ the automaton obtained from $\mathcal{A}'$ by allowing all states to be both initial and terminal. The automaton $\mathcal{A}$ recognizes S. Indeed any label of a path in $\mathcal{A}$ is clearly in F_S. Conversely, let $y \in F_S$. Since S is irreducible, there exists a word u such that $xuy \in F_S$. Thus $uy \in x^{-1}F_S$ showing that y is the label of some path in $\mathcal{A}$.

Let now $\mathcal{B}$ be any transitive deterministic automaton recognizing S. The automaton $\mathcal{B}'$ obtained from $\mathcal{B}$ by choosing as initial state a state i in $Q \cdot x$ and all states as terminal states, recognizes the set $x^{-1}F_S$. Indeed, it is clear that the language L recognized by $\mathcal{B}'$ is included in $x^{-1}F_S$.

Conversely, let y be a word of $x^{-1}F_S$. Let z be a word which has nonzero minimal rank in $\mathcal{B}$. As $\mathcal{B}'$ is a transitive automaton, there exists a word u such that zux is a label of a path of $\mathcal{B}'$ leading to state i. Since x is a synchronizing word,

$$zux \in F_S, xy \in F_S \Rightarrow zuxy \in F_S.$$

Then the ranks of $zuxy$, zux, and z are the same. Thus y is the label of a path of $\mathcal{B}'$ beginning at i, since otherwise $zuxy$ would have a nonzero rank strictly smaller than the rank of zux.

We now use the known result that a recognizable set of finite words has a unique minimal automaton. Let $\mathcal{A}'$ be the minimal automaton of the set $x^{-1}F_S$. Since $\mathcal{B}'$ recognizes $x^{-1}F_S$, there is a reduction from $\mathcal{B}'$ onto $\mathcal{A}'$. Thus there is a reduction from $\mathcal{B}$ onto $\mathcal{A}$.   □

The minimal automaton is used to characterize different classes of sofic systems like aperiodic systems [7], almost-of-finite-type shifts [42], or shifts of finite type. We give the result for shifts of finite type.

Proposition 5.4. *A sofic system is of finite type if and only if its minimal automaton is local.*

Proof. By Proposition 5.2, if the sofic system admits a local minimal automaton, it is of finite type. Conversely, by the same proposition, a shift of finite type is recognized by a deterministic local automaton $\mathcal{A}$. The minimal automaton $\mathcal{A}_S$ itself is obtained by a reduction from the local automaton $\mathcal{A}$. A reduction transforms a local automaton into a local one since a fixed number of symbols determine the current state in $\mathcal{A}$ and thus in $\mathcal{A}_S$.   □

There is a definition of deterministic automaton which is more abstract and which we introduce now.

Let S and T be two subshifts with S of finite type. Let $f : S \to T$ be a sequential local function and let $\mathcal{T}$ be a transducer realizing f with a local input automaton recognizing S. The morphism f is said to be *right-resolving* if the output automaton is deterministic.

The following statement, whose proof is straightforward, shows that one can define a right-resolving function directly, without reference to the transducer.

Proposition 5.5. *A sequential k-local function f is right-resolving iff for $y = f(x)$, the values of the block $x_{-k+1} \cdots x_{-1}$ and of the symbol y_0 determine the value of the symbol x_0.*

□

Right-resolving morphisms belong to a broader family of almost one-to-one morphisms called finite-to-one and introduced in Section 6.

A *right-resolving cover* of a sofic system S is a right-resolving morphism $f : T \to S$ from a shift of finite type T onto S. The minimal automaton of a sofic system S defines a right-resolving cover $f : T \to S$ of S which is

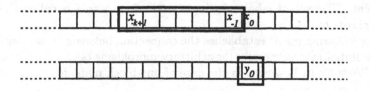

Fig. 5.1. A right-resolving map

minimal in the sense that for any other right-resolving cover $g : U \to S$ the subshift T is a factor of U.

6. Codes and finite-to-one maps

In this section, we are going to study the relationship between two notions: finite-to-one maps on the one hand and codes on the other hand (on codes, see also the chapter by Helmut Jürgensen in this handbook). We will show that the close connection between both notions allows one to prove new results and also to give new and simpler proofs of old ones. We begin with the definition of a finite-to-one map.

A morphism $f : S \to T$ between two subshifts S and T is said to be *finite-to-one* if, for all $y \in T$, the set $f^{-1}(y)$ is finite.

We shall see below that when S is an irreducible shift of finite type, a finite-to-one map is actually *bounded-to-one* in the sense that there is a constant n such that each point of T has at most n pre-images.

We now come to the concepts of codes and unambiguous automata, which are related to the notion of finite-to-one maps.

For a set X of finite words, we denote by X^* the set of all concatenations $x_1 x_2 \ldots x_n$ with $n \geq 0$ and $x_i \in X$.

A set X of finite words is called a *code* if no non-trivial equality holds between the words of X^*. In more precise terms, X is a code iff

$$u_1 u_2 \ldots u_n = v_1 v_2 \ldots v_m,$$

where $u_i, v_j \in X$, implies $n = m$ and $u_i = v_i$ for each index i.

As an example of a code, a *prefix code* is a set X such that no element of X is a prefix of another element of X.

An automaton is said to be *unambiguous* if two paths with the same origin state, the label, and the same final state, are equal. A deterministic automaton is unambiguous since the origin state and the label are sufficient to determine the path. As another particular case, a transitive local automaton is also unambiguous. Indeed, if two distinct paths have the same origin and end, and the same label, we can build two distinct cycles with the same label. This contradicts the hypothesis that the automaton is local.

Let $\mathcal{A} = (Q, E)$ be a transitive automaton and let $i \in Q$ be a particular state. A path from i to i is called *simple* if it does not use i between its endpoints. The set of labels of simple paths from i to i is called the *set of first returns to i*.

The following result establishes the connection between these concepts. It can be stated more generally for arbitrary morphisms between shifts of finite type. We state it, however, in the case of one-block maps for simplicity.

Proposition 6.1. *Let $\mathcal{A} = (Q, E)$ be a transitive automaton, $i \in Q$ and let X be the set of first returns to i. The following conditions are equivalent.*

1. *The automaton $\mathcal{A}$ is unambiguous.*
2. *X is a code and distinct simple paths from i to i are labeled by distinct elements of X.*
3. *The map going from bi-infinite paths in the automaton to their labels is finite-to-one.*

Proof. It is clear that 1 and 2 are equivalent.

Further, if the automaton is ambiguous, the map f going from paths in the automaton to their labels is not finite-to-one. Thus $3 \Rightarrow 1$.

Finally, if the map f is not finite-to-one, there exists an infinite number of infinite paths having the same image by f. As there is only a finite number of states, an infinity of these paths go through a same state p after the edge of index 0. We can assume that an infinity of them are distinct after the index 0. Let us take $n+1$ of them, where n is the number of states of the automaton. Let us assume that they can all be two by two distinguished before the edge of index m, where m is a positive integer. At least two of them go then through the same state q after this edge and we get two equally labeled paths going from p to q. Thus $1 \Rightarrow 3$. □

An easy consequence of this result is that a finite-to-one map f from an irreducible shift of finite type S to T is really bounded-to-one.

Proposition 6.2. *Let $f : S \to T$ be a finite-to-one map realized by a transducer with n states. The number of pre-images of an element of T is bounded by n^2.*

Proof. Let x be an element of T. For each $m \geq 1$ there are at most n^2 paths $p_m \xrightarrow{w_m} q_m$ labeled by $w_m = x_{-m} \cdots x_m$ since such a path is determined by the pair (p_m, q_m). Thus x has at most n^2 pre-images. □

The connection between codes and subshifts can be considered independently of automata. Let indeed X be a code and let S be the subshift $S_{F(X^*)}$ formed by all bi-infinite words having all its factors in the set $F(X^*)$. One then has the following statement.

Proposition 6.3. *Let $\mathcal{A}$ be an unambigous automaton such that X is the code of first returns to some state of $\mathcal{A}$. Then S is the subshift recognized by $\mathcal{A}$. If X is finite, then S is the set of bi-infinite words having at least one factorization in words of X.* □

It is of course still true that the set of bi-infinite words having a factorization in words of X is a subshift, even if X is a set of words which is not a code, provided it is finite. Such a system is sometimes called the *renewal system* generated by X.

Example 6.1. If $X = \{a, ba\}$, then S is the golden mean system of Figure 4.2

A code $X \subset A^+$ is said to be *maximal* if it is maximal for set-inclusion, that is if $X \subset Y$ for a code Y implies $X = Y$. It is known that a rational code $X \subset A^*$ is maximal iff it is *complete*, i.e. if the set $F(X^*)$ of factors of words of X^* is equal to A^* (see [13] p. 68). A result, due to Ehrenfeucht and Rozenberg, says that any rational code is included in a maximal one (see [13] p. 62).

An analogous proof can be used to obtain the following result [4].

Theorem 6.1. *If S, T are irreducible shifts of finite type, and $f : S \to T$ is a finite-to-one morphism, then there is an irreducible shift of finite type U containing S and a finite-to-one morphism from U onto T extending f.*

We make two comments about this statement before indicating its proof.

First, the fact that the larger subshift U is required to be irreducible is essential in the statement which would be otherwise trivial: it would be enough to take U to be a disjoint union of S and a copy of T, and to define f on U as being the identity on T.

Second, the link with the theorem of Ehrenfeucht and Rozenberg is the following. Consider the particular case of the map f: paths $\mapsto$ labels in a transitive unambiguous automaton $\mathcal{A}$. Thus f is a finite-to-one morphism from the edge shift S on $\mathcal{A}$ into the full shift $T = A^{\mathbf{Z}}$. Let X be the code defined by the first returns to some state q of $\mathcal{A}$. An embedding $X \subset Y$ of X into a maximal rational code Y can always be obtained by adding states and edges to $\mathcal{A}$ in such a way that Y is the set of first returns to q in the new automaton $\mathcal{B}$.

The set of labels in $\mathcal{B}$ is thus equal to A^* since it is the set of factors of the complete code Y. Thus embedding X into a rational maximal code Y corresponds to an extension of the finite-to-one map f to a larger subshift in such a way that it becomes surjective.

We now give an indication of the proof of Theorem 6.1.

Proof of Theorem 6.1. We first make the hypothesis that $S \subset A^{\mathbf{Z}}$ and $T \subset B^{\mathbf{Z}}$ are Markov shifts. This is true up to conjugacy and thus we may make this hypothesis. We also suppose that f is a one-block map. This is true again up to a conjugacy. Under these assumptions, we may realize f as the map paths $\mapsto$ labels in an automaton $\mathcal{A}$ whose set of states is $Q = A$. Also, T is recognized by an automaton $\mathcal{B}$ with set of states equal to B.

Let $q \in Q$ be a particular state of $\mathcal{A}$. Let X be the image under f of the set of first returns to q in $\mathcal{A}$. Let $b = f(q)$ and Y be the set of first returns to b in $\mathcal{B}$. Thus $X \subset Y^*$. Moreover Y^* and X^* are rational subsets of B^*.

If $T = f(S)$, there is nothing to prove. Otherwise, there is a word $y \in Y^*$ such that

1. y is unbordered, i.e. y has no non-trivial prefix which is also a suffix.
2. $y \notin F_{f(S)} = f(F_S)$

The construction of such a word y follows the same lines as the analogous construction in [13], p. 64.

Let $Z \subset B^*$ be the set

$$Z = Y^* - X^* - B^*yB^*$$

We build an irreducible shift of finite type U containing S and an extension g of f by considering the automaton C of Figure 6.1 where the component called Z is actually a finite automaton recognizing the set Z (with i as initial state and t as final state). The subshift U is the set of infinite paths in G and the function g is the map paths $\mapsto$ labels in C.

Then the extension g of f to U satisfies the following properties.

1. g maps U into T.
2. g is finite-to-one.
3. g is onto.

Thus g is a finite-to-one morphism from U onto T extending f. This completes the sketch of the proof of Theorem 6.1. □

The theorem of Ehrenfeucht and Rozenberg corresponds to the case where T is the full shift. The more general case where a code X is constrained to be included in a fixed factorial set F has been studied by A. Restivo who has shown that the results known previously extend to this case [53].

The paper [4] contains other results of the same kind. One of them deals with right-closing morphisms, a notion intermediate between finite-to-one and right-resolving which is defined precisely as follows.

A function $f : S \to T$ is *right-closing* if whenever $x, y \in S$ have a common left-infinite tail and $f(x) = f(y)$ then $x = y$ (see Figure 6.2).

Right-closing morphisms correspond to automata with bounded delay in the same way as right-resolving morphisms correspond to deterministic automata. The result on right-closing morphisms proved in [4] corresponds to the result on codes with bounded deciphering delay proved by V. Bruyère, L.

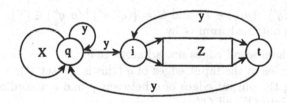

Fig. 6.1. The resulting transducer

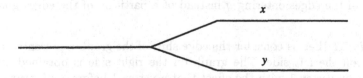

Fig. 6.2. Two left-asymptotic words x and y

Wang, and L. Zhang in [21]: any rational code with finite deciphering delay can be embedded into a rational maximal one (see also [20]).

Finally, the paper [4] contains an analogous result on extending morphisms belonging this time to the class of *biclosing* morphisms, i.e., morphisms that are both left and right closing. This is related to a result proved recently by L. Zhang: any rational biprefix code can be embedded into a rational maximal one [62].

7. State splitting and merging

We have seen that one may associate with every irreducible sofic system a minimal automaton that recognizes it. The computation of such a minimal automaton may be performed using one of the standard algorithms for automaton minimization, such as Moore's algorithm or Hopcroft's algorithm (see [3] for example). All minimization algorithms consist in some kind of state identification since the states of the resulting minimal automaton are equivalence classes of states (equivalent states are those with the same future). In this section, we introduce an operation on symbolic systems, called state merging which allows one to identify states of the automaton recognizing a sofic system S in such a way that the resulting system is conjugate to S. The inverse operation is called state splitting. These concepts are due to F. Williams [61]. We first define the operation on the edge shift of a graph.

Let $G = (Q, E)$ be a graph. Let $q \in Q$ and let I (resp. O) be the set of edges entering q (resp. going out of q). Let $O = O' + O''$ be a partition of O. The operation of *(output) state splitting* relative to (O', O'') transforms G into the automaton $G' = (Q', E')$ where $Q' = Q \cup q'$ is obtained from Q by adding a new state q' and E' is defined as follows.

$$E' = E - O' + I' + U$$

with $I' = \{(p, q') \mid (p, q) \in I\}$ and $U = \{(q', q'') \mid (q, q'') \in O'\}$.

Thus G' is obtained from G by

1. leaving unchanged all edges not adjacent to q.
2. giving q' copies of the input edges of q (this is the set I').
3. distributing the output edges of q between q' and q according to the partition of O into O' and O''.

The operation of input state splitting is analogous. It uses a partition $I = I' + I''$ of the edges entering q instead of a partition of the edges going out of q.

Example 7.1. Let us consider the edge shift of the graph represented on Figure 7.1 on the left side. The graph on the right side is obtained by state splitting on vertex 1 with the effect that a vertex 1 before a 2 is transformed into a 3 . The edges going out of 1 are split into two parts: the loop on 1 on

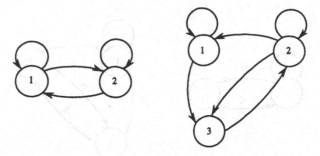

Fig. 7.1. An output split of state 1

the first hand remains unchanged. The edge going from 1 to 2 becomes an edge from 3 to 2. The edges incoming at 1 and 3 are identical.

The operation of *(output) state merging* is the inverse of that of (output) state splitting. Formally, let $G = (Q, E)$ be a graph and let $q, q' \in Q$ be two states such that the edges coming into q and q' are the same (except for the end). We merge q and q' in a single state q having the same input edges as the former state q (with loops (q, q) for the edges of the form (q', q)). The output edges are obtained as the union of those of q and q'.

Finally, the operation of input state merging is the inverse of input state splitting. It is the operation linked with the minimization of automata, as we shall see shortly.

The following statement is easy to prove.

Proposition 7.1. *Let G' be obtained from G by state splitting. Then the edge graphs S_G and $S_{G'}$ are conjugate. More precisely, a state splitting is a 2-block map whose inverse (the merge) is a 1-block map.*

Proof. It is clear that the merge is a 1-block map. Consider now an output split at q according to a partition $O = O' + O''$ of the set O of edges going out of q. The image of an edge is itself unless it is in I or O'. In the first case it is transformed into an edge of I' if it is followed by an edge of O'. In the second case, it is transformed into an edge of U. Thus a sliding window of length 2 is enough to perform the transformation.   □

Let S be a sofic system and let $\mathcal{A} = (Q, E)$ be an automaton recognizing S. The operations of state splitting and merging on the graph transfer to operations on the automaton. In an output split, the labels of the edges coming into and going out of q are transferred to the edges incident to the new state q'. More precisely, we have

$$I' = \{(p, a, q') \mid (p, a, q) \in I\} \text{ and } U = \{(q', a, q'') \mid (q, a, q'') \in O'\}.$$

Example 7.2. Let $\mathcal{A}$ be the automaton with two states represented on the left of Figure 7.2.

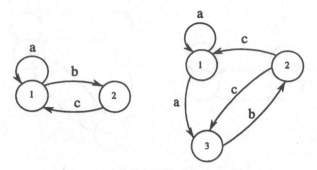

Fig. 7.2. An output split

There are two edges going out of state 1: $(1, a, 1)$ and $(1, b, 2)$. We transfer the second one to a new state called 3 which receives the same input edges as state 1. The result is represented on the right side of Figure 7.2.

Let $\mathcal{A}$ be a deterministic automaton. A sequence of input state mergings produces a deterministic automaton $\mathcal{B}$ with fewer states than $\mathcal{A}$ and still equivalent, i.e. recognizing the same subshift. However, it is not true in general that the minimal automaton $\mathcal{C}$ can be reached in this way. This is illustrated by the following example.

Example 7.3. Let us consider the automaton $\mathcal{A}$ of Figure 7.3. The two states cannot be merged since they have distinct output edges (taking the labels into account).

Actually, the map from paths to labels is 2-to-1 and the subshift recognized is the full shift. If it were possible to reach the minimal automaton with splits and merges, the result would be a conjugacy between paths and labels.

There is however one interesting case where the minimization can be obtained by merges.

Proposition 7.2. *Let $\mathcal{A}$ be a transitive and deterministic automaton. If $\mathcal{A}$ is local, there is a sequence of state merges that transforms $\mathcal{A}$ into a minimal equivalent automaton.*

Proof. We recall from Section 5 that the minimal automaton can be computed by an identification of states having the same future i.e. the same set

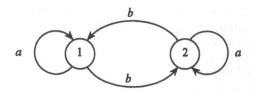

Fig. 7.3. A non minimal automaton

$F_q = \{w \in A^* \mid q \cdot w$ is well-defined$\}$.We suppose that $\mathcal{A}$ is not minimal and thus that there are distinct states q, q' with the same future. Let x be a word of maximal length such that $q \cdot x \neq q' \cdot x$. Then for all $a \in A$, we have $(q \cdot x) \cdot a = (q' \cdot x) \cdot a$ and thus $q \cdot x, q' \cdot x$ can be merged. $\square$

The following result is due to F. Williams [61].

Theorem 7.1. *Any conjugacy between shifts of finite type can be obtained, up to a renaming of the symbols, as a composition of splits and merges.*

Proof. We first consider the particular conjugacy which is the coding by overlapping blocks of fixed length, say k. This particular map can certainly be obtained by a series of splits. This allows us to obtain the shift map itself since it can be obtained through a coding in blocks of length 2.

It is therefore enough to prove that we can obtain a 1-block map whose inverse is sequential as a composition of splits and merges. Such a map is the map from paths to labels in a deterministic local automaton $\mathcal{A}$. Let k be such that k symbols in the past determine the current state. We can, up to a coding by blocks of length $k + 1$, make the labels all distinct. The result is a 1-block map whose inverse is also 1-block. Such a map is a renaming of the symbols. $\square$

Theorem 7.1 shows in particular that the group of automorphisms of a shift of finite type is generated by splits and merges, but through possibly larger shifts. For a primitive shift of finite type, the automorphism group contains every finite group [33]. It is not known whether, on a finite alphabet, it is generated by the shift and its elements of finite order, although this has been conjectured (on automorphisms see [19] or [46]).

The operation of state splitting plays an important role in the applications of symbolic dynamics to coding (see [41], [2]). In the next section, we shall see how it is related to the isomorphism of shifts of finite type.

8. Shift equivalence

In this section, we discuss the problem of the conjugacy of shifts of finite type. In particular, we shall give an algebraic formulation of the equivalence in terms of matrices. In fact, R. F. Williams [61] introduced two equivalence relations on matrices allowing one to formulate the relation of conjugacy on the subshifts in algebraic terms.

Two square matrices M, N with nonnegative coefficients are said to be *elementary shift equivalent* if there exist two nonnegative integral matrices U, V such that

$$M = UV, \qquad N = VU \tag{8.1}$$

Note that M and N may have different dimensions.

Then M and N are called *strong shift equivalent* if there is a chain of elementary shift equivalences between M and N.

Let $S = S_G, T = T_H$ be two edge shifts given by the adjacency matrices M, N of the graphs G, H. We then have the following result.

Theorem 8.1. (Williams [61]) *Two shifts of finite type S and T given by the matrices M, N as above are conjugate iff the matrices M, N are strong shift equivalent.*

Proof. Let first S and T be conjugate. By Theorem 7.1, there exists a sequence of splits and merges transforming S into T. It is therefore enough to prove that splits and merges correspond to shift equivalences on matrices. We consider an output split of a state q. We suppose that q corresponds to the last index of M. Then

$$M = \left[\begin{array}{c} M' \\ x+y \end{array} \right], \qquad N = \left[\begin{array}{ccc} N' & z & z \end{array} \right]$$

where x, y are row vectors, z is a column vector and the decomposition of the last row of M corresponds to the partition of the edges going out of q. We have $M = UV$ and $N = VU$ where

$$U = \left[\begin{array}{ccccc} 1 & 0 & & & 0 \\ 0 & \ddots & & & 0 \\ & & 1 & & 0 \\ & & & \ddots & 0 \ 0 \\ & & & 0 & 1 \ 1 \end{array} \right], \qquad V = \left[\begin{array}{c} M' \\ x \\ y \end{array} \right] = \left[\begin{array}{cc} N' & z \end{array} \right]$$

Thus conjugate subshifts have strong shift equivalent matrices. We prove the converse in the case where the matrices M and N have coefficients 0 or 1 or, equivalently when G and H are ordinary graphs. The general case is not substantially more difficult but the notation is more cumbersome.

Since neither G or H has multiple edges, we may consider S and T as formed by sequences of vertices instead of sequences of edges. Let $f : S \to T$ be the function defined as follows. For $x \in S$ and $n \in Z$, if $x_n = i, x_{n+1} = j$ then (i, j) is an edge of the graph G. Since $M = UV$ there is exactly one vertex k of H such that $U_{ik} = V_{kj} = 1$. We define a 2-block map by $f(i, j) = k$. The inverse is obtained in the same way and S, T are thus conjugate.   □

Example 8.1. For instance, in Example 7.1, the adjacency matrices M and N of the graphs are elementary shift equivalent since

$$M = \left[\begin{array}{cc} 1 & 1 \\ 1 & 1 \end{array} \right] = \left[\begin{array}{ccc} 1 & 0 & 1 \\ 0 & 1 & 0 \end{array} \right] \left[\begin{array}{cc} 1 & 0 \\ 1 & 1 \\ 0 & 1 \end{array} \right],$$

$$N = \left[\begin{array}{ccc} 1 & 0 & 1 \\ 1 & 1 & 1 \\ 0 & 1 & 0 \end{array} \right] = \left[\begin{array}{cc} 1 & 0 \\ 1 & 1 \\ 0 & 1 \end{array} \right] \left[\begin{array}{ccc} 1 & 0 & 1 \\ 0 & 1 & 0 \end{array} \right]$$

The relation of strong shift equivalence is not easy to compute because in a chain $(M_1, M_2, \ldots, M_n)$ of elementary equivalences, the dimensions of the matrices M_i are not a priori bounded. It is not known whether it is recursively computable or not.

We now come to the second equivalence relation on square matrices. Two square matrices M and N are called *shift equivalent*, denoted $M \sim_k N$, if there exist two nonnegative integral matrices U, V and an integer k such that

$$
\begin{array}{ccccc}
MU & = & UN, & NV & = & VM \\
M^k & = & UV, & N^k & = & VU
\end{array}
\tag{8.2}
$$

The relation $\sim_k$ is transitive. Let indeed $M \sim_k N$ and $N \sim_l P$, let $MU = UN, NV = VM, M^k = UV, N^k = VU$ and $NR = RP, PS = SN, N^l = RS, P^l = SR$. Then

$$
\begin{array}{ccccc}
MUR & = & URP, & PSV & = & SVM, \\
M^{k+l} & = & (UR)(SV), & P^{k+l} & = & (SV)(UR)
\end{array}
$$

The integer k is called the *lag* of the equivalence.

It is clear that $k = 1$ corresponds to an elementary shift equivalence and thus that strong shift equivalence implies shift equivalence.

The converse was posed by Williams as a problem: does shift equivalence imply strong shift equivalence? It was shown by Kim and Roush in [35] that the answer is negative in general. However the subshifts of their counterexample are reducible and the conjecture is still pending for irreducible shifts of finite type.

The definition of shift equivalence given above asks for the existence of nonnegative integral matrices U, V such that equations 8.2 are satisfied. If we only require that U, V have integer coefficients (possibly negative), we get the a priori weaker notion of *shift equivalence over* $\mathbb{Z}$.

Both notions coincide however for primitive matrices, i.e. matrices such that the associated shifts are primitive (Parry and Williams [48]).

Proposition 8.1. *Two primitive integral matrices are shift equivalent iff they are shift equivalent over* $\mathbb{Z}$.

It should be noted that the index k in equation (8.2) can be larger over $\mathbb{N}$ than over $\mathbb{Z}$ as shown in the following example.

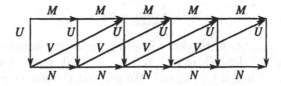

Fig. 8.1. Shift equivalence (lag 2)

Example 8.2. Let

$$M = \begin{bmatrix} 1 & 3 \\ 2 & 1 \end{bmatrix}, \qquad N = \begin{bmatrix} 1 & 6 \\ 1 & 1 \end{bmatrix}$$

Then M and N are similar over $\mathbb{Z}$ since if

$$P = \begin{bmatrix} 2 & 3 \\ 1 & 1 \end{bmatrix}, \qquad P^{-1} = \begin{bmatrix} -1 & 3 \\ 1 & -2 \end{bmatrix}$$

then $M = P^{-1}NP$. Thus M, N are shift equivalent over $\mathbb{Z}$. They are thus shift equivalent. We may indeed choose $k = 3$ and $U = P^{-1}N^3, V = P$.

It has been shown by Kirby Baker that the matrices M and N are indeed strong shift equivalent (see [38], p. 238). However the least number of pairs of elementary shift equivalent matrices used to go from M to N is 7.

The matrices M and N are particular cases of the more general case:

$$M = \begin{bmatrix} 1 & k \\ k-1 & 1 \end{bmatrix}, \qquad N = \begin{bmatrix} 1 & k(k-1) \\ 1 & 1 \end{bmatrix}$$

It is easy to see that these matrices are shift equivalent over $\mathbb{Z}$ and thus over $\mathbb{N}$. However, it is not known whether they are strong shift equivalent.

It is interesting to note that shift equivalence has been proved decidable by Kim and Roush [34]. Also, the problem of comparing, inside a given semigroup, the various relations generalizing the group conjugacy has been studied by several authors (see [22]). It is however a different problem here since the relation is defined among square matrices of different dimensions and not inside a semigroup.

9. Entropy

The notion of entropy in information theory has its root in the work of Shannon. It is defined as a measure of uncertainty and depends on the use of probabilities. Its use in symbolic dynamics, under the name of *topological entropy*, is independent of probabilities. It is an invariant under conjugacy as we shall now see.

The *entropy* of a nonempty subshift S is the limit

$$h(S) = \lim_{n \to \infty} \frac{1}{n} \log s_n$$

where s_n is the number of blocks of length n appearing in the elements of S.

This limit is well defined. In fact, if S is a subshift, we have $s_{n+m} \leq s_n s_m$. Then $\log(s_{n+m}) \leq \log(s_n) + \log(s_m)$. We get that the sequence $(\log(s_n))_{n>0}$ is a subadditive sequence of strictly positive integers and, as a consequence of this fact, that the sequence $(\log(s_n)/n)_{n>0}$ converges.

The entropy of a set of finite words X is the superior limit

$$h(X) = \limsup_{n \to \infty} \frac{1}{n} \log \alpha_n$$

where α_n is the number of words of length n belonging to X.

The following statement gives a method to compute the entropy of a irreducible sofic system provided we can compute the entropy of a set of the form X^* where X is a code.

Proposition 9.1. *Let S be an irreducible sofic system recognized by a transitive unambiguous automaton $\mathcal{A}$. Let X be the code of first returns to some state of $\mathcal{A}$. The entropy of S is equal to the entropy of X^*.*

Proof. Let us denote by l_n the number of words of X^* of length n and let s_n be the number of blocks of length n in S. For any positive integer n, we have $l_n \leq s_n$ since any word of X^* is a block of an element of S. This proves that $h(X^*) \leq h(S)$.

Let k be the number of states of $\mathcal{A}$. Let w_n be a block of length n of an element of S. As the graph of $\mathcal{A}$ is strongly connected, there exist two words u and v, of lengths $|u|, |v|$ satisfying $|u| + |v| \leq k$, such that $uw_n v$ belongs to X^*. This allows us to associate to each block of length n of a sequence of S, a word of length at most $(n + k)$ of X^*, which admits the block as factor. As the number of positions of the block in the word is at most $k + 1$, we get $s_n \leq (k+1)(l_n + l_{n+1} + \cdots + l_{n+k})$. It follows from this that $h(S) \leq h(X^*)$. $\square$

Let L be a set of finite words over an alphabet A and let $f_n = \mathrm{card}(L \cap A^n)$ be the number of words of length n in L. Then

$$f_L(z) = \sum_{n \geq 0} f_n z^n$$

is the generating series of the sequence f_n. One can show (see [13] p. 42) that a set X is a code iff the following equality holds.

$$f_{X^*} = \frac{1}{1 - f_X}$$

The following result allows one to compute the entropy of a set of the form X^* where X is a code.

Theorem 9.1. *Let S be an irreducible sofic system. Let $\mathcal{A}$ be a transitive unambiguous automaton recognizing S and let X be the code of first returns to some state of $\mathcal{A}$. The entropy of S is $\log(1/r_X)$ where r_X is the unique positive root of $f_X(r) = 1$.*

Proof. By Proposition 9.1 we have $h(S) = h(X^*)$. The entropy of X^* is equal to $\log(1/r)$ where r is the convergence radius of f_{X^*}. Since X is a code, we have $f_{X^*} = (1 - f_X)^{-1}$. As any $\mathbb{R}^+$-rational series which is not a polynomial has its convergence radius as a pole, we get that r_X is the unique positive root of $f_X(z) = 1$. $\square$

Theorem 9.1 gives a method to compute the entropy of an irreducible sofic system. One may compute the number r by solving the equation $f_X(z) = 1$. This is effective when X is a rational code or equivalently when S is a sofic system.

An alternative method to compute r is to use the fact that $1/r$ is the maximal eigenvalue of the matrix associated to any unambiguous automaton recognizing S. In fact, let M be a matrix with real coefficients. The spectral radius ρ of M is the maximal modulus of its eigenvalues. One has (see [29] for example)

$$\rho = \limsup \sqrt[n]{\| M^n \|}$$

where $\| M \|$ is any norm of the matrix M. If we choose the particular norm equal to the sum of modulus of all coefficients, then the number of blocks of length n in the edge shift of the automaton is $\| M^n \|$. Thus the entropy of S is $\log \rho$.

This is true also when the automaton is not transitive. Thus, the entropy of S is the maximum of the entropies of the irreducible components of S.

Example 9.1. Let S be the even system represented on Figure 4.1. We have $X = \{a, bb\}$ and $f_X(z) = z + z^2$. Thus $r = 1/\varphi$ where φ is the golden mean. Accordingly, the maximal eigenvalue of the matrix

$$M = \begin{bmatrix} 1 & 1 \\ 1 & 0 \end{bmatrix}$$

is φ.

We prove the following result which implies in particular that the entropy is invariant under conjugacy (a fact that could also be proved directly).

Proposition 9.2. *Let S and T be two irreducible shifts of finite type and let $f : S \to T$ be a morphism. The following conditions are equivalent.*

1. f is finite-to-one.
2. $h(S) = h(T)$

Proof. $1 \Rightarrow 2$. Let $\mathcal{A}$ be a transducer realizing f with a local input automaton. By Proposition 6.1, the output automaton is unambiguous. Let q be a state of $\mathcal{A}$, let X be set of first returns to q in the input automaton and let Y be the set of first returns to q in the output automaton. Then X and Y have the same number of words of each length and thus $h(X^*) = h(Y^*)$. Hence $h(S) = h(T)$ by Proposition 9.1.

$2 \Rightarrow 1$. Let $\mathcal{A}$ be a transducer realizing f. We suppose that the output automaton of $\mathcal{A}$ is ambiguous. If there exist two edges in $\mathcal{A}$ which only differ by the input, then we can remove one of these edges without changing the map realized. This removal decreases the entropy of the set of returns X^* in the input automaton since it increases strictly r_X.

To handle the general case, we consider two paths u, v of length k with the same label x. We shall consider the automaton $\mathcal{A}^k$ which has the same set of states as $\mathcal{A}$ but the set of words of length k as alphabet with the transitions induced by those of $\mathcal{A}$. Obviously, the entropies of the systems S^k, T^k recognized by the input and output automata of $\mathcal{A}^k$ satisfy $h(S^k) = k\, h(S), h(T^k) = k\, h(T)$. We may choose k to be prime to the gcd of the cycle lengths of the automaton. In this way the automaton $\mathcal{A}^k$ is still transitive. We are thus in the situation considered at the beginning. $\square$

If the alphabet A has k elements, then $r \geq 1/k$ or equivalently

$$f_X(1/k) \leq 1 \tag{9.1}$$

which is Kraft's inequality.

It is well-known that one has equality in (9.1) iff the code X is maximal (see [13]). This can be seen as equivalent to the fact that the sofic system S associated to X is equal to the full shift on k symbols.

There are actually several results for which one may indifferently use either the vocabulary of subshifts or that of codes and automata.

As an example, we have the following result, due to Hedlund [33].

Proposition 9.3. *Let S and T be irreducible sofic systems and let $f : S \to T$ be a morphism. Any two of the following conditions imply the third.*

1. *f is finite-to-one.*
2. *f is onto.*
3. *$h(S) = h(T)$*

This statement is the direct counterpart of the following one for codes (see [13], p. 69).

Proposition 9.4. *Let X be a recognizable subset of A^* and let $k = Card(A)$. Any two of the three following statements imply the third.*

1. *X is a code.*
2. *$f_X(1/k) = 1$.*
3. *X is complete.*

For any series f with positive coefficients satisfying (9.1), it is well-known that there exists a prefix code X on a k-symbol alphabet such that $f = f_X$.

Recall that a series $f = \sum_{k \geq 0} f_k z^k$ is said to be $\mathbb{N}$-rational if there exists a nonnegative integral $n \times n$ matrix M, and two vectors $i \in \mathbb{N}^{1 \times n}, t \in \mathbb{N}^{n \times 1}$ such that identically $f_k = iM^k t$.

If X is a rational code, then f_X satisfies (9.1) and is additionally an $\mathbb{N}$-rational series. What can be said conversely? It is tempting to conjecture that for any $\mathbb{N}$-rational series $f = \sum_{n \geq 0} \alpha_n z^n$, such that $f(1/k) \leq 1$, there exists a rational prefix code X over a k-letter alphabet such that $f = f_X$.

A particular case of this is proved in [49].

We recall that an algebraic number is a root r of a monic polynomial whose coefficients are rational numbers. Among these polynomials there is a unique one $p(z)$ of minimal degree, called the *minimal polynomial* of r. The algebraic conjugates of r are the roots of $p(z)$.

When X is a recognizable code, $f_X(z)$ is a rational series and thus r_X is an algebraic number. It is indeed the largest root of the numerator of $1 - f_X(z)$. In the case of a finite code, one has additional properties of this algebraic number.

Proposition 9.5. *If X is a finite code, then r_X has no other real positive algebraic conjugate.*

Proof. Since $f_X(z)$ has positive coefficients, the function $z \mapsto f_X(z) - 1$ is strictly increasing from -1 to $+\infty$ for $z \in [0, +\infty[$. Thus there can be only one positive real number r such that $f_X(r) = 1$. Hence, the algebraic integer r_X has the property that it has no other positive real algebraic conjugate.   □

Thus r_X is the only real positive root of its minimal polynomial. It is also its root of minimal modulus since for any other root ρ of $1 - f_X(z)$, one has $f_X(r) = 1 \leq f_X(|\rho|)$ whence $r \leq |\rho|$.

The above property can be used to prove that some systems cannot be obtained as a renewal system generated by a finite code, as shown in the following example.

Example 9.2. Let $X = \{a, bc, ab, c\}$. The set X is not a code since abc has two factorizations. Let S be the renewal system generated by X. An automaton recognizing S is represented in Figure 9.1.

The determinization and further minimization of the automaton of Figure 9.1 gives the automaton of Figure 9.2.

The minimal automaton of S is local and therefore S is a shift of finite type (although the automaton of Figure 9.1 is not local). Let M be the

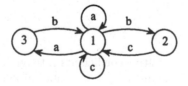

Fig. 9.1. The renewal system S

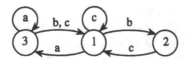

Fig. 9.2. The minimal automaton of S

adjacency matrix of $\mathcal{A}$. We have

$$M = \begin{bmatrix} 1 & 1 & 1 \\ 1 & 0 & 0 \\ 2 & 0 & 1 \end{bmatrix}$$

The characteristic polynomial of M is $p(z) = z^3 - 2z^2 - 2z + 1 = (1 + z)(1 - 3z + z^2)$. The roots of $p(z)$ are $-1, \varphi^2, \hat{\varphi}^2$ where φ is the golden mean and $\hat{\varphi}$ its conjugate. The entropy of S is $\log \varphi^2$. Since φ^2 has a positive real conjugate which is $\hat{\varphi}^2$, the system S cannot be generated by a finite code. This example is due to J. Ashley (unpublished). It answers negatively a conjecture formulated by A. Restivo asserting that a finitely generated renewal system which is at the same time a shift of finite type can be generated by a finite code.

The following result, due to D. Handelman gives a converse to Proposition 9.5.

Theorem 9.2. (Handelman [32]) *Let r be an algebraic integer stricly less in modulus than any of its conjugates. The number r is a root of a polynomial of the form $1 - zp(z)$ with p a polynomial with non-negative coefficients iff it has no other real positive algebraic conjugate.*

The theorem does not cover the case where r has conjugates of the same modulus. In this case, the set of roots of modulus r is of the form $r\epsilon$ where ϵ is any of the p-th roots of 1. The generalization of Handelman's theorem to this case has been obtained by F. Bassino ([6], [5]).

The polynomials of the form $1 - zp(z)$ with p non-negative are a particular case of *polynomials with one sign change*, i.e. such that, after deleting the zero coefficients, the sequence $(a_0, a_1, \ldots)$ of coefficients has exactly one sign change. The result of D. Handelman is actually stated in this more general case.

The proof of Theorem 9.2 uses properties of *log concave* polynomials, also called unimodal, which are polynomials such that the sequence $(a_0, a_1, \cdots)$ of coefficients satisfies $a_i^2 > a_{i+1}a_{i-1}$. It is also related to a result of Poincaré according to which, if a polynomial p with real coefficients has exactly one positive real root, then there exists a polynomial P such that the product pP has one sign change.

Theorem 9.2 has been extended to study the star-height of one-variable rational series. It is known that the star-height of a one-variable N-rational series is at most two (see [56]). F. Bassino has used Theorem 9.2 to obtain a characterization of the series of star-height one under the assumption that they have a unique pole of minimal modulus ([6], [5]).

Much of the study of shifts of finite type is linked to that of *positive matrices*. Indeed, a shift of finite type is given by a finite graph which in turn is given by its adjacency matrix. The shift of finite type itself corresponds to a class of equivalent matrices. The study of this equivalence has motivated a lot of research (see [18] for a survey).

One aspect of this research is the study of the cone of positive matrices inside the algebra of all integer matrices. The basic properties of positive matrices were obtained long ago by Perron and Frobenius. The theorem says essentially that a nonnegative real matrix has an eigenvalue of maximal modulus which is a positive real number. If it is irreducible, it has a corresponding eigenvector with positive coefficients. And if it is primitive, there is only one eigenvalue of maximal modulus.

In a more recent work, Handelman [31] has proved a kind of converse of the Perron Frobenius theorem. A matrix is said to be *eventually positive* if some power has only strictly positive coefficients. A *dominant eigenvalue* is an eigenvalue α such that $\alpha > |\beta|$ for any other eigenvalue.

Theorem 9.3. (Handelman [31]) *A matrix with integer coefficients is conjugate to an eventually positive matrix iff it has a dominant eigenvalue of multiplicity one.*

This result is very close to one due to M. Soittola (see [56]) characterizing N-rational series in one variable among Z-rational series. We quote it for series having a *minimal pole*, i.e., a unique pole with minimal modulus.

Theorem 9.4. (Soittola) *A Z-rational series with nonnegative integer coefficients*

$$f = \sum_{n \geq 0} \alpha_n z^n$$

having a minimal pole is N-rational.

In [50] it is shown that both theorems can be deduced from the construction of a basis in which the matrices have the appropriate properties. It also gives at the same time a proof that a one-variable N-rational series has at most star-height 2.

10. The road coloring problem

A classical notion in automata theory is that of a synchronizing word. We recall that, given a deterministic and complete automaton $\mathcal{A}$ on a state set Q, a word w is said to be synchronizing if the state reached from any state $q \in Q$ after reading w is independent of q. The automaton itself is called *synchronizing* if there exists a synchronizing word.

A maximal prefix code X is called *synchronizing* if it is the set of first returns to the initial state in a synchronizing automaton.

It is clear that a necessary (but not sufficient) condition for an automaton to be synchronizing is that the underlying graph is primitive (i.e., strongly connected and the gcd of the cycle lengths is 1). The same holds for a prefix code (which is called aperiodic if it is maximal and the gcd of the word lengths is 1).

The following problem was raised in [1]. Let G be a finite directed graph with the following properties:

1. All the vertices of G have the same outdegree.
2. G is primitive.

The problem is to find, for any graph G satisfying these hypotheses, a labeling turning G into a synchronizing deterministic automaton. The problem is called the *road-coloring problem* because of the following interpretation: a labeling (or coloring) making the automaton synchronizing allows a traveler lost on the graph G to follow a path which is a succession of colors leading back home regardless of where he actually started.

In terms of symbolic dynamics, such a labeling defines a right-resolving map $f : S_G \to A^{\mathbb{Z}}$ from the shift of finite type S_G onto a full shift which is 1-to-1 almost everywhere. In this context, a synchronizing word is called a *resolving block* and we say that f has a resolving block if there exists such a synchronizing word.

The road-coloring problem itself remains still open but some results have been obtained that we describe now.

The following result is proved in [1]. It shows that if the road coloring problem can perhaps not be solved on a given graph G, it can be solved on a subshift conjugate to S_G.

Theorem 10.1. *If G is a primitive graph with all vertices of the same outdegree, there exists a conjugate T of S_G and a right-resolving map $f : T \to A^{\mathbb{Z}}$ from T onto the full shift on k symbols having a resolving block.*

The proof consists in considering the subshift $S_{G^{(n)}}$, for large enough n, where $G^{(n)}$ is the graph having as edges the paths of length n in G.

Actually, Theorem 10.1 can also be obtained using a result on codes that we now describe.

A set $X \subset A^*$ is called *thin* if there exists a word $w \in A^*$ such that $A^* w A^* \cap X = \emptyset$, that is to say that w does not appear as a block in the words of X. It is known that every rational code is thin (see [13], p. 69).

The following result is due to Schützenberger [57] (see also [52]).

Theorem 10.2. (Schützenberger) *Let $k \geq 2$ be an integer and let $\alpha = (\alpha_n)_{n \geq 1}$ be a sequence of integers. Let $f = \sum_n \alpha_n z^n$ and let ρ_α denote the radius of convergence of the series f.*

There exists a thin maximal and synchronizing prefix code X on a k-letter alphabet such that $f_X = f$ iff the following conditions are satisfied.

1. $\sum_{n \geq 1} \alpha_n k^{-n} = 1$,
2. $\rho_\alpha > 1/k$,
3. the integers n such that $\alpha_n \neq 0$ are relatively prime,

Actually, conditions (1) and (2) are equivalent to the existence of a thin maximal prefix code such that $f_X = f$ and condition (3) holds then iff X is aperiodic. It is shown in [57] that, under the hypotheses of the theorem, one may choose an integer n and two symbols $a, b \in A$ such $a^n \in X$ and that the set

$$Y = a^n \cup (X \cap a^* b a^*) \tag{10.1}$$

contains a synchronizing word for X.

Theorem 10.2 can be used to prove Theorem 10.1. Indeed, let us consider a particular vertex i of the graph G and let α_n be the number of simple paths from i to i in G. Then the sequence $\alpha = (\alpha_n)_{n \geq 1}$ satisfies the conditions of Theorem 10.2. We can use a state splitting to be able to label $n + 1$ paths of the resulting graph by the words of the set Y of Eq (10.1).

The following result is proved in [52], providing a partial answer to the problem. For a prefix code X, we denote by T_X the usual (unlabeled) tree whose leaves correspond to the elements of X. Several prefix codes may thus correspond to the same tree according to the choice of a labeling of the sons of each node.

Theorem 10.3. *Given a finite aperiodic prefix code X, there is a synchronizing prefix code Y such that $T_X = T_Y$.*

The proof uses heavily the theorem of C. Reutenauer ([54]) on the noncommutative polynomial of a code.

In terms of symbolic dynamics, Theorem 10.3 solves positively the road-coloring problem for those graphs G satisfying the following additional assumption:

(3) All vertices except one have indegree 1.

Other results on the road coloring propblem have been obtained and in particular, by G. O'Brien [47] and by J. Friedman [27].

11. The zeta function of a subshift

Besides the entropy, there is an invariant of symbolic dynamical systems which takes into account the number of elements of a given period.

Let (S, σ) be a subshift and let

$$P_n = \{x \in S \mid \sigma^n(x) = x\}$$

be the set of points of period dividing n.

The *zeta function* of a subshift S is the series

$$\zeta_S(z) = \exp \sum_{n>0} \frac{p_n}{n} z^n$$

with $p_n = \mathrm{card}(P_n)$.

Two subshifts have the same zeta function iff they have the same number of elements of each period. Since a conjugacy preserves the period of the points, the zeta function is an invariant under conjugacy. This information is useful for separating non equivalent systems. It is actually stronger than entropy for sofic systems.

In fact, let S be an irreducible sofic subshift recognized by an unambiguous automaton $\mathcal{A}$. Let X be the code of first returns to some state of $\mathcal{A}$. Let t_n be the number of words of length n in X^*. Then

$$t_n \leq p_n \leq s_n$$

By definition, we have $h(S) = \lim \frac{1}{n} \log s_n$ and $h(X^*) = \lim \frac{1}{n} \log t_n$. By Proposition 9.1 we have $h(X^*) = h(S)$ and thus

$$\lim \frac{1}{n} \log t_n = \lim \frac{1}{n} \log p_n = \lim \frac{1}{n} \log s_n$$

Hence $h(S)$ is determined by $\zeta(S)$.

Another invariant related to the number of minimal forbidden blocks of each length is studied in [10].

It was proved by R. Bowen and O. Lanford in [17] that the zeta function of a shift of finite type S is a rational series. They proved actually the following proposition.

Proposition 11.1. *Let S be the edge shift of a graph G. If M is the adjacency matrix G,*

$$\zeta_S(z) = det(I - Mz)^{-1}$$

Proof. As we can associate bijectively to each sequence x of S such that $\sigma^n(x) = x$, a cycle of the graph of length n, we get that $s_n = \text{trace}(M^n)$. The computation of the zeta function of S can now be done as follows, where I is the identity matrix of the same size as M.

$$
\begin{aligned}
\zeta_S(z) &= \exp \sum_{n>0} \frac{1}{n} \text{trace}(M^n) z^n = \exp \text{trace}(\sum_{n>0} \frac{1}{n}(Mz)^n) \\
&= det \exp(\sum_{n>0} \frac{1}{n}(Mz)^n) = det \exp \log(I - Mz)^{-1} \\
&= det(I - Mz)^{-1} \qquad \qquad \qquad \qquad \qquad \square
\end{aligned}
$$

It was proved later by A. Manning [40] and also by R. Bowen [17] that the zeta function of a sofic system is also rational.

This has motivated further investigations in several directions. On one hand, J. Berstel and C. Reutenauer have extended the result of Manning to the case of a generalization of the zeta function to some formal series [14] and proved the rationality of the generalized zeta function for cyclic languages.

On the other hand, M. P. Béal has introduced an operation on finite automata, the *external power*, allowing one to obtain the generalized zeta function of a sofic system as a combination of the values obtained on the different external powers, (see [7] and [8]). This gives a proof of the formula of Bowen [17] and this proof can be extended to the case of cyclic languages [9].

More recently, C. Reutenauer [55] has obtained new results showing that the zeta function of a sofic system is not only rational but even N-rational. He has also extended his results to more general symbolic systems, introduced by D. Fried under the name of *finitely presented systems* [26].

12. Circular codes, shifts of finite type and Krieger embedding theorem

There is a close connection between shifts of finite type and a particular class of codes called circular codes (see [13] for a more comprehensive introduction).

A set $X \subset A^+$ is called a *circular code* if any circular word over A has at most one decomposition as a product of words from X. More precisely, X is a circular code if for any $x_1, x_2, \ldots, x_n, y_1, y_2, \ldots, y_m \in X$ and $p \in A^*, s \in A^+$ the equalities

$$sx_2x_3 \ldots x_np = y_1y_2 \ldots y_m, \tag{12.1}$$

$$x_1 = ps \tag{12.2}$$

imply $n = m$, $p = \epsilon$ and $x_1 = y_1, \ldots, x_n = y_n$.

Indeed, Equalities 12.1,12.2 corresponds to two decompositions of a word written on a circle as represented in Figure 12.1.

The following statement relates circular codes and local automata (see [7] p. 65).

Proposition 12.1. *Let X be a finite code and let $\mathcal{A}$ be an unambiguous strongly connected automaton such that X is the set of first returns to some state of $\mathcal{A}$. The following conditions are equivalent.*

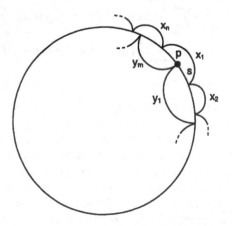

Fig. 12.1. Two circular factorizations

1. X is a circular code.
2. The automaton $\mathcal{A}$ is local.

Proof. Two cycles in $\mathcal{A}$ with equal labels define two factorizations of a circular word and conversely. Thus the result follows from Proposition 4.1.   □

The next result relates circular codes and shifts of finite type.

Proposition 12.2. *The renewal system generated by a finite circular code is a shift of finite type.*

Proof. Let S be the renewal system generated by the circular code X. By the previous proposition, there exists a local automaton recognizing S. By Proposition 4.3, S is a shift of finite type.   □

Let, as in Section 9., $\alpha_n = \text{Card}(X \cap A^n)$ and $f_X(z) = \sum_{n \geq 0} \alpha_n z^n$. Let S be the system generated by X, which is the set of all bi-infinite words having a factorization in words of X. The following statement shows that the zeta function of S can be easily computed.

Proposition 12.3. *Let X be a finite circular code. The zeta function of S is given by*

$$\zeta_S = (1 - f_X)^{-1} \qquad (12.3)$$

Proof. Since X is circular, S is a shift of finite type (Proposition 12.1). Let M be the adjacency matrix of the graph of the flower automaton of X. By Proposition 11.1, we have

$$\zeta_S(z) = \det(I - Mz)^{-1}$$

It is well-known for any graph made of n cycles of lengths $(\alpha_1, \ldots, \alpha_n)$ with one common vertex that

$$\det(I - Mz) = 1 - f_X$$

The result follows from the two above equations.   □

The number p_n of points of S of period dividing n can be computed from Formula (12.3). Indeed, we have

$$\sum_n \frac{p_n}{n} z^n = \log(\zeta_S)$$

and thus, by Formula (12.3)

$$\sum_n \frac{p_n}{n} z^n = -\log(1 - f_X)$$
$$= -\log(1 - \sum_n \alpha_n z^n)$$
$$= \sum_n s_n z^n$$

with

$$s_n = \sum_{k=1}^{n} \frac{1}{k} \alpha_n^{(k)}, \qquad \alpha_n^{(k)} = \sum_{i_1 + \ldots + i_k = n} \alpha_{i_1} \cdots \alpha_{i_n}$$

Thus we have

$$p_n = \sum_{i=1}^{n} \frac{n}{i} \alpha_n^{(i)} \tag{12.4}$$

The number of points having period exactly n is denoted by $q_n(S)$ or simply q_n. Obviously p_n and q_n are related by $p_n = \sum_{d|n} q_d$. The following inequalities are then satisfied for all $n \geq 1$.

$$q_n \leq l_n(k) \tag{12.5}$$

where $l_n(k)$ is the number of points of period exactly n in the full shift over k symbols. Indeed, the number of points in S having period exactly n cannot exceed the total number of points of period n in the full shift on k symbols.

The numbers $l_n(k)$, sometimes called Witt numbers, satisfy $\sum_{d|n} d l_d(k) = k^n$ or equivalently, by Möbius inversion formula,

$$l_n(k) = \frac{1}{n} \sum_{d|n} \mu(d) k^{n/d}$$

The length distribution $(\alpha_n)_{n \geq 1}$ of a circular code satisfies inequalities stronger than (9.1) which are obtained after expressing in (12.5) the integers p_n in terms of the α_n using Formula (12.4).

The first inequalities are, in explicit form:

$$\alpha_1 \leq k$$
$$\alpha_2 + \frac{1}{2}(\alpha_1^2 - \alpha_1) \leq \frac{1}{2}(k^2 - k)$$
$$\cdots \leq \cdots$$

It was shown by Schützenberger (see [13] p. 343) that these inequalities characterize the length distributions of circular codes.

Theorem 12.1. (Schützenberger) *A sequence α_n of integers is the length distribution of a circular code over a k-letter alphabet iff it satisfies the above inequalities.*

This is linked in a very interesting way with a theorem of Krieger which gives a necessary and sufficient condition for the existence of a strict embedding of a shift of finite type into another one.

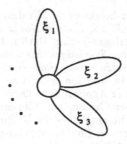

Fig. 12.2. A renewal graph

Theorem 12.2. (Krieger [36]) *Let S and T be two shifts of finite type. Then there exists an isomorphism f from S into T with $f(S) \neq T$ iff*

1. $h(S) < h(T)$.
2. *for each* $n \geq 1$, $q_n(S) \leq q_n(T)$

A proof of Krieger's theorem can be found in the book of D. Lind and B. Marcus [38].

We explain here the connection between Krieger's theorem and the theorem of Schützenberger on circular codes.

Given a finite sequence $\xi = (\xi_i)_{1 \leq i \leq n}$, let G be the *renewal graph* made of n simple cycles of lengths $\xi_1, \ldots, \xi_n$ with exactly one common point (see Figure 12.2). Any circular code on the alphabet A with length distribution ξ defines an isomorphism from the edge shift S_G into $A^{\mathbf{Z}}$. Indeed, there is a labeling of G which defines a flower automaton $\mathcal{A}$ for X. By Proposition 12.1, the subshift recognized by $\mathcal{A}$ is of finite type. The map from paths to labels is therefore an embedding of S_G into the full shift $A^{\mathbf{Z}}$.

Thus Theorem 12.1 gives a proof of Theorem 12.2 in the particular case where S is a the edge shift defined by a renewal graph and T is a full shift.

References

[1]   Roy Adler, I. Goodwin, and Benjamin Weiss. Equivalence of topological Markov shifts. *Israel J. Math.*, 27:49–63, 1977.

[2]   Roy L. Adler, D. Coppersmith, and M. Hassner. Algorithms for sliding block codes. *IEEE Trans. Inform. Theory*, IT-29:5–22, 1983.

[3]   Alfred V. Aho, John E. Hopcroft, and Jeffrey D. Ullman. *The Design and Analysis of Computer Algorithms*. Addison Wesley, 1974.

[4]   Jonathan Ashley, Brian Marcus, Dominique Perrin, and Selim Tuncel. Surjective extensions of sliding block codes. *SIAM J. Discrete Math.*, 6:582–611, 1993.

[5]   Frédérique Bassino. Non-negative companion matrices and star-height of N-rational series. *Theoret. Comput. Sci.*, 1996. (to appear).

[6]   Frédérique Bassino. Star-height of an N-rational series. In C. Puech and R. Reischuk, editors, *STACS 96, Lecture Notes in Computer Science*, volume 1046, pages 125–135. Springer-Verlag, Berlin, 1996.

[7]   Marie-Pierre Béal. *Codage Symbolique*. Masson, 1993.

[8]   Marie-Pierre Béal. Puissance extérieure d'un automate déterministe, application au calcul de la fonction zêta d'un système sofique. *R.A.I.R.O-Informatique Théorique et Applications*, 29:85–103, 1996.

[9]   Marie-Pierre Béal, Oliver Carton, and Christophe Reutenauer. Cyclic languages and strongly cyclic languages. In C. Puech and R. Reischuk, editors, *STACS 96, Lecture Notes in Computer Science*, volume 1046, pages 49–59. Springer-Verlag, Berlin, 1996.

[10]   Marie-Pierre Béal, Filippo Mignosi, and Antonio Restivo. Minimal forbidden words and symbolic dynamics. In C. Puech and R. Reischuk, editors, *STACS 96, Lecture Notes in Computer Science*, volume 1046, pages 555–566. Springer-Verlag, Berlin, 1996.

[11]   Danièle Beauquier. Minimal automaton for a factorial, transitive rational language. *Theoret. Comput. Sci.*, 67:65–73, 1989.

[12]   Tim Bedford, Michael Keane, and Caroline Series, editors. *Ergodic Theory, Symbolic Dynamics and Hyperbolic Spaces*. Oxford University Press, 1991.

[13]   Jean Berstel and Dominique Perrin. *Theory of codes*. Academic Press, 1985. (also available on http://www-litp.ibp.fr/berstel/LivreCodes).

[14]   Jean Berstel and Christophe Reutenauer. Zeta functions of formal languages. *Trans. Amer. Math. Soc.*, 321:533–546, 1990.

[15]   Garett D. Birkhoff. Quelques théorèmes sur le mouvement des systèmes dynamiques. *Bull. Soc. Math. France*, 40:305–323, 1912.

[16]   François Blanchard and Alejandro Maass. On dynamical properties of generalized cellular automata. In Ricardo Baeza-Yates and Eric Goles, editors. *LATIN 95: Theoretical Informatics, Lecture Notes in Computer Science*, volume 911, pages 84–98, Springer-Verlag, Berlin, 1995.

[17]   Rufus Bowen and O. Lanford. Zeta functions of restrictions of the shift transformation. *Proc. Symp. Pure Math.*, 14:43–50, 1970.

[18]   Michael Boyle. Symbolic dynamics and matrices. In S. Friedland, V. Brualdi, and V. Klee, editors, *Combinatorial and Graph-Theoretic Problems in Linear Algebra, IMA Volumes in Mathematics and its Applications*, volume 50, Springer-Verlag, Berlin, 1993.

[19]   Michael Boyle, Douglas Lind, and D. Rudolph. The automorphism group of a shift of finite type. *Trans. of Amer. Math. Soc.*, 306:71–114, 1988.

[20]   Véronique Bruyère and Michel Latteux. Variable-length maximal codes. In *ICALP 96, Lecture Notes in Computer Science*, Springer-Verlag, Berlin, 1996. (to appear).

[21]   Véronique Bruyère, L. Wang, and Liang Zhang. On completion of codes with finite deciphering delay. *European J. Combin.*, 11:513–521, 1990.

[22]   Christian Choffrut. Conjugacy in free inverse monoids. *Int. J. Alg. Comput.*, 3:169–188, 1993.

[23]   Aldo De Luca and Stefano Varricchio. *Combinatorics on Words and Regularity Conditions*. Springer-Verlag, Berlin, 1996. (to appear).

[24]   Samuel Eilenberg. *Automata, Languages and Machines*, volume A. Academic Press, 1974.

[25]   R. Fischer. Sofic systems and graphs. *Monatshefte Math.*, 80:179–186, 1975.

[26]   David Fried. Finitely presented dynamical systems. *Ergod. Th. Dynam. Sys.*, 7:489–507, 1987.

[27]   Joel Friedman. On the road coloring problem. *Proc. Amer. Math. Soc.*, 110:1133-35, 1990.

[28] Christiane Frougny and Boris Solomyak. Finite beta-expansions. *Ergod. Th. & Dynam. Syst.*, 12:45–82, 1992.

[29] F. R. Gantmacher. *The Theory of Matrices.* Chelsea, 1960.

[30] Andrew Gleason. Semigroups of shift register matrices. *Mathematical Systems Theory*, 25:253–267, 1992. (notes of a course given at Princeton in 1960).

[31] David Handelman. Positive matrices and dimension groups affiliated to C^*-algebras and topological Markov chains. *J. Operator Theory*, 6:55–74, 1981.

[32] David Handelman. Spectral radii of primitive integral companion matrices and log-concave polynomials. In Peter Walters, editor, *Symbolic Dynamics and its Applications, Contemporary Mathematics*, volume 135, pages 231–238. Amer. Math. Soc., 1992.

[33] George Hedlund. Endomorphisms and automorphisms of the shift dynamical system. *Math. Syst. Theory*, 3:320–375, 1969.

[34] K. H. Kim and F. W. Roush. Decidability of shift equivalence. In *Symbolic Dynamics, Lecture Notes in Mathematics*, volume 1342, pages 374–424. Springer-Verlag, Berlin, 1988.

[35] K. H. Kim and F. W. Roush. Williams's conjecture is false for reducible matrices. *J. Amer. Math. Soc.*, 5:213–215, 1992.

[36] Wolfgang Krieger. On the subsystems of topological Markov chains. *Ergod. Th. & Dynam. Syst.*, 2:195–202, 1982.

[37] Joseph Kruskal. The theory of well-quasi-ordering: a frequently rediscovered concept. *J. Comb. Theory (ser. A)*, 13:297–305, 1972.

[38] Douglas Lind and Brian Marcus. *An Introduction to Symbolic Dynamics and Coding.* Cambridge University Press, 1996.

[39] M. Lothaire. *Combinatorics on Words.* Cambridge University Press, 1983.

[40] Anthony Manning. Axiom a diffeomorphisms have rational zeta function. *Bull. London Math. Soc.*, 3:215–220, 1971.

[41] Brian Marcus. Factors and extensions of full shifts. *Monats. Math.*, 88:239–247, 1979.

[42] Brian Marcus. Sofic systems and encoding data. *IEEE Trans. Inf. Theory*, IT-31:366–377, 1985.

[43] J. C. Martin. Substitution minimal sets. *Amer. J. Math.*, 93:503–526, 1971.

[44] Marston Morse. Recurrent geodesics on a surface of negative curvature. *Trans. Amer. Math. Soc.*, 22:84–110, 1921.

[45] Marston Morse and George Hedlund. Symbolic dynamics. *Amer. J. Math.*, 3:286–303, 1936.

[46] Masakazu Nasu. *Textile Systems for Endomorphisms and Automorphisms of the Shift*, volume 546 of *Memoirs of Amer. Math. Soc.* Amer. Math. Soc., 1995.

[47] G. L. O'Brien. The road coloring problem. *Israel J. Math*, 39:145–154, 1981.

[48] William Parry and R. F. Williams. Block coding and a zeta function for finite markov chains. *Proc. London Math. Soc.*, 35:483–495, 1977.

[49] Dominique Perrin. Arbres et séries rationnelles. *C. R. Acad. Sci. Paris*, 309:713–716, 1989.

[50] Dominique Perrin. On positive matrices. *Theoretical Computer Science*, 94:357–366, 1992.

[51] Dominique Perrin. Symbolic dynamics and finite automata. In Jiri Wiedermann and Petr Hajek, editors, *Mathematical Foundations of Computer Science 1995, Lecture Notes in Computer Science*, volume 969, pages 94–104. Springer-Verlag, Berlin, 1995.

[52] Dominique Perrin and Marcel-Paul Schützenberger. Synchronizing prefix codes and automata and the road coloring problem. In Peter Walters, editor, *Symbolic Dynamics and its Applications, Contemporary Mathematics*, volume 135, pages 295–318. Amer. Math. Soc. 1992.

[53] Antonio Restivo. Codes and local constraints. *Theoret. Comput. Sci.*, 72:55–64, 1990.

[54] Christophe Reutenauer. Non commutative factorization of variable length codes. *J. Pure and Applied Algebra*, 36:157–186, 1985.

[55] Christophe Reutenauer. N-rationality of zeta functions. *Advances in Mathematics*, 1996. (to appear).

[56] Arto Salomaa and Matti Soittola. *Automata-Theoretic Aspects of Formal Power Series*. Springer-Verlag, New York, 1978.

[57] Marcel-Paul Schützenberger. On synchronizing prefix codes. *Inform and Control*, 11:396–401, 1967.

[58] Wolfgang Thomas. Automata on infinite objects. In J. van Leeuwen, editor, *Handbook of Theoretical Computer Science*, volume vol. B, Formal models and semantics, pages 135–191. Elsevier, 1990.

[59] Peter Walters, editor. *Symbolic Dynamics and its Applications, Contemporary Mathematics*, volume 135, Amer. Math. Soc., 1992.

[60] Benjamin Weiss. Subshifts of finite type and sofic systems. *Monats. Math.*, 77:462–474, 1973.

[61] Frank Williams. Classification of subshifts of finite type. *Ann. of Math.*, 98:120–153, 1973. (Errata ibid. **99**:380–381, 1974).

[62] Liang Zhang and Zhonghui Shen. Completion of recognizable bifix codes. *Theoret. Comput. Sci.*, 145:345–355, 1995.

Cryptology: Language-Theoretic Aspects

Valtteri Niemi

1. Introduction

Cryptology is the science and art of secret writing. The basic idea is to convert a meaningful text to another text in which the meaning is concealed. This starting point seems to imply that cryptology is essentially connected to *semantics*. The reason why this conclusion is not true is the following.

The conversion process is not useful unless it is also possible to reveal the original meaning somehow. Moreover, the revealing should be possible (in case suitable secret information is known) without knowing the original meaning *beforehand*. It follows that the revealing process must be based on syntactical properties of the concealed text.

Indeed, all modern cryptologic systems are built by syntactical rules. Although semantical issues are important when the quality of the system is concerned, cryptology can be rooted in formal language theory. On the other hand, a language-theoretic framework has not been widely used in cryptologic research. This fact constitutes a challenge for scientists in both fields.

We continue the discussion about connections between cryptology and language theory after introducing some cryptologic terminology in the next section. More background in cryptology can be obtained by consulting, e.g., the books [11], [38], [40], [41], [42], [44]. As for basic language-theoretic terminology we refer to other chapters of this handbook.

2. Basic notions in cryptology

The original meaningful message is called a *plaintext* while the converted message is called a *cryptotext*. The conversion process is called *encryption* and the revealing process is called *decryption*. All possible plaintexts constitute a *plaintext space* which can be defined as a language over some alphabet Σ. For instance, the plaintext space could be the set of all meaningful texts in English. As mentioned in the introduction, encryption does not usually depend on the meaning of the plaintext, hence all elements of Σ^* could be encrypted.

In the encryption algorithm there are two parameters, namely the plaintext p and an *encryption key* k which can be defined as a word over an alphabet K. If the algorithm produces a cryptotext c this relation is denoted by

$$E_k(p) = c.$$

Typically, E_k is a function for a fixed encryption key k but this is not always the case because the encryption algorithm can be *nondeterministic*.

Respectively, the decryption algorithm uses two parameters, the cryptotext c and a *decryption key* k' which is a word over an alphabet K'. If the algorithm produces a result p, this is denoted by

$$D_{k'}(c) = p.$$

Typically, D is a deterministic algorithm and $D_{k'}$ is a function. The pair (E, D) is called a *cryptosystem* if for each encryption key k there exists a decryption key k' such that

$$D_{k'}(E_k(p)) = p$$

for each plaintext p.

A cryptosystem is sometimes called a *cipher*.

The basic requirements for a *secure* cryptosystem are the following:

1) If k (resp., k') is known then the algorithm E_k (resp., $D_{k'}$) can be executed effectively.

2) If k (resp., k') is *not* known then it is impossible in practice to find out the outcome of the algorithm E_k (resp., $D_{k'}$).

The art of designing secure cryptosystems is called *cryptography*. (Note, however, that many authors use the term "cryptography" as a synonym for the more general term "cryptology".) It is usually quite easy to meet the requirement (1) but the other requirement (2) is much trickier. If it is shown that a cryptosystem *does not* meet (2) then the system is *broken*. The art of trying to break cryptosystems is called *cryptanalysis*.

Often the encryption key is equal to the decryption key, i.e., $k = k'$, and then the cryptosystem is called *classical* (or *one-key* or *symmetric*). In the mid-1970s Whitfield Diffie and Martin Hellman noted that it is possible to separate k and k' essentially from each other [12]. More precisely, they suggested that it is possible to construct such cryptosystems in which there is a third key k'' from which both k and k' can be derived but without knowing k'' it is impossible in practice to find out k' if k is given and/or vice versa. This kind of cryptosystem is called a *public-key* (or *two-key* or *asymmetric*) system.

Secure public-key systems are closely related to the following more general notions. A function is called *one-way* if it is easy to compute but, on the other hand, it is intractable to find any preimage of a given value. Furthermore, if there exists some additional information, called the *trapdoor*, by which it is easy to find preimages, then the function is called *trapdoor* function. (Without the trapdoor information it is, again, intractable to find preimages.)

Clearly, a public-key cryptosystem can be turned to a classical one by simply using k'' as both encryption and decryption key, thus including the derivation of k and k' into the encryption and decryption algorithms. The following example shows what are the benefits if this is *not* done. The example also illustrates some differences between the two types of cryptosystems.

Assume we have an open network of n users and each user is connected to one single server. It is possible that both insiders and outsiders can read any message sent in the network. In order to be able to change *confidential* messages over the network some kind of encryption is introduced. The biggest problem in a situation of this kind is the management of the keys. If a classical cryptosystem is used we have two options:

1) The server generates n keys and sends one of them to each user via a *secure channel* (e.g., ordinary mail). Suppose now that a user A wants to send a message to another user B. First A encrypts the message using her key and sends it to the server. The server decrypts and forwards the message encrypted by B's key.

An obvious disadvantage is the huge load on the server. Also, all messages appear temporarily in plaintext form inside the server.

2) The server generates $\frac{n(n-1)}{2}$ keys, one for each pair of users, and sends the appropriate $n-1$ keys to each user via a secure channel. Now A and B can change messages using the key designated to this specific pair.

An obvious disadvantage is the huge number of keys in the case n is large.

Suppose now that a public-key cryptosystem is used instead of a classical one. The server generates n pairs of encryption and decryption keys, one pair for each user. The decryption key is sent to each user via secure channel while the encryption key is made public! If A wants to send a message to B she first looks up B's encryption key in the server's public list. Then A encrypts the message using this key and, at the other end, B decrypts the message. It follows from the definition of a public-key cryptosystem that the public list of encryption keys does not give away any of the decryption keys.

Neither of the two previous disadvantages appear in this case. On the other hand, there is a new drawback which is not so obvious. It is possible in principle for a third user C to impersonate A in the process and send a message to B signed by A because B's encryption key is public. Hence, *authenticity* of the messages cannot be guaranteed. Fortunately, the concept of a public-key system offers a solution also to this problem.

Let the server reverse the roles of encryption and decryption keys in the set-up phase of the system. That means encryption keys are dealt to the users via secure channels while decryption keys are made public.

Now A sends the message to B using A's secret encyption key. At the other end, B can decrypt the message by the public decryption key. If the resulting text is meaningful it must be coming from A because only A knows the encryption key. Of course, in this reversed scenario also the drawback mentioned above is reversed. This means *confidentiality* of the messages is not guaranteed any more, since anyone in possession of the cryptotext can decrypt it by the public key. Messages encrypted by the second method are called *digital signatures.*

To save both authenticity and confidentiality a double system can be used in which two different sets of pairs of keys are constructed by the server. The user A encrypts first, along the first scenario, using B's public encryption key and after that encrypts the resulting text along the other scenario using her own secret encryption key. Of course, a double system like this decreases efficiency. Also, public-key cryptosystems are typically less efficient than classical systems. For these reasons, the most practical scenario seems to be as follows.

The user A sends a request "I want to talk to B" to the server using the system that guarantees authenticity. Then the server generates a *session key* in some classical cryptosystem and sends this key to both A and B using the double system that was described above. Now A and B can change authentic and confidential messages using the session key while the server need not be active any more.

This example also illustrates the fact that in building secure communication systems one usually needs both classical and public-key cryptosystems together with some amount of physically secured communication.

3. Connections between cryptology and language theory

Almost all cryptosystems can be modelled by *generalized sequential machines* (i.e., finite automata with output) in the following way. The machine takes first a preliminary input which contains the encryption key. After the key the input contains the plaintext stream. The first portion of the cryptotext depends only on some fixed-length prefix of the plaintext and the whole cryptotext can be computed effectively. Subsequent parts of the cryptotext may depend on earlier parts of cryptotext or plaintext but typically the memory needed for this purpose is of fixed size. Altogether, the input is read in a linear sweep and a fixed amount of memory is needed to produce the output. Decryption is done similarly: the encryption key is replaced by the decryption key and the plaintext and the cryptotext change roles.

Many famous mechanical cryptographic machines are concrete examples of this interpretation. Some of these machines are still used as software versions. A cryptanalysis of one of these versions can be found in [57].

Unfortunately, the automaton model of a cryptosystem is seldom useful, mainly because the number of states in the machine is often astronomical. Despite this drawback there are many connections between cryptology and language theory. Some of the connections are presented in subsequent sections of this survey.

Sections 4 and 5 show how language theory and automata theory can be used in design of both classical and public-key cryptosystems. In Section 6 it is shown that there are cases in which theoretical cryptologic research can benefit from theory of languages. Section 7 discusses how cryptanalysis may apply language-theoretic and automata-theoretic methods. In Section 8 we present results in language theory which are motivated by cryptology. Finally, in Section 9, a few examples are given of results in different areas making use of both cryptology and formal language theory.

4. Public-key systems based on language theory

The best known public-key cryptosystems (e.g., the *RSA* system [35]) are based on number theory. A very crucial assumption associated with these systems is that some specific number-theoretic problems (like factoring a big integer) are intractable in practice.

Many experts feel that public-key cryptology is dangerously dependent on a few problems whose complexity is not exactly known. This is the main reason why new systems based on other areas of mathematics are wanted. In this section we present some public-key cryptosystems the security of which depends on formal language theory. There are also classical cryptosystems based on language theory (e.g., [1]) but we do not discuss them here.

There are two general problems that seem to be difficult to overcome when language-theoretic cryptosystems are constructed. First, the cryptotext tends to be much longer than the plaintext. In many practical situations this *data expansion* cannot be tolerated. Secondly, while it is intractable (or at least difficult) to find the decryption key when the encryption key is given, the same is not true in the opposite direction. This means the systems cannot be used in the digital signature mode (see Section 2). The fundamental reason for these problems might be the fact that the structural richness of the set of integers is not met in objects of language theory.

In the sequel we present the main ideas of some specific systems. Many important details are omitted for the sake of simplicity.

4.1 Wagner-Magyarik system

This public-key cryptosystem [56] is based on an undecidable *word problem for finitely presented groups*. Consider generators $x_1, \ldots, x_n$ and their inverses $x_1^{-1}, \ldots, x_n^{-1}$. Assume further that we have *relators* $r_1 = \lambda, \ldots, r_m = \lambda$ where each r_i is a word over the alphabet $\{x_1, \ldots, x_n, x_1^{-1}, \ldots, x_n^{-1}\}$ and λ is the empty word. Two words are said to be *equivalent* if one of them can be derived from the other by the following rewriting rules:

$$r_i \to \lambda, \lambda \to r_i; x_j^{-1} x_j \to \lambda, x_j x_j^{-1} \to \lambda, \lambda \to x_j x_j^{-1} \mid x_j x_j^{-1}$$

where $i = 1, \ldots, m$ and $j = 1, \ldots, n$. It turns out that there are *specific* sets of relators for which the problem whether or not two words are equivalent (the word problem) is undecidable.

Encryption key: Generators, relators and two words w_0, w_1 which are known to be inequivalent.

Encryption: This is performed *bit-by-bit* and nondeterministically. Suppose next bit in the plaintext is i. Then a word equivalent to w_i is generated using the rewriting rules above. This word is added as a new entry to the cryptotext.

Decryption key: More relators $s_1 = \lambda, \ldots, s_p = \lambda$ which are chosen in such way that

1) w_0 and w_1 are still inequivalent,
2) the problem whether or not two words are equivalent can now be decided effectively.

Decryption: The effective decision procedure is used to check whether the word in the cryptotext is equivalent to w_0 or w_1.

Similar cryptosystems are constructed in [15] based on word problems for groups that are *not* finitely presented. The reference [3] contains another method to extend the Wagner–Magyarik cryptosystem.

4.2 Salomaa–Welzl system

In this system [37] we consider two morphisms $h_0, h_1 : \Sigma^* \to \Sigma^*$ and a nonempty word $w \in \Sigma^*$. The quadruple $G = (\Sigma, h_0, h_1, w)$ is said to be *backward deterministic* if the condition

$$h_{i_1}(h_{i_2}(\cdots (h_{i_n}(w)) \cdots)) = h_{j_1}(h_{j_2}(\cdots (h_{j_m}(w)) \cdots))$$

always implies

$$i_1 \cdots i_n = j_1 \cdots j_m.$$

A system of this kind can obviously be used as a classical cryptosystem by encrypting $i_1 \cdots i_n$ as $h_{i_1}(h_{i_2}(\cdots (h_{i_n}(w)) \cdots))$. To construct a public-key system let us proceed as follows.

Let Δ be an alphabet of much greater cardinality than Σ and $g : \Delta^* \to \Sigma^*$ be a morphism that maps each letter either to a letter or to the empty word. Construct now nondeterministically another quadruple $H = (\Delta, \sigma_0, \sigma_1, u)$ where $u \in g^{-1}(w)$ while σ_0 and σ_1 are *finite substitutions* on Δ such that the following condition holds: if $y \in \sigma_i(d)$ then $g(y) = h_i(g(d))$.

(In fact, H is a *T0L* system and G is a *DT0L* system .)

Encryption key: The quadruple H.

Encryption: A bit sequence $i_1 \cdots i_n \in \{0,1\}^*$ is encrypted by choosing at random one element c from $\sigma_{i_1}(\cdots(\sigma_{i_n}(u)\cdots)$.

Decryption key: The quadruple G and the morphism g.

Decryption: The word $g(c)$ equals $h_{i_1}(h_{i_2}(\cdots(h_{i_n}(w)\cdots)$. Since G is backward deterministic the sequence $i_1 \cdots i_n$ can be found. The process is effective if G is chosen in an appropriate manner.

One possible way to make the decryption process effective is to begin with a *strongly* backward deterministic G in which the condition

$$h_{i_1}(h_{i_2}(\cdots(h_{i_n}(w)\cdots) = h_t(x)$$

always implies the conditions $t = i_1$ and $x = h_{i_2}(\cdots(h_{i_n}(w)\cdots)$. Now it is possible to parse without any look-ahead.

A polynomial-time algorithm was given in [18] which breaks the cryptosystem if strongly backward deterministic G is used. The essential point in the algorithm is the fact that the image of a regular language under an inverse of a finite substitution is accepted by a finite nondeterministic automaton of the *same size* as the automaton accepting the original language. This fact makes parsing without any look-ahead possible also in the case where only H is known.

Other cryptanalytic observations concerning Salomaa–Welzl system can be found in [19] and [39]. There are also several extensions of the system.

4.3 Subramanian et al. system

This system [45] combines some ideas of the two previous systems. Consider two words u_0, u_1 over an alphabet Σ and an *injective* morphism $h : \Sigma^* \to \Sigma^*$. By using a similar morphism $g : \Delta^* \to \Sigma^*$ as in the previous section several finite substitutions $\sigma_1, \ldots, \sigma_m$ are defined with the following property: if $y \in \sigma_i(d)$ then $g(y) = h(g(d))$.

Encryption key: Two words $x_i \in g^{-1}(u_i)$ $(i = 1,2)$ and the substitutions σ_i for $i = 1, \ldots, m$.

Encryption: To encrypt a bit i we compute nondeterministically a word $c \in \sigma_{j_1}(\cdots(\sigma_{j_k}(x_i)\cdots)$ where each $j_l \in \{1, \ldots, m\}$ is chosen randomly.

Decryption key: Morphisms g and h.

Decryption: The word $g(c)$ is equal to $h^k(u_i)$. Since h is injective it is possible to find u_i.

The cryptanalytic method of [18] breaks also this system. More precisely, let us combine the different substitutions $\sigma_1, \ldots, \sigma_m$ into a single substitution σ with

$$\sigma(a) = \bigcup_{i=1}^{n} \sigma_i(a).$$

(In other words, we obtain a $0L$ system.) Clearly, $c \in \sigma^k(x_i)$. This condition already implies $g(c) = h^k(u_i)$. Hence, exactly one of the two words x_0, x_1 belongs to the language $\sigma^{-k}(c)$.

4.4 Siromoney–Mathew system

In this public-key cryptosystem [43] *Lyndon words* play a central role. Suppose $(\Sigma, <)$ is a totally ordered alphabet. The order $<$ is extended to a total *lexicographical* order in Σ^* by the condition: $u < v$ iff
i) u is a proper prefix of v or
ii) there exist words $r, s, t \in \Sigma^*$ and letters $a, b \in \Sigma$ with $a < b$ such that $u = ras$ and $v = rbt$.

Two words $u, v \in \Sigma^*$ are *conjugate* if there exist $x, y \in \Sigma^*$ such that $u = xy$ and $v = yx$. A word w is a *Lyndon word* if $w < v$ whenever v is a conjugate of w (and $v \neq w$). By a well-known theorem (of Chen, Fox, and Lyndon, see [23]) every word $w \in \Sigma^*$ admits a unique factorization $w = w_1 \cdots w_m$ such that each w_i is a Lyndon word and $w_1 \geq w_2 \geq \ldots \geq w_m$.

A morphism $g : \Delta^* \to \Sigma^*$ of the Salomaa–Welzl type is introduced again. Let us choose two disjoint sets L_1, L_2 consisting of Lyndon words in Σ^*. Furthermore, we construct a set $E = \{w_1, \ldots, w_n\} \subseteq \Delta^*$ such that $|w_i| < |w_j|$ implies $g(w_i) < g(w_j)$. Finally, define some rewriting rules $u_j \to v_j, v_j \to u_j$ such that $g(u_j) = g(v_j)$.

Encryption key: Two sets $E_i = g^{-1}(L_i) \cap E$ for $i = 1, 2$ and the rewriting rules above.

Encryption: Suppose the plaintext is $i_1 \cdots i_p$ with $i_j \in \{0, 1\}$. Choose words $v_1, \ldots, v_p$ such that $|v_j| \geq |v_{j+1}|$ and $v_j \in E_{i_j}$ for $j = 1, \ldots, p$. Finally, repeated use of the rewriting rules completes the encryption producing a word c.

Decryption key: The morphism g and the sets L_i which may be regular languages.

Decryption: The unique factorization theorem gives us the equation $g(c) = r_1 \cdots r_p$ where $r_j \in L_{i_j}$.

4.5 Niemi system

The following public-key cryptosystem [27] is based on *hiding regular languages*. Consider two arbitrary grammars G_0, G_1 and two finite automata A_0, A_1 such that $L(A_0) \cap L(A_1) = \emptyset$. By standard *triple construction* we may construct grammars G_0', G_1' with $L(G_i') = L(G_i) \cap L(A_i)$ for $i = 1, 2$. Let us, once again, introduce similar morphism g as in Salomaa–Welzl system. Now we build two more grammars G_0'', G_1'' in such way that after the morphism g is applied to all elements (also to productions) of G_i'' the result is a subgrammar of G_i'.

Encryption key: The grammars G_0'' and G_1''.

Encryption: Done bit-by-bit. To encrypt the bit i we (nondeterministically) generate a word $c \in L(G_i'')$.

Decryption key: The automata A_0 and A_1, the morphism g.

Decryption: Now $g(c) \in L(G_i') \subseteq L(A_i)$. We simply check whether A_0 or A_1 accepts $g(c)$.

4.6 Oleshchuk system

In this system [28] we consider a *word problem* for *semigroups*, similarly as in Section 4.1. Let us define an alphabet Σ and some *equations*

$$l_1 = r_1, \ldots, l_m = r_m$$

where l_i, r_i are words over Σ for $i = 1, \ldots m$. Now two words are *equivalent* if one of them can be derived from the other by the rewriting rules $l_i \rightarrow r_i, r_i \rightarrow l_i$.

The set of equations has the *Church–Rosser* property if the following condition holds: two words are equivalent iff some third word can be derived from both of them using only *length-decreasing* rewriting rules. In this case the word problem, i.e., the problem whether or not two words are equivalent, can be solved in linear time.

The basic idea of the cryptosystem is the same as in the Wagner–Magyarik system: the encryption key consists of two inequivalent words w_0, w_1 and some equations with a difficult word problem while the decryption key adds more equations which possess the Church–Rosser property.

There is one additional trick in the system: w_0 and w_1 are chosen in such way that

1) all words in $\{w_0, w_1\}^*$ are pairwise inequivalent and
2) $\{w_0, w_1\}$ is a *code*: the condition $w_{i_1} \cdots w_{i_n} = w_{j_1} \cdots w_{j_m}$ implies $i_1 = j_1$.

Suppose the plaintext is $i_1 \cdots i_n$. In the encryption, the rewriting rules can be applied directly to the long word $w_{i_1} \cdots w_{i_n}$ and decryption is still possible. Of course, this trick can be introduced similarly to the system of Section 4.1.

This addition is useful mainly because in bit-by-bit encryption the *integrity* of the message cannot be guaranteed easily. For instance, an active intruder can modify the cryptotext in such way that the order of bits in the decrypted message is changed. This can be done without breaking the cryptosystem at all.

5. Cryptosystems based on automata theory

Automata theory offers a natural basis for cryptosystem design. As explained in Section 3 all cryptosystems can be viewed as finite automata (or as sequential machines). There are also many purely automata-theoretic problems suitable for constructing cryptosystems. The basic model is simple: the encryption key consists of an automaton and its inverse is the decryption key. In many cases it is difficult to find a machine that inverts the function of a given machine.

Automata-theoretic cryptosystems seem to avoid the biggest drawbacks of the language-theoretic systems, i.e., data expansion and lack of digital signatures. On the other hand, both topics suffer from the lack of systematic and massive cryptanalytic research.

As in the previous section, we survey the key ideas of a few specific systems. Let us begin with a couple of systems based on *cellular automata*.

5.1 Wolfram system

One popular method in constructing classical cryptosystems is to produce an *unpredictable pseudorandom* bit sequence which is added bitwise to the plaintext stream. The key of the system is the *seed* of the pseudorandom sequence. For instance, *nonlinear feedback shift registers* are often used in this manner.

Wolfram system [58] uses a one-dimensional cellular automaton to generate a pseudorandom stream. The value of each cell (which is a bit) is updated synchronously in discrete time steps and the new value depends on the old values of the cell itself and its two neighbours according to the rule

$$a_i' = a_{i-1} \oplus (a_i \vee a_{i+1}).$$

The values of a particular cell through time serve as a pseudorandom sequence while the key (or the seed) is the initial state of the automaton.

A cryptanalytic algorithm for the cryptosystem was developed in [25]. It is based on an equivalent description of the system in which the key space is considerably reduced. As a conclusion, a key size of at least one thousand bits is needed to use the system in a secure way.

5.2 Guan public-key system

Another way to build a cryptosystem upon a cellular automaton is to use the plaintext as the initial state while the rules of the automaton constitute the encryption key. The final state of the automaton is the cryptotext. Decryption is possible if the CA is *invertible*, i.e., there exists another set of rules which reverses the process. These rules are used as the decryption key. It is easy to construct classical cryptosystems along these guidelines because both keys are kept secret. To obtain a public-key cryptosystem one must find such rules for the CA that even if they are given it is impossible in practice to find the rules of the inverse automaton.

Guan [16] uses easily invertible rule sets as basic building blocks in such way that a *composition* of these blocks is difficult to invert. More specifically, a composite of so-called *partially linear* functions is not necessarily partially linear.

The resulting cryptosystem can be used for both confidentiality and authenticity purposes because the encryption and decryption keys are essentially in symmetric positions.

Other cryptologic results concerning cellular automata can be found in, e.g., [10], [9], [17], [26].

Next we present some public-key systems based on sequential machines. They serve as a rare example of a public-key *stream cipher*. This means a stream of plaintext bits can be encrypted without any division into blocks of bits beforehand. All systems presented so far, as well as the best-known systems like *RSA*, need block division before encryption.

5.3 Tao-Chen public-key system

Let us consider a sequential machine in which the output sequence determines uniquely the input sequence provided that the initial state of the machine is fixed. Such machine is called *weakly invertible* and it can be proved that there exists another machine which reverses function of the original machine. It seems to be possible to construct cryptosystems based on sequential machines analogously to the case of cellular automata. Unfortunately, it is quite easy to find the inverse machine of the given weakly invertible machine. Hence, some modifications are needed.

A sequential machine is called *weakly invertible with delay* τ if the output sequence determines uniquely the input sequence *except* possibly the last τ letters. It turns out that in this case there exists another machine which reverses the action of the original machine with the exception that the first τ letters of the input sequence may be changed.

Already for relatively small values of τ it becomes very difficult to find an inverse machine of a given machine which is weakly invertible with delay τ. The original Tao-Chen system [51] is built on this phenomenon. In the case where each output symbol is a linear function of the input symbol and a few previous input and output symbols the machine is called *linear* and the problem to find an inverse machine is still easy even for larger delay values τ. If a *nonlinear* weakly invertible machine M_0 with delay 0 and a linear weakly invertible machine M_1 with delay τ are combined to operate sequentially, then a composed machine $M_0 \times M_1$ can be defined and it is weakly invertible with delay τ. Moreover, the inverse machine of $M_0 \times M_1$ is not easily found if M_0 and M_1 are not given.

Encryption key: A submachine M of $M_0 \times M_1$.

Encryption: Feed the plaintext with τ random symbols in the end to the machine M and obtain cryptotext as the output.

Decryption key: Machines M_0 and M_1.

Decryption: Feed the cryptotext first to the inverse of M_1. Remove then τ first symbols from the output and feed the rest to the inverse of M_0. The plaintext is obtained as the output.

Like the CA systems also this system may be applied for both confidentiality and authenticity purposes. The system is easily implemented and it has been in practical use in China [22], [53], [59].

An attempt to find out the decryption key given the encryption key leads to a problem of finding special common factors of two given *matrix polynomials*. No fast algorithms solving this problem are known in the general case.

However, the system was cryptanalyzed in [6] by showing that it is in fact enough to solve the problem for *modular* matrix polynomials which can be done effectively. Other cryptanalytic attacks can be found in [5] and [33]. There exists also an attack by Dai et al. referred to in [55].

Both the designers of the system [52], [55] and the cryptanalysts [6] have shown that the original system can be modified to avoid the attacks.

Tao et al. have also constructed classical systems based on sequential machines [48]. They include the nice features of no data expansion during encryption and bounded effect of an error in cryptotext to the (rest of) the decryption process. Other classical cryptosystems based on sequential machines can be found in, e.g., [4], [34].

6. Theoretical cryptologic research based on language theory

As explained in Section 3 almost all cryptosystems can be studied from the point of view of automata theory. Examples of research in general level that uses sequential machines as models of cryptosystems can be found in [30] and [31]. Altogether, the model has not turned out to be very fruitful.

On the other hand, some important parts of various cryptosystems can be modelled by sequential machines more successfully. The generation of a pseudorandom sequence is a good example. As already mentioned earlier, feedback shift registers have been heavily used in this sense (see, e.g., [42]).

The area of modern cryptologic research contains much more than mere design and analysis of cryptosystems. Active research is done on cryptologic *protocols*. For example, in an *interactive proof system* for membership in a language L one party called the *prover* convinces another party called the *verifier* that words $w \in L$ are actually in L. The concept is interesting only if the verifier has a limited computational power. The system is said to be *zero knowledge* if the only information which the verifier can get during the protocol is the single bit of information that w belongs to L. In [14] the subject is studied in the case the verifier is a *two-way probabilistic finite automaton*. It is worth mentioning that zero knowledge protocols have direct applications in, for instance, *user identification* and *computer voting*.

In practice, cryptosystems are typically used as building blocks of a larger system which can be called a *security architecture*. An example of other type of building block is *access control* which can also be modelled by finite automata [7]. The whole security architecture is often a complex object. In [36] it is modelled as a word in a language. The language is generated by a grammar in which the building blocks of the system are terminals and different specific *security tasks* are nonterminals. Examples of basic security tasks are authenticity and confidentiality of messages received/sent by some specific users of the computer system. Productions of the grammar are general rules which describe how different tasks can be reduced to other tasks and building blocks (e.g., concrete implementations of cryptosystems). The analysis of a specific security architecture transforms in the model into a *parsing* problem.

As already explained in Section 2, authenticity is as important ingredient of security as is confidentiality. A secure public-key cryptosystem as such is typically effective only against *passive* eavesdroppers who merely tap the lines and try to cryptanalyze the messages. On the other hand, an *active* saboteur may be able to read confidential messages even if the cryptosystem used is secure. A simple example of this was given in Section 2 where one user impersonates another user. To resist all possible attacks of active intruders the protocols used in communication must be designed carefully. In [13] a formalism is developed by which it is possible to analyze and charaterize security of protocols. The protocols are modelled by series of words over an alphabet. Algorithms constructed in the model involve, e.g., similar word problems as those introduced in Section 4.

7. Cryptanalysis based on language theory

Cryptanalysis is as important part of cryptology as design of cryptosystems. Usually cryptanalysis of a specific cryptosystem applies results of the same theory on which the system itself relies. In earlier sections we have mentioned examples of cryptanalytic methods which use language theory and automata theory to break cryptosystems based on these theories [18], [6]. The next example shows how automata theory can be used in cryptanalysis of a well known classical cryptosystem [29].

In a *monoalphabetic substitution cipher* each element of the plaintext alphabet is uniformly encrypted to some element of the cryptotext alphabet.

Hence, the encryption algorithm computes the value of an injective *letter-to-letter morphism*. If the plaintext is known to be in some natural language and a relatively long sequence of cryptotext is also known there is usually only one morphism that could have been used in the encryption. The traditional cryptanalytic method for finding this morphism is to consult some statistical information about the plaintext language. For instance, there exist tables of frequencies of all different n-letter subwords in the English language for $n = 1, 2, 3$.

In [29] the problem of breaking substitution ciphers involves the use of a *stochastic learning automaton*. Candidates for the encryption morphism are coded in the states of the automaton. A fixed *dictionary list* of the most common words in, say, English are used in order to measure the "fitness" of the current state when the cryptotext is read as input to the automaton. The transition to the new state depends on the fitness of the previous state. Experimental results show that the automaton reaches quite rapidly a state where a "correct" amount of decrypted words are found in the dictionary list. The method seems to be faster than other methods in breaking substitution ciphers.

8. Language-theoretic research inspired by cryptology

When results of a scientific theory are applied to solve problems in the area of another theory the interaction is seldom one-way. Those problems and their solutions usually generate further research in both fields. This phenomenon has occurred also between cryptology and formal language theory. There are many language-theoretic contributions in cryptology but cryptologic concepts have given rise to new ideas and notions also from the point of view of pure language theory. In the sequel we briefly present a few examples.

In the cryptosystem customarily attributed to *Richelieu* the cryptotext is obtained from the plaintext by adding irrelevant text which is scattered over the plaintext. This method referred to as *garbage-in-between* is also an essential part of many modern cryptosystems (see the morphism g in several systems presented in Section 4). The notions of *sequential* (resp., *parallel*) *insertion* and *deletion* studied in [20] are directly connected to this idea.

In [2] the method of garbage-in-between is modelled by *guided filtering*. Consider a word $w \in \{0, 1\}^*$ whose length equals the length of the cryptotext. Now w can be used as a decryption key that shows in which positions letters of the plaintext can be found. To be able to decrypt cryptotexts of different lengths a whole language of *key words* is introduced. For each length n only one w with $|w| = n$ is needed and the concepts of *thin* and *slender* languages are introduced.

A cryptosystem of [1] uses the plaintext as a *control word* in a *Szilard language*. The result of a derivation controlled by this word is the cryptotext.

In this framework a single control word may correspond to several derivation trees and interesting language-theoretic problems arise which are further studied in [24].

Cryptologic applications of rewriting systems (see cryptosystems in Section 4 and, also, [13]) have brought new insights to this area. An example of a paper in rewriting systems inspired by cryptology is [8]. Similarly, the cryptosystems based on generalized sequential machines have been a starting point for a great amount of research on invertible automata, e.g., [46], [47], [50], [54].

9. Research associated with language theory and cryptology

There are a number of papers in complexity theory which have direct connections to both cryptology and language theory. An example of such a paper is [15]. Because another chapter in this handbook is devoted to complexity theory we omit closer discussion of these connections here. Instead, we mention a couple of examples of other topics that have both language-theoretic and cryptologic aspects.

Let us consider a public-key cryptosystem (E, D). Breaking the system is related to *learning theory* in the following way. Anybody may generate as many pairs of the form $(x, E_k(x))$ as wanted. These pairs are also of the form $(D_{k'}(y), y)$, hence learning D from examples must be intractable if the cryptosystem is secure. Starting from this observation it is proved in [21] that, for instance, learning deterministic finite automata is intractable if some well known cryptosystems like RSA are secure. Other results in the area can be found in, e.g., [32].

A *Latin array* is a combinatorial notion which is motivated by cryptology and automata theory. It is a generalization of a *Latin square*: each element occurs exactly once in every column but exactly k times in each row. (This means a Latin array is an $n \times nk$-matrix.) The reference [49] is a survey of results in this area.

References

[1]   Andrasiu, M., Atanasiu, A., Păun, G. and Salomaa, A.: A new cryptosystem based on formal language theory. Bull. Math. de la Soc. Sci. Math. de Roumanie, Tome **36**, (1992).

[2]   Andrasiu, M., Păun, G., Dassow, J. and Salomaa, A.: Language-theoretic problems arising from Richelieu cryptosystems. Theoretical Computer Science, **116**, (1993) 339–357.

[3]   Anshel, I. M. and Anshel, M.: From the Post-Markov theorem through decision problems to public-key cryptography. American Mathematical Monthly 100, (1993) 835–844.

[4]   Atanasiu, A.: A class of coders based on gsm. Acta Informatica, 29, (1992) 779–791.

[5]   Bao, F. and Igarashi, Y.: A randomized algorithm to finite automata public key cryptosystem. Proceedings of ISAAC'94, Lecture Notes in Computer Science 834, Springer-Verlag, Berlin, 1994, 678–686.

[6]   Bao, F. and Igarashi, Y.: Break finite automata public key cryptosystem. Proceedings of ICALP'95, Lecture Notes in Computer Science 944, Springer-Verlag, Berlin, 1995, 147–158.

[7]   Bell, D. E. and LaPadula, L. J.: Secure computer system: unified exposition and Multics interpretation. Technical report MTR-2997, Mitre Corp., Bedford, Mass., 1976.

[8]   Benois, M.: Descendants of regular language in a class of rewriting systems: algorithm and complexity of an automata construction. Proceedings of Rewriting Techniques and Applications, Lecture Notes in Computer Science 256, Springer-Verlag, Berlin, 1987, 121–132.

[9]   Daemen, J., Govaerts, R. and Vandewalle, J.: A framework for the design of one-way hash functions including cryptanalysis of Damgård's one-way function based on a cellular automaton. Advances in cryptology-ASIACRYPT'91, Lecture Notes in Computer Science 739, Springer-Verlag, Berlin, 1993, 82–96.

[10]   Damgård, I.: A design principle for hash functions. Advances in cryptology-CRYPTO'89, Lecture Notes in Computer Science 435, Springer-Verlag, Berlin, 1990, 416–427.

[11]   Denning, D. E.: Cryptography and data security. Addison-Wesley, Reading, Mass., 1982.

[12]   Diffie, W. and Hellman, M.: New directions in cryptography. IEEE Transactions on Information Theory IT-22 (1976), 644–654.

[13]   Dolev, D. and Yao, A. C.: On the security of public key protocols. IEEE Transactions on information theory IT-29 (1983) 198-208.

[14]   Dwork, C. and Stockmeyer, L.: Zero-knowledge with finite state verifiers. Advances in cryptology-CRYPTO'88, Lecture Notes in Computer Science 403, Springer-Verlag, Berlin, 1990, 71-75.

[15]   Garzon, M. and Zalcstein, Y.: The complexity of Grigorchuk groups with application to cryptography. Theoretical Computer Science 88, (1991), 83–98.

[16]   Guan, P.: Cellular Automaton public-key cryptosystem. Complex Systems 1, (1987) 51–56.

[17]   Gutowitz, H.: Cryptography with dynamical systems. Proceedings of Cellular Automata and Cooperative Systems, NATO Advanced Institute, Kluwer, Dordrecht, 1993, 237–274.

[18]   Kari, J.: A cryptanalytic observation concerning systems based on language theory. Discrete Applied Mathematics 21, (1988) 265–268.

[19]   Kari, J.: Observations concerning a public-key cryptosystem based on iterated morphisms. Theoretical Computer Science 66, (1989) 45–53.

[20]   Kari, L.: On insertion and deletion in formal languages. Ph. D. thesis, Univ. of Turku, Turku, 1991.

[21]   Kearns, M. and Valiant, L.: Cryptographic limitations on learning boolean formulae and finite automata. Journal of the ACM 41, (1994) 67–95.

[22]   Li, J. and Gao, X.: Realization of finite automata public key cryptosystem and digital signature. Proceedings of CRYPTOCHINA'92, 110–115. (in Chinese)

[23]   Lothaire, M.: Combinatorics on Words. Addison-Wesley, Reading, MA, 1982.

[24] Matei, C. and Tiplea, F. L.: (0,1)-total pure context-free grammars. Proceedings of Developments in Language Theory '95, World Scientific, Singapore, 1996, 148–153.

[25] Meier, W. and Staffelbach, O.: Analysis of pseudo random sequences generated by cellular automata. Advances in cryptology-EUROCRYPT '91, Lecture Notes in Computer Science **547**, Springer-Verlag, Berlin, 1991, 186–199.

[26] Nandi, S., Kar, B. K. and Pal Chaudhuri, P.: Theory and applications of cellular automata in cryptography. IEEE Transactions on Computers **43**, (1994) 1346–1357.

[27] Niemi, V.: Hiding regular languages: a public-key cryptosystem. Manuscript, 1989.

[28] Oleshchuk, V. A.: On public-key cryptosystem based on Church-Rosser string-rewriting systems. Proceedings of Computing and Combinatorics '95, Lecture Notes in Computer Science **959**, Springer-Verlag, Berlin, 1995, 264–269.

[29] Oommen, B. J. and Zgierski, R.: Breaking substitution cyphers using stochastic automata. IEEE Transactions on Pattern Analysis and Machine Intelligence **15**, (1993) 185–192.

[30] Pichler, F.: Finite state machine modelling of cryptographic systems in LOOPS. Advances in cryptology-EUROCRYPT '87, Lecture Notes in Computer Science, Springer-Verlag, Berlin, 1988, 65–73.

[31] Pichler, F.: Application of automata theory in cryptology. Elektrotechnik und Informationstechnik **105**, (1988) 18–25. (in German)

[32] Pitt, L. and Warmuth, M. K.: Prediction-preserving reducibility. Journal of Computer and System Sciences **41**, (1990) 430–467.

[33] Qin, Z. and Zhang, H.: The attack algorithm A tau M to finite automaton public key cryptosystems. Chinese Journal of Computers, **18**, (1995) 199–204.

[34] Rayward-Smith, V. J.: Mealy machines as coding devices. In: Beker, H. J. and Piper, F. C. (eds.): Cryptography and codings, Clarendon Press, Oxford, 1989.

[35] Rivest, R. L., Shamir, A. and Adleman, L.: A method for obtaining digital signatures and public-key cryptosystems. Communications of ACM, **21**, (1978), 120–126.

[36] Rueppel, R.: A formal approach to security architecture. Advances in cryptology – EUROCRYPT'91, Lecture Notes in Computer Science **547**, Springer-Verlag, Berlin, 1991, 387–398.

[37] Salomaa, A.: A public-key cryptosystem based on language theory. Computers and Security **7**, (1988) 83–87.

[38] Salomaa, A: Public-key cryptography. EATCS Monographs in Theoretical Computer Science, vol. 23, Springer-Verlag, Berlin, 1990.

[39] Salomaa, A. and Yu, S.: On a public-key cryptosystem based on iterated morphisms and substitutions. Theoretical Computer Science **48**, (1986) 283–296.

[40] Schneier, B.: Applied cryptology: protocols, algorithms and source code in C. John Wiley & Sons, New York, 1994.

[41] Seberry, J. and Pieprzyk, J.: Cryptography: an introduction to computer security. Prentice Hall, New York, 1989.

[42] Simmons, G. J. (ed.): Contemporary cryptology: the science of information integrity. IEEE Press, 1992.

[43] Siromoney, R. and Mathew, L.: A public key cryptosystem based on Lyndon words. Information Processing Letters **35**, (1990) 33–36.

[44] Stinson, D. R.: Cryptography: theory and practice. CRC Press, 1995.

[45] Subramanian, K. G., Siromoney, R. and Abisha, P. J.: A D0L-T0L public key cryptosystem. Information Processing Letters **26**, (1987), 95–97.

[46]   Tao, R.: Some results on the structure of feedforward inverses. Scientia Sinica (Series A) **27**, (1984) 157-162.

[47]   Tao, R.: Invertibility of linear finite automata over a ring. Proceedings of ICALP'88, Lecture Notes in Computer Science **317**, Springer-Verlag, Berlin, 1988, 489–501.

[48]   Tao, R.: On finite automaton one-key cryptosystems. Proceedings of Cambridge Security Workshop, Fast Software Encryption, Lecture Notes in Computer Science **809**, Springer-Verlag, Berlin, 1994, 135–148.

[49]   Tao, R.: On Latin arrays, Proceedings of International Workshop on Discrete Mathematics and Algorithms, Jinan University Press, 1994, 1–14.

[50]   Tao, R.: On invertibility of some compound finite automata. Technical report No. ISCAS-LCS-95-06, Laboratory for Computer Science, Institute of Software, Chinese Academy of Sciences, Beijing, 1995.

[51]   Tao, R. and Chen, S.: Finite automata public key cryptosystem and digital signature. Computer Acta **8**, (1985) 401–409 (in Chinese).

[52]   Tao, R. and Chen, S.: Two varieties of finite automata public key cryptosystem and digital signature. Journal of Computer Science and Technology **1**, (1986) 9–18.

[53]   Tao, R. and Chen, S.: An implementation of identity-based cryptosystems and signature schemes by finite automaton public key cryptosystems. Proceedings of CRYPTOCHINA'92, 87–104. (in Chinese)

[54]   Tao, R. and Chen, S.: Generating a kind of nonlinear finite automata with invertibility by transformation method. Technical report No. ISCAS-LCS-95-05, Laboratory for Computer Science, Institute of Software, Chinese Academy of Sciences, Beijing, 1995.

[55]   Tao, R., Chen, S. and Chen, X.: FAPKC3: a new finite automaton public key cryptosystem. Technical report No. ISCAS-LCS-95-07, Laboratory for Computer Science, Institute of Software, Chinese Academy of Sciences, Beijing, 1995.

[56]   Wagner, N. R. and Magyarik, M. R.: A public-key cryptosystem based on the word problem. Advances in cryptology-CRYPTO'84, Lecture Notes in Computer Science **196**, Springer-Verlag, Berlin, 1985, 19–37.

[57]   Wichmann, P.: Cryptanalysis of a modified rotor machine. Advances in cryptology-EUROCRYPT'89, Lecture Notes in Computer Science **434**, Springer-Verlag, Berlin, 1990, 395–402.

[58]   Wolfram, S.: Cryptography with cellular automata. Advances in cryptology-CRYPTO '85, Lecture Notes in Computer Science **218**, Springer-Verlag, Berlin, 1986, 429–432.

[59]   Zhang, H., Qin, Z. et al.: The software implementation of FA public key cryptosystem. Proceedings of CRYPTOCHINA'92, 105–109. (in Chinese)

Index

Printing and Binding: Strauss GmbH, Mörlenbach